FISHES
of the
TROPICAL EASTERN PACIFIC

The research upon which this volume is based was sponsored by the
Smithsonian Tropical Research Institute
with additional financial support from the
W. Atherton Seidell Endowment Fund of the Smithsonian Institution.

FISHES

of the

TROPICAL EASTERN PACIFIC

Gerald R. Allen – D. Ross Robertson

UNIVERSITY OF HAWAII PRESS
HONOLULU

A CHP Production

Produced and published by
Crawford House Press Pty Ltd
PO Box 1484
Bathurst NSW 2795 Australia

Published in the United States of America by
University of Hawaii Press
2840 Kolowalu Street
Honolulu, Hawaii 96822

Designed by David H. Barrett

Library of Congress Cataloguing-in-Publication Data
Allen, Gerald R.
Fishes of the tropical eastern Pacific / Gerald R. Allen
and David R. Robertson.
p. cm.
Includes bibliographical references (p.) and index.
ISBN 0-8248-1675-7
1. Fishes – Pacific Ocean. 2. Marine fishes – Tropics.
I. Robertson, David R. (David Ross), 1946- . II. Title.
QL623.6.A58 1994
597.092'61 – dc20 94-11879
CIP

Printed in Hong Kong by Colorcraft Ltd

10 9 8 7 6 5 4 3 2 1

CONTENTS

ACKNOWLEDGEMENTS

We could not have completed this volume without the assistance of numerous individuals. We are particularly grateful to Dr Ira Rubinoff, Director of the Smithsonian Tropical Research Institute (STRI), for his continuous support and encouragement. This project was financed from STRI general research funds and a grant from the Smithsonian Institution's Atherton Siedell Endowment Fund. The governments of Colombia, Costa Rica, Ecuador, Mexico, and Panama granted us permission to conduct field studies. Travel arrangements were expedited for us by Mercedes Arosemena, Gloria Maggiori, and Hector Guzman, all of STRI. We thank Dave West, skipper of the STRI research vessel *Benjamin* and first mate Luiz Cruz for their assistance on a collecting and photographic voyage to the Gulf of Chiriqui and Perlas Islands. Gustavo Justines of Departamento de Recursos Marinos, Panama arranged for us to go aboard trawling vessels and assisted with fish collections. Peter Glynn (STRI) enabled D. Ross Robertson to join his cruise (number 90-053) on RV *Gyre*, sponsored by the United States National Science Foundation, to Isla del Cocos and the Galapagos. Les Knapp of the Smithsonian Oceanographic Sorting Center provided collection chemicals, plastic bags, and storage drums. We are grateful to Daniel Evans, Director of the Charles Darwin Research Station at the Galapagos, and his assistant Fionnuala Walsh for their logistic assistance. Collecting assistance in the Galapagos was rendered by Eldridge Birmingham, Harris Lessios, Benjamin Victor, and Jerry Wellington. We thank the Centro de Investigaciones Oceanograficos y Hidrograficas de la Armada Nacional de Colombia for use of their research vessel ARC *Malpelo* and the Division de Parques Nacionales de INDERENA for permission to conduct research and collecting activities at Isla Malpelo. Funds for this work were provided by the Smithsonian Tropical Research Institute (from their ROF fund). J.D. Lopez assisted with field collections at Malpelo. We are grateful to Ursula Schober (STRI) for diving assistance at Isla del Cocos and the Galapagos. Special thanks are due Richard Cooke (STRI) who accompanied us on trips to the Panama City fish market and also made it possible for us to obtain valuable collections of Panamanian fishes. We were also assisted on trawling trips fish market visits by Conrado Tapia (STRI). We received valuable field assistance in Panama from STRI staff members Anibal Velarde, Alcibiades Cedeño, Pancho Sanchez, and Ismael Gonzalez. Tita and Ana Tapia kindly provided accomodation for us during collecting trips to Aguadulce, Panama. Rex and Joan Allen (G. Allen's parents) provided accomodation and hospitality during a week-long stay at Golfito, Costa Rica. Kirstie Kaiser helped with collections in the Gulf of California. Tim Means and Gary Cotter of Baja Expeditions (San Diego, California) were instrumental in organizing our month-long visit to the Gulf of California in 1990. We also thank John Fox and Almei Moehl of Cabo Aquadeportes for providing logistic assistance at Cabo San Lucas. Vernon Scholey provided working facilities at the Achotines Laboratory of the Interamerican Tropical Tuna Commission. The first author wishes to thank Victor and Shirley Springer and the Norman Duke family for providing accomodation while studying eastern Pacific fishes at the Smithsonian Institution. Richard Rosenblatt provided valuable assistance in identifying photographs of eastern Pacific fishes.

We are especially indebted to the following people who helped to identify our specimens and photos or served as critical reviewers of the manuscript: N. Labbish Chao of Bio-Amazonia Conservation, Talahassee, Florida (Sciaenidae); Bruce Collette of the National Marine Fisheries Service, Washington, DC (Ammodytidae, Batrachoididae, Belonidae, Exocoetidae, and Hemiramphidae); Richard Cooke (Ariidae); William Eschmeyer of the

California Academy of Sciences, San Francisco (Scorpaenidae), Martin Gomon of the Museum of Victoria, Melbourne (Labridae); Phil Hastings of the University of Arizona (Chaenopsidae); Phil Heemstra of the J.L.B. Institute of Ichthyology, Grahamstown, South Africa (Serranidae); Doug Hoese of the Australian Museum, Sydney (Gobiidae); Dave Johnson of the Smithsonian Institution, Washington, DC (Anomalopidae); Patricia Kailola of the Bureau of Rural Resources, Canberra, Australia (Ariidae); John McCosker of the California Academy of Sciences (Muraenidae, Ophichthidae); John McEachran of Texas A & M University (batoid fishes); Rolly Mckay of the Queensland Museum, Brisbane (Haemulidae); Tom Munroe of the US National Marine Fisheries Service (Achiridae, Bothidae, Cynoglossidae); Stuart Poss of the Gulf Coast Research Laboratory, Ocean Springs, Mississippi (Scorpaenidae); Jack Randall of the Bishop Museum, Honolulu (Acanthuridae, Mullidae and Scaridae); Dave Smith of the Smithsonian Institution (Chlopsidae and Congridae); Bill Smith-Vaniz, formerly of the Academy of Natural Sciences, Philadelphia (Carangidae and Opistognathidae); Victor Springer of the Smithsonian Institution (Labrisomidae and Blenniidae); Wayne Starnes of the Smithsonian Institution (Priacanthidae); and the late Peter Whitehead of the Centro de Investigaciones Biologicas, La Paz, Mexico (Clupeidae).

Special thanks are due Roger Steene (Cairns, Australia) for his collecting assistance and underwater photographic expertise at the Galapagos and Panama. Alex Kerstitch (Tuscon, Arizona) performed similar servics for us in the Gulf of Californa and allowed us to use many of his excellent photographs to illustrate this work. We also obtained valuable photographs from Jack Randall (Bishop Museum) and Mark Conlan (San Marcos, California). Francisco Neira (Murdoch University, Perth, Australia) prepared the line drawings. Sketches of the tooth plates of ariid catfishes were made by Conrado Tapia (STRI). We are very thankful to Roger Swainston (Perth, Australia) for his superb color paintings of eastern Pacific fishes. Our photographic endeavors were greatly assisted by the dark room expertise of Carl Hansen (STRI) and Lorie Aceto and her staff (Smithsonian Institution Photographic Services, Washington, DC). William Bussing of the University of Costa Rica provided valuable information about the fishes of Isla del Coco.

We are extremely grateful to Richard Cooke and John McCosker for critically reviewing our manuscript.

Finally, we thank Connie Allen for preparing the typescript, index, and bibliography.

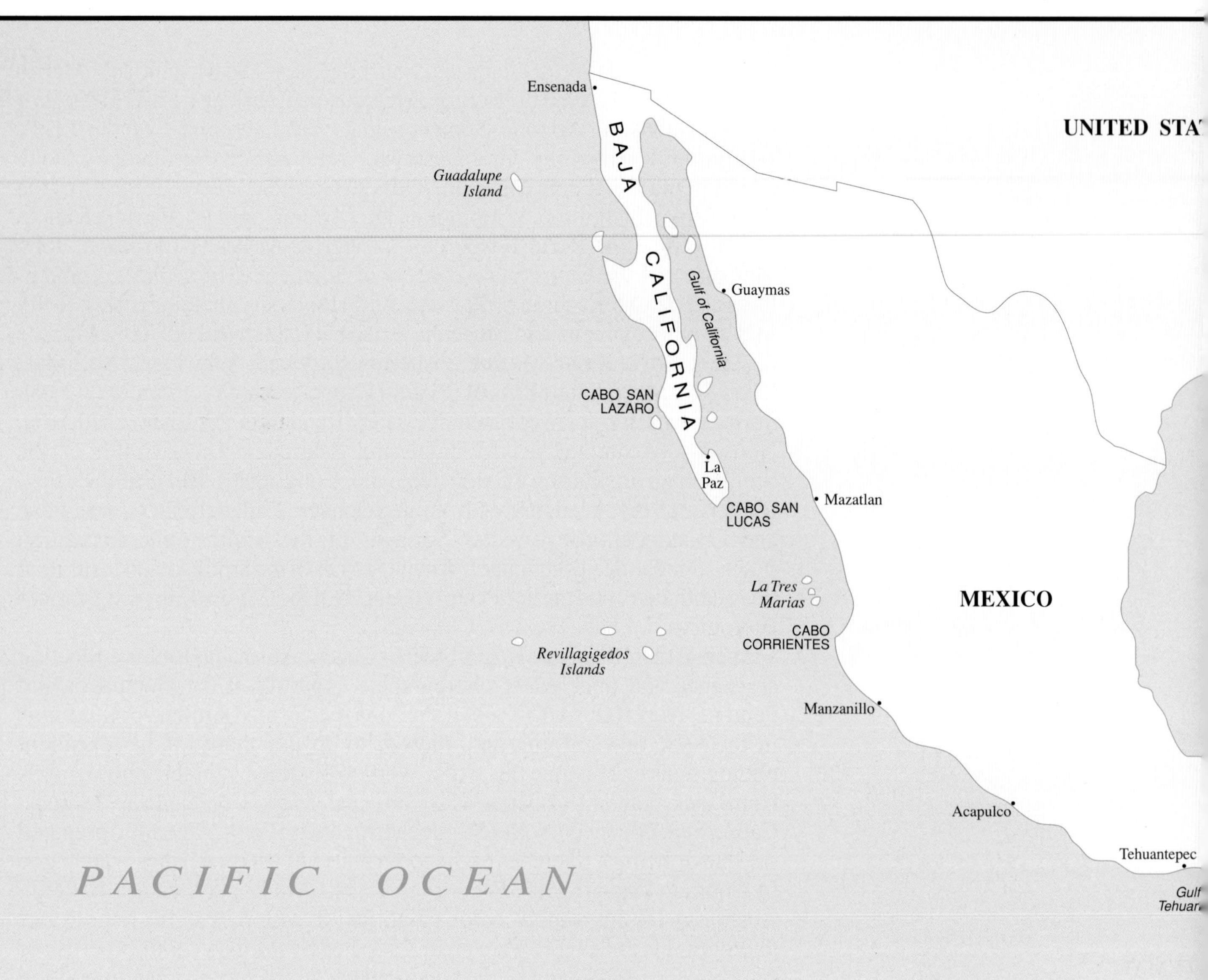

THE TROPICAL EASTERN PACIFIC REGION

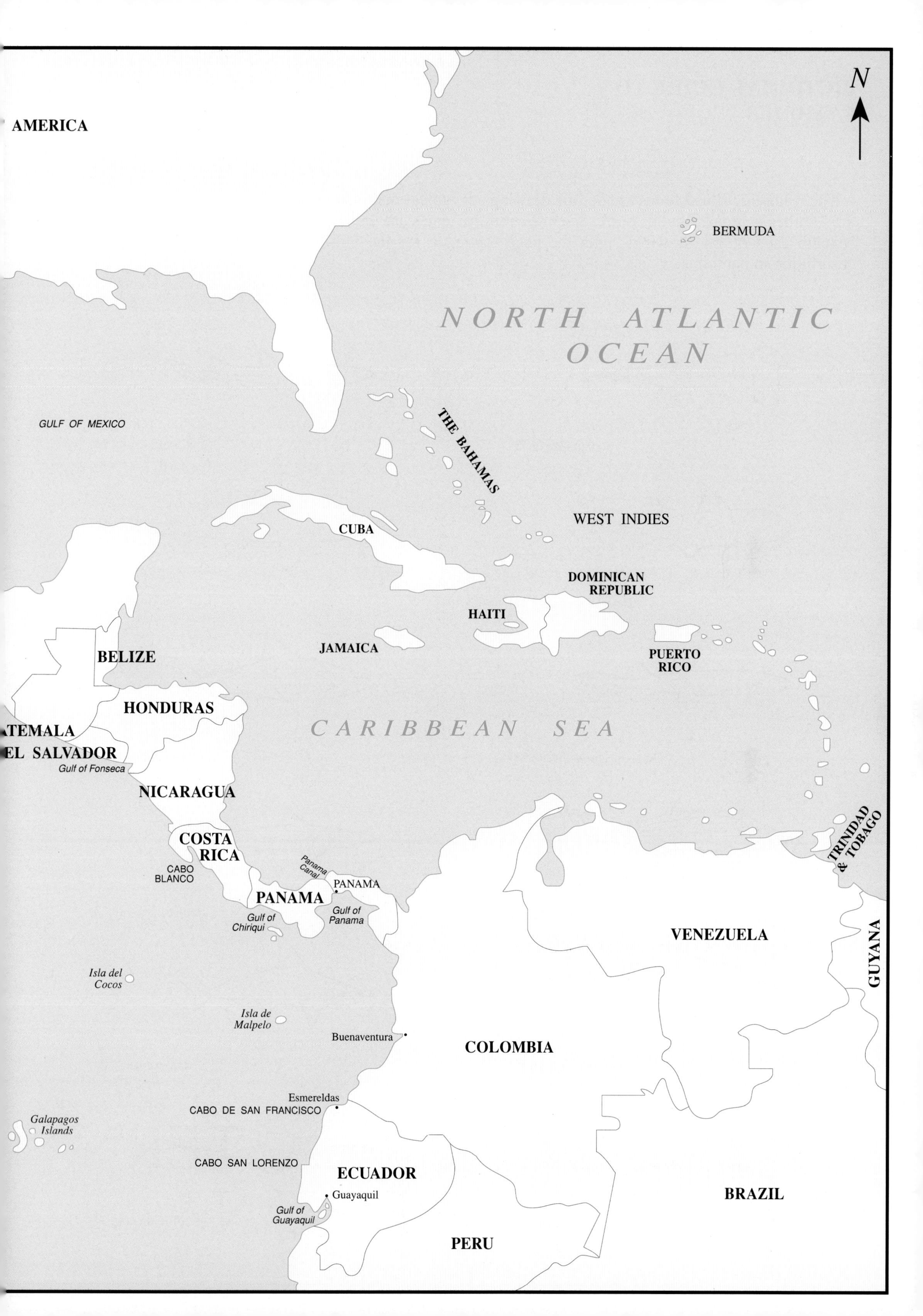

N
AMERICA
BERMUDA
NORTH ATLANTIC OCEAN
GULF OF MEXICO
THE BAHAMAS
WEST INDIES
CUBA
DOMINICAN REPUBLIC
HAITI
JAMAICA
PUERTO RICO
BELIZE
HONDURAS
TEMALA
EL SALVADOR
Gulf of Fonseca
CARIBBEAN SEA
NICARAGUA
COSTA RICA
CABO BLANCO
Panama Canal
PANAMA
PANAMA
Gulf of Chiriqui
Gulf of Panama
TRINIDAD & TOBAGO
VENEZUELA
GUYANA
Isla del Cocos
Isla de Malpelo
Buenaventura
COLOMBIA
Esmereldas
CABO DE SAN FRANCISCO
Galapagos Islands
CABO SAN LORENZO
ECUADOR
Guayaquil
Gulf of Guayaquil
BRAZIL
PERU

PICTORIAL GUIDE TO FAMILIES

The following illustrations are outline drawings of typical members of the families contained in this book. Scientific family names are indicated beneath each of the drawings, with the page reference for the family description in parentheses.

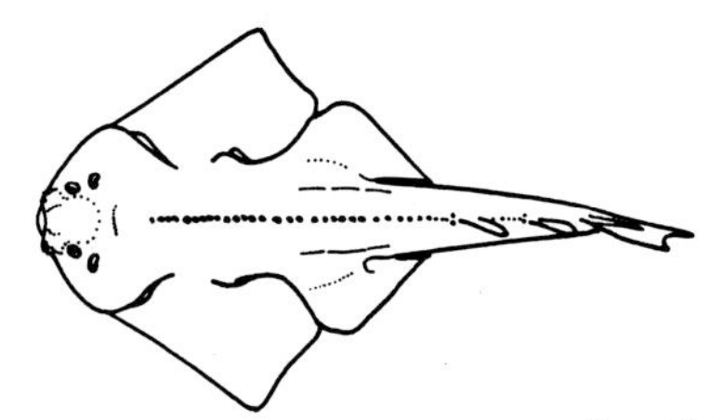

Squatinidae (p 18)

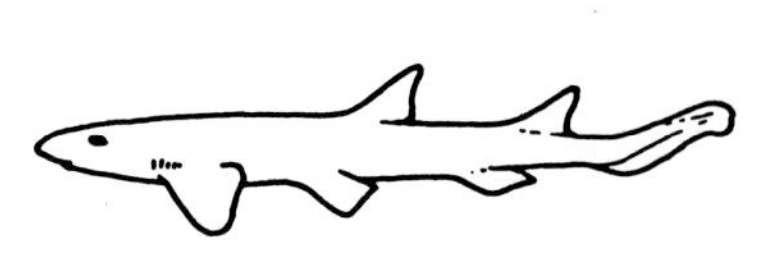

Scyliorhinidae (p 20)

Heterodontidae (p 18)

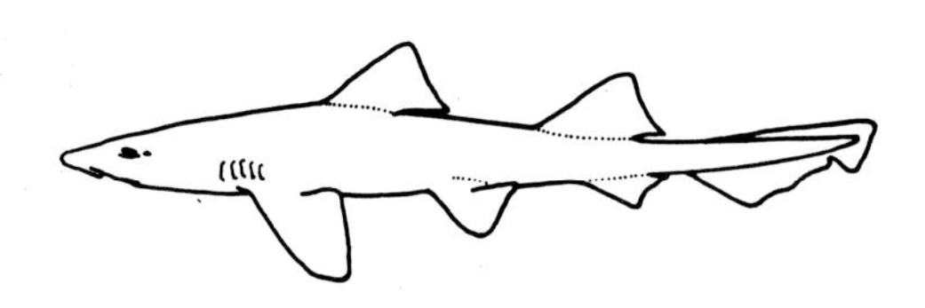

Triakidae (p 20)

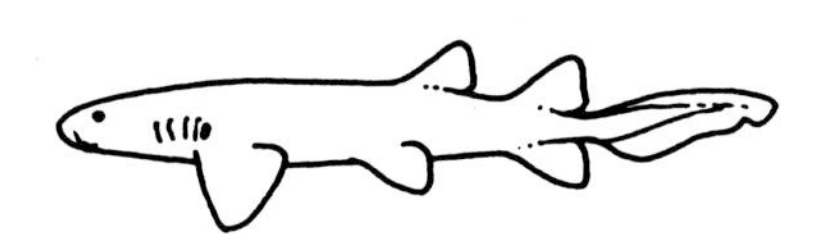

Ginglymostomatidae (p 19)

Carcharhinidae (p 21)

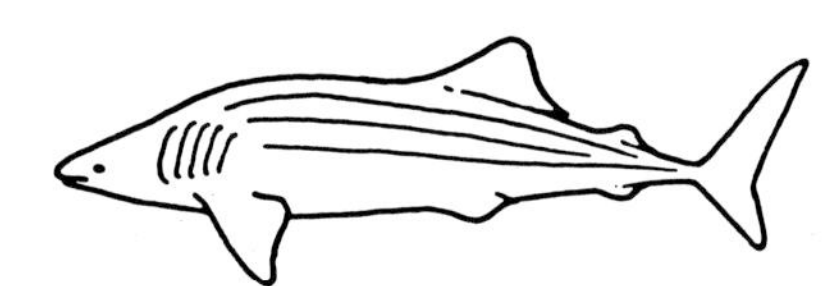

Rhincodontidae (p 19)

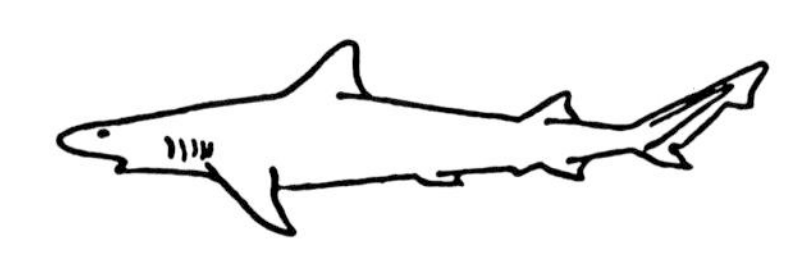

Hemigaleidae (p 23)

Odontaspidae (p 19)

Sphyrnidae (p 24)

Lamnidae (p 20)

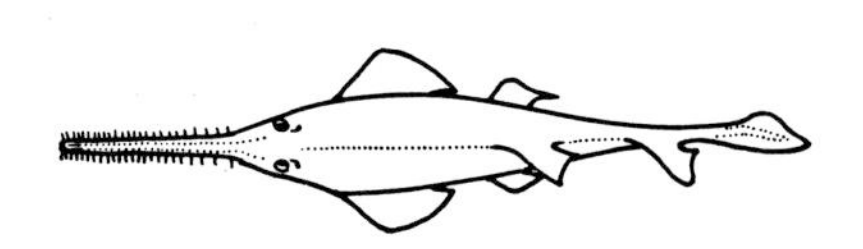

Pristidae (p 31)

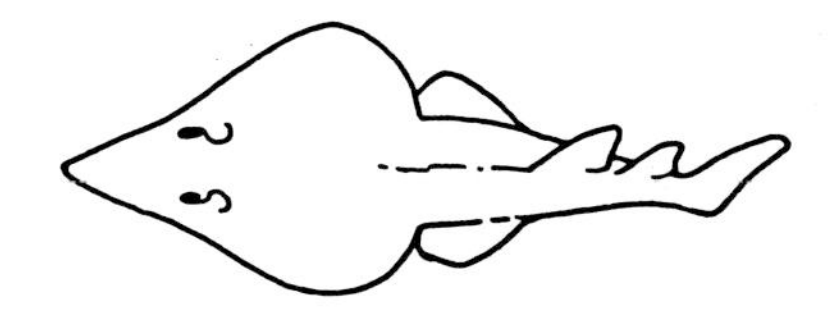

Rhinobatidae (p 31)

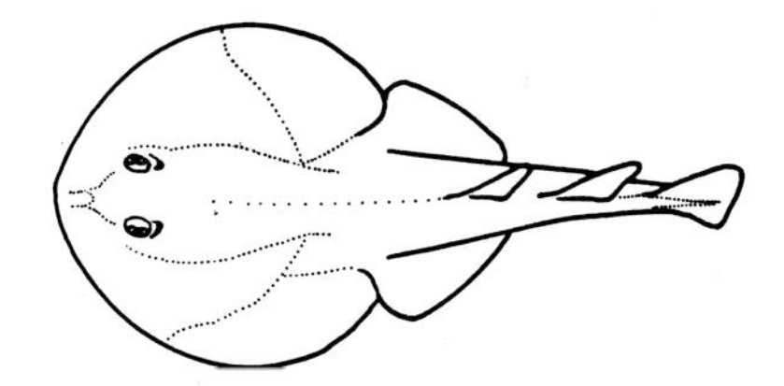

Narcinidae (p 32)

Torpedinidae (p 33)

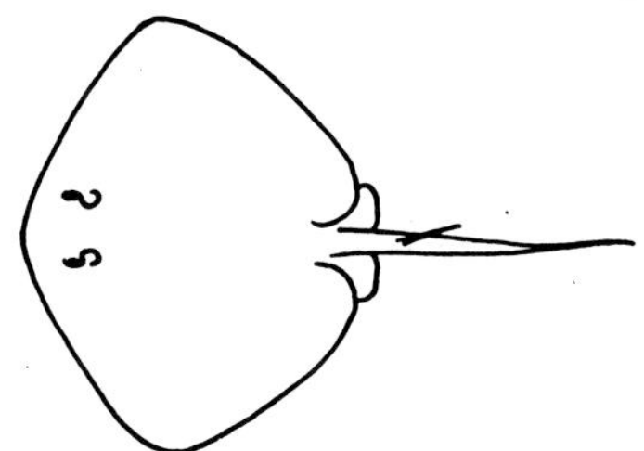

Dasyatididae (p 33)

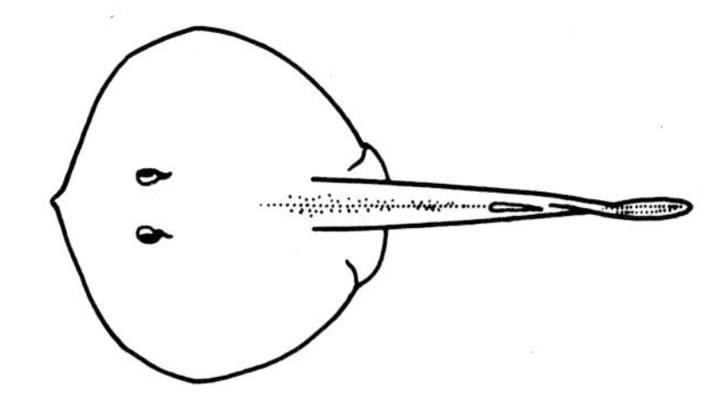

Urolophidae (p 34)

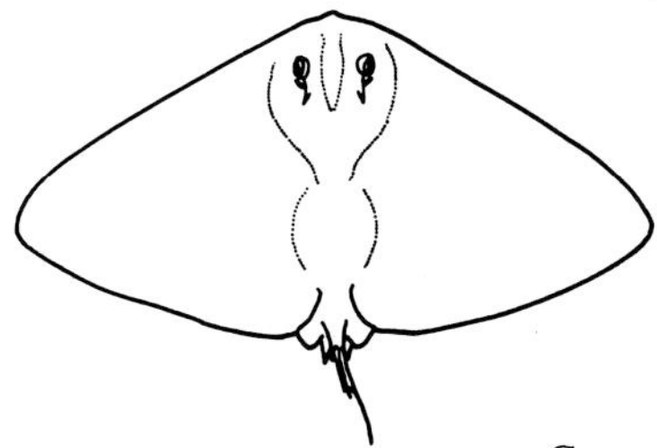

Gymnuridae (p 36)

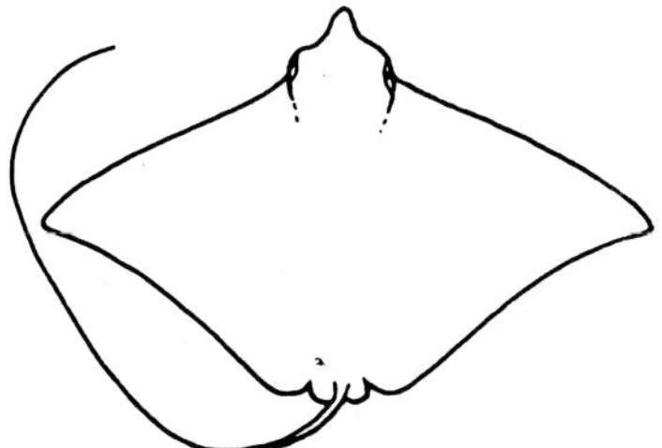

Myliobatidae (p 36)

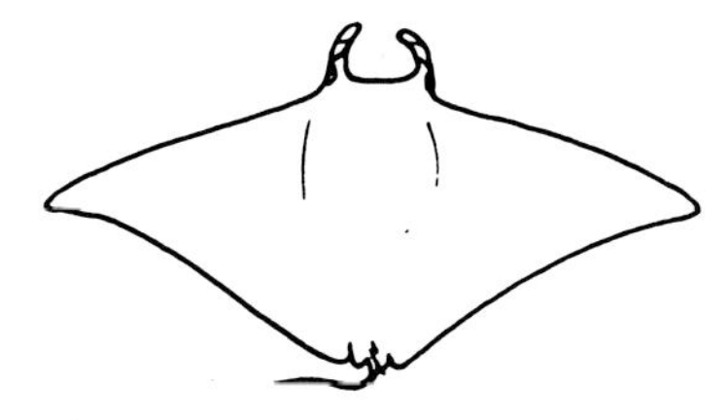

Mobulidae (p 37)

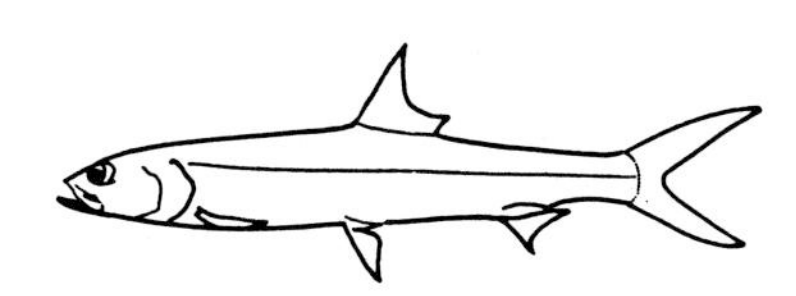

Elopidae (p 41)

Albulidae (p 41)

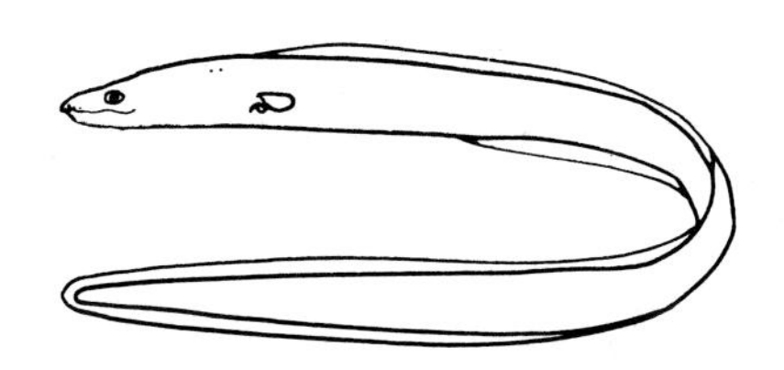

Chlopsidae (p 42)

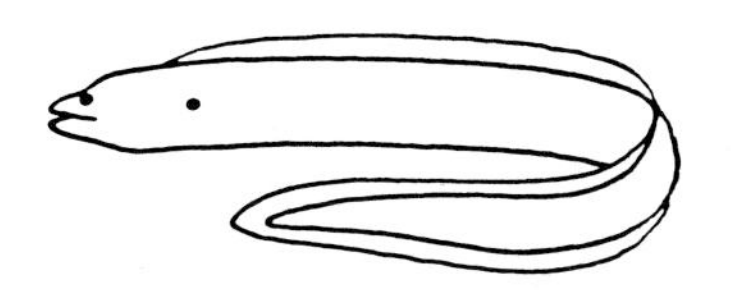

Muraenidae (p 42)

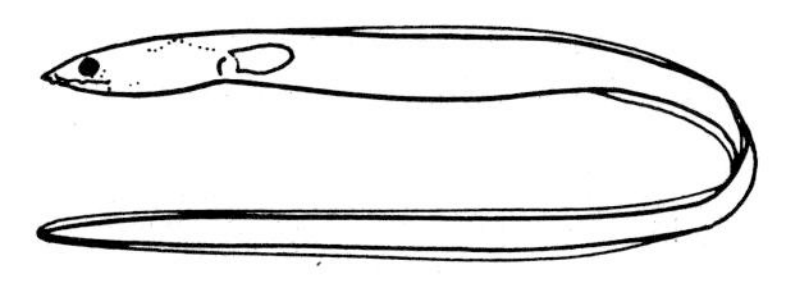

Ophichthidae (p 50)

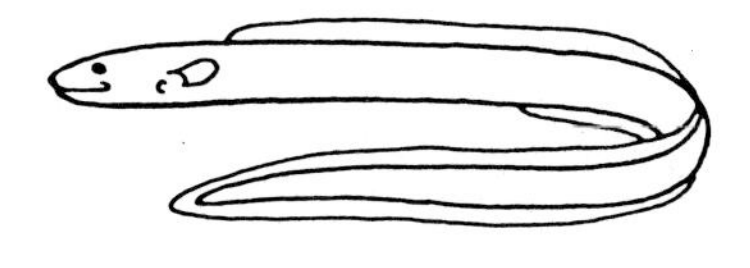

Congridae (p 56)

Clupeidae (p 60)

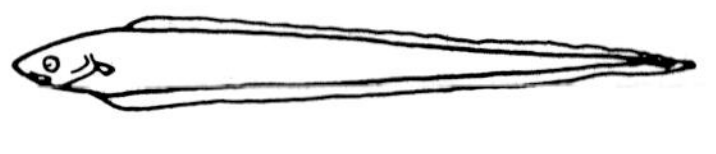

Carapidae (p 76)

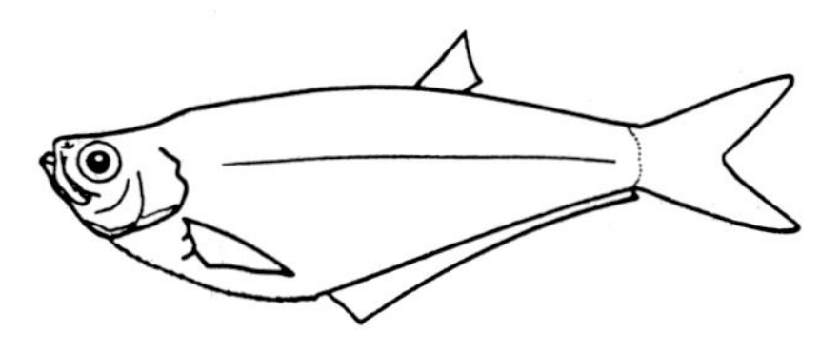

Pristigasteridae (p 61)

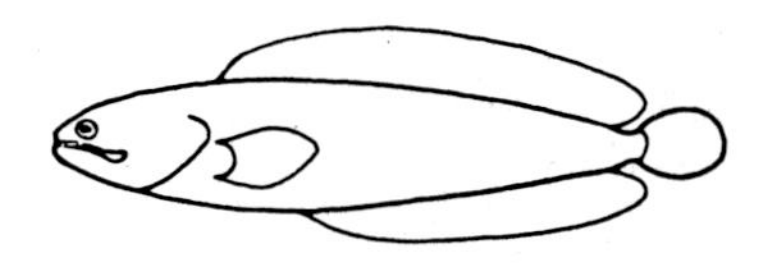

Bythitidae (p 76)

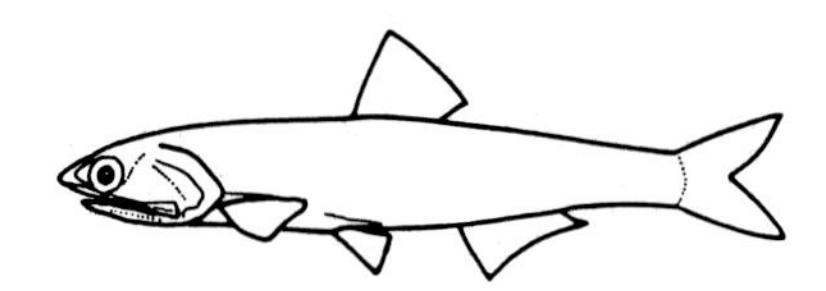

Engraulidae (p 62)

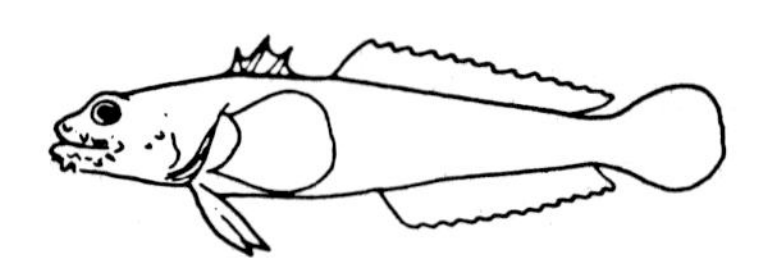

Batrachoididae (p 78)

Chanidae (p 62)

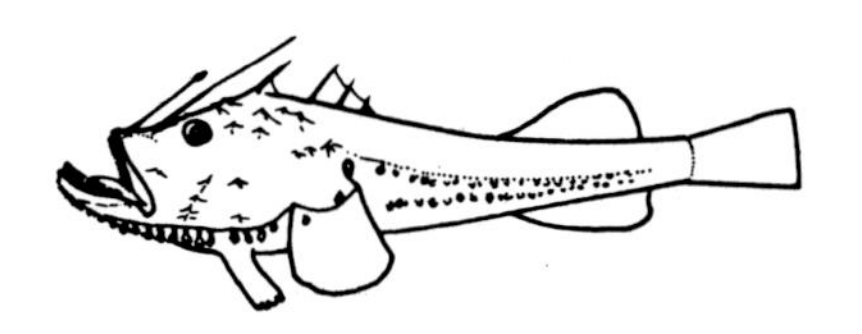

Lophiidae (p 81)

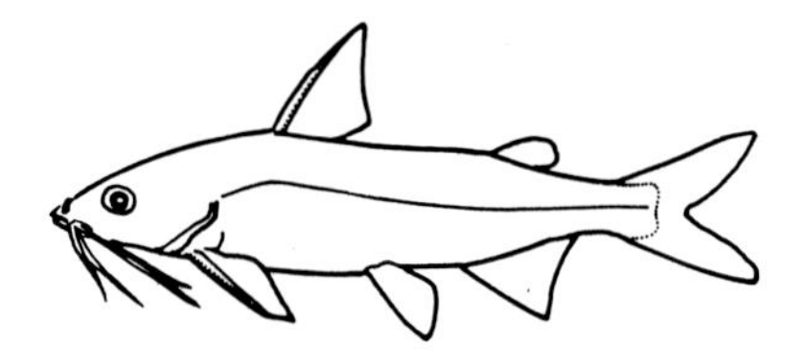

Ariidae (p 66)

Antennariidae (p 81)

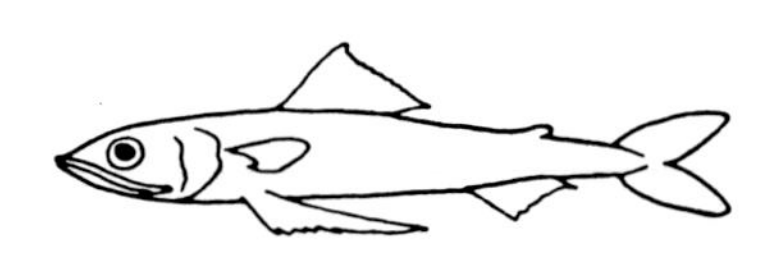

Synodontidae (p 72)

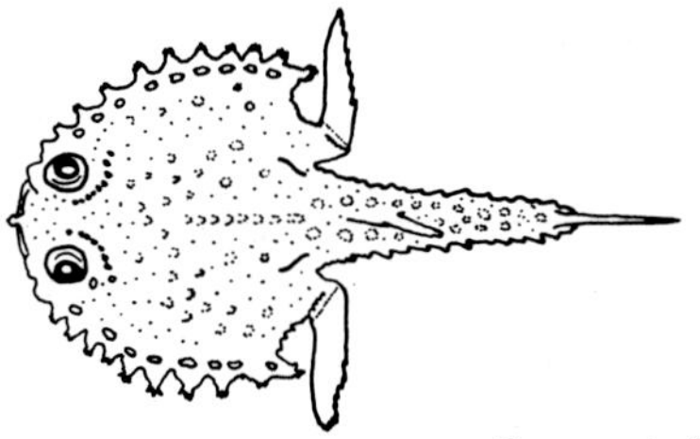

Ogcocephalidae (p 83)

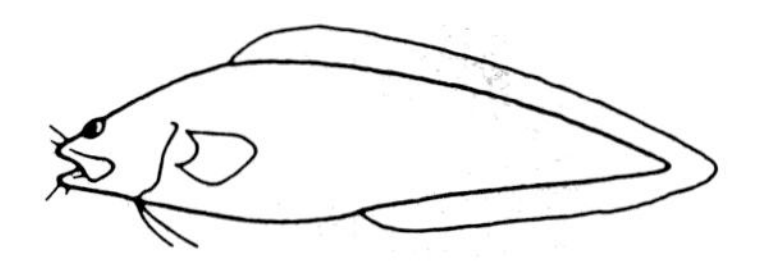

Ophidiidae (p 73)

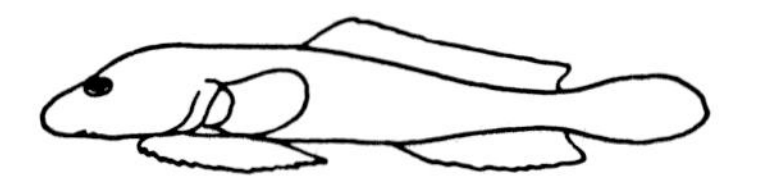

Gobiesocidae (p 84)

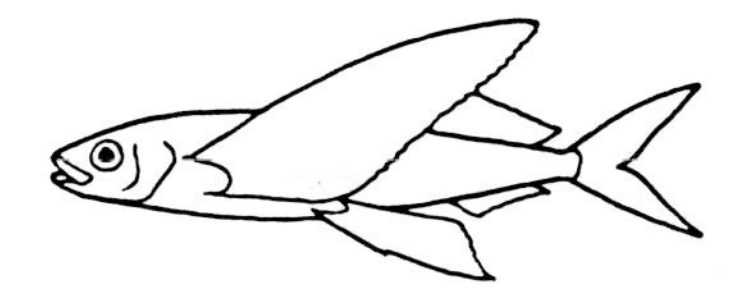

Exocoetidae (p 88)

Hemiramphidae (p 89)

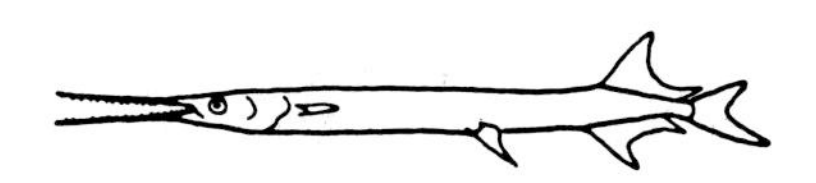

Belonidae (p 90)

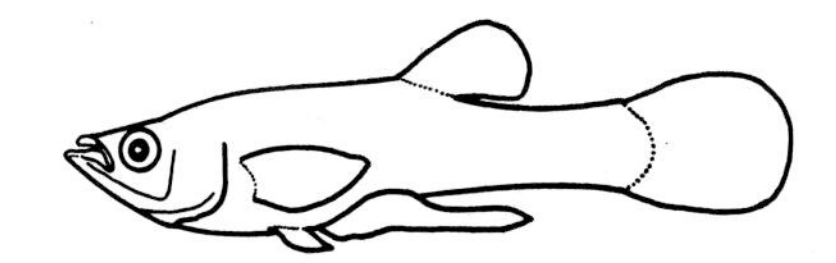

Poeciliidae (p 90)

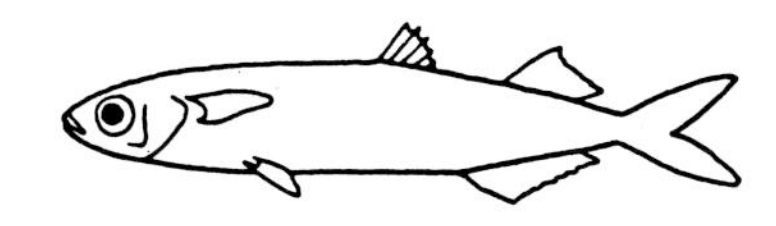

Atherinidae (p 91)

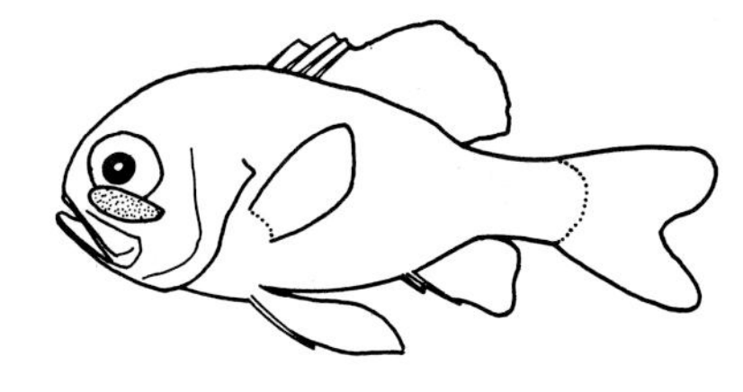

Anomalopidae (p 94)

Holocentridae (p 94)

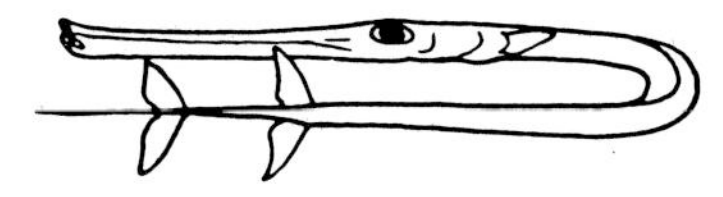

Fistulariidae (p 96)

Aulostomidae (p 96)

Syngnathidae (p 96)

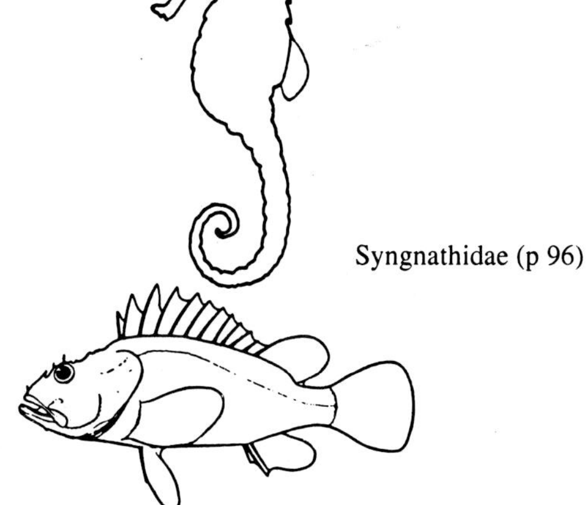

Syngnathidae (p 96)

Scorpaenidae (p 98)

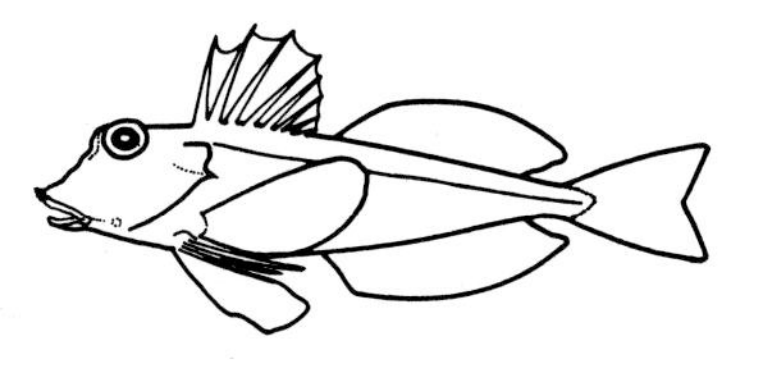

Triglidae (p 101)

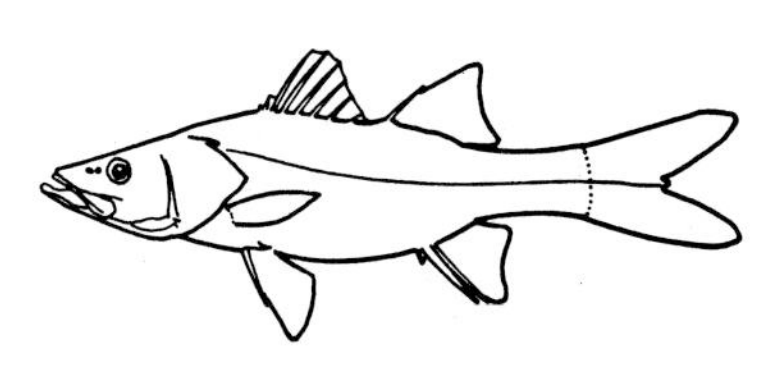

Centropomidae (p 104)

Serranidae (p 106)

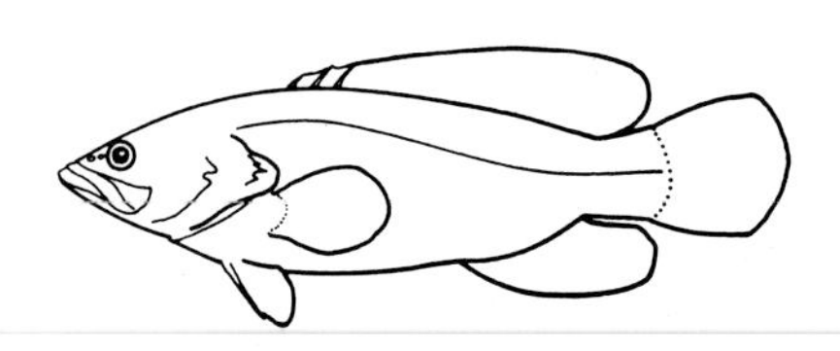

Grammistidae (p 118)

Kuhliidae (p 119)

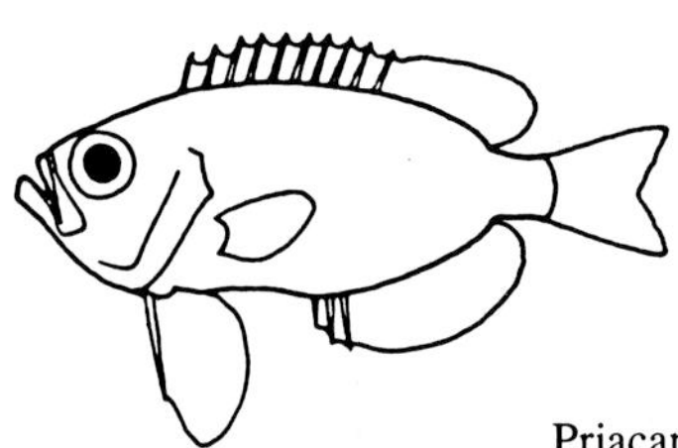

Priacanthidae (p 120)

Apogonidae (p 121)

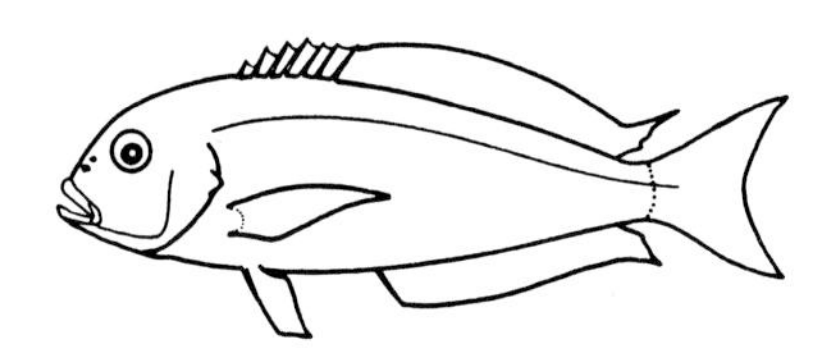

Malacanthidae (p 123)

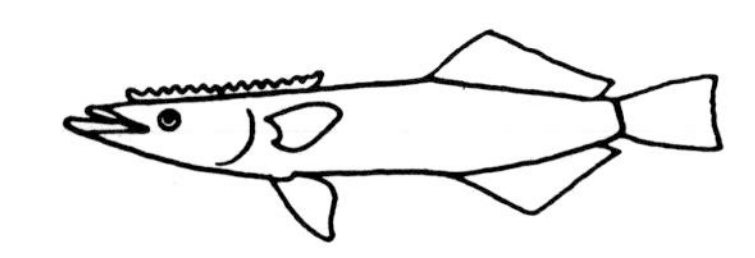

Echeneidae (p 124)

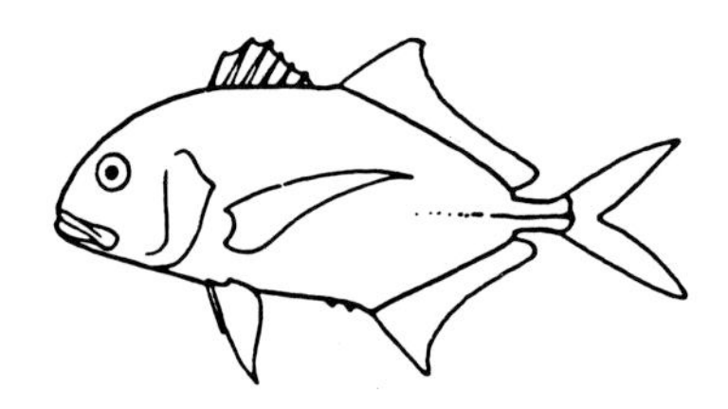

Carangidae (p 125)

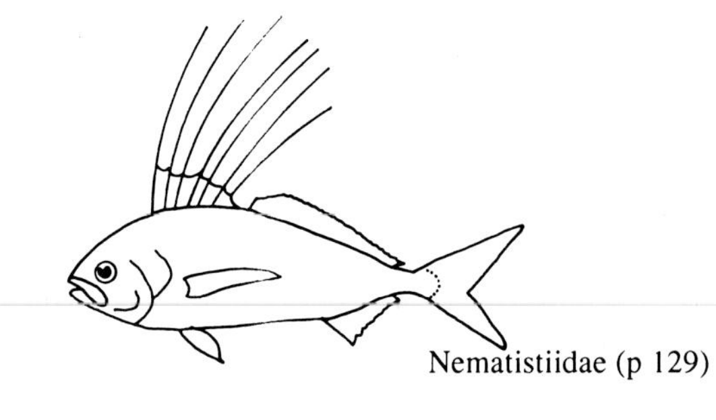

Nematistiidae (p 129)

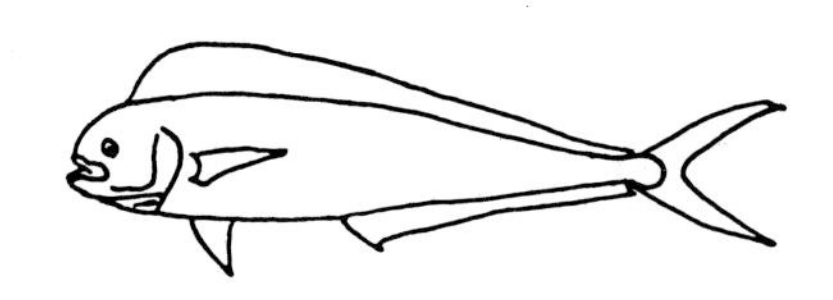

Coryphaenidae (p 129)

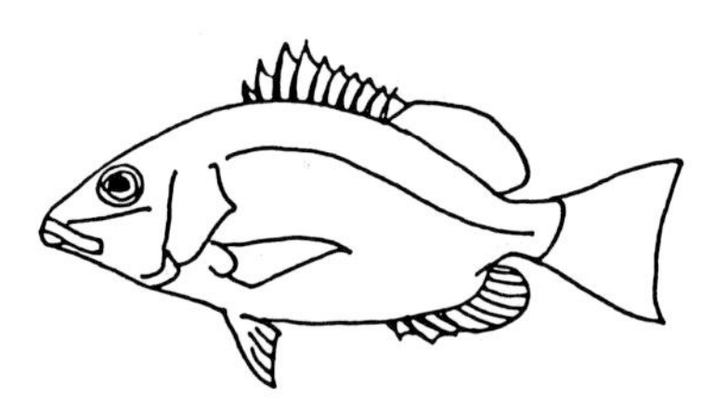

Lutjanidae (p 136)

Lobotidae (p 141)

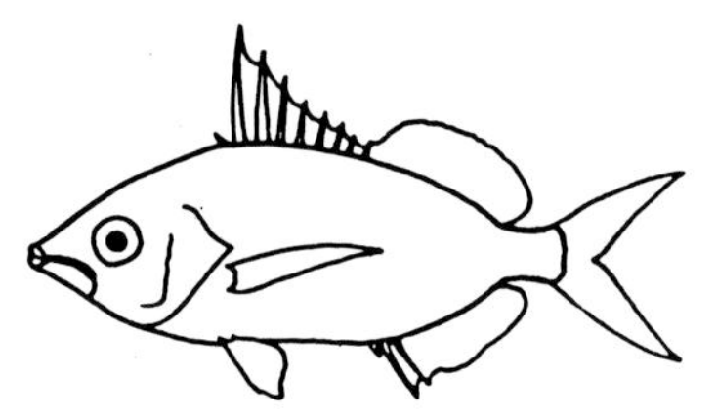

Gerreidae (p 141)

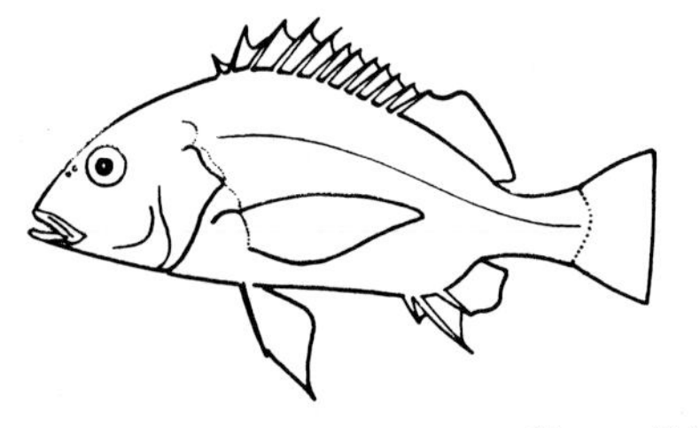

Haemulidae (p 144)

Sparidae (p 155)

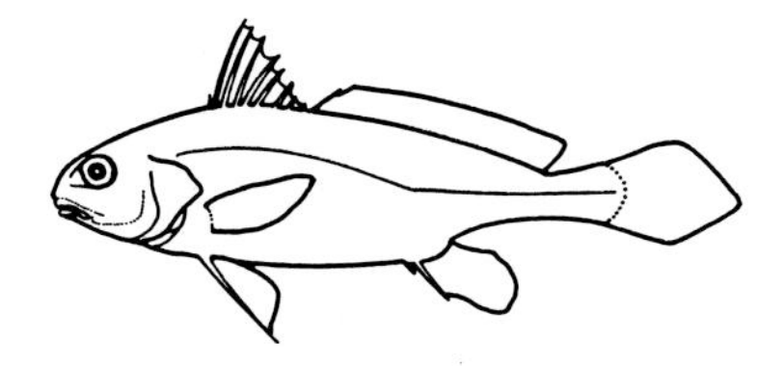

Sciaenidae (p 157)

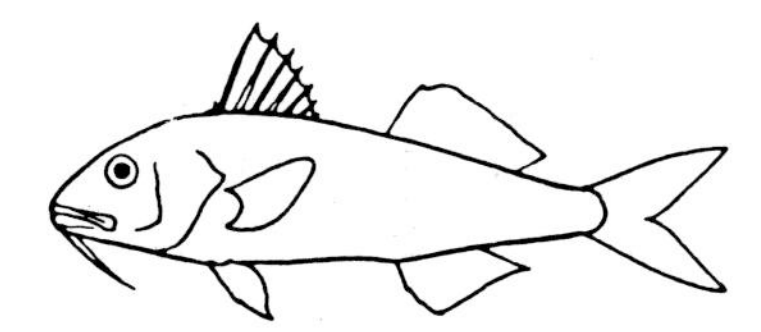

Mullidae (p 174)

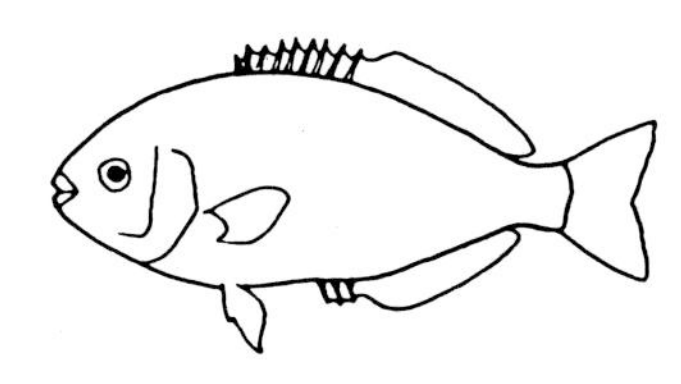

Kyphosidae (p 175)

Ephippididae (p 178)

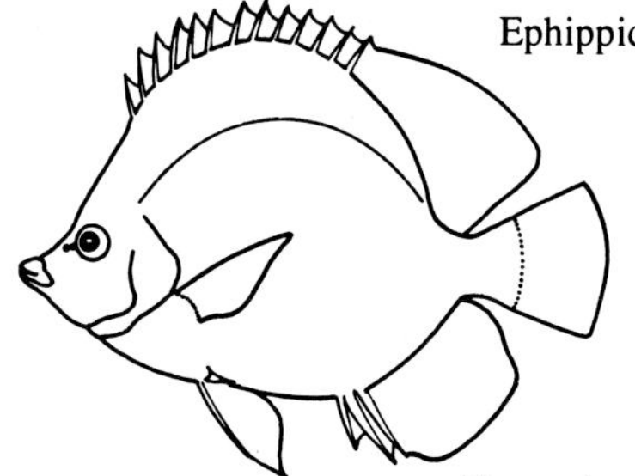

Chaetodontidae (p 179)

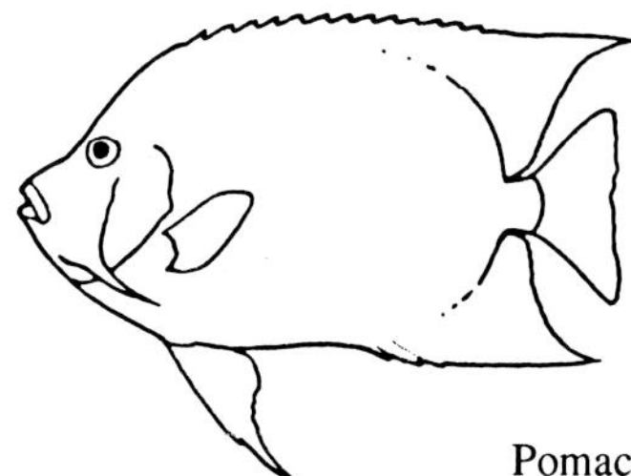

Pomacanthidae (p 181)

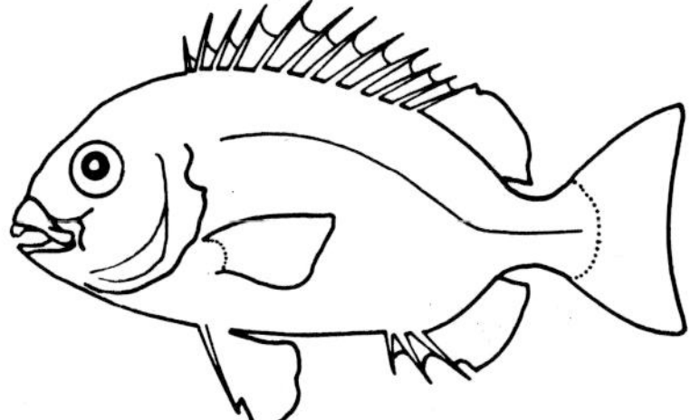

Oplegnathidae (p 183)

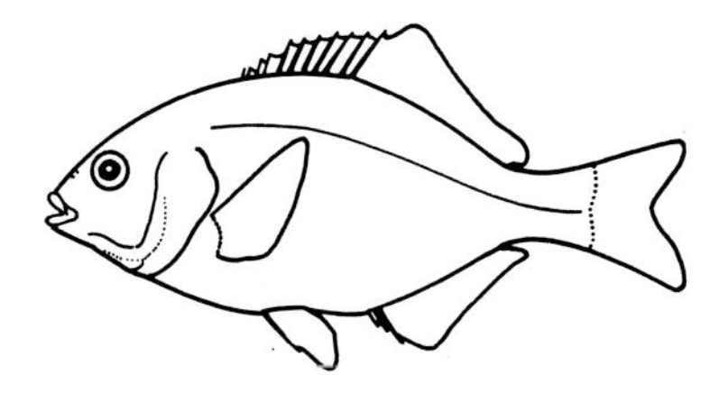

Embiotocidae (p 184)

Pomacentridae (p 185)

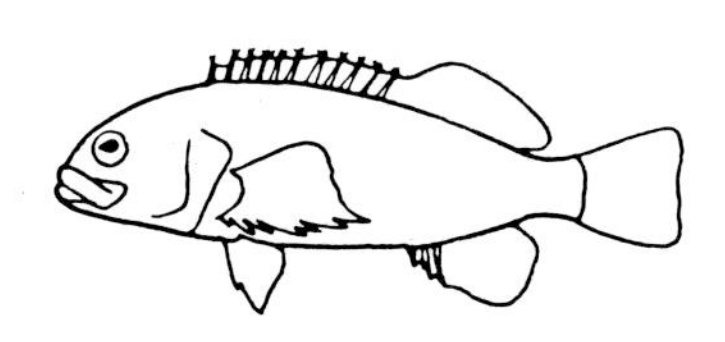

Cirrhitidae (p 192)

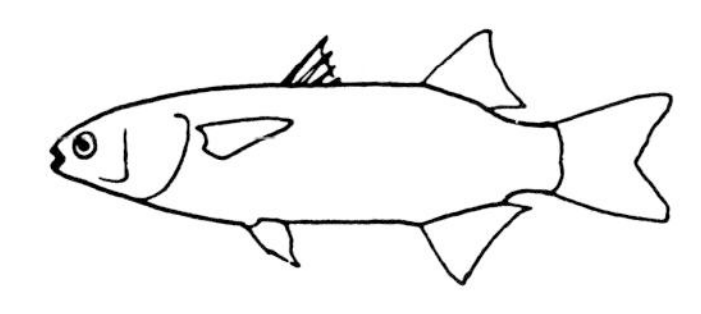

Mugilidae (p 194)

Sphyraenidae (p 195)

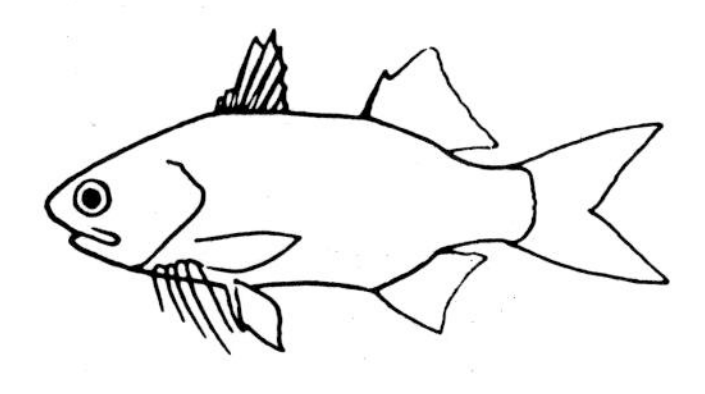

Polynemidae (p 195)

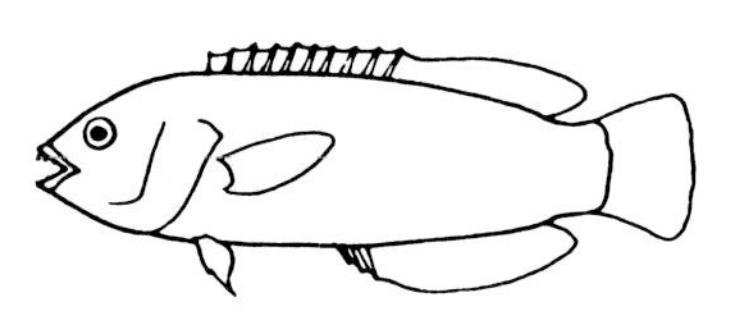

Labridae (p 196)

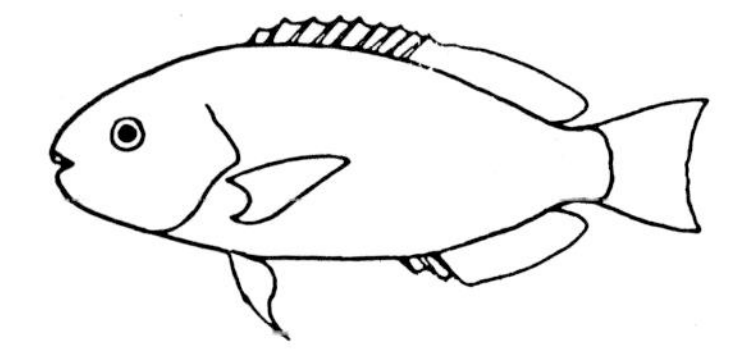

Scaridae (p 209)

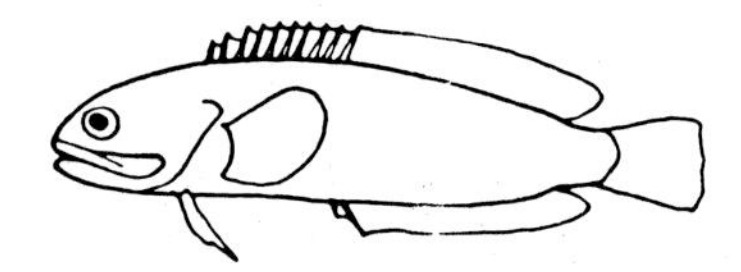

Opistognathidae (p 212)

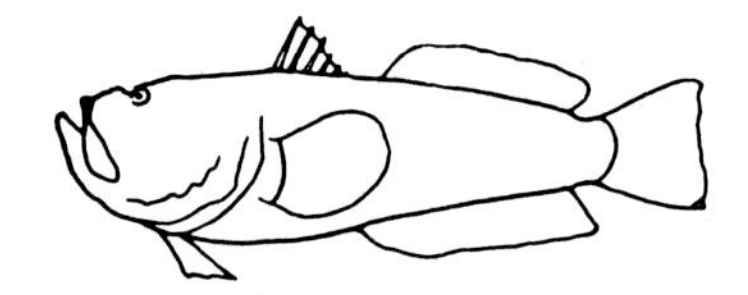

Uranoscopidae (p 215)

Tripterygiidae (p 218)

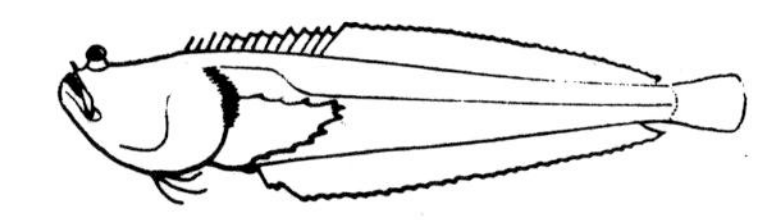

Dactyloscopidae (p 223)

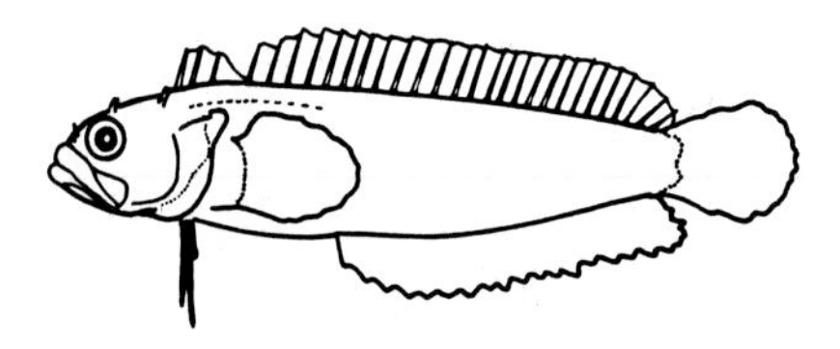

Labrisomidae (p 226)

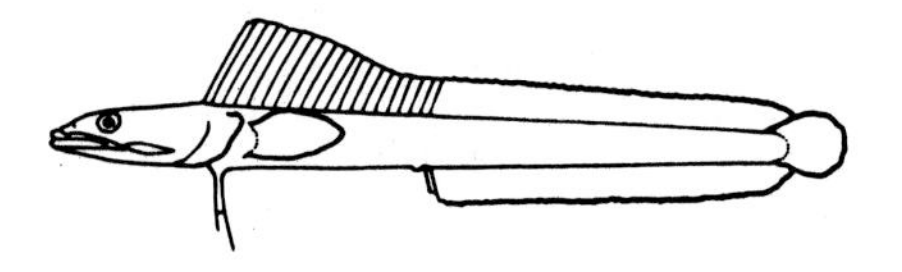

Chaenopsidae (p 238)

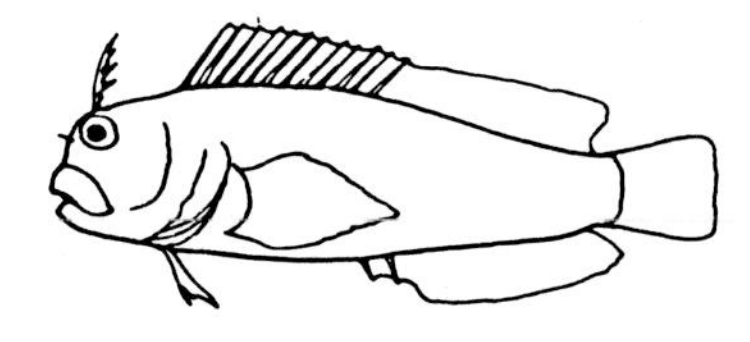

Blenniidae (p 252)

Ammodytidae (p 256)

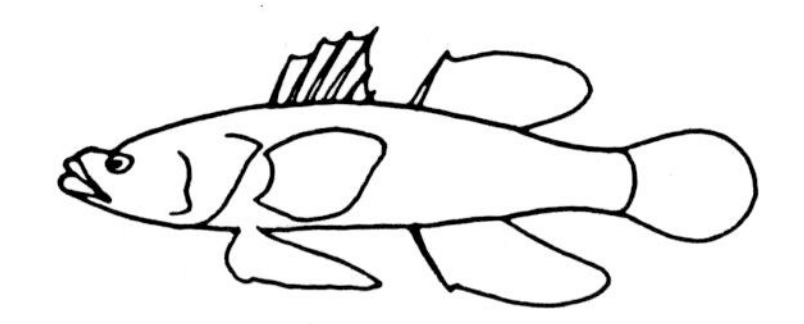

Eleotrididae (p 256)

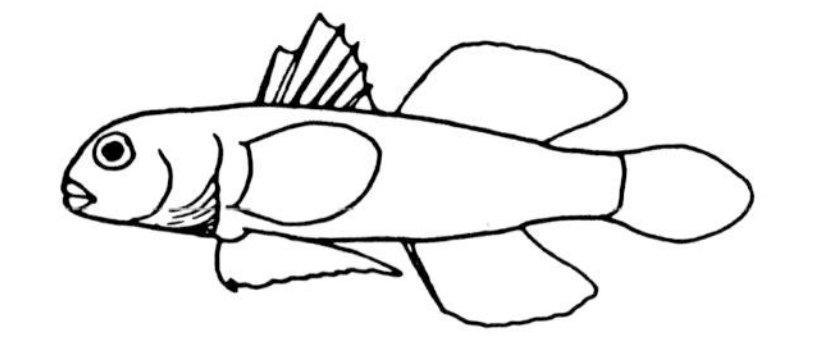

Gobiidae (p 257)

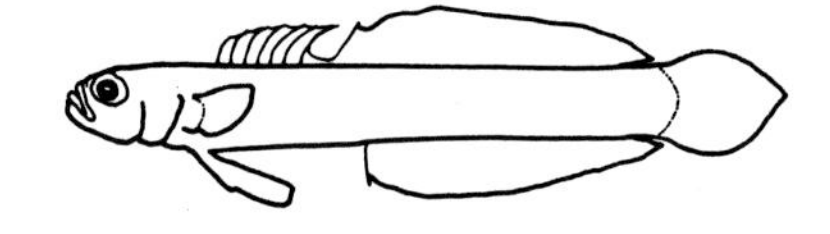

Microdesmidae (p 270)

Acanthuridae (p 271)

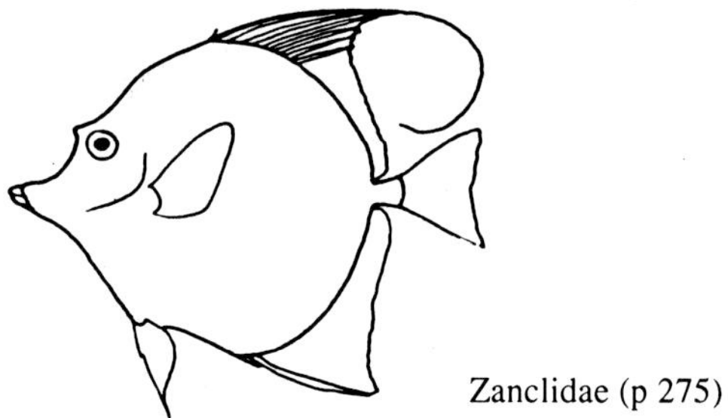

Zanclidae (p 275)

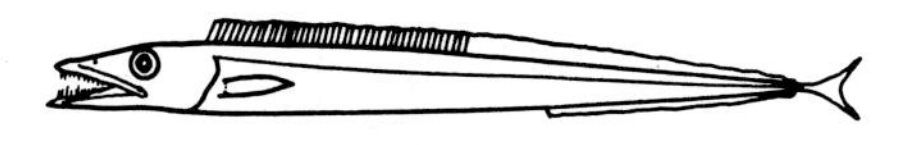

Trichiuridae (p 275)

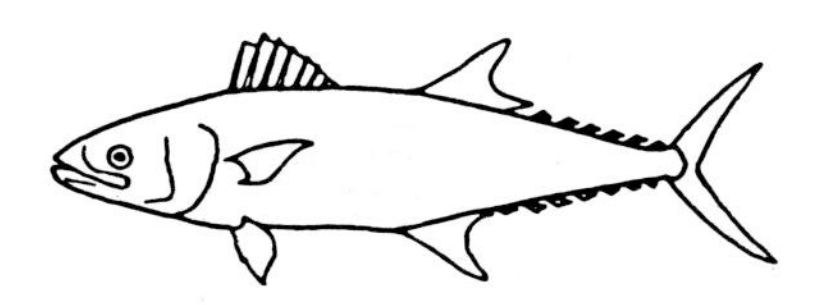

Scombridae (p 276)

Xiphiidae (p 278)

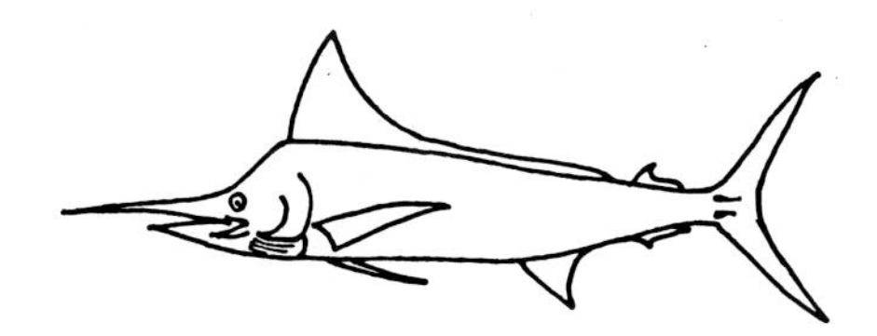

Istiophoridae (p 278)

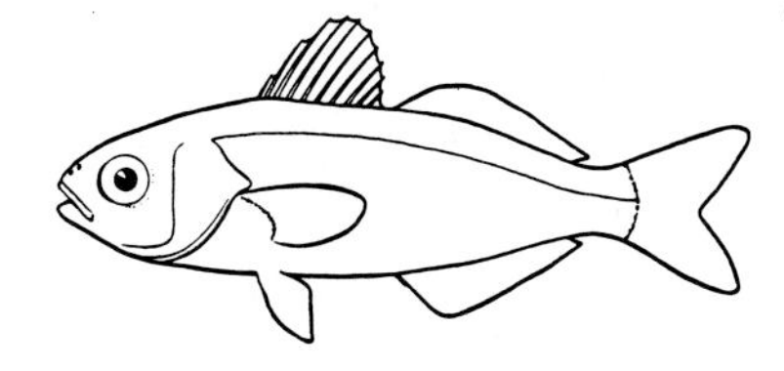

Nomeidae (p 284)

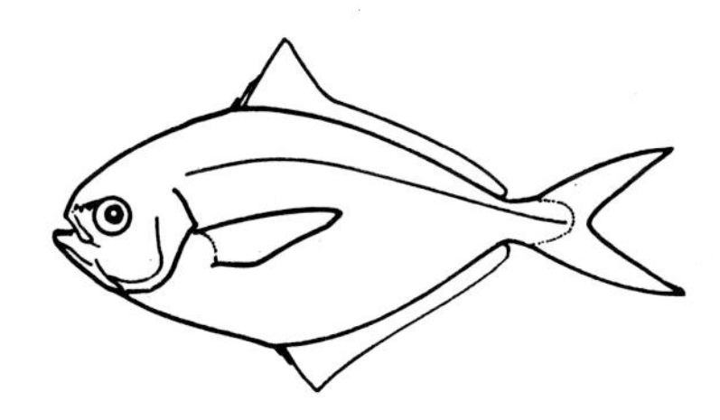

Stromateidae (p 285)

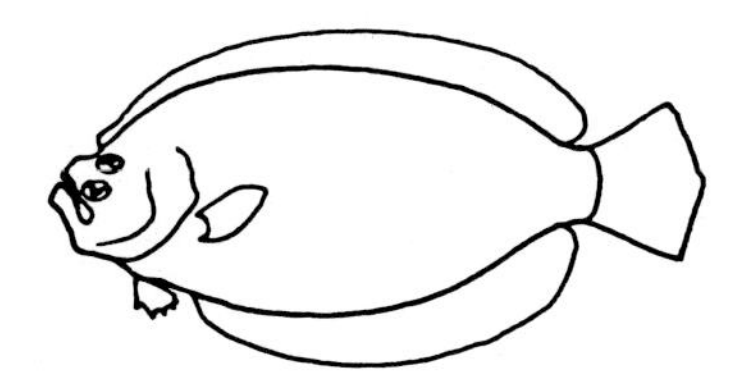

Bothidae (p 285)

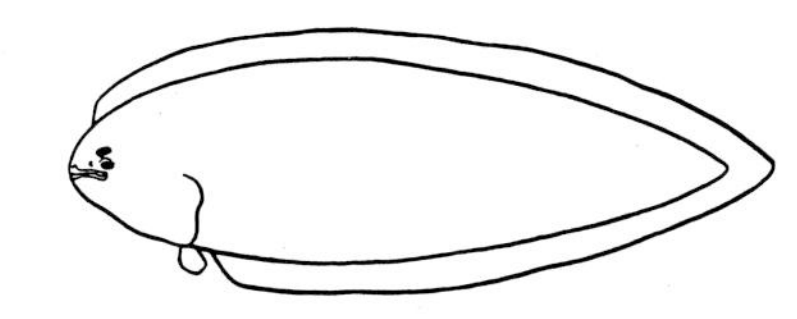

Cynoglossidae (p 290)

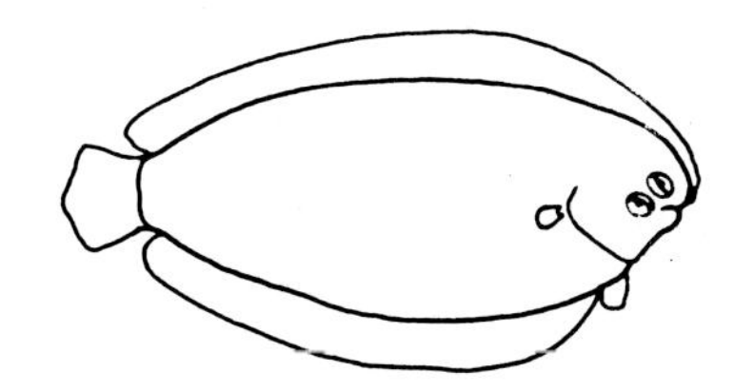

Achiridae (p 292)

Balistidae (p 294)

Monacanthidae (p 296)

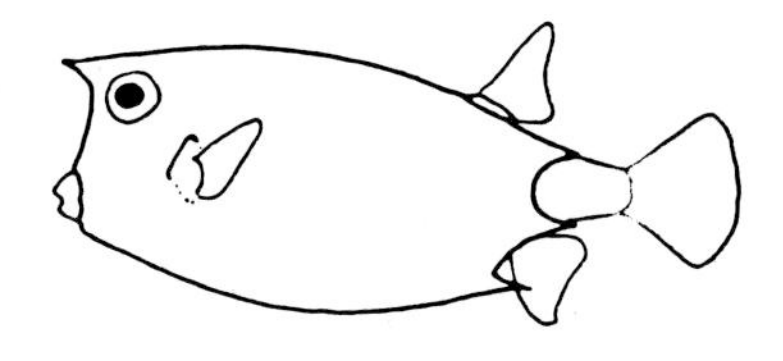

Ostraciidae (p 298)

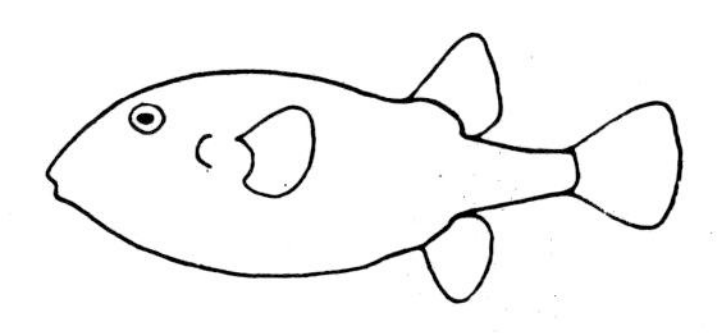

Tetraodontidae (p 299)

Diodontidae (p 301)

INTRODUCTION

Our coverage encompasses the region extending from approximately the central Gulf of California southward to Ecuador, including the offshore Revillagigedos, Cocos, and Galapagos island groups. Primary emphasis is placed on species that inhabit inshore reefs and adjacent sand, rubble, and weed habitats. We also provide coverage of some of the more common members of coastal sand and mud-bottom communities, such as those frequented by shrimp trawlers, and those found in river mouths and cstuarics. Also included are pelagic families of interest to commercial and recreational fishermen; for example, the tunas, mackerels, and billfishes.

There is a rich history of ichthyological exploration in the tropical eastern Pacific. The first scientific collections were obtained during Darwin's voyage to the Galapagos aboard the *Beagle*. Fishes from this expedition were described by Jenyns in 1842. Since that time a number of major works have been published. An extensive listing is presented in the reference section at the end of the book, but several are worth mentioning because of their considerable depth and overall contribution to our knowledge of the region's fish fauna. The four-volume work of Jordan and Evermann (1896-1900), *The Fishes of North and Middle America*, presented an invaluable synopsis of the known fauna as it stood at the turn of the century. This was a solid foundation upon which Jordan, his colleagues, and their students continued to build on during the early part of the twentieth century. Charles H. Gilbert, a collaborator of Jordan's, also had a considerable impact. He made large collections in Mexico and Panama, and along with Edwin C. Starks (1904) produced *The Fishes of Panama Bay*. Another landmark publication was Meek and Hildebrand's (1923-1928) *The Marine Fishes of Panama*. These are just a few of the larger works, but there exists considerable taxonomic literature devoted to the region's fish fauna. For the most part these consist either of new species descriptions or of important generic revisions. The Scripps Institution of Oceanography (La Jolla, California)

A protected anchorage at Golfito, Costa Rica.

has earned a reputation as a leading center for studies on tropical eastern Pacific fishes. This reputation is largely the result of the work of the late Carl Hubbs, Richard Rosenblatt, and their many students. The institute maintains a large collection of regional fishes, and many important publications are based on this material.

It is somewhat surprising that a comprehensive illustrated guide to the region's fishes has not been published in recent times, given the large body of knowledge that has accumulated and the highly unique nature of the fauna. One exception, although its coverage is limited to the northernmost part of the region, is *Reef Fishes of the Sea of Cortez* (Thomson et al., 1979). This book provides good coverage of the Gulf of California. In addition, Chirichigno's (1974) illustrated guide to Peruvian marine fishes is very useful for species identification throughout the region. Two other important works are currently in preparation. These include a guide to Galapagos fishes by Lavenberg and Grove, and a multivolume species identification guide to the tropical eastern Pacific region to be published by the Food and Agriculture Organization of the United Nations (FAO).

The eastern Pacific has a relatively impoverished inshore fish fauna, compared with other tropical areas, such as the Caribbean Sea and the Indo-West Pacific. The reasons for this impoverishment are complex, but several are major contributors. One involves the physical size of the region; although the mainland coastline seems extensive, there are few offshore islands, and due to cool upwellings and currents that sweep into the region from adjacent temperate areas, the tropics are significantly constricted. Although the warm-water region extends northward to the Gulf of California, it terminates rather abruptly at about Cabo Blanco, Peru, due to the influence of the cool Peru Current. There is also a significant reduction in reef habitat diversity compared with the

(Opposite) An imposing rock spire at Isla Santiago, Galapagos.

The busy fishing port of Guaymas, Sonora, Mexico.

Hordes of pelicans attracted to shrimp trawlers in Panama Bay.

Caribbean and Indo-West Pacific. Very few species of reef-building corals are present, hence a lack of the numerous microhabitats offered on a fully developed coral reef. Lastly, the region is isolated, with minimal recruitment of fauna from the neighboring central western Pacific, apparently due to contrary current patterns for larval transport, and the great distances involved to the nearest island "stepping stones" to the east.

The tropical eastern Pacific fish fauna exhibits a very strong regional endemicity at both generic and specific levels. There is also a strong relationship with the western Atlantic shore fauna. Rosenblatt (1967) noted that about 25 percent of a total of 479 genera found in the New World tropics occurred on both sides of the Central American Barrier. He further noted that 14 percent, or about 67 genera, were endemic to the eastern Pacific. Of the shore fishes (excluding the pelagic tunas, billfishes, and nomeiids) treated in the present volume, fully 84 percent of the species are endemic to the eastern Pacific. The remaining species are either shared with the Indo-West Pacific region (7 percent) or the western Atlantic (2.5 per cent), or have a circumglobal distribution (6.5 percent). The low percentage of species shared with the western Atlantic is somewhat deceiving as there are numerous species pairs composed of a close relative occurring on either side of the Central American isthmus. The members of these pairs are very similar in appearance, and are often referred to as geminate or twin species. The spadefishes of the genus *Chaetodopterus* present a good example of this phenomenon; *C. zonatus* of the Pacific is nearly identical to the Atlantic *C. faber*. Numerous examples of similar twins occur in many other families. The high number of shared genera and the occurrence of twin species reflect the common ancestry of the eastern Pacific and Atlantic faunas. The two areas were interconnected until uplifting of the Central American land bridge closed the gap, probably in the early Pliocene epoch.

Although relatively small compared to its Indo-West Pacific and Atlantic neighbors, the shore-fish fauna of the eastern Pacific tropics has a special "flavor" imparted by its large number of endemic species and the unique faunal "mix" or community of fishes that is present. Many coral-reef families that are well represented in other seas, particularly the Indo-West Pacific, have very few species. Examples of this phenomenon are found in a number of reef families including Holocentridae, Scorpaenidae, Apogonidae, Mullidae, Chaetodontidae, Pomacanthidae, Scaridae, Blenniidae, and Acanthuridae. However, at least one family, the Sciaenidae, exhibits the opposite trend, and has undergone a tremendous radiation in the eastern Pacific. The region also has a large indigenous representation among the reef-dwelling blenny families Labrisomidae and Chaenopsidae. Both are absent from the vast Indo-West Pacific, but are well represented in the western Atlantic.

The exact limit of zoogeographic subregions or provinces in the eastern Pacific has generated considerable debate. However, most biogeographers would agree that the region is divisible into two major provinces, based on the distribution patterns displayed by the fish inhabitants. The northern or Mexican province extends southwards from the Gulf of California and about Magdalena Bay on the outer coast of Baja California, to the Gulf of Tehuantepec, southern Mexico. Special mention should be made of the Gulf of California, which has a high level of endemism (estimated at about 15 percent). Springer (1958) proposed the name "Pacific Central America Faunal Gap" for the area between the Gulf of Tehuantepec and the Gulf of Fonseca, Honduras. He noted that this coast, occupying some 1 000 km, was apparently devoid of rocky reefs, consisting instead of sand, mud, or mangroves. This gap appeared to coincide with the distributional limits of certain species of clinid fishes (now grouped in Labrisomidae). The southern or Panamanian subregion extends from the Gulf of Fonseca southward (and westward) to the lower portion of the Gulf of Guayaquil, or about to the border between Ecuador and Peru.

In addition to the mainland zones, there are several discrete island groups that, because of their isolation and degree of endemism, need to be treated as separate entities outside of the major provinces. The volcanic Galapagos Archipelago, lying nearly 1 000 km off the coast of Ecuador, is composed of thirteen major islands and numerous small islets. The Galapagos fauna, containing approximately 300 inshore species, is an interesting blend from the tropical eastern Pacific, temperate South America, and Indo-West Pacific, and there is a significant percentage of endemism, estimated at about 17 percent. The faunal

Golfo Dulce, Costa Rica.

complexity of the Galapagos is partly explained by its diversity of habitat, including mangroves, extensive sandy areas, rocky reefs, and limited areas of good coral growth. There is also a complex interaction of warm tropical seas and cold waters of the Peru or Humboldt Current. In recent years, an increasing number of species from the Indo-West Pacific has been reported in the archipelago. It is thought that these fishes are periodically transported as larvae by unusual current gyres that flow in from the west. These "vagrant" species are omitted from our coverage because they are extremely rare, in most cases having been reported on the basis of only one or two individuals. Among the species involved are the butterflyfishes *Chaetodon auriga, C. kleini,* and *C. lunula*, and the sharpnose puffers *Canthigaster amboinensis* and *C. janthinoptera* (Grove and Lavenberg, in press).

Malpelo Island is situated about 400 km off the Colombian coast. This rocky outpost is volcanic in origin, largely devoid of vegetation, and rises steeply to an elevation of approximately 450 m. The main island is 2 km in length and about 1 km at its widest point. In addition, there are a number of rocky outcrops and pinnacles in the immediately adjacent waters. McCosker and Rosenblatt (1975) recorded 70 fish species, including several possible endemics. Most of the fauna is eastern Pacific mainland in composition, although five species are shared only with the Galapagos and Isla del Coco. Endemism is estimated to be less than 5 percent.

Isla del Coco lies about 560 km northeast of the Galapagos, and some 480 km west of Costa Rica. It is a small, steep-sided, well-vegetated island, with a land area of about 47 square km. Approximately 100 shore fishes are known from the island, of which about 88 percent are Panamic, and the remainder are a mix of endemics (about 5 percent), and either Indo-West Pacific or circumglobal species.

Tiny Clipperton is the only true atoll in the eastern Pacific. This small reef complex is only about 3 km in diameter and lies slightly more than 1 000 km from the mainland southwest of Acapulco, Mexico. The fishes of this remote outpost remain poorly studied, but there appears to be a larger percentage of Indo-West Pacific species than at other offshore islands in the region.

The Revillagigedo Group, consisting of three small volcanic islands and several adjoining rocky islets, lie about 400 km south of the tip of Baja California. Although the fauna has not been thoroughly investigated, it appears to consist of about 120 species, including about 10 percent endemics.

Biology of Shore Fishes

The diversity of shore fishes found in the tropical eastern Pacific is reflected in a wide variety of reproductive habits and life-history strategies. The following discussion is intended to give an overview of the most common patterns. More detailed information is available in the scientific literature, or semipopular works such as Thresher's (1984) *Reproduction in Reef Fishes.*

The majority of shore fishes are egg layers that employ external fertilization. Relatively few species bear live young that are prepared to fend for themselves at birth. Included in the latter category are sharks, rays, and cuskeels. Basically, two patterns of oviparous or egg-laying reproduction are evident in most reef species. Females of many fishes, including the highly visible wrasses and parrotfishes, scatter relatively large numbers of small, positively buoyant eggs into open water, where they are summarily fertilized by the male. The spawning event is typically preceded by nuptial chasing, temporary color changes, and courtship displays in which fins are erected. This behavior is generally concentrated into a short period, often at sundown or shortly afterwards. This pattern is seen in diverse groups such as lizardfishes, angelfishes, wrasses,

A mantle of snow-white guano is typical of rocky islets in the Gulf of California, such as Isla Farralon, shown here.

Tranquil waters near Golfito, Costa Rica.

parrotfishes, and boxfishes. Typically, either pair or group spawning occurs, in which the participants make a rapid dash towards the surface, releasing their gonadal products at the apex of the ascent.

The fertilized eggs float near the surface and are dispersed by waves, winds, and currents. Hatching occurs within a few days, and the young larvae are similarly at the mercy of the elements. Recent studies of the daily growth rings found on the ear bones (otoliths) of reef fishes indicate that the larval stage generally varies from about one to eight weeks, depending on the species involved. The extended larval period no doubt accounts for the wide dispersal of many reef species. For example, many fishes that occur in other Indo-Pacific regions have geographic ranges that extend from East Africa to Polynesia, or range even further eastward to the Americas.

A second reproductive pattern involves species that lay their eggs on the bottom, frequently in rocky crevices, empty shells, sandy depressions, or on the surface of invertebrates such as sponges, corals, or gorgonians. Among the best-known fishes in this category are the damselfishes, blennies, gobies, and triggerfishes. Prior to egg deposition, these fishes often prepare the surface by cleaning away detritus and algal growth. Bottom spawners also exhibit elaborate courtship rituals, which involve much aggressive chasing and displaying. This behavior has probably been best studied amongst the damselfishes. In addition, one or both parents may exhibit a certain degree of nest-guarding behavior, in which the eggs are kept free of debris and guarded from potential egg-feeders, such as wrasses and butterflyfishes. A very specialized mode of parental care is seen in cardinalfishes, in which the male broods the egg mass in its mouth. Similarly, male pipefishes and seahorses brood their eggs on a highly vascular region of the belly or underside of the tail. As a rule, the eggs of benthic nesting fishes are less numerous, larger, have a longer incubation period, and are at a more advanced developmental stage when hatched, compared with the eggs and larvae of pelagic spawning fishes. Incubation may last several days, and the larvae then lead a pelagic existence for up to several weeks before settling on the bottom in a suitable reef habitat.

Arid hills meet the sea at Isla Cerralvo, Gulf of California.

There is very little information on the longevity of most reef fishes. Perhaps one of the longest life spans is that of the Lemon Shark, which may reach 50 years or more. Most of the larger reef sharks probably live to an age of at least 20 to 30 years. In general, the larger reef fishes such as groupers, snappers, and emperors tend to live longer than smaller species. Otolith aging techniques indicate that large groupers may live at least 25 years, and some snappers approximately 20 years. Most of our knowledge of smaller reef fishes has resulted from aquarium studies. The values obtained from captive fishes may exceed the natural longevity due to lack of predators and the protective nature of the artificial environment. Species of *Platax* from the Indo-West Pacific are known to survive for 20 years, and even small species such as damselfishes and angelfishes may reach an age of 10 years or more.

Ecology of Reef Fishes

The fishes included in this book are mainly inhabitants of inshore environments, particularly reefs and adjacent habitats. In general, tropical shore fishes are finely synchronized with their environment. Each species exhibits very precise habitat preferences that are dictated by a combination of factors, including the availability of food and shelter, and various physical parameters such as depth, water clarity, currents, and wave action.

The majority of inshore fishes in the region – at least the majority of those likely to be encountered by divers and anglers – are associated with rocky habitats. Rock outcrops and ledges provide necessary shelter, and a source of benthic invertebrates and algal growth, important dietary components for many reef fishes. Rocky reefs often have a dense covering of various types of seaweed, which provides additional shelter for a variety of small fishes. Some of the more prominent families found on rocky reefs include moray eels, snappers, grunts, damselfishes, wrasses, parrotfishes, blennies (three families), gobies, and surgeonfishes.

Reef environments in the eastern Pacific can be broadly classified into two major categories: coastal mainland reefs, and offshore or island reefs. Inshore or coastal reefs may be strongly influenced by freshwater run-off and resultant siltation. Underwater visibility on these reefs is usually poor or nonexistent, particularly during the wet season, when rivers are flowing at their maximum. Offshore reefs are usually more pristine and less likely to be affected by freshwater run-off and consequent siltation. Hence, underwater visibility is sometimes greater than 20 m. Conditions are more suitable for coral, and indeed, the best growth is found well away from the mainland. Coral growth is most abundant between about 5 and 15 m depth. In shallower water, corals are inhibited by the pounding surge, and in deeper water they are inhibited by the much reduced penetration of light.

Although coral reefs represent the main habitat for reef fishes in most other parts of the vast Indo-Pacific tropics, they are poorly represented in the eastern Pacific. In most areas, only scattered, somewhat isolated coral formations are found, but one notable exception is the Gulf of Chiriqui, just south of the westernmost part of Panama. This area is not subjected to cold seasonal upwellings, unlike the nearby Gulf of Panama. Because of the more stable thermal regime, there is extensive development of certain hermatypic corals to depths of 10 to 15 m. Although coral diversity is relatively low, several species of *Pocillopora* are present, as well as species of *Porites*, *Pavona*, and the hydrocoral *Millepora*. Several fishes that are primarily distributed on coral reefs of the Indo-West Pacific (for example, the Convict Surgeonfish, *Acanthurus triostegus*) have only been reported from islands in the Gulf of Chiriqui, and other offshore areas such as Clipperton, Revillagigedos, and the Galapagos.

Another large and important segment of the eastern Pacific fauna is associated with soft-bottom habitats. These range from clean, white sand in clear waters, to soft, thick mud, characteristic of turbid bays and estuaries. Although this habitat is often considered to be low in fish diversity, the total number of species utilizing this environment is surprising. The soft-bottom community includes such fishes as rays, snake eels, catfishes, lizardfishes, croakers, sand stargazers, and

Lush tropical vegetation shrouds many of the islands in the Gulf of Chiriqui, Panama. Isla Uva is shown here.

Los Islotes, a group of rocky islets off La Paz, Baja California.

flatfishes.

Tropical shore fishes generally exhibit a high degree of habitat partitioning. The tube blennies (family Chaenopsidae) provide a good example of the fine scale on which this principle operates. Many of the species live in a narrow zone which is determined by the presence of particular types of barnacle shells or polychaete worms, whose empty shells and tubes are utilized for shelter and the site of egg deposition. Water depth is also an important partitioning factor, and there are numerous examples of reef fishes that have well-defined depth ranges. In the broadest sense, there are three main depth categories: shallow (0 to 4 m), intermediate (5 to 19 m), and deep (20 m+). The depth limits of these zones may vary locally, depending largely on the degree of shelter and sea conditions. The shallow environment is typified by wave action, which, in highly protected areas such as coastal bays, may exert its effect down to only a few centimeters. However, the effect of surface waves may sometimes be felt below 10 m on reefs exposed to heavy ocean swells. The intermediate depth zone generally has the greatest abundance of fishes and live corals. Here, wave action is minimal, although currents are often strong.

How to Use this Book

This book is intended as an identification guide to the common inshore fishes, and pelagic species of interest to sport anglers and commercial fishermen. We provide coverage of at least 90 percent of the region's inshore fauna. Coverage for the two areas at the extreme limits of the region, namely the Gulf of California and the Galapagos Islands, is less complete. Both of these areas encompass faunal elements from cooler, temperate waters, which we do not include. However, both of these border regions are covered in separate publications (Thomson et al., 1979, for the Gulf of California, and Lavenberg and Grove, in press, for the Galapagos).

The fishes are presented in phylogenetic order by family (the scientific names of which all end in -idae); in other words, the most primitive fishes are presented first, followed by more advanced forms. With few exceptions, we follow the sequence presented in Nelson's (1984) *Fishes of the World*. At the beginning of each family section we include a general discussion of diagnostic characters, overall geographic distribution, usually an estimate of the number of genera and species, and basic biological facts. Within each family section, the species accounts appear alphabetically by scientific name. This is the italicized name in two (rarely three, in the case of subspecies) parts, the first being the genus name (capitalized), and the second the species name. This is immediately followed by the name of the author(s) who gave the fish its scientific name, and the year in which the description was published. If the author's name is in parentheses, it indicates that the fish was originally placed in a genus that is different from the one that is currently accepted. A few species treated in this book are still undescribed, and are therefore indicated by sp. (an abbreviation of species) following the genus name.

The common names of fishes used in this book have been taken from a variety of sources. In many cases they are names that have been previously published in other books, such as *Reef fishes of the Sea of Cortez* (Thomson et al., 1979). Unfortunately, there is a scarcity of popular fish literature for this region, hence many of the smaller reef fishes previously lacked a common name. In these cases we have introduced a name, often after consultation with appropriate family specialists. We regret that it is not possible in this edition to include local Spanish-language names, but we do provide Spanish family names. Hopefully this problem can be rectified in a subsequent Spanish-language edition.

Most of the fishes which occur in the tropical eastern Pacific were given their names by ichthyologists in the latter half of the nineteenth century or the first half of the present century. In some cases, the same fish was given a different scientific name by different researchers. By the law of priority, the oldest name is the acceptable one, provided it is binomial, was accompanied by a description, and was published on or after 1758, the date of the tenth edition of Carl Linnaeus' *Systema Naturae*, the starting point of our current system of biological nomenclature. Subsequent, invalid names for previously described organisms are called synonyms. Most of the problems involving synonyms have been sorted out, but some older names are still being discovered, which means they must replace names in current usage. We occasionally mention synonyms of species that have undergone a relatively recent name change.

The port of San Carlos, just north of Guaymas, Sonora, Mexico

To accommodate the large number of species in this volume, we give only a brief diagnosis for each. *Features that are particularly useful in distinguishing the species from close relatives are italicized.* Often, the number of spines and soft rays in the fins, and the number of scales in a lengthwise series along the middle of the body, or the number of pored lateral-line scales are important for species determination. *We rely primarily on the color illustrations as the basis for identification.*

We give the maximum length in centimeters (cm) attained by each species. This is frequently an approximation, due to a general lack of information in the literature.

The geographical distribution of each species is given in general terms; for example, Gulf of California to Peru, which is a common pattern for many of the region's fishes.

We include more than one illustration for a number of species which have a different color pattern at different stages of the life cycle (for example, the juveniles of damselfishes). Nearly all the wrasses (Labridae) and parrotfishes (Scaridae) undergo a female to male sex change, often accompanied by striking change in color. Therefore, we include illustrations of both initial phase (either male or female) and terminal phase (male) fish for most species.

Formulas for fin ray counts are given at the beginning of the species accounts of most bony fishes. To differentiate between spines (not branched, nor segmented, and usually sharp-tipped) and soft (segmented, flexible, and often branched) rays, we use the standard ichthyological convention of giving the number of spines in Roman numerals and the number of soft rays in Arabic numerals. A comma between Roman and Arabic numerals indicates that both spines and soft rays are contained in the same fin; a plus sign in the formula indicates that there are two separate fins. For example, the dorsal fin formula X + I,15 indicates there are two separate fins, the first containing 10 spines, and the second with a single spine and 15 soft rays.

The scale count that is most often diagnostic for fishes is the number of pored or tubed scales in the lateral line, which frequently extends from the upper end of the gill opening to the base of the caudal fin. When the lateral line is not apparent, the longitudinal series of scales is counted between the same points.

Occasionally, the count of gill rakers is needed to differentiate closely related species. These are protuberances along the inner edge of the first gill arch (see illustration on page 15). For many fishes it is necessary to utilize a microscope to obtain this count.

The shape of the rear margin of the caudal fin is frequently useful in distinguishing a species. It may be forked, emarginate (inwardly concave), lunate (very deeply concave), truncate (the edge vertical) or rounded (outwardly convex). In addition, the caudal fin may be unusual in shape; for example, lanceolate (long and pointed; most common in gobies and their relatives), or with a rear margin that is double emarginate and pointed in the middle to S-shaped (found mainly in sciaenids).

Various body and fin proportions are often used to characterize a fish. These are presented as percentages of a larger measurement such as the standard length (straight-line distance from tip of snout to base of caudal fin). More often, the proportions are expressed as the number of times the smaller measurement can be divided into the larger. For example, the depth of the body may be expressed as the number of times this measurement can be divided or "stepped into" the standard length. A body depth range of 5-7 would indicate a slender species, whereas a range of 1.3-2 would indicate a relatively high-bodied fish.

In describing bands of color, a stripe refers to a horizontal marking, and a bar is vertically oriented.

External Features of Fishes

Cartilaginous Fishes (Sharks and Rays)

The two illustrations below and the four on the facing page are labelled to show the principal external parts of fishes.

Silvertip Shark
(*Carcharhinus albimarginatus*)

Panamic Stingray
(*Urotrygon aspidura*)

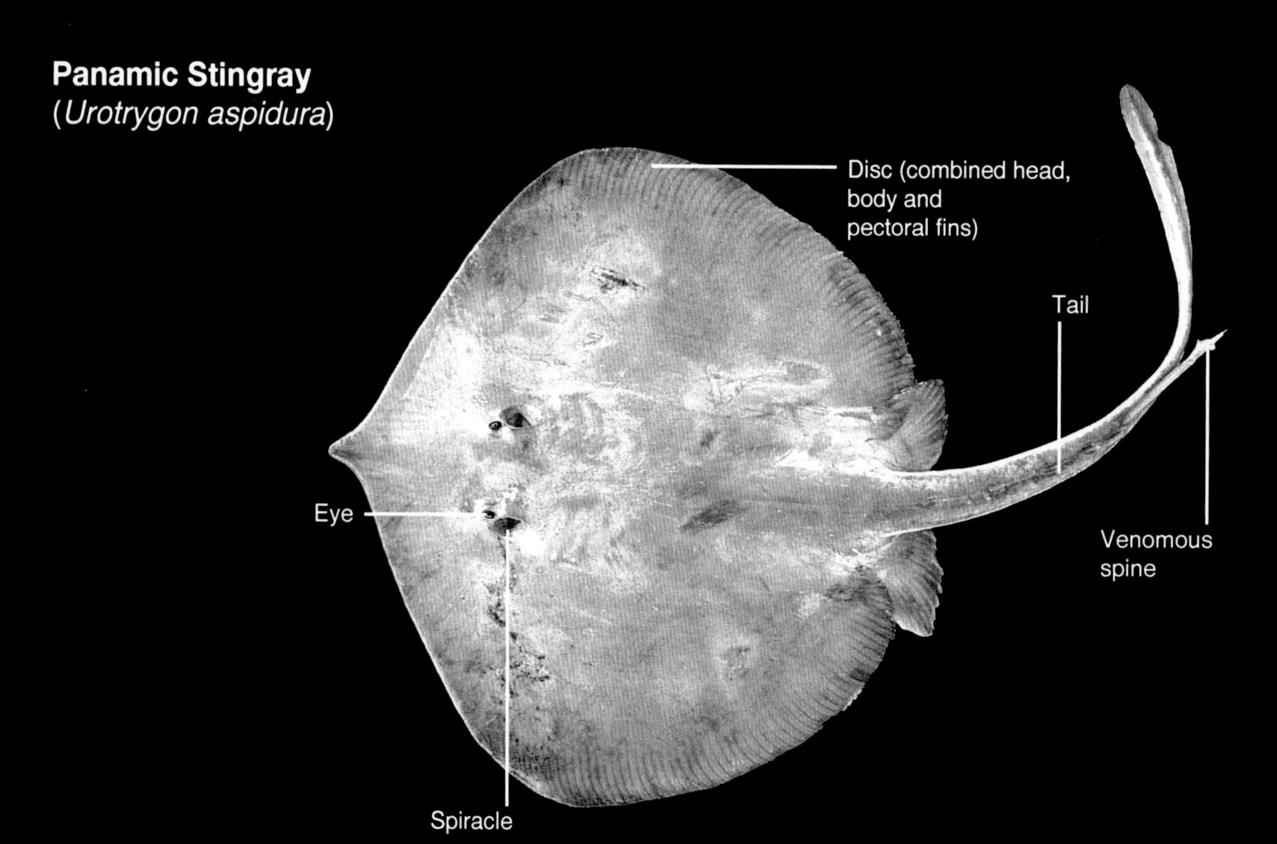

Bony Fishes

Broomtail Grouper
(*Mycteroperca xenarcha*)

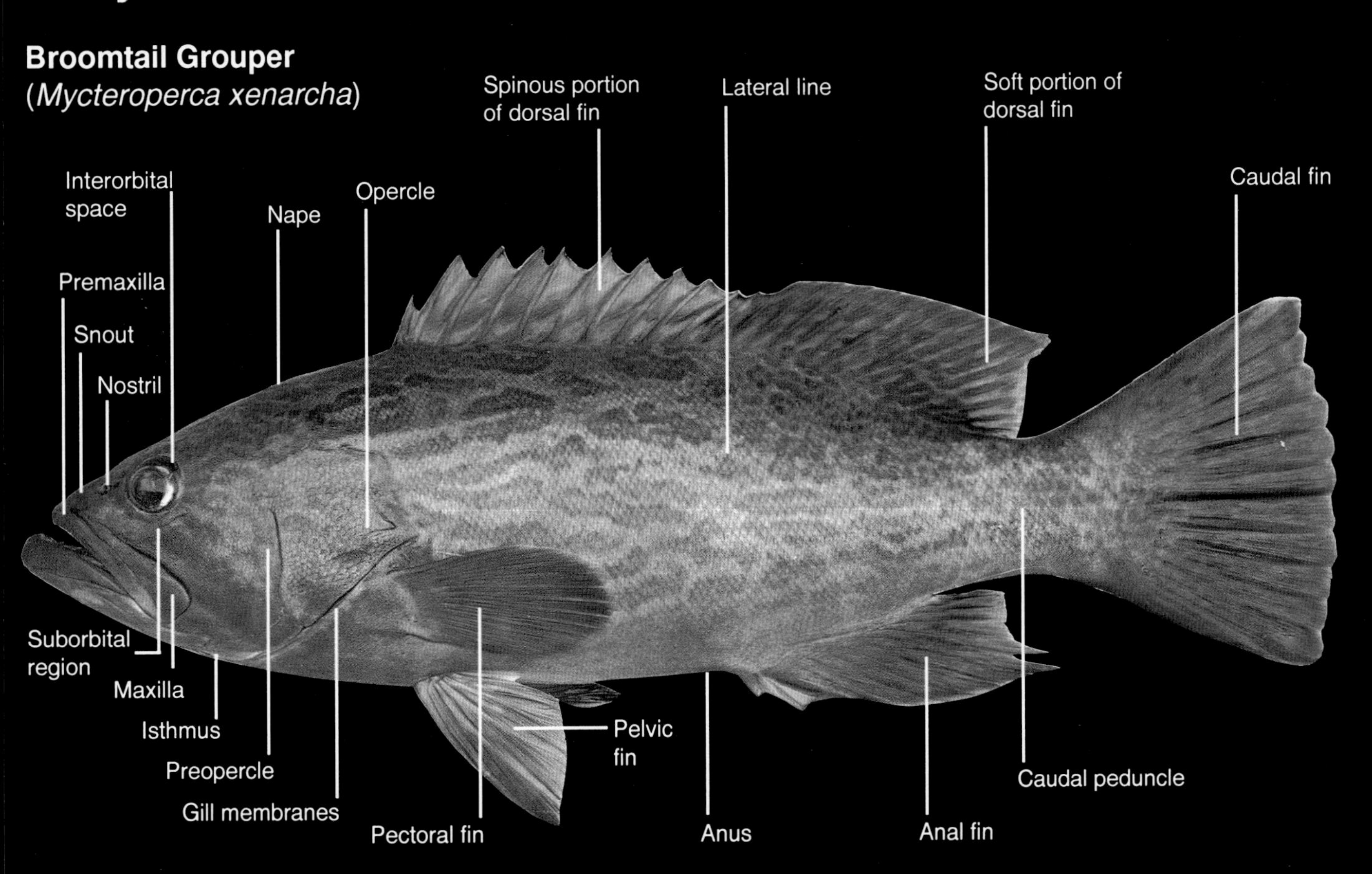

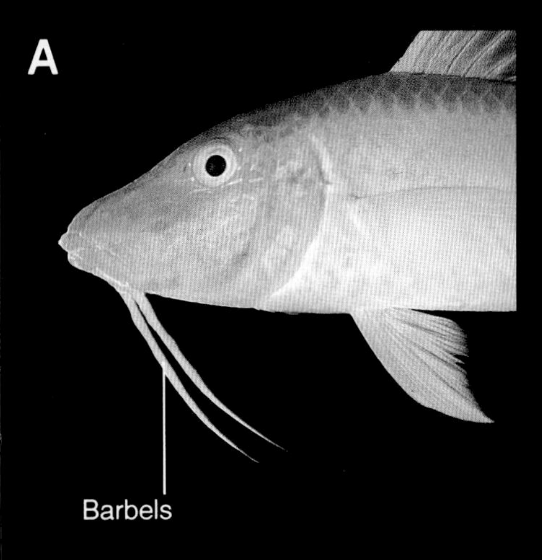

The picture labelled **A** is the head of a goatfish (Mullidae) and shows the pair of barbels on the chin. These are moved over the bottom or thrust into the sediment during feeding to assist the fish in finding its food.

B shows the tail of a trevally (Carangidae) which has a falcate caudal fin; this shape is often found on fishes capable of swimming very rapidly. Because of the stress placed on the narrow caudal peduncle, fishes such as jacks and tunas usually reinforce it with scutes and/or keels.

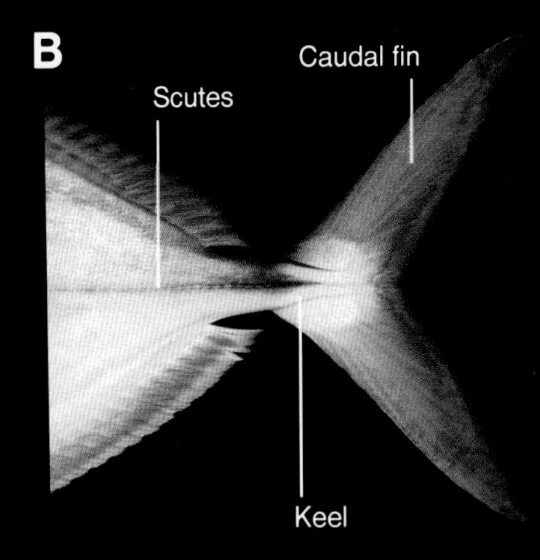

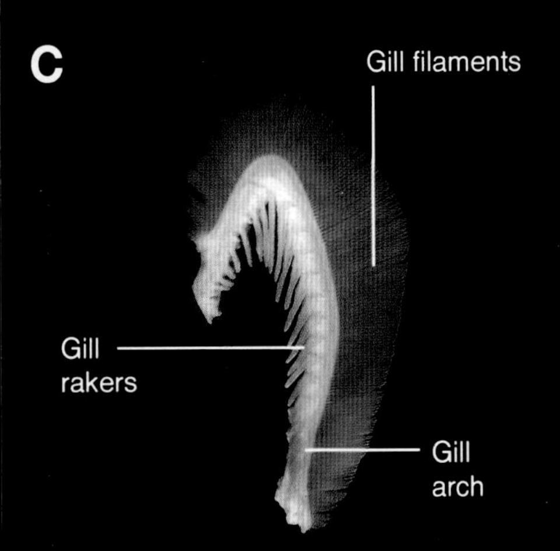

C depicts one of the gills (respiratory organs of fishes). The gill arch is the structural part. Gaseous exchange takes place in the gill filaments and the gill rakers keep food items from passing out of the gill opening along with expired water.

D is the roof of the mouth of a percomorph fish and shows the typical dentition of the premaxilla, vomer and palatine bones.

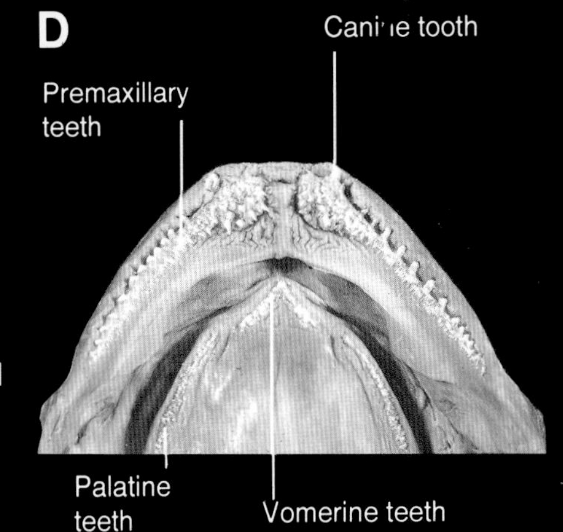

SHARKS (TOLLOS, TIBURONES)

SUPERORDER SELACHIMORPHA

Perhaps no other group of fishes has captured the interest of humankind to the same extent as sharks. The habits and reputations of a relatively small number of species have shaped the popular notion that sharks are basically evil, menacing predators. This is definitely not true, as the majority of the approximately 340 species are no more threatening than most other fishes. Unfortunately a few species such as the Tiger Shark and Great White Shark are known to fatally attack man. Even though the incidence of attacks is small, the danger represented by these animals is consistently blown out of proportion by overzealous journalists. Certainly the automobile is much more menacing than all the sharks in all the world's seas could ever be. Sharks exhibit a remarkable diversity of body shapes and habits. They range in size from the gargantuan Whale Shark that grows to more than 15 m to species that are less than 0.5 m. Sharks are characterized by a cartilaginous skeleton; 5 to 7 lateral gill openings; upper jaw not fused to the cranium; usually numerous sharp, conspicuous teeth in the jaws, that are replaced by replacement teeth from intact rows behind when broken or worn; some short ribs present, but not protecting the body cavity as in most bony fishes; a spiracle opening usually present; males with claspers used to transfer sperm to the cloaca of females. The closely related rays share most of these features, but are generally separated by the body shape (disc-like in rays) and placement of the gill slits. The gill openings are located on the underside of the head, rather than on the sides of the head as in sharks. However, the differences are not absolute, and certain groups, such as sawfishes, guitarfishes, sawsharks, and angelsharks, have intermediate characters. Sharks utilize a variety of reproductive modes; all fertilization is internal. Depending on the species, the embryos may develop freely, attach to a placenta, or be sealed in leathery egg cases. Relatively few species produce egg cases; these are generally deposited in bottom vegetation. Most species bear their young alive in broods that range from a few individuals to nearly 100. The newborn are called pups and usually have the appearance of miniature adults. Sharks are flesh eaters that feed mainly on fishes, crustaceans and molluscs. A few of the larger species take such prey as marine mammals, seabirds, sea turtles, and other sharks. Fish-eating species generally have well developed sharp teeth, often with lateral cusps or incisors, that are designed for seizing and tearing. The tooth shapes are very useful for helping to identify individual species. Sharks have keen vision and an excellent sense of smell. Also, their highly developed lateralis system (composed of special fluid-filled sensory canals) enables them to detect low frequency vibrations at considerable distances, thus facilitating prey detection and the avoidance of predation. They also have a sense to detect an electric field such as that which surrounds a sleeping fish at night. Sharks generally feed at night and have a remarkable adaptation called the tapetum lucidum that is also found in cats and other vertebrates that are nocturnal hunters. This structure, located behind the retina, increases the sensitivity of the shark's eye to the available light.

Heterodontus quoyi (see description on p. 18)

ANGELSHARKS (ANGELOTES, PECES ANGEL)
FAMILY SQUATINIDAE

Angelsharks are unusually shaped creatures characterized by a flattened head and body, greatly expanded pectoral and pelvic fins, and very small dorsal, anal, and caudal fins. The family contains a single genus, *Squatina*, with 13 species. They occur in cool temperate to tropical seas at depths ranging from inshore shallows to at least 1 300 m. Most species are relatively small, usually under about 160 cm, and they are not considered dangerous to humans. They feed on a variety of small fishes, crustaceans, cephalopods, gastropods, and bivalves. At least some species are nocturnal. During the day they are sometimes seen resting, partly buried on mud or sand bottoms.

PACIFIC ANGELSHARK
Squatina californica Ayres, 1859

Anterior half of body and head (including pectoral and anal fins) roughly diamond-shaped and greatly flattened dorsoventrally; eyes dorsally positioned with prominent spiracles behind them; pectoral and pelvic fins greatly enlarged; dorsal fins very small, of about equal size, and situated near base of tail; generally mottled brown or grayish without distinguishing marks. Alaska to Gulf of California, and Ecuador to Chile; occurs in 3-183 m depth, usually seen resting on sand bottoms. Has litters of approximately 10 pups. Maximum size about 152 cm; size at birth 21-26 cm.

HORNSHARKS (DORMILONES, TIBURONES CORNUDOS)
FAMILY HETERODONTIDAE

This family contains a single genus (*Heterodontus*) with eight species that occur in temperate and tropical seas along the fringe of the Indian and Pacific oceans. They are small, easily recognizable sharks distinguished by a squarish head and a stout, sharp spine at the beginning of each dorsal fin. They are slow-moving animals, often seen resting on the bottom among rocks or weeds. Hornsharks are oviparous. They lay unusual, large, spiral-flanged egg cases, usually among rocky crevices. The young are generally over 14 cm long at hatching. The Port Jackson Shark of Australia breeds during late

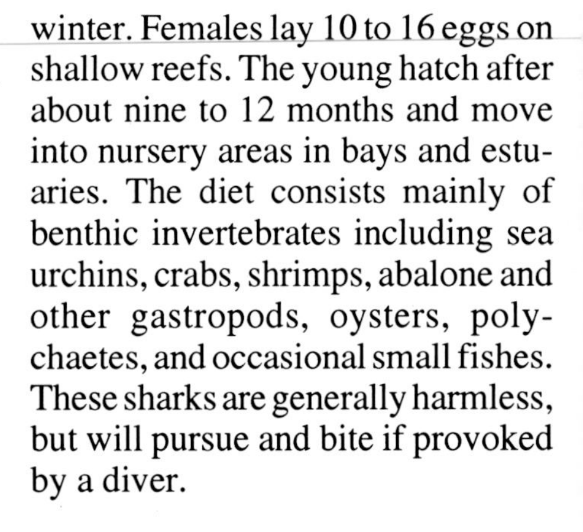

winter. Females lay 10 to 16 eggs on shallow reefs. The young hatch after about nine to 12 months and move into nursery areas in bays and estuaries. The diet consists mainly of benthic invertebrates including sea urchins, crabs, shrimps, abalone and other gastropods, oysters, polychaetes, and occasional small fishes. These sharks are generally harmless, but will pursue and bite if provoked by a diver.

Squatina californica

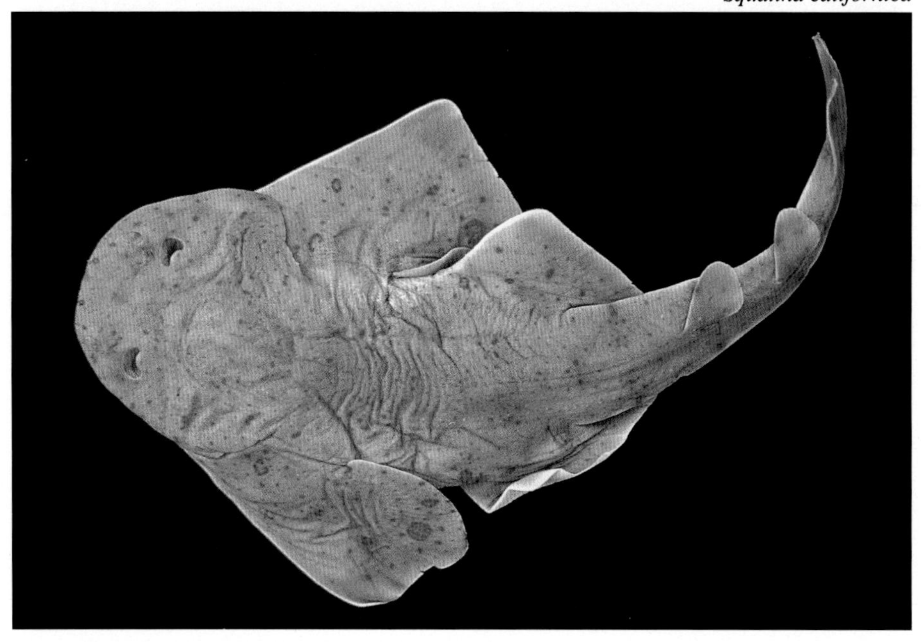

MEXICAN HORNSHARK
Heterodontus mexicanus Taylor & Castro-Aguirre, 1972
(Plate I-4, p. 27)

A slender-bodied shark with an enlarged squarish head and pig-like snout; a low bony ridge above each eye; nostrils without barbels; with nasoral and circumnarial grooves that are connected to the mouth; anterior nasal flaps elongate posteriorly; small mouth with enlarged molariform teeth posteriorly; *both dorsal fins with a sharp spine; origin of first dorsal fin over pectoral fin bases; gray-brown with scattered large black spots on fins and body*. Gulf of California to Peru; occurs on rocky to sandy bottoms to at least 20 m depth. To 70 cm. *H. francisci* occurs in the northernmost part of our area (Gulf of California to central California) and is similar in appearance but has much smaller, more numerous spots.

GALAPAGOS BULLHEAD SHARK
Heterodontus quoyi (Freminville, 1840)
(Photograph on p. 17)

A slender-bodied shark with an enlarged squarish head and pig-like snout; a low bony ridge above each eye; nostrils without barbels; with nasoral and circumnarial grooves that are connected to the mouth; anterior nasal flaps elongate posteriorly; small mouth with enlarged molariform teeth posteriorly; *both dorsal fins with a sharp spine; origin of first dorsal fin behind pectoral fin bases; gray or light brown with blackish, leopard-like spotting*; a series of diffuse dark gray bars may be present across back and sides. Galapagos Islands and Peru; often seen resting on sand adjacent to rocky reefs in about 3-20 m depth. Reaches 83 cm.

NURSE SHARKS (GATAS, TIBURONES DE BARBILLA)

FAMILY GINGLYMOSTOMATIDAE

This small family contains three species in two genera and has a circumglobal distribution in tropical and subtropical seas. These sharks occur on coral and rocky reefs, usually close to shore. Diagnostic features include two relatively close-set dorsal fins of about equal size on posterior half of body, a pair of barbels below the snout, and a groove between each nasal opening and the corner of the mouth. Knowledge of the reproduction is incomplete, but at least one species (*Ginglymostoma cirratum* from the Atlantic and eastern Pacific) is ovoviviparous with young that are nourished mainly by yolk while in the uterus; litters of 20 to 30 young have been reported. These sharks cruise around the bottom in search of food, with their mouth and barbels close to the substrate. They are nocturnally active and ingest prey with a powerful sucking action. The diet includes mainly fishes, crabs, shrimps, lobsters, other crustaceans, and cephalopods. Nurse sharks are generally considered harmless, but may bite and inflict serious injury if provoked.

NURSE SHARK
Ginglymostoma cirratum (Bonnaterre, 1788)
(Plate I-7, p. 27)

A large brownish shark with 2 dorsal fins of similar size; a pair of barbels below snout; nasoral grooves present; mouth inferior with multicusped teeth; caudal fin moderately long, about one-third of total length; yellow-brown to gray-brown, with or without small dark spots and obscure saddle markings on back. Both sides of the tropical Atlantic Ocean and eastern Pacific from the Gulf of California to Peru; occurs in a variety of shallow inshore habitats from the intertidal zone to at least 15 m depth. Maximum total length reported as 430 cm, but usually under 300 cm.

WHALE SHARK (TIBURONES BALLENA)

FAMILY RHINCODONTIDAE

See discussion of the single species of the family below.

WHALE SHARK
Rhincodon typus Smith, 1828
(Plate I-1, p. 27)

Head broad and flat with terminal mouth situated just in front of eyes; minute, extremely numerous teeth; *prominent ridges on sides of body with the lowermost expanding into a prominent keel on each side of caudal peduncle*; first dorsal fin relatively large; a small second dorsal and anal fin; a somewhat lunate caudal fin without a prominent subterminal notch; *generally blackish with "checkerboard" pattern of whitish spots, stripes, and bars; ventral parts whitish.* Circumglobal in tropical and warm temperate seas; pelagic in habits and often seen far offshore, as well as inshore near reefs; encountered as single individuals, but aggregations are sometimes seen. Whale sharks appear to be highly migratory and may occur in certain areas at more or less predictable intervals. Their movements are probably dependent on availability of their planktonic food source and also on changes in sea temperatures. They are sometimes associated with schools of pelagic scombrids or other fishes. The exact mode of reproduction is unknown, but they are thought to be ovoviviparous, wherein the egg cases are retained in the uterus for most of the embryonic development. An adult female was recorded to have 16 egg cases in its uteri. Whale sharks are considered harmless and are often examined at close range by divers. They are suction filter feeders that consume a wide variety of organisms including small crustaceans, squids, and fishes, especially sardines, anchovies, mackerels, and tunas. Maximum size uncertain, perhaps to 18 m, but lengths above 12 m are rare; most reported specimens between 4 and 12 m. It is the world's largest fish; the smallest free-living whale sharks are about 55 cm in length.

SAND TIGER SHARKS (SOLRAYOS, TOROS)

FAMILY ODONTASPIDAE

These are large, bulky sharks with a pointed snout, a long mouth extending well past the eyes, a pair of similar-sized dorsal fins, and large dagger-like teeth. The family contains two genera, each with two species. They are distributed in all tropical and temperate seas. The best-known member of the family is the Grey Nurse Shark (also called the Sand Tiger Shark). It is often common on rock and coral reefs, sometimes forming schools. The body is

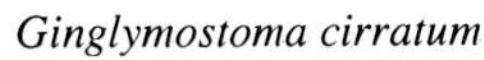
Ginglymostoma cirratum

denser than water, but it swallows air at the surface to provide buoyancy. Normally there are two young per litter, one from each uterus, although considerably more egg cases are produced. It is believed that the larger embryo eats fertilized eggs and smaller embryos within the uterus until the time of birth. These sharks are active predators of a variety of small to moderately large fishes, small sharks, rays, squids, crabs, and lobsters. Although they are mainly inoffensive and non-aggressive if left alone, they have a reputation as maneaters. It would appear this notoriety is undeserved and probably stems from confusion with certain requiem sharks (Carcharhinidae).

SMALLTOOTH SAND TIGER
Odontaspis ferox (Risso, 1810)
(Plate I-3, p. 27)

A large, grayish shark with a robust body; head with a long, bulbously conical snout; eyes moderately large, without nictitating eyelids; mouth relatively long, extending behind eyes; *teeth moderately large with prominent dagger-like median cusp and 2-3 small cusplets on either side; second dorsal fin only slightly smaller than first dorsal and situated well in front of anal fin*; caudal fin asymmetrical with moderate subterminal notch and strong ventral lobe; medium gray on back, light gray below; young with black tips on dorsal fins. Circumglobal in mainly subtropical waters; in the eastern Pacific it is known from Southern California to the Gulf of California; depth range 13-420 m. To 360 cm.

MACKEREL OR MAKO SHARKS (JAQUETONES, MARRAJOS)
FAMILY LAMNIDAE

This family is distributed in all temperate and tropical seas. It contains three genera and five species, including the notorious Great White Shark or White Pointer. They are generally large, powerful sharks, further distinguished by their spindle-shaped body, pointed snout, a slender, keeled caudal peduncle, lunate caudal fin, and sharp, dagger-like teeth (or in the case of the White Shark, triangular and serrate). Mackerel sharks are partly warm-blooded, and have a circulatory system that enables them to retain a body temperature that is warmer than the surrounding sea. Development of the young is ovoviviparous, without a yolk-sac placenta. They exhibit a remarkable phenomenon known as oophagy in which developing foetuses feed on fertilized eggs and possibly smaller siblings within the uterus. These rapid swimmers are capable of spectacular leaps when chasing their prey. They feed on a variety of fishes, other sharks, rays, seabirds, turtles, marine mammals, squids, and benthic crustaceans. They are aggressive, dangerous sharks that are sometimes responsible for fatal attacks on swimmers and surfers. The Great White (*Carcharodon carcharias*) ranges into the northern and southernmost portions of our region, but is omitted because its main distribution lies in cooler seas.

SHORTFIN MAKO
Isurus oxyrinchus Rafinesque, 1809
(Plate I-2, p. 27)

Body moderately slender; *snout long and pointed; teeth smooth-edged, long and slender at front of jaws, blade-like and triangular posteriorly*; a large first dorsal fin and very small second dorsal and anal fins positioned near base of caudal fin; pectoral fins relatively long and narrow; *strong lateral keel on each side of caudal peduncle*; caudal fin crescent-shaped; dark blue to nearly white on ventral parts. Circumglobal in tropical and temperate seas. Attains nearly 400 cm; size at birth between 60 and 70 cm.

CATSHARKS (PEJEGATOS, PINTARROJOS)
FAMILY SCYLIORHINIDAE

This is the largest family of sharks, with approximately 90 species. The group has successfully penetrated a wide range of habitats from tropical shallows to arctic waters and oceanic depths over 2 000 m. Catsharks are particularly well represented in African and Australian seas. Of the nine species that occur in our region, only the Swellshark occurs inshore on shallow reefs.

SWELLSHARK
Cephaloscyllium ventriosum (Garman, 1880)
(Plate I-5, p. 27)

A large, stocky member of the catshark family; stomach inflatable with air or water; second dorsal fin considerably smaller than first; no labial furrows; no barbels; snout broadly rounded when viewed from below, extremely short; anterior nasal flaps broadly lobate or subtriangular, overlapping mouth posteriorly; origin of second dorsal fin over origin of anal fin; teeth at front of jaws with dagger-like central cusp and 1-2 small cusplets on each side; *yellow-brown with variegated pattern of dark brown blotches, saddles, and spots; underside of head and abdomen spotted.* Central California to Chile; a sluggish, bottom-dwelling shark occurring from shallow depths to over 400 m; found most often on algae-covered rocky reefs. Grows to at least 100 cm.

HOUNDSHARKS (MUSOLAS, VIUDAS)
FAMILY TRIAKIDAE

This is one of the largest shark families with approximately 35 species inhabiting tropical and temperate seas. They are small to moderate-sized sharks (up to about 200 cm) characterized by a rather slender body; a long, pointed snout (when viewed laterally); horizontally oval eyes; and a long, angular or arched mouth that reaches past the anterior margins of the eyes. They lack nasal barbels and there are no spines in front of the dorsal and anal fins. The teeth are highly variable, ranging from low molars to more typical shark-like teeth that are blade-like or cuspidate. Houndsharks are generally seen close to the bottom, ranging from coastal shallows to depths in excess of 2 000 m. Many of the species are mainly active at night, feeding primarily on crustaceans, cephalopods, and fishes. None of the species are considered harmful. Only one species, *Triakis maculatus*, sometimes seen in shallow waters at the Galapagos, is treated here. A total of nine species occur in the eastern Pacific, most of which belong to the genus *Mustelus*, commonly known as smooth-hounds. The Leopard Shark (*Triakis semifasciatus*), a popular resident in California public aquariums, is also a member of this family.

SPOTTED HOUNDSHARK
Triakis maculata Kner & Steindachner, 1866
(Plate I-6, p. 27)

A medium-sized, stout shark with a short, broadly rounded snout; eyes in a dorsolateral position; anterior nasal flaps lobate, not reaching mouth and well separated from each other; long upper labial furrows that reach the lower symphysis of the mouth; teeth not blade-like, with straight erect cusps and cusplets absent or poorly developed; *dorsal fins of similar size; origin of second dorsal fin well forward of anal fin origin*; pectoral fins broadly falcate in adults; generally light gray, becoming white on ventral surface; *usually scattered black spots on body, but sometimes plain-colored.* Galapagos Archipelago and Peru to northern Chile. Reaches 200 cm.

REQUIEM SHARKS (CAZONES, TINTORERAS, GAMBUSOS, CHATOS)
FAMILY CARCHARHINIDAE

The requiems are one of the largest and best-known families of sharks. Worldwide there are 48 species in 12 genera. They are active, strong swimmers that occur singly or in small to large groups. They span a considerable range of sizes. For example, some milk sharks (genus *Rhizoprionodon*) reach a maximum length of less than 100 cm, whereas the Tiger Shark (*Galeocerdo cuvier*) is among the largest of sharks, with a maximum size of at least 7.4 m. Except for the Tiger Shark, which is ovoviviparous, all species are viviparous with a yolk-sac placenta. They have litters of young that number from one or two to as many as 135. The pups resemble miniature versions of the adults and are able to fend for themselves moments after birth. Requiem sharks are responsible for about half of all reported shark attacks on humans. However, less than 100 attacks are reported each year worldwide and no more than 30 of these are fatal. Although attacks are extremely rare, the potential danger of the more aggressive species, such as the Tiger and Bull sharks, should be recognized. Spearfishermen are more prone to shark attack than other divers because fishes struggling on a spear emit low-frequency vibrations which can attract sharks unerringly to the site. Once sharks are in the vicinity, the presence of blood, which is readily detected by the keen olfactory sense, is apt to add to their aggressive behavior. A few species have been shown to exhibit threat posturing, involving exaggerated lateral swimming movements (that is, arc of movement of the head increases noticeably), arching the back, holding the pectoral fins downward, and snapping of jaws. If these motions are evident, one should move slowly away from the shark and exit the water. They are voracious predators that feed mainly on a variety of fishes, other sharks, rays, squid, octopuses, cuttlefish, crabs, lobsters, and shrimps. Lesser items include seabirds, turtles, sea snakes, marine mammals, molluscs, carrion, and garbage debris.

Carcharhinus albimarginatus

SILVERTIP SHARK
Carcharhinus albimarginatus (Rüppell, 1837)
(Plate II-5, p. 29)

Snout moderately long and broadly rounded, the preoral length 6.8-9.2 percent of total length; *interdorsal ridge present*; origin of first dorsal fin over or slightly anterior to inner pectoral corner; height of first dorsal 7.1-10.6 percent of total length; apex of first dorsal obtusely pointed to acute; origin of second dorsal over or slightly behind origin of anal fin; gray, darker on back and shading to white ventrally; *distinctive white tips or margins on first dorsal, caudal, and pectoral fins*. Tropical Indo-Pacific, including eastern Pacific. Eats fish; may be aggressive; common on outer reef slopes below about 18 m. Usually has 5 or 6 pups per litter, but up to 11. Grows to 300 cm; size at birth 55-80 cm.

BIGNOSE SHARK
Carcharhinus altimus (Springer, 1950)
(Plate II-6, p. 29)

Snout long and slightly pointed, the preoral length 7.5-10.0 percent of total length; interdorsal ridge present on back; origin of first dorsal fin over pectoral axil or behind it to almost as far back as halfway along inner pectoral margin; height of first dorsal 8.3-11.9 percent of total length; apex of first dorsal bluntly pointed; origin of second dorsal in front of anal fin origin; gray, becoming whitish below; *distal ends of all fins except pelvics dusky (pigment on tips of pectorals darker on underside of fins)*. Circumglobal distribution in temperate and tropical seas. Feeds on fishes (including sharks and rays) and cephalopods; usually found between 90 and 430 m depth. Has litters of 3-11 pups; young occur in estuaries. Reaches 300 cm; 65-80 cm at birth.

SILKY SHARK
Carcharhinus falciformis (Bibron, 1839)
(Plate II-4, p. 29)

Slender-bodied; body depth 11.5-17.5 percent of total length; snout moderately long and slightly pointed, the preoral length 6.9-9.3 percent of total length; a low, narrow interdorsal ridge; origin of first dorsal fin behind inner pectoral corner by not less than one-third length of inner pectoral margin; *first dorsal fin small, its height 5.2-9.9 percent of total length*; apex of first dorsal narrowly rounded; trailing edge of first dorsal very falcate; origin of second dorsal over or slightly behind anal origin; gray to dark gray dorsally, shading to white ventrally, sometimes with faint band of white invading gray on upper abdomen; first dorsal fin unmarked; second dorsal, anal, lower caudal lobe, and pectoral fins may have dusky tips. Circumtropical, oceanic and coastal; found inshore as shallow as 18 m, but more abundant offshore from the surface to at least 500 m. Eats mainly fishes; a potentially dangerous shark, but no attacks have been reported. Has litters of 2-14 pups. Grows to 230 cm; size at birth 70-87 cm. *C. floridanus* is a synonym.

GALAPAGOS SHARK
Carcharhinus galapagensis (Snodgrass & Heller, 1905)
(Plate II-7, p. 29)

A large, relatively slender, gray shark; snout moderately long and broadly rounded; *a low interdorsal ridge*; origin of first dorsal fin about opposite rear edge of pectoral fins; *first dorsal fin moderately large and falcate with a short rear tip, to height 9.1-12.1 percent or more of total length*; apex of first dorsal fin pointed or narrowly rounded; origin of second dorsal fin about over anal fin origin; pectoral fins large with narrowly rounded to pointed tips; brownish gray on back and sides, white below; *tips of most fins dusky*. Circumtropical distribution, but usually associated with oceanic islands; in the eastern Pacific, reported from Galapagos, Cocos, Revillagigedo, Clipperton, and Malpelo islands, and also the coasts of Baja California, Guatemala, and Colombia. Often occurs in aggregations to at least 180 m depth. Feeds mainly on bottom fishes; is known to attack humans and should be regarded as potentially dangerous. Has 6-14 pups per litter. Attains 350 cm; size at birth 57-80 cm. *Very similar in appearance and easily confused with* C. obscurus; *best means of separation is the taller dorsal fin of* C. galapagensis *(9.1-12.1 percent of TL vs 5.8-9.9 percent for* C. obscurus*)*.

BULL SHARK
Carcharhinus leucas (Valenciennes, 1839)
(Plate II-8, p. 29)

Heavy-bodied; snout short and broadly rounded, the preoral length 4.6-6.7 percent of total length; no interdorsal ridge; origin of first dorsal fin usually over or just posterior to pectoral axil; first dorsal fin moderately large, its height 7.0-11.3 percent of total length; apex of first dorsal fairly pointed; origin of second dorsal distinctly in front of anal origin; gray, becoming white ventrally, often with faint, pale gray horizontal band extending into the white of the upper abdomen; *fins of small individuals with dusky tips or edges, adults plain*. Continental coasts of all tropical and subtropical seas; often travels far up rivers. Feeds on about anything edible; a dangerous shark responsible for fatal attacks. Has litters of 3-13 pups. May reach 300 cm; size at birth 56-81 cm.

BLACKTIP SHARK
Carcharhinus limbatus (Valenciennes, 1839)
(Plate II-9, p. 29)

Snout moderately long and pointed, its preoral length 6.3-9.0 percent of total length; no interdorsal ridge; origin of first dorsal fin over or slightly posterior to pectoral fin axil; first dorsal fin moderately large and falcate, its height 8.2-13.8 percent of total length; apex of first dorsal acute; origin of second dorsal about over or slightly in front of anal fin origin; gray-brown dorsally, shading to white ventrally, with nearly horizontal band of gray on midside extending into white of upper abdomen; *black tips on dorsal fins, lower lobe of caudal, pelvic fins, and pectoral fins*. Circumtropical distribution. Feeds mainly on fishes, but also cephalopods and larger crustaceans; usually not aggressive, but has been known to attack. Has litters of 1-10 pups. Reported to reach 250 cm; size at birth 38-72 cm.

OCEANIC WHITETIP SHARK
Carcharhinus longimanus (Poey, 1861)
(Plate II-11, p. 29)

Snout moderately short and broadly rounded, the preoral length 5.4-7.1 percent of total length; interdorsal ridge usually present; origin of first dorsal fin slightly anterior to inner posterior corner of pectoral fins; first dorsal fin very large with apex broadly rounded, its height 9.2-15.2 percent of total length; *pectoral fins extremely long, broad, and distally rounded, length of anterior edge 20.2-27.1 percent of total length*; origin of second dorsal in front of or over anal fin origin; brownish gray on back, becoming white ventrally; *tips of first dorsal fin, paired fins, and caudal fin lobes broadly mottled white; anal fin usually blackish at tip, and second dorsal fin may be dusky at tip; juveniles with most fins tipped with black*. Circumtropical; primarily oceanic-epipelagic. Feeds on fishes, squids, birds, turtles, pelagic crustaceans and gastropods, and occasional marine mammals; a dangerous species responsible for human attacks. Has litters of 1-15 pups. Grows to 396 cm; 60-65 cm at birth. *C. maou* (Lesson) is an

Carcharhinus limbatus

earlier name, but due to long-term usage, *longimanus* has been placed on the Official List of Specific Names in Zoology.

DUSKY SHARK

Carcharhinus obscurus (Lesueur, 1818)
(Plate II-10, p. 29)

Snout of moderate size, rounded, the preoral length 5.7-8.4 percent of total length; *interdorsal ridge present*; origin of first dorsal fin over or slightly posterior to inner rear corner of pectoral fins; first dorsal fin of moderate size, the height 5.8-10.4 percent of total length; apex of first dorsal pointed to narrowly rounded; origin of second dorsal above anal fin origin; gray, shading to white ventrally, with a faint, near-horizontal, gray band invading the white of upper abdomen; tips of fins dusky. Circumglobal in tropical and warm temperate seas; primarily on continental shelves from shallow water to 400 m. Feeds on wide variety of marine animals; a dangerous shark that has attacked humans. Has litters of 6-14 pups. Reaches 362 cm; size at birth 69-100 cm. *Similar to Galapagos Shark (see comments under that species).*

SMALLTAIL SHARK

Carcharhinus porosus (Ranzani, 1839)
(Plate II-12, p. 29)

A small, slender, gray shark; snout moderately long and pointed; eyes circular and large; no interdorsal ridge; origin of first dorsal fin over rear margin of pectoral fins; first dorsal fin large and falcate with bluntly pointed apex; origin of second dorsal fin over or slightly behind middle of anal fin; pectoral fins small, falcate with narrowly rounded to pointed tips; generally gray on back and sides, white on lower parts; *tips of pectoral, dorsal, and caudal fins may be dusky or blackish, but are not prominently marked.* Western Atlantic and eastern Pacific (Gulf of California to Peru); a common inshore shark ranging from shallow estuaries to at least 36 m depth. Feeds primarily on fishes, including small sharks; considered to be harmless. Has litters of 2-7 pups. Grows to 150 cm; size at birth 31-40 cm.

TIGER SHARK

Galeocerdo cuvier (Peron & Lesueur, 1822)
(Plate I-14, p. 27)

Head, thorax and abdomen stout, but body becoming very attenuate posteriorly, with *a low lateral keel on each side of the narrow caudal peduncle*; snout very short and slightly rounded, its preoral length 3.7-4.8 percent of total length; spiracle a narrow slit behind eye; *interdorsal ridge present*; origin of dorsal fin over posterior corner of pectoral fin; first dorsal fin not very large, its height 7.5-9.3 percent of total length; apex of first dorsal fin pointed; origin of second dorsal fin distinctly in front of anal fin origin; *adults gray with vertical bars on upper half of sides (sometimes faint or absent); young with large dark spots, some coalescing to form bars.* Circumtropical; retires to deeper water during the day, but feeds on shallow reefs at night. Feeds on a wider variety of items than any other shark, including bony fishes, sharks, rays, turtles, birds, marine mammals, cephalopods, spiny lobsters, crabs, gastropods, jellyfishes, and carrion; an extremely dangerous shark responsible for attacks on divers and swimmers. Has from 10 to over 80 pups per litter. Attains at least 7.4 m and possibly 9 m; size at birth 50-75 cm.

WHITENOSE SHARK

Nasolamia velox (Gilbert, 1898)
(Plate II-3, p. 29)

A small slender gray shark; snout very long and pointed; preoral length much greater than distance between nostrils; nostril slits very large and nearly transverse; labial furrows very short; no interdorsal ridge; origin of first dorsal fin just behind rear insertion of pectoral fins; first dorsal fin moderate in size and erect with concave posterior margin; second dorsal fin slightly smaller than anal fin; origin of second dorsal fin over anal fin origin or slightly anterior; pectoral fins moderately broad and triangular; gray or brownish gray on back and sides, whitish below; *a black spot outlined with white on the dorsal surface of snout tip; no distinguishing marks on fins.* Gulf of California to Peru; a relatively common shark usually seen in 15-25 m depths, but ranging to at least 190 m. Feeds on small fishes and crabs; considered harmless. Gives birth to around 5 pups per litter. Reaches at least 150 cm; size at birth 50-55 cm.

LEMON SHARK

Negaprion brevirostris (Poey, 1868)
(Plate II-1, p. 29)

A large, stocky shark; snout short and broad, somewhat rounded; narrow, dagger-like, smooth-cusped teeth in both jaws; spiracles usually absent (occasionally with very small spiracles); *no interdorsal ridge*; origin of first dorsal fin behind rear margin of pectoral fins; *second dorsal fin about same size or only slightly smaller than first dorsal fin*; pectoral fins broad and slightly falcate; *pale yellow-brown on back and sides*, yellowish or whitish below. Tropical waters of the Atlantic and eastern Pacific (Gulf of California to Ecuador); a common inshore shark seen in a variety of habitats including rocky reefs, estuaries, and river mouths (may enter fresh water). Feeds mainly on fishes, but also consumes molluscs and crustaceans; has been implicated in human attacks and is considered dangerous, especially if provoked. Litter size ranges from 4-17. Attains about 340 cm; size at birth 60-65 cm. *N. fronto* (Jordan & Gilbert) is a synonym.

PACIFIC SHARPNOSE SHARK

Rhizoprionodon longurio (Jordan & Gilbert, 1882)
(Plate II-2, p. 29)

A small, relatively slender, gray shark; *snout long and pointed with semitranslucent appearance when viewed from below; nostril slits oblique; labial furrows very long; interdorsal ridge absent or rudimentary*; origin of first dorsal fin over rear edge of pectoral fins; first dorsal fin moderate in size and erect with concave posterior margin; second dorsal fin slightly smaller than anal fin; origin of second dorsal fin over middle of anal fin; pectoral fins moderately broad and triangular; gray or gray-brown on back and sides, white below. Southern California to Peru; a relatively common inshore shark ranging from shallow estuarine waters to at least 30 m depth. Feeds on small fishes and crustaceans. Usually has 2-5 pups per litter. Maximum size to at least 110 cm, possibly 154 cm; size at birth 33-34 cm.

WEASEL SHARKS (TIBURONES COMADREJA)

FAMILY HEMIGALEIDAE

These are small to moderate-sized sharks that have a relatively slender body and horizontally oval eyes. Other features include absence of barbels; small to moderately large blade-like teeth usually with cusps (at least those in lower jaw); two dorsal fins of unequal size, the first dorsal base well ahead of the pelvic bases; and precaudal pit present. The members of this Indo-Pacific and eastern Atlantic family are often considered to be a subfamily of the Carcharhinidae. Only one of the species, *Triaenodon obesus*, is commonly encountered on reefs in our

Triaenodon obesus

area. Weasel sharks are viviparous and feed primarily on fishes and cephalopods. The largest species grows to about 250 cm.

WHITETIP REEF SHARK
Triaenodon obesus (Rüppell, 1837)

A slender shark, the depth about 11-16 percent of total length; head depressed, about twice as broad as deep; very short and blunt snout; spiracles absent or minute; labial furrows very short; tubular anterior nasal flaps; small, smooth-edged teeth with strong cusplets in both jaws; brownish gray, shading to whitish with a yellow cast ventrally, usually with a few scattered, roundish, dark gray spots on body (more on ventral half than dorsal); *tips of first dorsal fin and upper caudal lobe broadly white; tips of second dorsal fin and lower lobe of caudal fin also often white*. Indo-Pacific, including tropical eastern Pacific. Spends most of the day at rest on the bottom in caves or beneath ledges; a seemingly curious shark that often approaches divers at close range. Feeds mainly on fishes, also cephalopods; considered harmless, but a few attacks on spearfishermen have been reported. Has litters of 1-5 pups. Maximum size, 210 cm; 52-60 cm at birth.

HAMMERHEAD SHARKS (TIBURONES MARTILLO, CORNUDAS)
FAMILY SPHYRNIDAE

Hammerhead sharks are easily recognized by the hammer or mallet-shaped lateral expansions of the head. Otherwise the general body shape is very similar to that of carcharhinid sharks. The enlargement on either side of the head is thought to increase maneuvering capabilities and also increases their sensory capacities related to vision, smell, and pressure detection. The family is distributed worldwide in tropical and temperate seas. There are nine species; all except one belong to the genus *Sphyrna*. They range in maximum size from about 140 to 600 cm. Some of the larger species have been responsible for attacks on humans, but recent studies indicate that they are not particularly aggressive unless provoked or baited. Their natural diet includes a variety of items such as fishes, sharks, cephalopods, crustaceans, and turtles.

SCALLOPED BONNETHEAD
Sphyrna corona Springer, 1940
(Plate I-10, p. 27)

A small hammerhead shark with moderately broad but short mallet-shaped lateral extensions on head; width of hammer 24-29 percent of total length; anterior margin of head broadly arched with very shallow lateral and medial indentations; preoral length usually more than 40 percent of head width; first dorsal fin moderately large and erect, its posterior margin concave; second dorsal fin base about half length of anal fin base; origin of second dorsal fin above middle of anal fin; concavity of anal fin margin very slight; generally gray on back and sides, white below; fins without prominent markings. Gulf of California to Peru; an uncommon inshore shark. Reaches a maximum size of 92 cm; size at birth about 25 cm. This is the smallest species of hammerhead.

SCALLOPED HAMMERHEAD

Sphyrna lewini (Griffith & Smith, 1834)
(Plate I-9, p. 27)

A large hammerhead shark with broad, narrow-bladed lateral extensions on the head; width of hammer 24-30 percent of total length; anterior margin of head broadly convex with a prominent median indentation and another more conspicuous notch laterally near the end, setting off the terminal lobe bearing the eye; a slight indentation in anterior margin between the median and lateral notches on each side, the overall effect being a scalloped edge; first dorsal fin moderately large and erect, outer posterior margin concave, the height 11.9-14.5 percent of total length; free rear tip of second dorsal fin nearly reaching caudal fin; base of anal fin noticeably larger than that of second dorsal fin; brownish gray, shading to white ventrally; undersides of pectoral fins tipped with black. Worldwide in tropical and warm temperate seas. Has litters of 15-31 pups. Attains 420 cm; size at birth 42-55 cm.

Sphyrna media ▼

SCOOPHEAD

Sphyrna media Springer, 1940
(Plate I-13, p. 27)

A small hammerhead shark with moderately broad but short mallet-shaped lateral extensions on head; width of hammer 22-33 percent of total length; anterior margin of head broadly arched, with very shallow lateral and medial indentations; preoral length usually less than 40 percent of head width; first dorsal fin moderately large and erect, its posterior margin noticeably concave; second dorsal fin base about half length of anal fin base; origin of second dorsal fin above middle of anal fin; concavity of anal fin margin very pronounced; gray-brown on back and sides, lighter on ventral surfaces. Western Atlantic and eastern Pacific (Gulf of California to Ecuador); a poorly-known inshore species. Reported to attain 150 cm; size at birth 30-35 cm.

GREAT HAMMERHEAD

Sphyrna mokarran (Rüppell, 1837)
(Plate I-12, p. 27)

A large hammerhead shark with broad, narrow-bladed lateral extensions on the head; width of hammer 23-27 percent of total length; anterior margin of head nearly straight in adults, with shallow medial and lateral indentations; first dorsal fin very tall and falcate, its apex pointed; rear margin of second dorsal and anal fins strongly notched, their bases about equal; gray-brown on back and sides, whitish below; no prominent markings on fins. Circumtropical distribution; in the eastern Pacific from Gulf of California to northern Peru; a coastal pelagic and semioceanic species, it ranges from coastal shallows to depths over 80 m. Feeds on a wide range of fishes, with a preference for rays, groupers, and catfishes; considered dangerous, but there are few reported attacks by this species. Litters contain 13-42 pups. Said to reach 600 cm, but uncommon above 350 cm; size at birth 50-70 cm.

BONNETHEAD

Sphyrna tiburo (Linnaeus, 1758)
(Plate I-8, p. 27)

A small hammerhead shark with a very narrow, shovel-shaped head, thus lacking the prominent lateral-blade extensions of other hammerheads; width of head 18-25 percent (usually below 21 percent) of total length; anterior margin of head broadly arched without indentations; preoral length about 40 percent of head width; enlarged molariform teeth at back of jaws; first dorsal fin moderately large and erect, its posterior margin concave; second dorsal fin base about half length of anal fin base; origin of second dorsal fin above middle of anal fin; gray or gray-brown on back and sides, whitish below. Western Atlantic and eastern Pacific (Southern California to Ecuador); a common inshore hammerhead ranging from shallow estuaries and the intertidal zone to at least 80 m depth. Feeds mainly on crustaceans, including crabs, shrimps, isopods, and barnacles. Gives birth to 4-16 pups per litter. Maximum size about 150 cm; size at birth 35-40 cm.

SMOOTH HAMMERHEAD

Sphyrna zygaena (Linnaeus, 1758)
(Plate I-11, p. 27)

A large hammerhead shark with broad, narrow-bladed lateral extensions on the head; width of hammer 26-29 percent of total length; anterior margin of head broadly arched with prominent lateral indentations, but no medial indentation; first dorsal fin moderately large and erect, its posterior margin concave; anal fin slightly larger than second dorsal fin, its posterior margin prominently notched; origin of second dorsal fin slightly behind origin of anal fin; dark olive to dark gray-brown on back and sides, white below; undersides of pectoral fin tips dusky. Circumglobal in tropical and temperate seas; in the eastern Pacific from California to Chile, but most common in cooler temperate waters; a coastal-pelagic and semioceanic hammerhead often sighted near the surface. Feeds mainly on fishes, but also consumes cephalopods and crustaceans; considered dangerous, although there are few authenticated attacks on humans. Litter size ranges from 29-37. Grows to about 370-400 cm; size at birth 50-61 cm.

PLATE I

1 **WHALE SHARK** (*Rhincodon typus*)

2 **SHORTFIN MAKO** (*Isurus oxyrinchus*)

3 **SMALLTOOTH SAND TIGER** (*Odontaspis ferox*)

4 **MEXICAN HORNSHARK** (*Heterodontus mexicanus*)

5 **SWELLSHARK** (*Cephaloscyllium ventriosum*)

6 **SPOTTED HOUNDSHARK** (*Triakis maculata*)

7 **NURSE SHARK** (*Ginglymostoma cirratum*)

8 **BONNETHEAD** (*Sphyrna tiburo*)

9 **SCALLOPED HAMMERHEAD** (*Sphyrna lewini*)

10 **SCALLOPED BONNETHEAD** (*Sphyrna corona*)

11 **SMOOTH HAMMERHEAD** (*Sphyrna zygaena*)

12 **GREAT HAMMERHEAD** (*Sphyrna mokarran*)

13 **SCOOPHEAD** (*Sphyrna media*)

14 **TIGER SHARK** (*Galeocerdo cuvier*)

15 **SAWFISH** (*Pristis perotteti*)

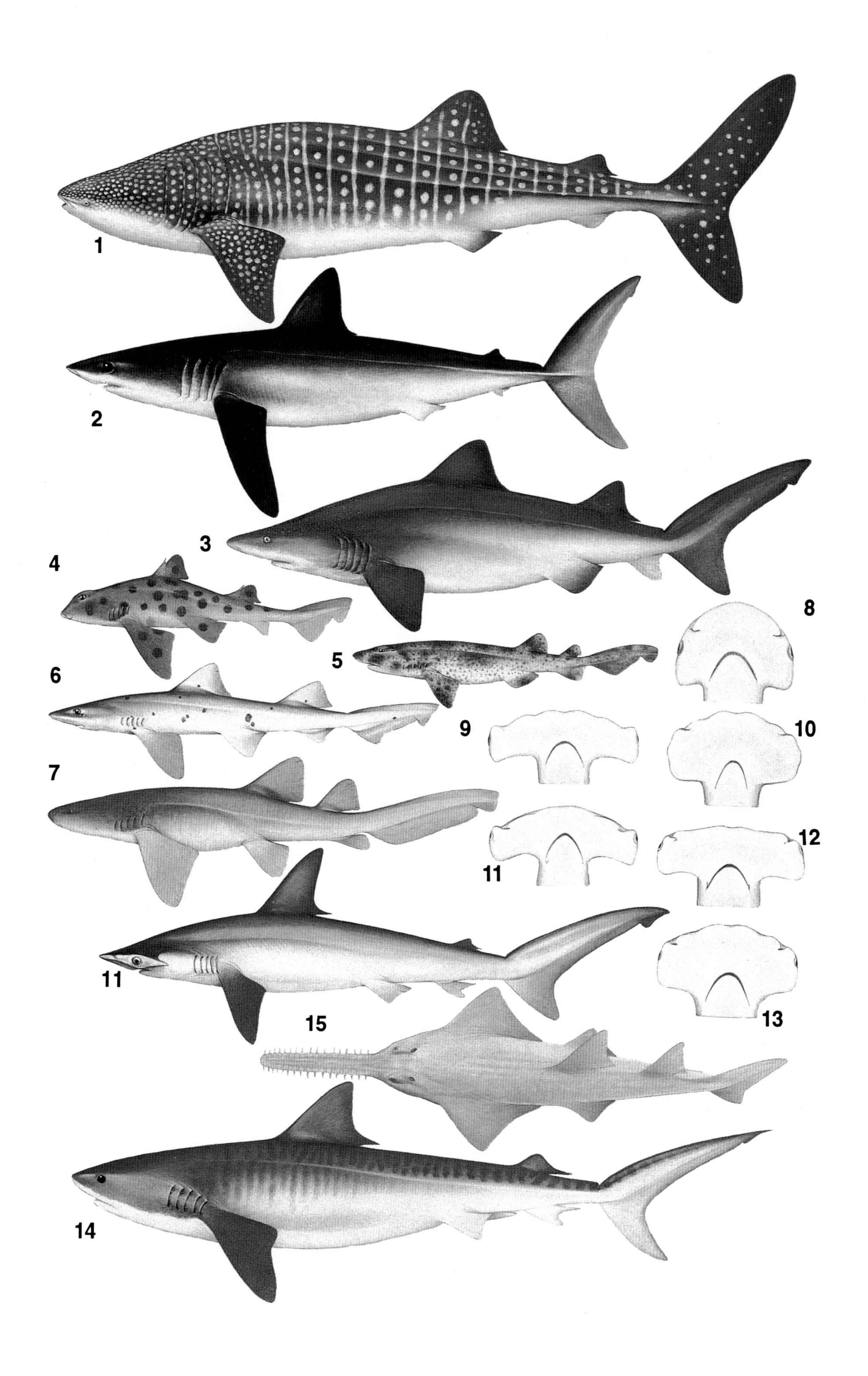
1
2
3
4
5
6
7
8
9
10
11
12
11
13
15
14

PLATE II

1 **LEMON SHARK** (*Negaprion brevirostris*)

2 **PACIFIC SHARPNOSE SHARK** (*Rhizoprionodon longurio*)

3 **WHITENOSE SHARK** (*Nasolamia velox*)

4 **SILKY SHARK** (*Carcharhinus falciformis*)

5 **SILVERTIP SHARK** (*Carcharhinus albimarginatus*)

6 **BIGNOSE SHARK** (*Carcharhinus altimus*)

7 **GALAPAGOS SHARK** (*Carcharhinus galapagensis*)

8 **BULL SHARK** (*Carcharhinus leucas*)

9 **BLACKTIP SHARK** (*Carcharhinus limbatus*)

10 **DUSKY SHARK** (*Carcharhinus obscurus*)

11 **OCEANIC WHITETIP SHARK** (*Carcharhinus longimanus*)

12 **SMALLTAIL SHARK** (*Carcharhinus porosus*)

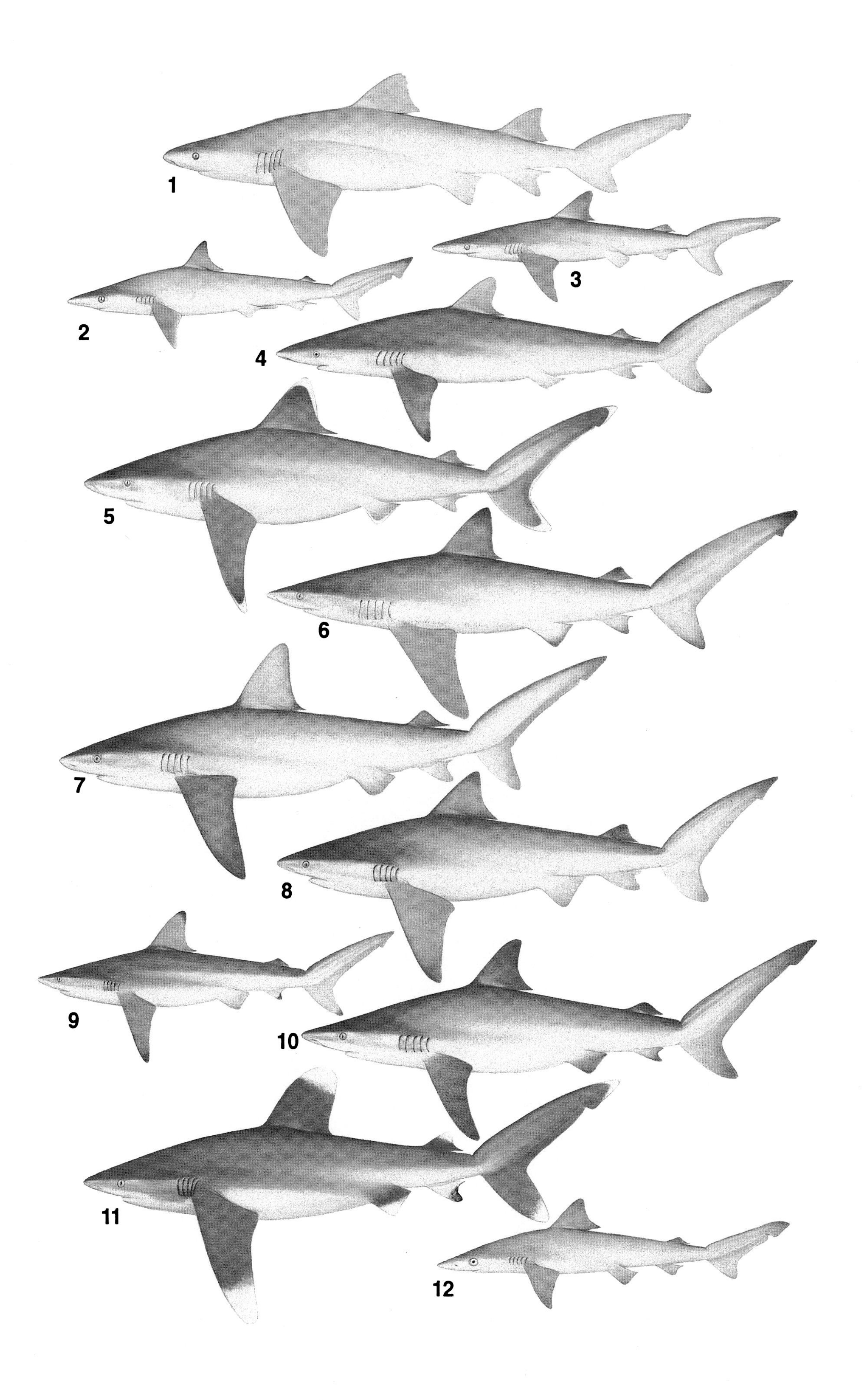
1
3
2
4
5
6
7
8
9
10
11
12

RAYS (RAYAS)
SUPERORDER
BATOIDIMORPHA

Rays, or batoid fishes, share many morphological similarities with sharks. Consequently they are grouped together in the subclass Elasmobranchii. Shared characteristics include a cartilaginous skeleton, presence of placoid scales (developed in the same manner as teeth, hence different from usual fish scales), a series of slit-like external gill openings, often a pair of spiracles (vestigial gill openings) on top of the head, no air or swim bladder, a unique spiral valve in the intestine that greatly increases the absorptive area of the digestive tract, and internal fertilization by means of elongate claspers located on the inner edge of the male pelvic fins. Although similar to sharks in many respects, most rays are easily differentiated by the ventral location of the gill slits and by the greatly enlarged pectoral fins, which are fused to the sides of the head. Bottom-dwelling species, including all skates and stingrays, draw in water to the gill chambers via the spiracles. This avoids bringing in bottom sediment that would result if the mouth was utilized as in most fishes. Free-swimming rays, such as mantas and eagle rays, respire in a normal way, bringing in water through the mouth and passing it out via the gill slits. Sawfishes and guitarfishes have a shark-like body, but the ventral position of the gill openings indicate they are indeed rays. There are 17 families of rays, with approximately 480 species currently recognized. Nearly half the species belong to the skate family Rajidae. This family is generally not associated with reefs in our area, nor are its members common in shallow water. Therefore we do not provide coverage of them.

Manta Ray (*Manta birostris*), Baja California, Mexico

Photograph: Alex Kerstich

SAWFISHES (PEJESIERRAS, PEJEPEINES, PECES RASTRILLO)

FAMILY PRISTIDAE

Sawfishes are primarily confined to marine and estuarine habitats in tropical regions, but several species regularly occur in fresh water and are believed to reproduce there. Currently two genera and four species are recognized from the tropical Atlantic, Indian and Pacific oceans. They are shark-like in many respects, particularly the shape of the fins, but are actually related to rays, as indicated by the ventrally situated gill slits. The elongate snout is equipped with a row of large, flattened teeth along the lateral margins. This saw apparatus is used to dig for food, which largely consists of molluscs and other benthic invertebrates. It has also been reported that it is used to kill fish, but there is insufficient documentation. Sawfishes are ovoviviparous; females give birth to about three to five pups at a time.

SAWFISH
Pristis perotteti Müller & Henle, 1841
(Plate I-15, p. 27)

Body shark-like with 2 nearly equal-sized dorsal fins; head flattened; *17-21 teeth on each side of saw-like snout*; a large spiracle behind each eye; pectoral fins broad and triangular; *caudal fin shark-like with large upper lobe and small but pronounced lower lobe*; origin of first dorsal fin well forward of pelvic fin origin; dark gray to golden brown, whitish on ventral surface. West Africa, West Atlantic, and eastern Pacific, including Mexico to Peru; inhabits shallow bays and estuaries, also entering rivers. Known from Lake Nicaragua and 720 kilometers from the sea in the Amazon River. Grows to about 600 cm and estimated weight of 635 kg. *P. zephyreus* Jordan & Starks is a synonym. Very similar to and possibly conspecific with the Indo-West Pacific *P. microdon* Latham. A second species of sawfish, *P. pectinatus* Latham, occurs in the eastern Pacific. It lacks a definite lower caudal lobe and has 25-32 teeth on each side of the saw.

GUITARFISHES (GUITARRAS)

FAMILY RHINOBATIDAE

Guitarfishes are batoids, characterized by a large triangular head, incorporating the fused pectoral fins which taper to a pointed or narrowly rounded snout tip. The eyes and spiracles are situated on the dorsal surface; the mouth and nostril openings, as well as the gill slits are directed ventrally. The remainder of the body is more or less shark-like, with a pair of similar-sized dorsal fins on the back. The teeth resemble those of skates, sawfishes, and primitive stingrays. They form low pavement-like bands in the jaws, but are not designed for crushing prey, as are those of the myliobatids and rhinopterids. They feed on benthic invertebrates and small fishes. Guitarfishes are commonly encountered on sand or mud bottoms. They sometimes bury themselves below the surface when resting. The distribution encompasses all temperate and tropical seas from shallow inshore waters to depths down to 370 m. The family contains about 38 species. Guitarfishes are considered harmless. All species are viviparous, without a placenta.

SHOVELNOSE GUITARFISH
Rhinobatos productus Girard, 1855

Body shark-like with 2 equal-sized dorsal fins; head and adjoining pectoral fins forming a distinctly triangular structure; snout cartilage forming a double ridge from front of interorbital to snout tip; a large spiracle behind each eye; a row of thorny projections along dorsal midline from behind eyes to origin of first dorsal fin; caudal fin asymmetrical without distinct lower lobe; generally yellowish brown, whitish ventrally; snout somewhat translucent. Central California to Gulf of California; inhabits shallow coastal waters, including bays and estuaries, to 15-20 m depth. Grows to 170 cm and 18.4 kg.

BANDED GUITARFISH
Zapteryx exasperata Jordan & Gilbert, 1880
(Plate III-8, p. 39)

Body slender and somewhat shark-like with 2 equal-sized dorsal fins; head and adjoining pectoral fins forming a distinctly triangular structure; snout cartilage forming a double ridge from front of interorbital to snout tip; a large spiracle behind each eye; a row of thorny projections along dorsal midline from behind eyes to origin of first dorsal fin; caudal fin asymmetrical, without distinct lower lobe; *snout shorter and head broader than* R. productus*; reddish brown with irregular blotching and dark bars; several dark-margined yellow spots scattered on back and head disc.* Southern California to Peru; inhabits rocky reefs and adjacent sand flats. Reaches 100 cm.

Rhinobatos productus

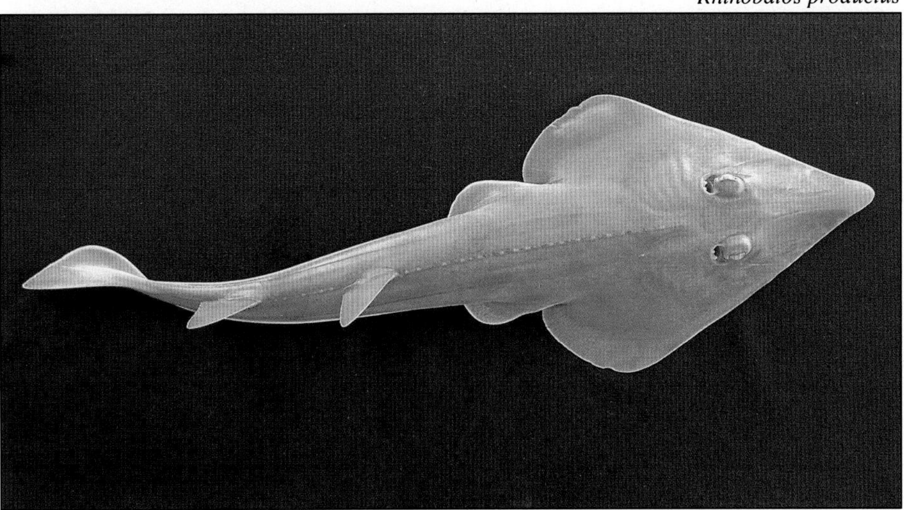

ELECTRIC RAYS (TEMBLADERAS, RAYAS ELÉCTRICAS, TORPEDOS)

FAMILY NARCINIDAE

These rays have a very flat and rounded disc composed of the head, trunk, and enlarged pectoral fins. There are two large, more or less equal-sized dorsal fins, and the upper and lower lobes of the caudal fin are continuous around the vertebral column. The skin is soft and loose, and is free of denticles. Perhaps their most noteworthy feature is the well-developed electrical organs that are derived from branchial musculature and are externally visible on either side of the head. These are used to stun prey and probably also act as a defensive weapon against enemies, including man. In addition, posterior to the main organ, the narcinids have an accessory electric organ with low-voltage discharge, which supposedly functions in intraspecific communication. Electric rays are generally sluggish fishes that occur in sandy or muddy areas. They sometimes conceal themselves by burying under the substratum. Young are born alive. The family is distributed in all tropical and temperate seas and contains about eight genera and 24 species.

Diplobatis ommata Normal variety ▲ Plain variety ▼

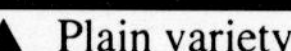

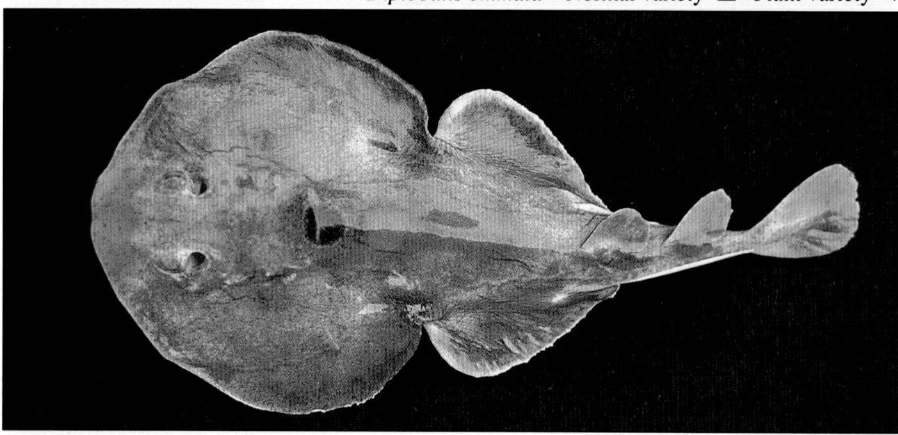

Narcine entemedor ▼

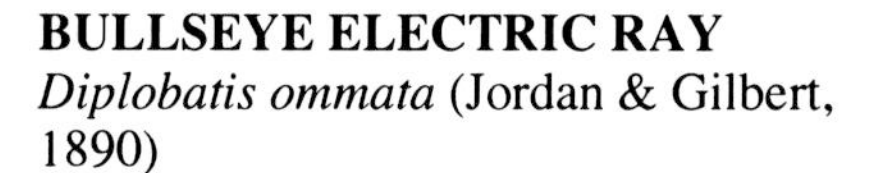

BULLSEYE ELECTRIC RAY
Diplobatis ommata (Jordan & Gilbert, 1890)

Disc rounded; *tips of dorsal fins and caudal rounded*; anterior margin of pelvic fins concealed by pectoral fins; *nostril of each side divided into 2 separate apertures; teeth entirely enclosed within closed mouth*; light brown, usually with profuse leopard-like spotting and diffuse, dark crossbars, also *a large "bullseye" ocellated marking at center of disc*; a specimen from Panama is plain brown except for a large black spot at center of disc. Gulf of California to Panama; inhabits sandy or rubble bottoms. Grows to about 20 cm total length.

LESSER ELECTRIC RAY
Narcine entemedor Jordan & Starks, 1895 (Plate III-9, p. 39)

Disc rounded; *tips of dorsal fins and caudal somewhat angular*; anterior margin of pelvic fins concealed by pectoral fins; *nostril of each side not divided into 2 separate apertures; teeth largely exposed when mouth closed; generally gray without distinctive markings*; posterior margins of fins narrowly white. Gulf of California to Panama; inhabits sandy bottoms. Reaches about 60 cm. Females give birth to 4-15 young; size at birth 11-12 cm. Closely related to *N. brasiliensis* (Olfers) from the western Atlantic.

TORPEDO ELECTRIC RAYS (TORPEDOS)
FAMILY TORPEDINIDAE

This family is closely related to the electric rays (Narcinidae). Both groups are characterized by the presence of a pair of well-developed kidney-shaped electric organs that are externally visible on either side of the head. These organs can discharge up to 45 volts and are used for defense or to stun prey. Both families have a roundish, soft, flaccid disc that is rather thick around the margin. The best means of separating the two families is the difference in size of the two dorsal fins. Those of narcinids are roughly equal in size, whereas the first dorsal fin of torpedinids is significantly larger than the second dorsal. In addition, torpedinids have a larger mouth that does not protract into a short tube as in narcinids. Torpedo rays inhabit most tropical and temperate seas, usually in depths less than 100 m, but some may occur to depths of at least 360 m. They are relatively inactive, soft-bottom dwellers that frequently cover themselves with sand or mud. Their diet consists of small, bottom-living invertebrates and fishes. They are viviparous, without a placenta; the young closely resemble their parents at birth. The family contains a single genus with several species.

TORPEDO ELECTRIC RAY
Torpedo tremens de Buen, 1959
(Plate III-10, p. 39)

Disc round, soft and flaccid, very thick near margin; tail very stout, not demarcated from disc, much shorter than disc length; first dorsal fin distinctly larger than second; caudal fin large and subtriangular; mouth moderate in size and slightly arched; grayish brown to blackish, without spotting on dorsal surface, white on underside of disc. Costa Rica to Chile; inhabits sand and mud bottoms. Grows to 58 cm.

STINGRAYS (RAYAS BATANAS, RAYAS LÁTIGO, BATEAS)
FAMILY DASYATIDIDAE

These rays are characterized by an angular to rounded disc, the width of which is not more than 1.3 times the length; the head is not separated or distinguishable from the rest of the disc; the floor of the mouth has fleshy papillae; the jaws have small blunt or cuspidate teeth in many series, forming bands; dorsal and caudal fins are absent, but some species have longitudinal skin folds on the upper and lower surface of the tail; the tail is moderately slender to very slender and whip-like, longer (considerably so in some species) than the disc, and usually with one or more large, serrate, venomous spines on its dorsal surface; the skin on the dorsal surface is smooth or armed with tubercles, thorns or dermal denticles. Stingrays occur in all tropical and subtropical seas; about six genera and 50 species are known. Most species are found in coastal waters, in estuaries, off beaches and river mouths, and on flat "trawl ground" bottoms on sand or mud. Relatively few species occur in the vicinity of coral reefs. Several are also known from freshwater habitats. The tail spine is extremely dangerous and capable of delivering an excruciating wound. Human fatalities have been reported. Caution should be exercised when wading on sandy bottoms. If a ray is stepped on, it has the ability to thrust its tail upward and forward, impaling the victim with remarkable speed. Immersion of the injured limb in hot water (about 50°C) for 30-90 minutes will often dramatically relieve the pain, but medical assistance should be obtained. Stingrays feed on a variety of sand- and mud-dwelling organisms, including crabs, shrimps, worms, molluscs, and fishes. They are livebearers, the young resembling miniature adults.

DIAMOND STINGRAY
Dasyatis brevis (Garman, 1880)
(Plate III-7, p. 39)

Disc ovate to roughly diamond-shaped, broader than long; anterolateral margins of disc nearly straight; snout pointed, very slightly projecting; *tail slightly greater than length of disc; caudal fin absent, but fin fold on upper and lower surface of tail*; upper surface of disc mainly smooth except for long median row of thorns on back and base of tail, flanked on each side by short row of thorns or spines on middle of disc; generally

Himantura pacifica (see description on p. 34)

dark brown or blackish, without distinctive markings. Mexico to Peru; inhabits sand-mud bottoms to 17 m depth. Maximum length 180 cm (disc width 103 cm).

LONG-TAILED STINGRAY
Dasyatis longus (Garman, 1880)
(Plate III-7, p. 39)

Disc ovate, broader than long; anterolateral margins of disc nearly straight; snout bluntly angular, but not protruding; *tail very long and slender, more than twice length of disc; caudal fin absent, but fin fold present midventrally on middle portion of tail*; a median row of blunt spines or thorns extending from head to shoulder girdle, 2 additional thorns much farther back and 1-2 thorns on each shoulder, skin otherwise smooth; generally gray to nearly blackish without distinctive marks. Gulf of California to Panama; inhabits sand-mud bottoms to at least 25 m depth; common in estuaries, and dangerous to humans. Attains a total length of at least 257 cm (disc width 117 cm).

PACIFIC STINGRAY
Himantura pacifica (Beebe & Tee-Van, 1941)
(Photograph on p. 33)

Disc with small, somewhat pointed *protuberance at snout tip*; "wings" of disc rounded; *tail without keel or fold on upper surface and with a low keel, but without fold, on lower surface*; upper surface of disc and basal portion of tail covered with coarse denticles. Mexico to Panama, also reported to be common around the Galapagos; inhabits silty or muddy bottoms in relatively shallow water. Width of disc to 62 cm and maximum total length of 152 cm.

Urobatis concentricus ▲

Urobatis halleri (see description on p. 35) Mottled variety ▲ Plain variety ▼

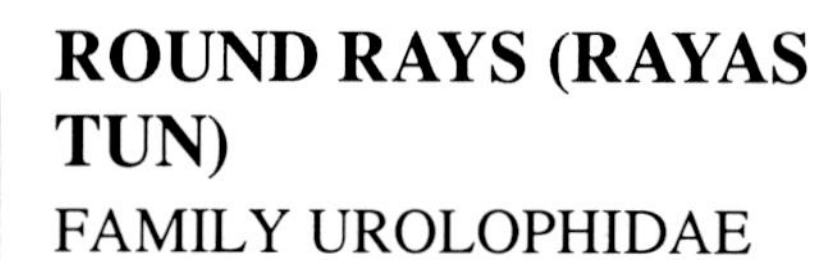

ROUND RAYS (RAYAS TUN)

FAMILY UROLOPHIDAE

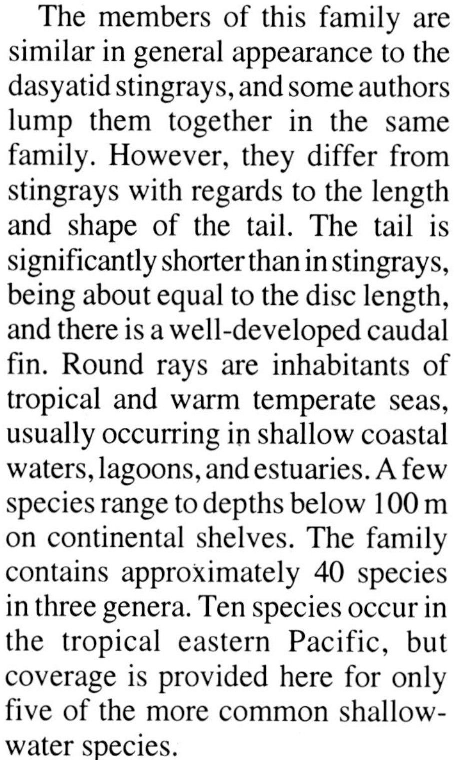

The members of this family are similar in general appearance to the dasyatid stingrays, and some authors lump them together in the same family. However, they differ from stingrays with regards to the length and shape of the tail. The tail is significantly shorter than in stingrays, being about equal to the disc length, and there is a well-developed caudal fin. Round rays are inhabitants of tropical and warm temperate seas, usually occurring in shallow coastal waters, lagoons, and estuaries. A few species range to depths below 100 m on continental shelves. The family contains approximately 40 species in three genera. Ten species occur in the tropical eastern Pacific, but coverage is provided here for only five of the more common shallow-water species.

CONCENTRIC STINGRAY
Urobatis concentricus Osburn & Nichols, 1916

Disc rounded, but with anterolateral margins nearly straight; snout rounded or bluntly angular; tail equal to or less than half total length; tooth rows in upper jaw 26-35; caudal fin rounded; skin smooth, without denticles or thorns; *generally light gray with large blackish blotches and smaller dark spots more or less arranged in concentric rows on disc*. Gulf of California; commonly seen on rubble bottoms adjacent to reefs in 5-20 m depth. Grows to about 40 cm (disc width 28 cm).

ROUND STINGRAY
Urobatis halleri (Cooper, 1863)
(Photograph on p. 34)

Disc rounded, slightly broader than long; snout somewhat angular; tail equal to or less than half total length; tooth rows in upper jaw 26-35; caudal fin rounded; *generally light brown or tan with dense, darker brown spotting and vermiculations; central portion of disc usually with large, brown, circular markings superimposed on spotted or vermiculated background.* Northern California to Panama; frequently sighted on sand, rubble, or rock bottoms in the vicinity of reefs; often completely or partially covers its back with sand when resting on the bottom. Grows to 55 cm (disc width 31 cm).

SPOTTED STINGRAY
Urobatis maculatus Garman, 1913

Disc rounded, but with anterolateral margins nearly straight; snout bluntly angular, tail equal to or less than half total length; tooth rows in upper jaw 35-37; caudal fin rounded; *generally brown or brownish gray, with irregular, variable-sized, relatively widely spaced, dark brown to blackish blotches and spots.* Gulf of California; frequently sighted on rubble bottoms in 2-20 m depth. Grows to about 42 cm length (disc width 26 cm).

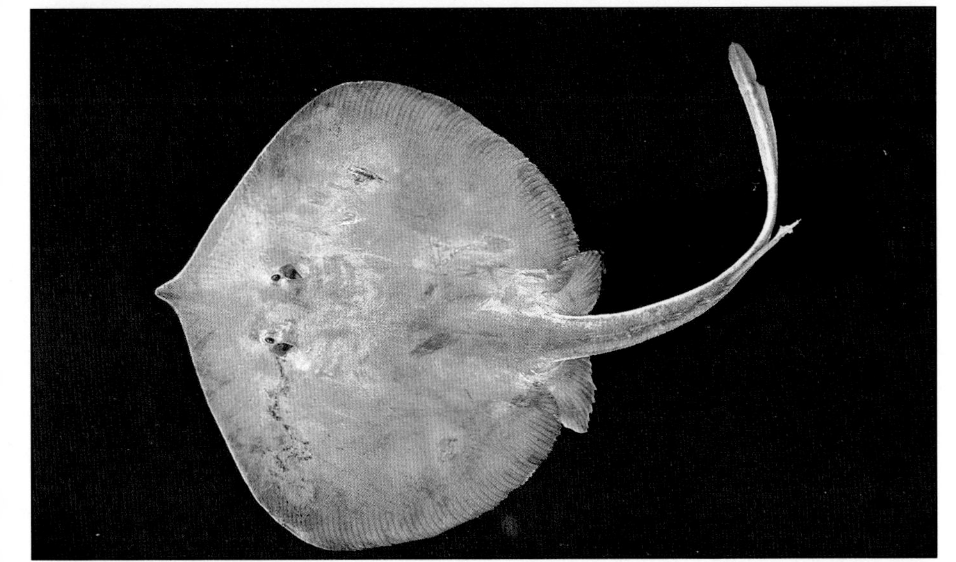

PANAMIC STINGRAY
Urotrygon aspidura (Jordan & Gilbert, 1881)

Disc ovate, somewhat broader than long; anterolateral margins of disc nearly straight; *tip of snout pointed, usually well produced*; tail length greater than half total length; caudal fin relatively rounded and narrow, its dorsal and ventral lobes not confluent; small, weak denticles sparsely covering disc and tail, and *large thorns on dorsal midline of tail*; brown or gray-brown without conspicuous markings. Mexico to Peru; depth range 5-20 m. Attains at least 42 cm (disc width 23 cm).

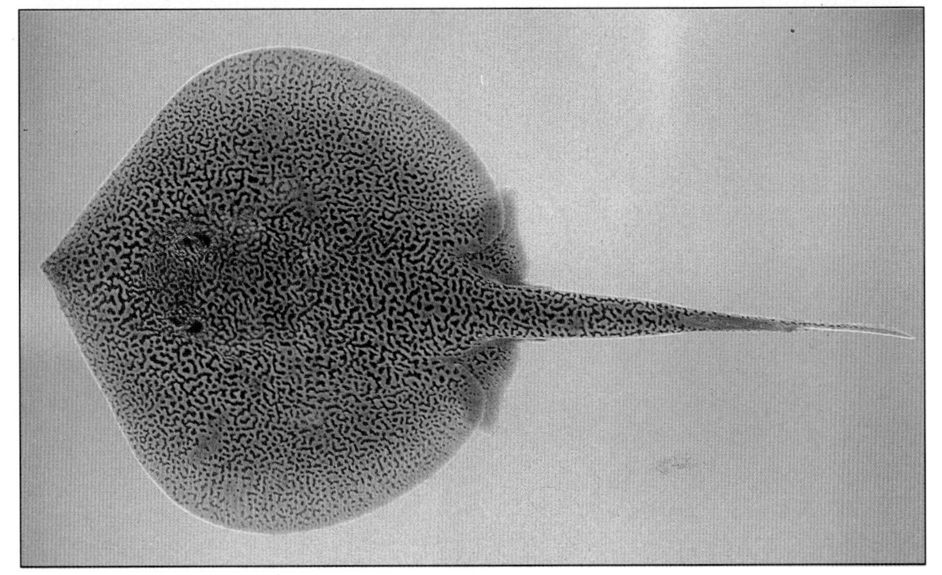

RETICULATED RAY
Urotrygon reticulata Miyake & McEachran, 1988

Disc oval-shaped, with anterolateral margins nearly straight; *snout pointed, but not forming projection*; tail greater than half total length; caudal fin robust and of equal height over most of length, its tip rounded; denticles in adults present on snout, in front of eyes, and margin of disc from snout to level of eyes, sparsely distributed on nuchal, scapular regions, and mid-portion of visceral cavity; *tan with dark brown to blackish vermiculations covering entire dorsal surface.* Known only from Panama; we captured a single specimen in 17 m on a flat sandy bottom. Largest specimen 24 cm total length.

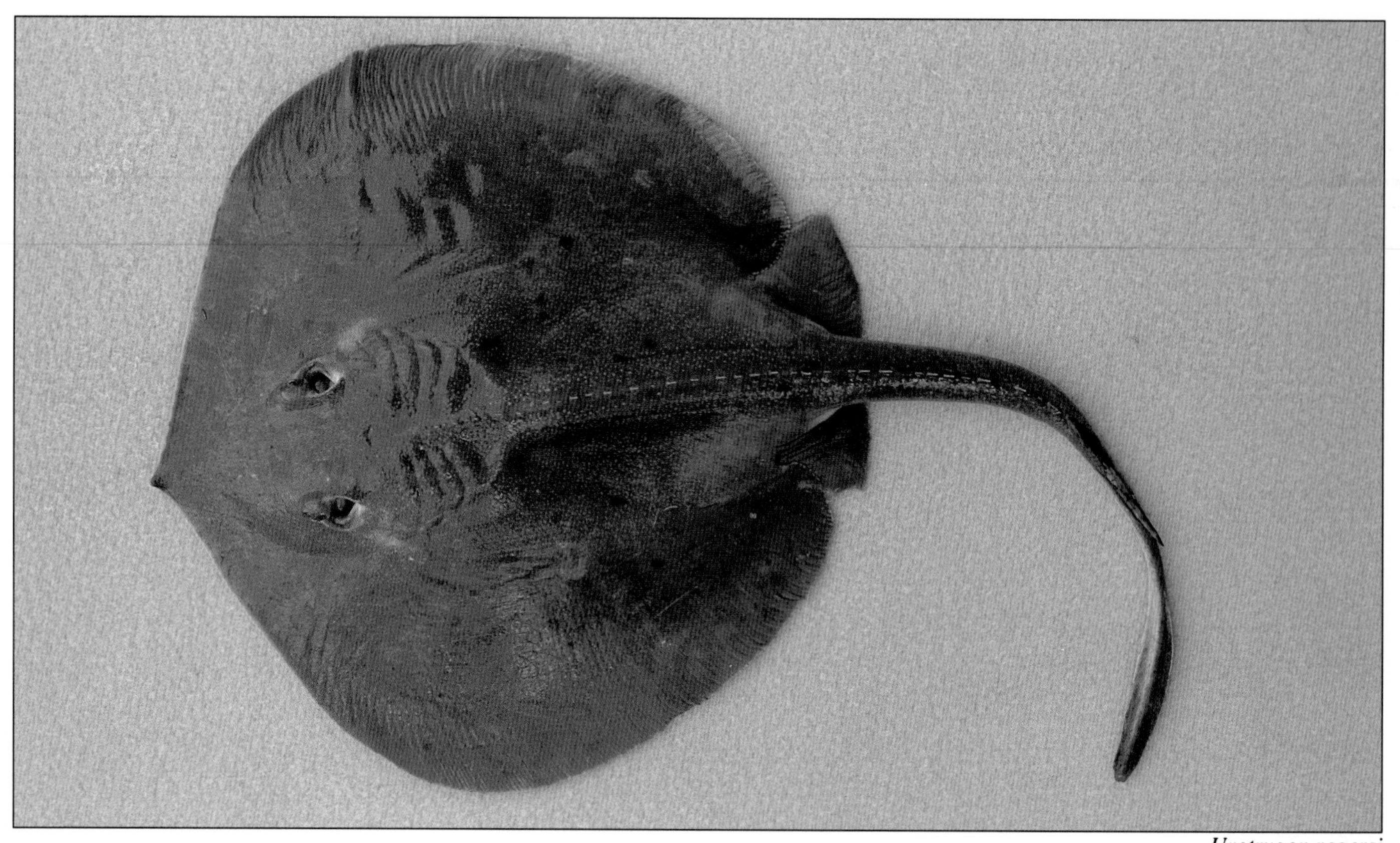

Urotrygon rogersi

ROGER'S STINGRAY
Urotrygon rogersi (Jordan & Starks, 1895)

Disc angular, somewhat diamond-shaped; anterolateral margins nearly straight; tip of snout pointed, slightly projecting; tail greater than half total length; caudal fin relatively narrow, pointed; small denticles on snout, along margin of disc, and on area behind scapular region at midline of disc; *3 rows of denticles on middle of back, the row on midline composed of large thorns from nuchal region to base of tail*; generally light brown, without distinguishing marks. Gulf of California to Chiapas, Mexico; we netted a single specimen in 12 m at San Carlos, Gulf of California. To 63 cm.

BUTTERFLY RAYS (RAYAS MARIPOSA, TUYAS)
FAMILY GYMNURIDAE

These rays are sometimes included as a subfamily of the Dasyatidae. They have a very distinctive appearance characterized by an extremely broad disc that is much wider (more than 1.5 times) than long. The tail is extremely short, lacks a caudal fin, and is sometimes equipped with a spine at its base. The habitat, as for most rays, is open sandy bottoms. The family occurs worldwide in tropical and subtropical seas. It contains two genera and about 12 species.

CALIFORNIA BUTTERFLY RAY
Gymnura marmorata (Cooper, 1863)
(Plate III-3, p. 39)

Disc more or less triangular with *huge angular "wings"*, ventrolateral margins gently rounded; anterolateral margins of disc concave; snout blunt, except for slightly protruding triangular tip; *tail very short, about half length of disc*; caudal fin absent; brown, sometimes mottled with small brown or blackish spots. Southern California to Peru; inhabits shallow bays and beaches. Attains a disc width of about 150 cm.

EAGLE AND COW-NOSED RAYS (RAYAS ÁGUILA, CHUCHOS, RAYAS BAGRE)
FAMILY MYLIOBATIDAE

These graceful rays are frequently sighted in the vicinity of reefs. Like manta rays, a few species can leap high above the ocean's surface. Characteristic features include a protruding head that is distinct from the disc; eyes and spiracles that are lateral on the head; the disc having triangular "wing" flaps; the tail being much longer than the disc; and most having one or more serrated venomous spines near the tail base. They have powerful jaws equipped with large, plate-like crushing teeth arranged in several rows. These are adapted for crushing molluscs, which form the main part of their diet. Young are born alive. The family occurs worldwide in tropical and subtropical seas; there are five genera containing about 27 species. The Cow-nosed rays represent a separate subfamily (Rhinopterinae) and are identified by their squarish, indented snout. They sometimes form large, spectacular aggregations.

SPOTTED EAGLE RAY
Aetobatus narinari (Euphrasen, 1790)
(Plate III-5, p. 39)

Head squarish, with protruding, rounded snout; large triangular "wings" (pectoral fin flaps); tail very long and slender with 2-

6 barbed spines at base; *dark gray to black with numerous white spots dorsally*, white ventrally. Cosmopolitan in tropical to warm temperate seas; feeds on clams, mussels, and oysters; often sighted alone or in small groups cruising on the perimeter of reefs; common in estuaries. Disc width to 250 cm.

GOLDEN RAY

Rhinoptera steindachneri Evermann & Jenkins, 1891
(Plate III-6, p. 39)

A diamond-shaped ray with head distinctly protruding from disc; width of disc about 1.7-1.8 times its length; tail with 1-2 serrated spines, their length about 2.5 times eye width; tail very slender, about 1.2-1.6 times length of disc; skin smooth; *overall pale yellowish brown*, white on ventral surface; tail blackish. Gulf of California to Galapagos; inhabits coastal lagoons, often forms large schools in mangrove areas. Grows to about 100 cm disc width.

MANTA OR DEVIL RAYS (MANTAS, DIABLOS)

FAMILY MOBULIDAE

These distinctive rays are easily recognized by the pair of large, protruding flaps in front of the mouth; the lateral eyes and spiracles; a large wing-like disc that is much wider than long, the ends pointed; no tail spine, or only a rudimentary spine, and a small dorsal fin at the base of the tail. The cephalic flaps are used to direct planktonic food items into the mouth. Some mantas grow to a width of nearly 7 m and weigh more than 1 300 kg; they are among the largest of fishes. Manta rays occur in all warm seas; the family contains two genera and about 10 species. They are frequently seen from boats far out to sea and also are encountered by divers in the vicinity of reefs. Mantas can make spectacular leaps above the water surface.

MANTA RAY

Manta birostris (Donndorff, 1798)
(Plate III-1, p. 39)

Head projecting with pair of paddle-like extensions; large triangular "wings" (pectoral fin flaps); *tail whip-like; disc above and below covered with small denticles*; mouth terminal (at front of head); teeth in lower jaw only; dark gray to black, sometimes with white patches on shoulders; white on ventral surface. Circumtropical. To at least 6.7 m disc width, and over 2 metric tons in weight.

DEVIL RAY

Mobula tarapacana (Philippi, 1892)

Head projecting with pair of paddle-like extensions; large triangular "wings" (pectoral fin flaps); *tail relatively short, without spine; skin smooth*; mouth on underside of head; teeth in both jaws, relatively large and tassellated; back brown to olivaceous green, ventral side white anteriorly and gray posteriorly. Indo-Pacific and eastern Atlantic, possibly circumtropical. Grows to a disc width of at least 305 cm and a weight of 350 kg. Also known as Box Ray.

THURSTON'S DEVIL RAY

Mobula thurstoni (Lloyd, 1908)
(Plate III-2, p. 39)

Head projecting with pair of paddle-like extensions; large triangular "wings" (pectoral fin flaps); *tail moderately long, about 60 percent of disc width in adults, without spine*; mouth on underside of head; teeth in bands in both jaws, forming low, pavement-like mosaic; dark blue to black on dorsal surface; ventral surface mainly white, becoming silvery towards pectoral fin tips, *a dark greenish patch usually evident on each side near posterior edge of pectoral fin*. Probably tropical circumglobal; known for certain from the eastern Pacific, eastern Atlantic, Gulf of Siam, and Indian Ocean; ranges from Gulf of California to Panama and possibly farther southwards; often forms huge schools containing several hundred individuals. Previously known as *M. lucasana*, but *M. thurstoni* is an earlier name. Grows to at least 180 cm disc width and 54 kg; size at birth 65-85 cm disc width.

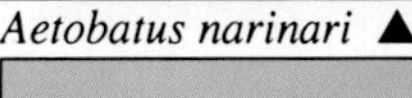

Aetobatus narinari ▲ *Mobula tarapacana* ▼

PLATE III

1 **MANTA RAY** (*Manta birostris*)

2 **THURSTON'S DEVIL RAY** (*Mobula thurstoni*)

3 **CALIFORNIA BUTTERFLY RAY** (*Gymnura marmorata*)

4 **DIAMOND STINGRAY** (*Dasyatis brevis*)

5 **SPOTTED EAGLE RAY** (*Aetobatus narinari*)

6 **GOLDEN RAY** (*Rhinoptera steindachneri*)

7 **LONG-TAILED STINGRAY** (*Dasyatis longus*)

8 **BANDED GUITARFISH** (*Zapteryx exasperata*)

9 **LESSER ELECTRIC RAY** (*Narcine entemedor*)

10 **TORPEDO ELECTRIC RAY** (*Torpedo tremens*)

1
2
3
4
5
6
7
8
9
10

BONY FISHES
CLASS OSTEICHTHYES

All of the remaining fishes treated in this book belong to the class Osteichthyes. This huge assemblage differs from the previously discussed sharks and rays by the possession of a skeleton composed of true bone rather than cartilage. Other diagnostic features include sutures between the skull bones; teeth that are fused to the bones; presence of segmented fin rays; usually a double nostril on each side of snout; biting edge of upper jaw formed by the maxillae and premaxillae (both are dermal bones); swim bladder or functional lung usually present; and fertilization external in the majority of species. The group is divided into four subclasses: Lungfishes (Dipneusti), Tassel-finned fishes (Crossopterygii, of which the Coelacanth is the only living example); Bichirs from fresh waters of Africa (Brachiopterygii); and Ray-finned fishes (Actinopterygii). All except the last-mentioned group have very few living representatives. The vast majority of modern fishes within the Actinopterygii belong to the subdivision Teleostei. Indeed, teleosts are by far the most abundant of all vertebrates, containing an estimated 21 000 species in 410 families, and 3 900 genera. This total represents about 96 percent of all known fishes. Teleosts are extremely diverse in size, shape, habits, and habitat. They occur from deep ocean trenches, kilometers below the surface, to high mountain streams. They range in size from tiny 2 cm gobies to giant groupers that weigh over 400 kg. The variety of shapes is seemingly endless ... eels, seahorses, flatfishes, anglerfishes, catfishes, flyingfishes ... to name just a few examples. Teleosts represent one of our most valuable commercial resources and provide sustenance to millions the world over. Some of the better-known fishes in this category include trout, salmon, herrings, sardines, anchovies, snappers, billfishes, tunas, and mackerels.

Galapagos Grunt (*Orthopristis forbesi*), Isla Pinzon, Galapagos

TENPOUNDERS (PECES TORPEDO, SÁBALOS, CHIROS)
FAMILY ELOPIDAE

Tenpounders are slender, silvery fishes which have a single dorsal fin on the middle of the back with pelvic fins directly below. The pectorals are ventrally located just behind the gill openings, and the caudal fin is strongly forked. They inhabit shallow inshore areas, frequently in estuaries, and occasionally enter fresh water. The diet includes small fishes and shrimps. Although often caught by anglers, the flesh is not very palatable due to the presence of numerous small bones. Tenpounders are close relatives of the tarpons (Megalopidae), a family that is absent from the eastern Pacific. Both have a characteristic ribbon-like, transparent larval stage known as a leptocephalus. It is very similar to the larvae of eels, but differs in having a distinctly forked caudal fin. Whitehead (1962) recognized six species, all in the genus *Elops*. The actual number of species has been somewhat controversial; some researchers suggest there is but a single circumtropical species, but there appear to be at least three to five. One occurs in our area.

MACHETE
Elops affinis Regan, 1909
(Plate IV-1, p. 65)

Dorsal rays 21-25; anal rays 14-17; pectoral rays 16-17; gill rakers on first arch (excluding rudiments) 10-11+16-20; body long and slender, somewhat oval in cross-section; mouth terminal or lower jaw projecting slightly; scales small, about 100-120 in lateral line; anal fin base slightly shorter than dorsal fin base; pelvic fins at middle of body, below level of dorsal fin origin or slightly in front of it; pectoral fins very low on side, just behind edge of gill cover; caudal fin deeply forked; generally silvery with bluish reflections. Southern California to Peru; inhabits shallow inshore waters, particularly in bays, lagoons, and mangrove areas; frequently enters fresh water. Reaches about 90 cm. Closely related to *E. saurus* Linnaeus of the western Atlantic.

BONEFISHES (ZORROS, LISONES, MACABIÉS)
FAMILY ALBULIDAE

This small family of three genera is one of the most primitive of teleost fishes. The young pass through a late larval stage called the leptocephalus which is also found in other primitive Teleostei such as the machete (*Elops*), the tarpon (*Megalops*), and true eels. This characteristic larva is transparent, ribbon-like, with a small head and forked tail. The body of bonefishes is elongate and only slightly compressed. Their most distinctive external feature is their overhanging snout and ventral mouth. There are very small teeth in bands anteriorly in the jaws and patches of small molariform teeth on the roof and floor of the mouth. The fins are all soft-rayed, the single dorsal in the middle of the body, and the anal fin far posterior; as in other primitive fishes, the pectoral fins are low on the body, and the pelvic fins abdominal in position; the caudal fin is deeply forked. Bonefishes get their name from the numerous fine bones in the flesh. *Albula vulpes* was formerly believed to be a wide-ranging circumtropical species, but a recent electrophoretic study in Hawaii by Shaklee and Tamaru has shown that it is actually divisible into two Indo-Pacific species, two from the western Atlantic, and one from the eastern Atlantic. Bonefishes root into sand with their conical snout for their usual food of small clams, various worms, and crustaceans; the hard parts of prey are crushed with their molariform teeth. These fishes are world renown as hard-fighting gamefishes.

PACIFIC BONEFISH
Albula neoguinaica Valenciennes, 1846
(Plate IV-3, p. 65)

Dorsal rays 18-19; anal rays 9; pectoral rays 16-18; lateral-line scales 62-72 (usually 67-70); in standard length; mouth distinctly ventral; *no prolonged rays at end of dorsal and anal fins; jaw not reaching posteriorly to below eye, usually well anterior of this point*; silvery with a blackish spot at tip of snout. Indo-Pacific; typically found on sandy substrata, sometimes coming into very shallow water on sand flats. Attains at least 100 cm. Frequently identified as *A. vulpes*, an Atlantic species.

SHAFTED BONEFISH
Albula nemoptera (Fowler, 1911)

Dorsal rays 18-19; anal rays 8; pectoral rays 16; lateral-line scales 76-84; gill rakers on first arch 5-7 + 10-11; mouth distinctly ventral; *a prolonged, filamentous last ray on the dorsal and anal fins (most pronounced on the dorsal fin); jaw reaches posteriorly to below eye*; overall silvery; yellowish wash on pelvic and pectoral fins; caudal fin dusky. Mexico to Panama; also occurs in the Caribbean Sea; coastal seas and estuaries. Reaches 51 cm.

Albula nemoptera

FALSE MORAYS (MORENAS FALSAS)

FAMILY CHLOPSIDAE

These small eels are characterized by having the pectoral fins present or absent; the dorsal fin originating above the gill opening or just behind; the median fins being continuous around the tail; the mouth reaching beyond the rear margin of the eye; and the posterior nostrils opening downwards through the lip. The teeth are small and pointed in about three rows on the jaws and two separate rows on the vomer. Most species are small (less than 20 cm). They occur in all tropical seas, but are seldom seen due to their cryptic habits. Many of the species burrow in sand or are found deep in crevices of the reef. Members of the family have sometimes been included in the Xenocongridae.

Chlopsis bicollaris

COLLARED EEL

Chlopsis bicollaris (Myers & Wade, 1941)

Body elongate, somewhat worm-like in appearance; dorsal and anal fins well developed, confluent around tip of tail; dorsal fin origin above level of gill opening; snout conical with rounded tip, its length about 4.2 in head length; lower jaw slightly shorter than upper; *pectoral fins absent*; anterior nostrils tubular; *posterior nostrils opening downward in upper lip*, below anterior margin of eye; teeth small and pointed in bands; vomerine teeth in 2 parallel rows; overall grayish brown, whitish ventrally; *a collar-like whitish bar at level of gill opening and dorsal fin origin.* Galapagos Islands. Attains 20 cm.

MORAY EELS (MORENAS)

FAMILY MURAENIDAE

Moray eels are characterized by a very elongate, muscular, compressed body and large mouth. This large assemblage (15 genera and about 200 species) is very diverse in head shape, dentition, color pattern, and maximum size. All species lack pectoral fins, but there is great variation in the height of the dorsal and anal fins, which are continuous with the caudal fin (or absent in the genus *Uropterygius*). Morays in the genus *Echidna* have blunt teeth (flat and molar-like in some species) that are adapted for crushing crabs and molluscs. Most of the other species have sharp needle-like fangs that, in larger morays, are capable of inflicting deep, painful wounds. Fortunately most species are not aggresive, but it is unwise for divers to thrust their arms blindly into crevices and holes in the reef. This action can sometimes provoke an attack. Likewise, the feeding of large morays with fish baits should be discouraged. We know of at least two cases in which fingers and arms were severely damaged by careless divers. Normal prey items of morays include fishes and a variety of invertebrates such as crabs, shrimps, and octopuses. The family is found worldwide in tropical and temperate seas.

Panamic Green Moray (*Gymnothorax castaneus*), Isla Uva, Panama

HARDTAIL MORAY
Anarchias galapagensis (Seale, 1940)

Dorsal and anal fins scarcely apparent, developed only as skin-covered ridges near tip of caudal fin; *tip of tail hard and pointed; posterior nostril closely associated with an enlarged interorbital pore, thus giving the appearance of a double nostril opening*; brown with stellate whitish blotches; tip of tail white. Gulf of California to Panama and the Galapagos. Reaches about 30 cm.

PALENOSE MORAY
Echidna nocturna (Cope, 1872)

Dorsal and anal fins moderately developed, clearly evident skin-covered ridges; origin of dorsal fin in front of level of gill openings; *teeth blunt, without differentiated canines, becoming molariform in adults; dark brown to nearly blackish with scattered white to yellowish spots; nostrils bright orange*. Gulf of California to Peru, including Galapagos. Maximum size, 75 cm.

Close-up of head showing molariform teeth

STARRY MORAY
Echidna nebulosa (Ahl, 1789)

Dorsal and anal fin rays developed as skin-covered ridges, but clearly evident; tip of tail blunt, with a skin-covered caudal; *short, stout, conical teeth at front of jaws; side of jaws with 1-2 rows of small, close-set, compressed, nodular teeth; white with 2 rows of dendritic black blotches containing small yellow spots, these blotches sometimes forming more or less complete bars on head region*; numerous small black spots between large blotches. Widespread in the tropical Indo-Pacific from East Africa to the Americas; in the eastern Pacific it ranges from the Gulf of California to Panama, including Clipperton Island and Isla del Coco. To 70 cm.

LICHEN MORAY

Enchelycore lichenosa (Jordan & Snyder, 1901)

Dorsal and anal fins developed as skin-covered ridges, but clearly evident; posterior nostril not tubular, at most with a raised rim; mouth large, *lower jaw curved, so that a gap is present and teeth visible when mouth is closed*; sharp, caniniform teeth of both jaws in 2-3 rows; *dark brown, head and throat with numerous irregular white to yellowish blotches and spots; a series of large pale blotches on side of body.* Known only from Japan and the Galapagos. Grows to 90 cm.

SLENDERJAW MORAY

Enchelycore octaviana (Myers & Wade, 1941)

Dorsal and anal fins developed as skin-covered ridges, but clearly evident; posterior nostril not tubular, at most with a raised rim; mouth large, *lower jaw curved, so that a gap is present and teeth visible when mouth is closed*; sharp, caniniform teeth of both jaws in 2-3 rows; *pores along upper lip are elongate slits with crenate margins; uniform brown to gray, without distinct markings.* Gulf of California to Colombia and the Galapagos. Reaches 70 cm.

Enchelycore octaviana ▲

LONGFANG MORAY

Enchelynassa canina (Quoy & Gaimard, 1824)

Dorsal and anal fins developed as skin-covered ridges, but clearly evident; *extremely long canine teeth at front of jaws, on intermaxilla, and inner row on side of upper jaw; anterior nostril a short tube with a long, bilobed flap at posterior end*; posterior nostril above front of eye, large and oval with a raised rim; skin with narrow, irregular vertical ridges; dark brownish gray with vertical, thin, blackish lines in grooves on skin; each pore along side of jaws is in a whitish spot. Widespread in the tropical Indo-Pacific; in the eastern Pacific it is known from Clipperton Island and the Gulf of Chiriqui, Panama. Reported to 150 cm.

Enchelynassa canina ▼

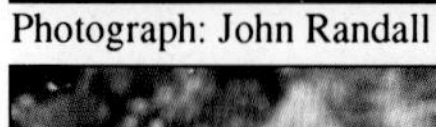

Photograph: John Randall

ZEBRA MORAY

Gymnomuraena zebra (Shaw, 1797)

Dorsal and anal fins scarcely apparent, developed as skin-covered ridges on posterior part of body; tip of tail blunt with a skin-covered caudal; *teeth blunt, molariform, covering jaws and palate like cobblestone pavement; brown to nearly blackish with narrow whitish bars.* Widespread in the tropical Indo-Pacific from East Africa to the Americas; in the eastern Pacific it ranges from the Gulf of California to Panama, but mainly found on offshore islands. Attains 150 cm, but common to 100 cm.

Gymnomuraena zebra ▼

LATTICETAIL MORAY
Gymnothorax buroensis (Bleeker, 1875)

Dorsal and anal fins covered with skin, but clearly evident; dorsal fin origin above level of gill opening; *posterior nostril not tubular, at most a raised rim; teeth pointed and well developed; 3 rows of large canines on lower jaw in an inner row on front portion of jaw*; light brown anteriorly, finely mottled with dark brown; *most of body dark brown with narrow white marks, sometimes forming diffuse crossbars; margin of tail yellowish.* Widespread in the tropical Indo-Pacific from East Africa to the Americas; in the eastern Pacific known from the Gulf of Chiriqui (Panama) and the Galapagos. Reaches 35 cm.

PANAMIC GREEN MORAY
Gymnothorax castaneus (Jordan & Gilbert, 1882)

Dorsal and anal fins developed as skin-covered ridges, but relatively tall and distinct; *posterior nostril not tubular, at most a raised rim*; dorsal fin origin on top of head, well in front of gill opening; *teeth of jaws caniniform; maxillary teeth uniserial; 3 longitudinal rows of teeth at front of upper jaw; brown to brownish green, usually plain*, but sometimes with a few white or yellow flecks, mostly on posterior half and on dorsal fin. Gulf of California to Panama and the Galapagos. Attains 120 cm.

FINE-SPOTTED MORAY
Gymnothorax dovii (Günther, 1870)

Dorsal and anal fins developed as skin-covered ridges, but relatively tall and distinct; *posterior nostril not tubular, at most a raised rim*; dorsal fin origin on top of head, well in front of gill opening; *teeth of jaws caniniform; maxillary teeth uniserial; 3 longitudinal rows of teeth at front of upper jaw; dark brown to black with peppering of numerous small white spots.* Costa Rica to Colombia and the Galapagos. To 140 cm.

EQUATORIAL MORAY
Gymnothorax equatorialis (Hildebrand, 1946)

Dorsal and anal fins very low, but forming distinct skin-covered ridges; anterior nostrils in simple, unbranched tubes; *posterior nostril not tubular, more or less flush with head profile; jaws closing completely when shut, not curved and elongate; teeth enlarged and somewhat shark-like, with serrate posterior margin, in a single row in each jaw*; vomerine teeth absent, or a few inconspicuous teeth hidden by folds of skin; brown, grading to light tan or whitish on ventral surface; *head and body covered with white spots, these very small on head, becoming larger posteriorly, those toward tip of tail very large and widely spaced.* Gulf of California to Peru. Maximum length about 100 cm.

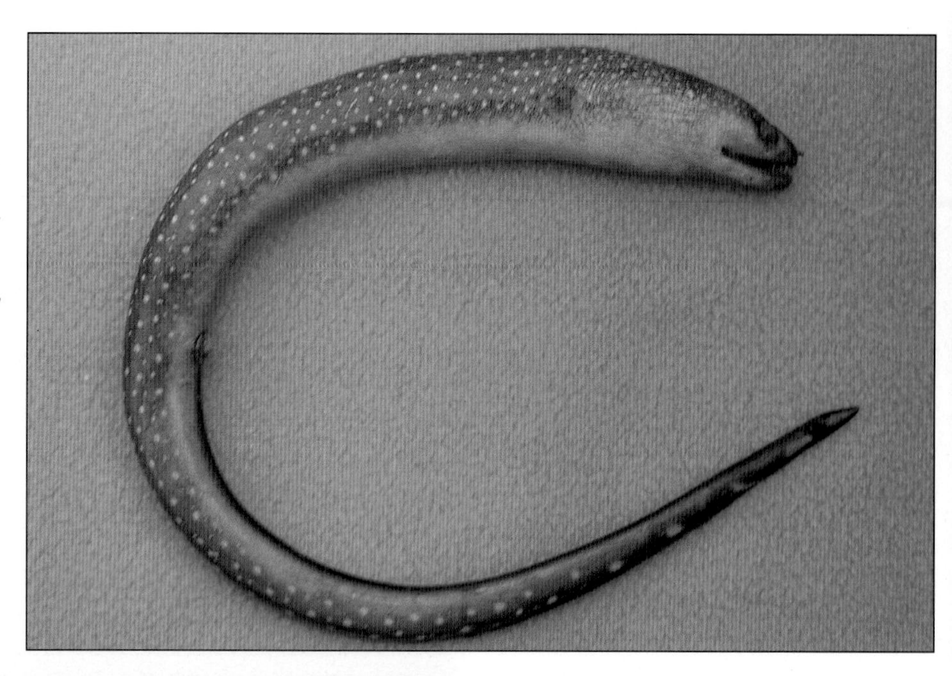

YELLOWMARGIN MORAY
Gymnothorax flavimarginatus (Rüppell, 1830)

Dorsal and anal fins covered with skin, but clearly evident; dorsal fin origin above level of gill opening; *posterior nostril not tubular, at most a raised rim; a single row at front of upper jaw; a pair of canines anteriorly in lower jaw; teeth on side of jaws in a single row; yellowish, densely mottled with dark brown; a prominent black blotch covering gill opening*; juveniles dark brown with yellow-green margin on dorsal and anal fins. Widespread in the tropical Indo-Pacific from East Africa to the Americas; in the eastern Pacific occurs mainly at offshore islands including Isla del Coco and the Galapagos. Reaches 120 cm.

MASKED MORAY
Gymnothorax panamensis (Steindachner, 1876)

Dorsal and anal fins covered with skin, but clearly evident; dorsal fin origin well in front of level of gill opening; posterior nostrils not tubular, at most a raised rim; teeth pointed and well developed; *outer series of teeth in jaws thickened, bent abruptly backwards at tips, their rear margins serrate*; posterior nostril not tubular, at most with a raised rim; medium brown; yellowish brown on head; *eye surrounded by ring of dark brown; pores on jaws surrounded by white.* Gulf of California to Panama and the Galapagos; also known from Juan Fernandez and Easter Islands. Grows to 30 cm.

UNDULATED MORAY
Gymnothorax undulatus (Lacepède, 1803)

Dorsal and anal fins covered with skin, but clearly evident; dorsal fin origin in front of level of gill opening; *posterior nostril not tubular, at most a raised rim; teeth pointed and well developed; a single row of long fang-like canines on intermaxilla, and the usual canines anteriorly in jaws*; young with a few canines in middle row of upper jaw; *head variable, often yellowish*, but also brown, gray or whitish; young brown with diffuse yellowish bars on body forming "chain-link" pattern; adults irregularly mottled. Widespread in the Indo-Pacific from East Africa to the Americas; in the eastern Pacific it occurs mainly on offshore islands including those in the Gulf of Chiriqui, Panama. Attains at least 100 cm.

ARGUS MORAY
Muraena argus (Steindachner, 1870)

Dorsal and anal fins covered with skin, but clearly evident; dorsal fin origin well ahead of level of gill opening; *posterior nostril tubular, above anterior margin of eye*; teeth pointed and well developed, those of adults uniserial; *teeth on vomer non-depressible*; brown, body with 3 rows of large, irregular yellow blotches and numerous small white spots; *intense black spot covering gill opening; a black spot at rear corner of mouth, not preceded by white; margin of dorsal and anal fin white*. Baja California to Peru, including the Galapagos. Attains 120 cm.

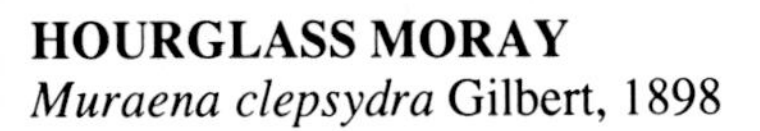

HOURGLASS MORAY
Muraena clepsydra Gilbert, 1898

Dorsal and anal fins covered with skin, but clearly evident; dorsal fin origin well ahead of level of gill opening; *posterior nostril tubular, above anterior margin of eye*; teeth pointed and well developed, those of adults uniserial; *all teeth on vomer depressible*; brown with dense covering of irregular, cream-colored spots; *a large (equal to 2.5 times size of eye diameter or more), ocellated black spot covering gill opening; a black spot at rear corner of mouth preceded by white area on the lower jaw*; young with 5-6 series of small hourglass-shaped spots. Baja California to Peru, including the Galapagos. Largest about 120 cm.

JEWEL MORAY

Muraena lentiginosa Jenyns, 1842

Dorsal and anal fins covered with skin, but clearly evident; dorsal fin origin well ahead of level of gill opening; *posterior nostril tubular, above anterior margin of eye*; teeth pointed and well developed, those of adults uniserial; *all teeth on vomer depressible; brown with dark-edged yellow spots, becoming more numerous with growth; other varieties are yellowish brown to nearly white with darker brown mottling*; a rather inconspicuous black area covering gill opening; black spot at rear corner of mouth preceded by a white area on lower jaw. Gulf of California to Peru, including the Galapagos. Reaches 60 cm.

Ocellated variety ▲ Mottled variety ▼

Reticulated variety ▲ Yellow variety ▼

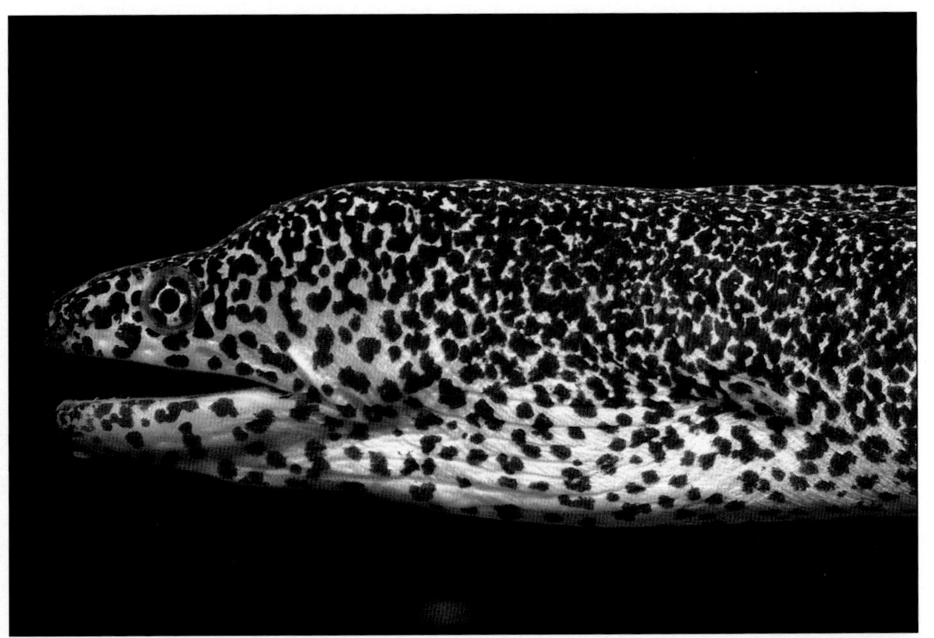

PEPPERED MORAY
Siderea picta (Ahl, 1789)

Dorsal and anal fins covered with skin, but clearly evident; dorsal fin origin above level of gill opening; posterior nostrils not tubular, at most a raised rim; *teeth pointed and well developed, but not of canine proportions; teeth in upper jaw in a single row, conical at front of jaw, becoming compressed along side of jaw; a single conical tooth on intermaxilla; 2 rows of teeth at front of lower jaw, the outer row small; no median depressible fangs in upper jaw; white with dense covering of small black spots*; juveniles with relatively larger black spots in about 3 longitudinal rows. Widespread in the tropical Indo-Pacific from East Africa to the Americas; in the eastern Pacific it occurs at offshore islands including the Galapagos. To 100 cm.

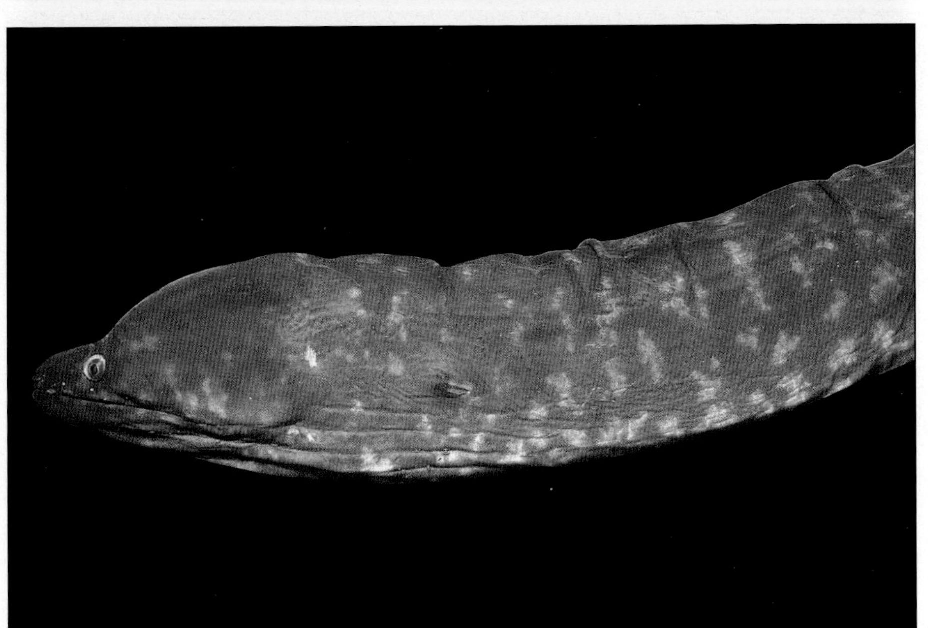

LARGEHEAD MORAY
Uropterygius macrocephalus (Bleeker, 1865)

Dorsal and anal fins mainly absent, developed only as skin-covered ridge at tip of tail; posterior nostril not tubular, only with a raised rim; *a single, minute branchial pore in front of gill opening*; teeth pointed and well developed; *teeth of jaws biserial, those of vomer uniserial*; overall brown, *usually with mottled pattern evident; tip of tail white to yellowish.* Christmas Island (eastern Indian Ocean) and Japan to the eastern Pacific; known in the latter region from offshore islands, including Isla del Coco. Reaches 47 cm. *U. nectura* (Jordan & Gilbert) is a synonym.

CRAFTY MORAY
Uropterygius versutus Bussing, 1991

Dorsal and anal fins mainly absent, developed only as skin-covered ridge at tip of tail; posterior nostril tubular; a pair of minute branchial pores in front of gill opening; teeth pointed and well developed; *teeth of jaws biserial, those of vomer uniserial; uniform brown or golden brown, sometimes with irregular whitish blotches on body.* Gulf of California to Colombia, including Isla del Coco and the Galapagos; ranges from tide pools to a depth of 40 m. Largest specimen, 56 cm. A third species of *Uropterygius* also occurs in the region. *U. tigrinus* is easily distinguished by its whitish body covered with large dark spots interspersed with small spots.

SNAKE EELS (ANGUILAS CULEBRA, ZAFIOS, TIESOS)

FAMILY OPHICHTHIDAE

These fishes have a typical eel-shaped body that is more or less rounded in cross-section. The body is scaleless; the eyes are usually small and situated just above the mouth; the snout is often pointed or nearly so; the lower jaw is frequently underslung; the posterior nostril is usually within or piercing the upper lip; the gill openings are midlateral to entirely ventral in position; the pectoral fins are present or absent; the tail tip is either hard and pointed or the caudal rays are conspicuous and confluent with the dorsal and anal fins. This large family is distributed in all tropical and temperate oceans; there are about 55 genera and approximately 250 species. Although very common, snake eels are seldom seen because they spend most of the time buried in the sand. The pointed snout is useful for burrowing. In addition, many have a bony, sharp tail and are equally adept at burrowing backward and forward. The species that have conspicuous bands or spotting are sometimes mistaken for sea snakes, but they are easily distinguished by their lack of scales and possession of a more or less pointed tail (paddle-like in sea snakes). The diet of most snake eels consists of small fishes, crabs and shrimps.

Equatorial Snake Eel (*Apterichtus equatorialis*), Perlas Islands, Panama

EQUATORIAL SNAKE EEL
Apterichtus equatorialis (Myers & Wade, 1941)

Tail tip a hard, finless point; *all fins absent; gill openings ventral, converging forward*; snout pointed; teeth of jaws uniserial, the largest on intermaxillary; *posterior nostril opening outside mouth, with a flap; anterior nostril tubular*; pale, creamy yellow with faint network of diffuse brownish spots. Mexico to Galapagos Archipelago. Grows to at least 30 cm.

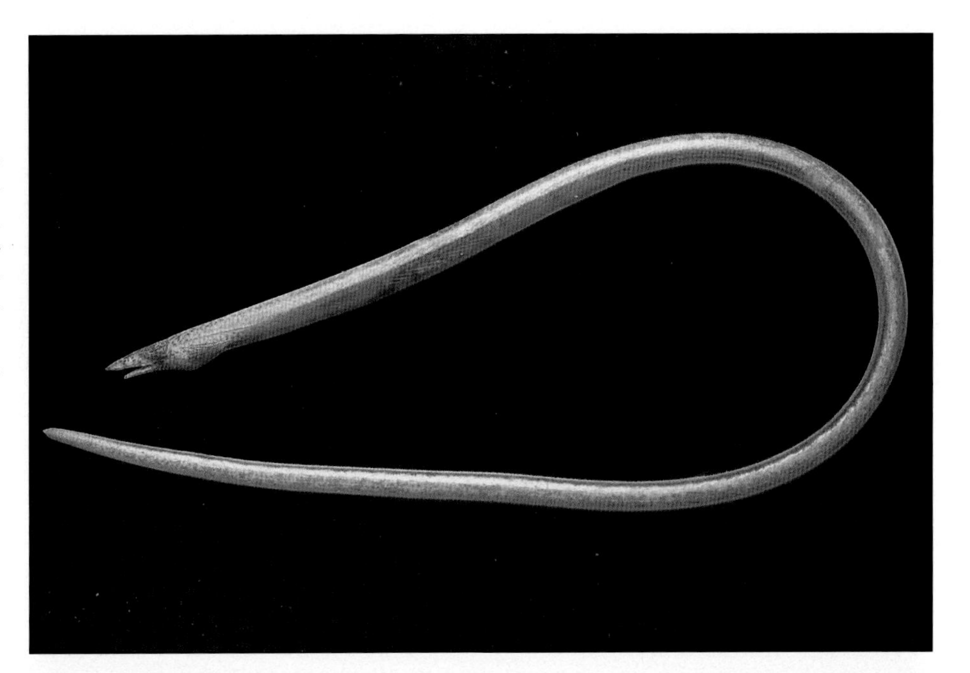

CYLINDRICAL SNAKE EEL
Bascanichthys cylindricus Meek & Hildebrand, 1923

Tail tip a hard, finless point; body very elongate; dorsal and anal fins low; dorsal fin origin well in front of gill opening, a little closer to snout tip than gill opening; *pectoral fin present, but very tiny*; snout short and blunt, its underside grooved; teeth of jaws small and conical, uniserial; *tail (anus to tip of tail) very long, about 1.0-1.2 in head and trunk (snout to anus)*; brown, grading to tan on ventral part; dorsal fin whitish or tan. Costa Rica and Panama. Reaches 90 cm.

PANAMIC SNAKE EEL
Bascanichthys panamensis Meek & Hildebrand, 1923

Tail tip a hard, finless point; body moderately robust, the trunk long, but *tail relatively short; tail (anus to tip of tail) about 1.3-1.4 in head and trunk (snout to anus)*; dorsal and anal fins low; dorsal fin origin well in front of gill opening, but slightly closer to gill opening than to snout tip; *pectoral fin present, but very tiny*; snout short and blunt, its underside grooved; teeth of jaws small and conical, uniserial; light brown with whitish dorsal fin; *head largely whitish with brown markings*. Baja California to Panama. To at least 80 cm.

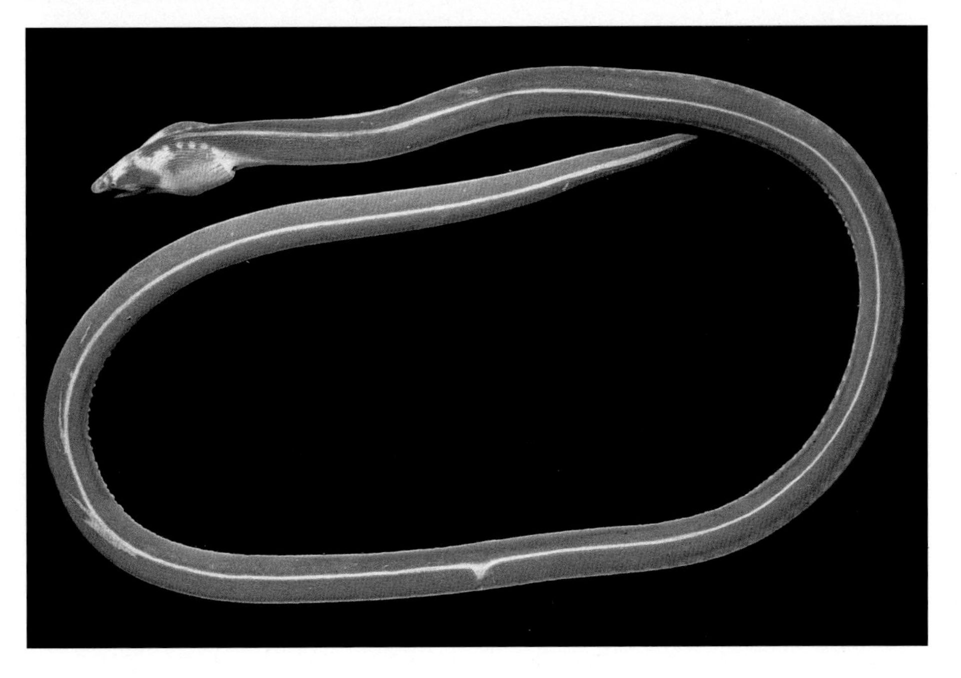

CLIFF'S SNAKE EEL
Callechelys cliffi Böhlke & Briggs, 1954

Tail tip a hard, finless point; dorsal and anal fins present; dorsal fin relatively well developed, originating well forward on head, slightly behind level of rear corner of mouth; *pectoral fins absent*; head and body laterally compressed; *anterior nostrils tubular*; a median groove on underside of snout; *gill openings low and lateral, converging forward*; teeth of jaws small and pointed, uniserial; tail (anus to tip of tail) nearly equal to head and trunk (i.e. snout to anus); *head creamy white; body yellowish, covered with numerous fine spots, not much larger than eye diameter*; dorsal fin with a distinct white edge. Gulf of California to Panama. To 50 cm. Two other species (not shown) of *Callechelys* occur in the region: *C. eristigmus* (Gulf of California to Panama) has the tail shorter than head and trunk and has large oval dark spots; *C. galapagensis* (Galapagos) has dark spots intermediate in size to the other species.

FANGJAW EEL
Echiophis brunneus (Castro-Aguirre and Suarez de los Cabos, 1983)

Tail tip a blunt, finless point; body elongate, cylindrical, pointed at both ends; tail (tail tip to anus) longer than head and trunk (snout tip to anus); *pectoral fins present; dorsal fin origin level with pectoral fin tips*; gill openings relatively elongate, vertical and lateral; snout short, subconical, slightly constricted near tip; *teeth very stong and pointed, biserial, largest anteriorly on jaws; teeth also well developed on roof of mouth;* posterior nostril in a short tube, before and beneath eye*; overall tan with numerous small, brown spots on back and sides; dorsal fin brown.* Gulf of California to Panama. Grows to at least 120 cm.

YELLOW FINLESS EEL
Ichthyapus selachops (Jordan & Gilbert, 1882)

Tail tip a hard, finless point; body elongate, cylindrical, pointed at both ends; tail (tail tip to anus) longer than head and trunk (snout tip to anus); *all fins absent; gill openings entirely ventral*; snout pointed and depressed, grooved and flattened on underside; teeth pointed, uniserial; *posterior nostril opening inside mouth; anterior nostril not tubular*; overall creamy yellow. Gulf of California to Ecuador. Grows to at least 40 cm.

BLACK SAILFIN EEL
Letharchus rosenblatti McCosker, 1974

Tail tip a hard, finless point; body moderately elongate and laterally compressed; snout pointed; dorsal fin moderately developed, originating well forward on head, a short distance behind level of rear corner of mouth; *pectoral and anal fin absent; anterior nostril a hole, without a raised rim; gill openings entirely ventral*; dark brown, nearly blackish, except anterior end of head, including snout, white. Gulf of California to Panama. Largest reported 35 cm.

SPOTTED SNAKE EEL
Myrichthys maculosus (Cuvier, 1817)

Tail tip a hard, finless point; head and trunk (snout tip to anus) shorter than tail (tail tip to anus); snout subconical, not acutely pointed, relatively short and broad when viewed from above; dorsal fin origin well in front of gill opening; *pectoral fin well developed and broad based (i.e. extends over the full length of gill opening); teeth molariform or granular, multiserial on jaws and vomer*; yellowish to white, with large black or dark gray oval spots. Widespread in the tropical Indo-Pacific from East Africa to the Americas; in the eastern Pacific it occurs from the Gulf of California to Peru, and at offshore islands including the Galapagos. Attains 50 cm.

ROCKPOOL WORM EEL
Myrophis vafer Jordan & Gilbert, 1881

Tail tip flexible; caudal fin rays conspicuous, confluent with dorsal and anal fins; body very long and slender; tail (tail tip to anus) much longer than head and trunk (snout tip to anus); snout broad and depressed; dorsal fin originating well posteriorly, closer to anus than to snout tip; pectoral fins well developed; *posterior nostril opens into mouth*; pale gray, nearly whitish, slightly darker on dorsal half. Baja California to Peru. Grows to 25 cm.

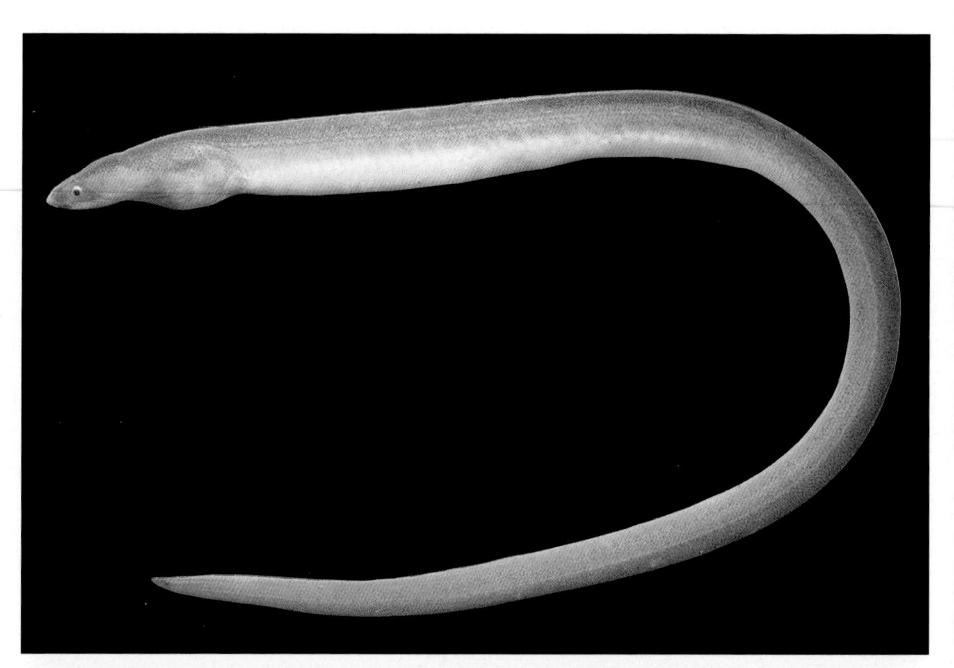

PACIFIC SAILFIN EEL
Paraletharchus pacificus (Osburn & Nichols, 1916)

Tail tip a hard, finless point; body moderately elongate and laterally compressed; snout short, rounded at tip; dorsal fin moderately developed, originating well forward on head, a short distance behind level of rear corner of mouth; *pectoral and anal fins absent; anterior nostrils tubular; gill openings entirely ventral*, each with a pronounced anterolateral pocket; head whitish with brown spots; remainder of body dark brown, nearly blackish, except white ventrally and also with white dorsal fin. Baja California to Panama. To 50 cm. *P. opercularis* (not shown) from the Galapagos is similar, but is cream-colored to tannish, with numerous brown spots.

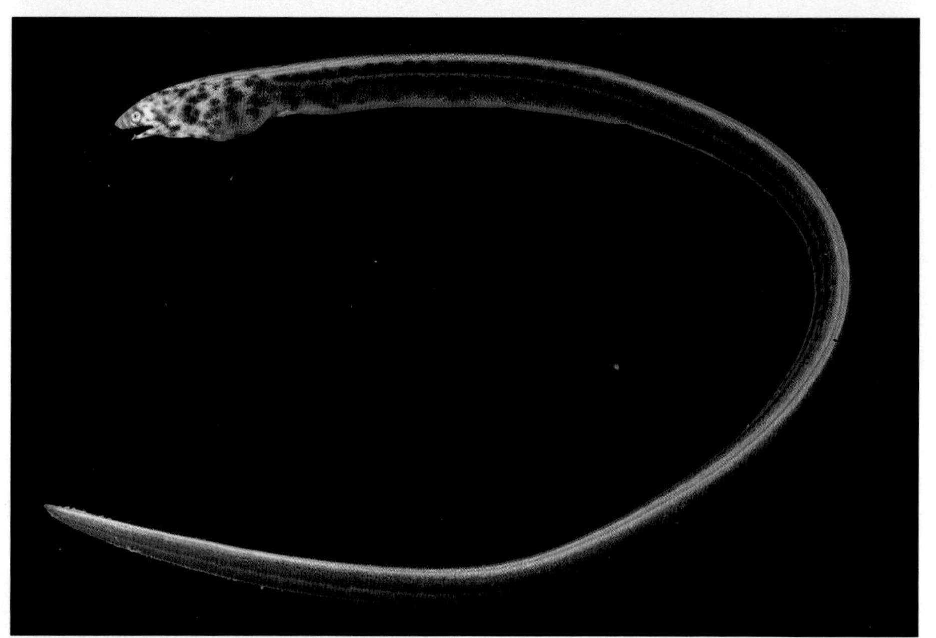

BLUNT-TOOTHED SNAKE EEL
Pisodonophis daspilotus Gilbert, 1898

Tail tip a hard, finless point; body long and slender; snout narrow and depressed; dorsal and anal fins relatively low; *dorsal fin origin behind level of gill opening; pectoral fin broad based; teeth molariform or granular, multiserial on jaws and vomer*; pale yellowish with small brown spots; dorsal and anal fins yellowish. Costa Rica to Ecuador. Maximum size to 66 cm.

RED-BANDED SNAKE EEL

Quassiremus nothochir (Gilbert, 1890)

Tail tip a hard, finless point; body moderately elongate, cylindrical; head and trunk (snout tip to anus) slightly longer than tail (tail tip to anus); snout conical, a short groove on underside; dorsal fin origin behind level of gill opening; *pectoral fin very tiny, smaller than or equal to eye*; teeth pointed, uniserial; ground color cream to tan; head with close-set black spots; *about 15 large black-edged, red, saddle-like markings on body,* with large and small black spots in the spaces between. Gulf of California to Costa Rica. Reaches about 60 cm.

Quassiremus nothochir ▼ Close-up of head ▲

FRECKLED SNAKE EEL

Quassiremus evionthas (Jordan & Bollman, 1889)

Tail tip a hard, finless point; body moderately elongate, cylindrical; head and trunk (snout tip to anus) slightly longer than tail (tail tip to anus); snout conical, a short groove on underside; dorsal fin origin behind level of gill opening; *pectoral fin very tiny, smaller than or equal to eye*; teeth pointed, uniserial; *white with numerous small black spots*; at regular intervals there is a yellowish ground color, and the dark spots are longer and darker, forming about 15 crossbars. Galapagos Islands. Grows to about 60 cm.

CONGER EELS (ANGUILAS CONGRIO, ANGUILAS JARDÍN, ZAFIROS)

FAMILY CONGRIDAE

Congers have a typical eel-shaped body with a small gill opening, and the caudal fin is continuous with the dorsal and anal fins. They differ noticeably from moray eels in having pectoral fins, and their body is nearly round anteriorly rather than compressed. The family is divisible into four subfamilies with approximately 42 genera and about 110 species worldwide: Congrinae (well developed pectoral fins; about 25 genera including *Rhynchoconger*); Heterocongrinae, commonly known as garden eels (pectoral fins tiny or absent; four genera); Muraenesocinae, or pike congers (pectorals very well developed and fang-like teeth; eight genera including *Cynoponticus*); and Bathymyrinae (pectorals well developed; about five genera including *Ariosoma* and *Paraconger*). The reef-dwelling congers are generally nocturnal predators of fishes and crustaceans. Garden eels are found in large colonies that feed exclusively on plankton that is picked from passing currents. The flesh of larger conger eels is considered good eating.

Galapagos Garden Eel (*Heteroconger klausewitzi*) colony, Isla Marchena, Galapagos

GILBERT'S CONGER
Ariosoma gilberti (Ogilby, 1898)

Dorsal and anal fins well developed; head cylindrical, body compressed posteriorly; tail (tail tip to anus) relatively short, only slightly longer than head and body (snout tip to anus); snout conical, slightly projecting; *pectoral fins long and narrow, about 2.8-3.1 in head length; teeth pointed, in villiform bands in jaws; vomer with triangular patch of teeth with long posterior projection*; dorsal fin origin above level of gill opening; overall tan or very pale brown. Gulf of California to Colombia; inhabits flat sandy bottoms in 10-40 m. To at least 20 cm.

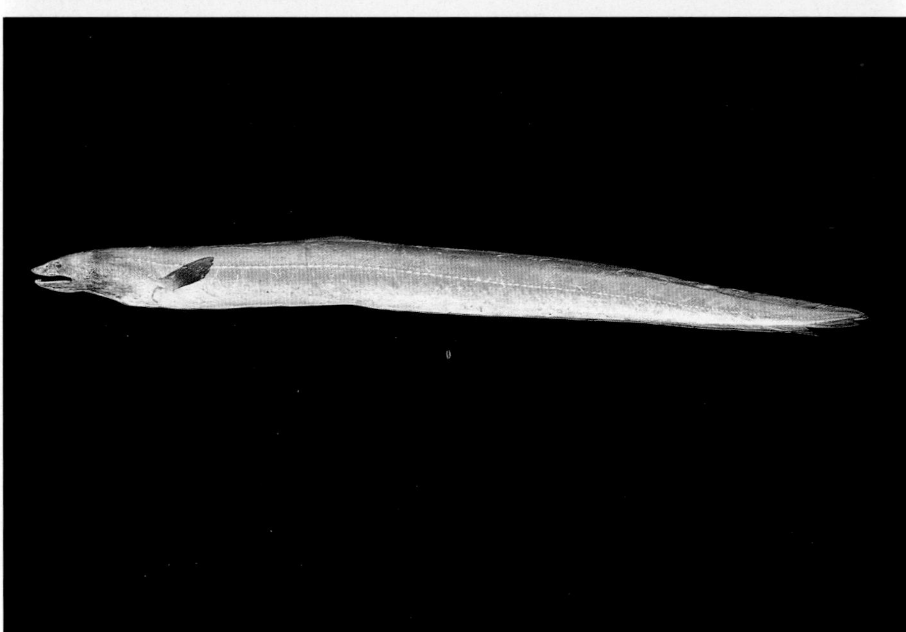

CONEHEAD EEL
Cynoponticus coniceps (Jordan & Gilbert, 1881)

Dorsal and anal fins well developed, continuous around tip of tail; dorsal fin origin about level with gill opening; pectoral fins well developed; *mouth large; teeth multiserial, including large fang-like canines at front of jaws; vomer with median row of about 15 very large teeth with their tips angled posteriorly*; olive-brown to grayish above and whitish below; dorsal and anal fins pale gray to brownish with dark margin. Baja California to Colombia. Reaches 90 cm.

SPOTTED GARDEN EEL
Gorgasia punctata Meek & Hildebrand, 1923

Body very elongate and cylindrical; snout very short; eye large, longer than snout and about 6-7 in head length; *mouth reaching below rear part of eye*; teeth pointed in narrow bands on anterior part of jaws, uniserial on sides, and directed backward; vomer with single series of large recurved teeth; *pectoral fin well developed, longer than eye; dorsal and anal fins low, not confluent around tail tip, which is fleshy*; generally pale tan to whitish, with numerous tiny brown spots. Baja California to Panama. Grows to at least 50 cm.

CAPE GARDEN EEL
Heteroconger canabus (Cowan & Rosenblatt, 1974)

Body very elongate and cylindrical; snout very short; eye large, longer than snout, 7.1-11.5 in head length; mouth reaching below front part of eye; teeth pointed, in narrow bands; vomerine teeth in 1-2 irregular longitudinal rows; teeth of jaws and vomer recurved; *pectoral fins very tiny*, much smaller than pupil of eye; *dorsal and anal fins low, confluent around tip of tail; dark chocolate brown,* grading to tan on extreme posterior portion of body; a narrow whitish bar below anterior portion of dorsal fin; black spot covering gill opening and pectoral fins also black; an oblique whitish bar on middle of head and white area in front of gill opening; *anterior pores of lateral line surrounded by halos of white pigment.* Gulf of California; occurs on clean sandy slopes in 3-18 m. To 77 cm.

CORTEZ GARDEN EEL
Heteroconger digueti Pellegrin, 1923

Body very elongate and cylindrical; snout very short; eye large, longer than snout, about 7.0-10.1 in head length; mouth reaching below front part of eye; teeth pointed, in narrow bands; vomerine teeth in 1-2 irregular longitudinal rows; teeth of jaws and vomer recurved; *pectoral fins very tiny,* much smaller than pupil of eye; *dorsal and anal fins low, confluent around tip of tail*; light brown, with a few whitish patches on head and several light bars on lower half of body; *anterior pores of lateral line surrounded by unpigmented areas, but not by white-pigmented halos.* Baja California to Panama; occurs on clean sandy bottoms in 4-25 m. Attains at least 65 cm.

GALAPAGOS GARDEN EEL
Heteroconger klausewitzi Eibl-Eibesfeldt & Koster, 1983

Body very elongate and cylindrical; snout very short; eye large, longer than snout, about 6.0-10.0 in head length; mouth reaching below front part of eye; teeth pointed, in narrow bands; vomerine teeth in 1-2 irregular longitudinal rows; teeth of jaws and vomer recurved; *pectoral fins very tiny*, much smaller than pupil of eye; *dorsal and anal fins low, confluent around tip of tail*; light to dark brown; several white blotches on head and on body below lateral line; *anterior pores of lateral line surrounded by white-pigmented areas*. Galapagos Islands; inhabits clean sand in 10-30 m. Reaches 70 cm.

CALIFORNIA CONGER
Paraconger californiensis Kanazawa, 1961

Dorsal and anal fins well developed; snout rounded; anterior nostrils tubular at tip of snout; posterior nostrils without tube, next to eye; mouth large, reaching below posterior part of eye; teeth in 1-2 rows, outer row compressed with pointed tips; upper edge of gill opening directly anterior to upper edge of pectoral fin base; *pores in lateral line 39-41; pectoral rays 15-18*; generally pale tannish, with *white margin on dorsal and anal fins*. Gulf of California to Peru. Grows to at least 50 cm.

SILVERBELLY CONGER
Rhynchoconger nitens (Jordan & Bollman, 1889)

Dorsal and anal fins well developed; snout conical, *lower jaw much shorter than upper*; anterior nostrils tubular at tip of snout; posterior nostrils without tube, next to eye; dorsal fin origin slightly anterior to level of gill opening; *pectoral fins relatively large, about 3.1-3.2 in head length*; teeth pointed, in broad bands; grayish, upper half of head and body with numerous tiny black spots, *silvery on lower parts*. Panama. Reaches at least 20 cm.

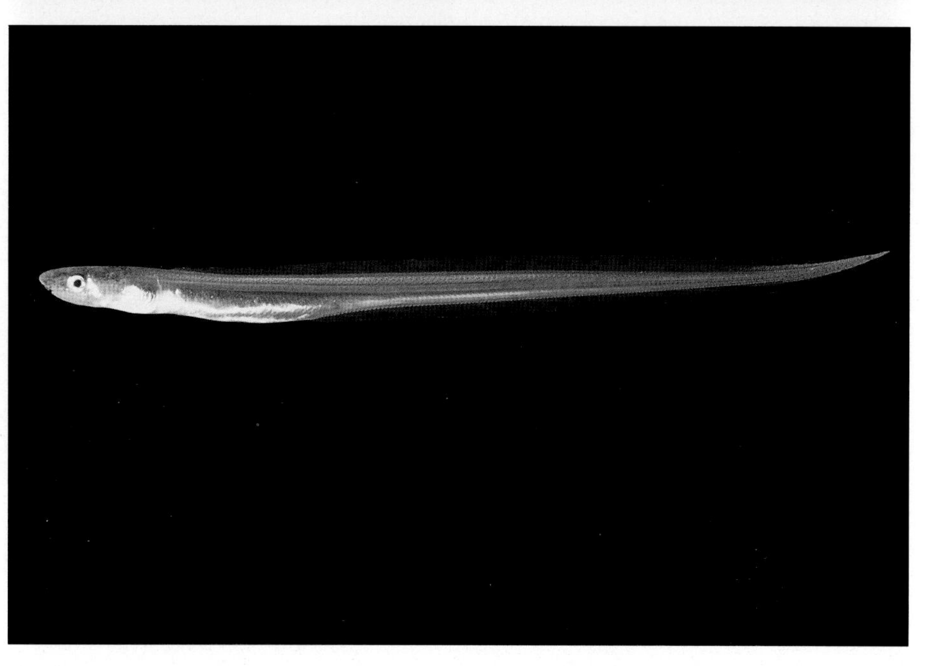

HERRINGS (SARDINAS, ARENQUES)
FAMILY CLUPEIDAE

Clupeids are generally small, silvery fishes that are easily recognized by their keel of scutes (special spiny scales) along the belly, and small, often poorly-toothed mouths. Other features include a single dorsal fin that is located over the middle of the body; a forked caudal fin; usually a short anal fin located well posteriorly; pelvic fins on abdomen below dorsal fin; pectoral fins low on the side just behind the head; and no fin spines. Herrings and their relatives are of major importance to fisheries; they constitute about half of the world's catch of fishes. Clupeids are found in most shallow-water habitats including fresh water, brackish estuaries, coastal embayments, and oceanic reefs. They occur in all warm and cold seas. Most are found in schools that may contain hundreds or thousands of individuals. Zooplankton, often crustaceans, is the major dietary item for most clupeids. The family was reviewed by Whitehead (1985).

School of herrings (*Opisthonema* sp.), Perlas Islands, Panama

RED-EYE ROUND HERRING
Etrumeus teres (DeKay, 1842)
(Plate IV-12, p. 65)

Dorsal rays 18; anal rays 13; body long and slender, its depth about 5.8-6.0 in standard length; branchiostegal rays 11-15; *base of pelvic fins behind dorsal fin base; no mid-ventral scutes on breast or belly*; blue on back, silvery on sides and belly. Southern Australia, Japan, East Africa and Red Sea, western Atlantic, eastern Pacific, and Hawaiian Islands; in the eastern Pacific occurs along the outer coast of Baja California, Galapagos, and Peru; forms pelagic schools along rocky shorelines. Reaches 25 cm.

FLATIRON HERRING
Harengula thrissina (Jordan & Gilbert, 1882)
(Plate IV-5, p. 65)

Dorsal rays 17-18; anal rays 15-17; gill rakers on lower limb of first arch 28-34; *body moderately deep, about 2.9-3.0 in standard length*; operculum smooth without bony striae; none of dorsal fin rays forming elongate filament; *rear margin of gill opening with 2 fleshy lobes*; irridescent blue on back, silvery on sides; a faint yellowish stripe on upper side. Gulf of California to Peru; forms large inshore schools. Grows to 8 cm.

PACIFIC PIQUITINGA
Lile stolifera (Jordan & Gilbert, 1881)
(Plate IV-7, p. 65)

Dorsal rays 11-15; anal rays 17; *body moderately slender, its depth 3.2-3.5 in standard length*; operculum smooth without bony striae; none of dorsal fin rays forming elongate filament; *rear margin of gill opening evenly rounded without fleshy lobes*; olive to bluish on back, white below, with broad silver midlateral stripe; tips of caudal fin black. Gulf of California to Peru; inhabits sandy or muddy shores, also entering brackish estuaries. Maximum size 15 cm.

PACIFIC THREAD HERRING
Opisthonema libertate (Günther, 1867)
(Plate IV-9, p. 65)

Dorsal rays 17; anal rays 19-20; body moderately deep, about 2.8-3.2 in standard length; *last dorsal fin ray long and filamentous, extending nearly to caudal fin base*; bluish on back, silvery white on sides and belly; upper part of body with narrow dark stripes, and scattered blackish spots often present on side; usually a black spot or blotch behind upper rear corner of gill cover. Gulf of California to Peru; a very common schooling species of herring found over soft bottoms close to shore. Three other *Opisthonema* in the eastern Pacific are virtually identical in appearance, and are difficult to separate. Lower limb gill raker counts of fishes in excess of 14 cm (counts increase with growth) provide the best means of separation. *O. bulleri* = 25-36; *O. libertate*=63-110; *O. medirastre*=41-69; *O. berlangai* (Galapagos only) = 50-87. Maximum length about 27 cm; common to 12-18 cm.

CALIFORNIA PILCHARD
Sardinops caeruleus (Girard, 1854)
(Plate IV-14, p. 65)

Dorsal rays 14; anal rays 17; body slender, its depth 4.5-5.0 in standard length; *lower half of gill cover with distinct bony striae radiating downward*; mouth relatively large, upper jaw reaching to below middle of eye; blue on back, silvery on sides, and white on belly; widely scattered blackish spots on side. Alaska to Gulf of California. The genus *Sardinops* (4 species) represents about 40 percent of the world clupeid catch; nearly 8 million tons were landed in 1982. Grows to about 38 cm.

LONGFIN HERRINGS (SARDINAS MACHETE)

FAMILY PRISTIGASTERIDAE

The members of this family were formerly included as a subfamily of the Clupeidae. They are very similar in body shape and overall coloration, but exhibit several important differences in their internal features (structure of the gill arches and skeletal system). Like the clupeids, they possess scutes along the belly, and are perhaps best differentiated from them by their long-based anal fin that contains more than 30 rays (60 or more in some species). The family has a worldwide, mainly tropical distribution, except it is absent in the eastern Atlantic; nine genera and 33 species are known. Most are inhabitants of shallow coastal seas and estuaries, but a few freshwater species occur in South America. The family was reviewed by Whitehead (1985).

PACIFIC ILISHA
Ilisha furthii (Steindachner, 1875)
(Plate IV-4, p. 65)

Dorsal rays 16; *anal rays 46-50*; body moderately slender, its depth about 2.8-3.3 in standard length; eye large; lower jaw projecting; *pelvic fins very tiny, below tip of pectoral fin*; breast and belly strongly compressed with 23-25 + 11-14 scutes; overall silvery; pectoral and caudal fins yellow. Costa Rica to Ecuador; inhabits coastal seas, frequently entering estuaries and rivers. Exceeds 28 cm.

PANAMA LONGFIN HERRING
Odontognathus panamensis
(Steindachner, 1876)
(Plate IV-8, p. 65)

Dorsal rays 12; *anal rays 61-68; pelvic fins absent*; body slender, its depth about 4.2-4.9 in standard length; breast and belly strongly compressed with uninterrupted keel of about 29-30 scutes; *upper jaw (maxillary) very long, reaching to or beyond gill opening*; overall silvery. Honduras to Panama; inhabits coastal seas, sometimes entering brackish estuaries. Maximum length to 18.5 cm.

TROPICAL LONGFIN HERRING
Neoopisthopterus tropicus (Hildebrand, 1946)
(Plate IV-6, p. 65)

Dorsal rays 13-15; *anal rays 43-48; pelvic fins absent; no gap between teeth-bearing outer edge of premaxilla and maxilla; lower jaw not projecting*; body slender, its depth about 3.8-4.0 in standard length; about 26 scutes on belly; greenish brown on back, silvery white below, with a silver midlateral band. Gulf of California to Peru; forms large schools in bays, estuaries, and in brackish river mouths. Grows to 11 cm.

DOVE'S LONGFIN HERRING
Opisthopterus dovii (Günther, 1868)
(Plate IV-13, p. 65)

Dorsal rays 11; anal rays 55-62; gill rakers on lower limb of first arch 15-19; *pelvic fins absent*; body elongate, its depth about 3.5-4.0 in standard length; belly with about 29 scutes; mouth upturned with lower jaw strongly projecting; *rear tip of premaxilla not contacting maxilla, thus resulting in a prominent gap; upper jaw relatively short, not reaching level of center of eye*; olive-brown to bluish on back, silvery on sides and belly; caudal fin slightly yellowish. Gulf of California to Peru; inhabits inshore waters over soft bottoms, sometimes entering estuaries. Grows to 21 cm. Three other *Opisthopterus* occur in the area: *O. equitorialis* (silver band on side, and 21-25 lower limb gill rakers), *O. macrops* (dorsal origin equidistant between caudal and pectoral fin bases), and *O. effulgens* (anal fin origin equidistant between front margin of eye and caudal base, and only 15 lower gill rakers).

YELLOWFIN HERRING
Pliosteostoma lutipinnis (Jordan & Gilbert, 1881)
(Plate IV-10, p. 65)

Dorsal rays 13; *anal rays 49-51; pelvic fins absent; body relatively deep, its depth about 3.0 in standard length; a small, toothed bone (hypo-maxilla) present between rear tip of premaxilla and edge of maxilla*; belly with about 27-28 scutes; *mouth upturned with lower jaw strongly projecting*; olive on back, silvery white below, with *silver midlateral band*; fins yellowish. Mexico (Mazatlan) to Colombia; inhabits coastal seas, possibly entering brackish estuaries. Attains 16 cm.

ANCHOVIES (ANCHOAS, ANCHOVETAS, BOQUERONES)
FAMILY ENGRAULIDAE

Anchovies are herring-like fishes that typically have a prominent shark-like snout that projects beyond the tip of the lower jaw. Many of the species possess scutes on the belly, but this feature is absent in all New World species, hence a valuable character in distinguishing clupeids and engraulids in our region. Anchovies frequently form dense shoals in coastal areas and are captured in large numbers by net fishermen. They are utilized fresh, frozen, or canned for human consumption or pet food. Other products resulting from the anchovy fishery include animal feeds (for cultured fishes, poultry, cattle, sheep, pigs), oil, fertilizers, and live tuna bait. The world anchovy catch in 1982 was 4 046 105 tons, but was considerably higher during the boom years of the Peruvian Anchovy, which yielded 13 059 900 tons alone in 1970. The family is found worldwide in all seas between about 60 degrees north and 50 degrees south latitude. A few species occur permanently in fresh water. They feed primarily on planktonic organisms, particularly crustaceans, which are filtered by the gill rakers. The family was reviewed by Whitehead et al. (1988), who recognized 139 species in 16 genera.

SPICULE ANCHOVY
Anchoa spinifer (Valenciennes, 1848)
(Plate IV-11, p. 65)

Dorsal rays 15; anal rays 34-40; gill rakers on lower limb of first arch 12-19; snout relatively pointed; *upper jaw (maxilla) elongate, reaching to edge of gill cover, its rear tip pointed; a few short gill rakers present on posterior surface of third gill arch; a small triangular projection on lower margin of gill cover (on suboperculum); pectoral fin reaching beyond pelvic fin origin*; bluish on back, silvery below; small specimens with silver midlateral stripe that disappears with age; caudal fin yellow. Atlantic (Panama and Trinidad to Brazil) and eastern Pacific (Costa Rica to Peru); the ony anchovy known from both the Pacific and Atlantic coasts of the Americas. To 20 cm. *Anchoa* is represented in the eastern Pacific by 17 species. Most are quite similar in appearance and difficult to distinguish.

BIGSCALE ANCHOVY
Anchovia macrolepidota (Kner & Steindachner, 1865)
(Plate IV-15, p. 65)

Dorsal rays 15; anal rays 29-32; *gill rakers on lower limb of first arch increasing with growth, about 70 at 5 cm standard length, 100 at 8-13 cm, and 120-135 in large fish*; snout relatively pointed; *upper jaw (maxilla) elongate, but not reaching to edge of gill cover, its rear tip pointed; no gill rakers on posterior surface of third gill arch*; bluish on back, silvery below; a silvery stripe sometimes present along middle of side; caudal fin yellowish with blackish rear margin. Gulf of California to Peru; inhabits sandy shores and brackish river mouths, forming large schools. Grows to about 15 cm.

BALBOA ANCHOVY
Anchoviella balboae (Jordan & Seale, 1926)
(Plate IV-18, p. 65)

Dorsal rays 13-15; anal rays 20-26; *gill rakers on lower limb of first arch increasing with growth, ranging from about 20 in juveniles to 28-35 in adults*; body depth 3.5-4.0 in standard length; *snout short and bluntly rounded; upper jaw (maxilla) short, barely reaching beyond level of rear edge of eye, its rear tip bluntly rounded*; bluish on back, silvery below. Known only from Panama, but probably more widespread; occurs along sandy shores forming large schools. To 8 cm. A second species of *Anchoviella, A. analis*, occurs along the coast of northern Mexico; it is more slender (depth 4.0-4.5 in standard length) and has more anal fin rays (31-37).

PACIFIC ANCHOVETA
Centengraulis mysticetus (Günther, 1867)
(Plate IV-17, p. 65)

Dorsal rays 15-17; anal rays 20-27; gill rakers on lower limb of first arch increasing with growth, about 25 at 5 cm standard length, about 60 at 12 cm; snout short and relatively pointed; *upper jaw (maxilla) moderately long, reaching well beyond rear margin of eye, but not reaching rear margin of preopercle, its rear tip bluntly rounded; gill membranes joined across isthmus (throat region), almost completely covering it*; bluish on back, silvery white below; a silver midlateral stripe in smaller fish, disappearing at about 6-9 cm length. Gulf of California to Peru; forms large schools over muddy inshore areas. Reaches about 18 cm, common to 10-12 cm.

PACIFIC SABRETOOTH ANCHOVY
Lycengraulis poeyi (Kner & Steindachner, 1865)
(Plate IV-16, p. 65)

Dorsal rays 13-16; anal rays 24-27; gill rakers on lower limb of first gill arch 18-23; snout short and blunt; *upper jaw (maxilla) moderately long, reaching well beyond rear margin of eye, but not reaching rear margin of opercle, its rear tip rounded; lower jaw with small canine-like teeth, becoming larger posteriorly*; blue on back, silvery white below; several small black dots on upper gill cover; a silver midlateral stripe present on smaller fish. El Salvador to Panama; forms large schools over muddy inshore areas; also found in brackish estuaries and river mouths. To 23 cm.

MILKFISH (CHANOS, ABUELAS)
FAMILY CHANIDAE

See discussion of the single species of the family, which follows.

MILKFISH
Chanos chanos (Forsskål, 1775)
(Plate IV-2, p. 65)

Dorsal rays 13-17; anal rays 8-10; pectoral rays 15-17; pelvic rays 10-12; lateral-line scales 78-90; *gill rakers on first arch 147-160 + 107-165; caudal fin strongly forked; scales cycloid, except head scaleless; a large axillary scale above pectoral and pelvic fins; no fin spines; mouth small without teeth; eye covered with thick layer of gelatinous tissue*; silvery blue-green on back, silvery on sides, and white below. East Africa to the Americas; in the eastern Pacific it ranges from Baja California to the Galapagos. Spawns in the sea, but spends part of the life cycle in estuaries; highly esteemed as food in several regions, they form the basis of an extensive fish-farming industry in southeastern Asia; often encountered in schools. Grows to 180 cm.

Large flocks of pelicans and other seabirds frequently signal the presence of huge concentrations of anchovies and herrings. Gulf of Panama.

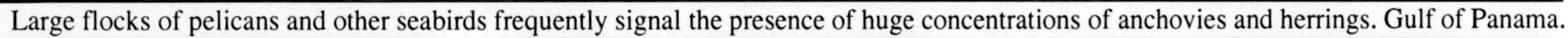

PLATE IV

1 **MACHETE** (*Elops affinis*)

2 **MILKFISH** (*Chanos chanos*)

3 **PACIFIC BONEFISH** (*Albula neoguinaica*)

4 **PACIFIC ILISHA** (*Ilisha furthii*)

5 **FLATIRON HERRING** (*Harengula thrissina*)

6 **TROPICAL LONGFIN HERRING** (*Neoopisthopterus tropicus*)

7 **PACIFIC PIQUITINGA** (*Lile stolifera*)

8 **PANAMA LONGFIN HERRING** (*Odontognathus panamensis*)

9 **PACIFIC THREAD HERRING** (*Opisthonema libertate*)

10 **YELLOWFIN HERRING** (*Pliosteostoma lutipinnis*)

11 **SPICULE ANCHOVY** (*Anchoa spinifer*)

12 **RED-EYE ROUND HERRING** (*Etrumeus teres*)

13 **DOVE'S LONGFIN HERRING** (*Opisthopterus dovii*)

14 **CALIFORNIA PILCHARD** (*Sardinops caeruleus*)

15 **BIGSCALE ANCHOVY** (*Anchovia macrolepidota*)

16 **PACIFIC SABRETOOTH ANCHOVY** (*Lycengraulis poeyi*)

17 **PACIFIC ANCHOVETA** (*Centengraulis mysticetus*)

18 **BALBOA ANCHOVY** (*Anchoviella balboae*)

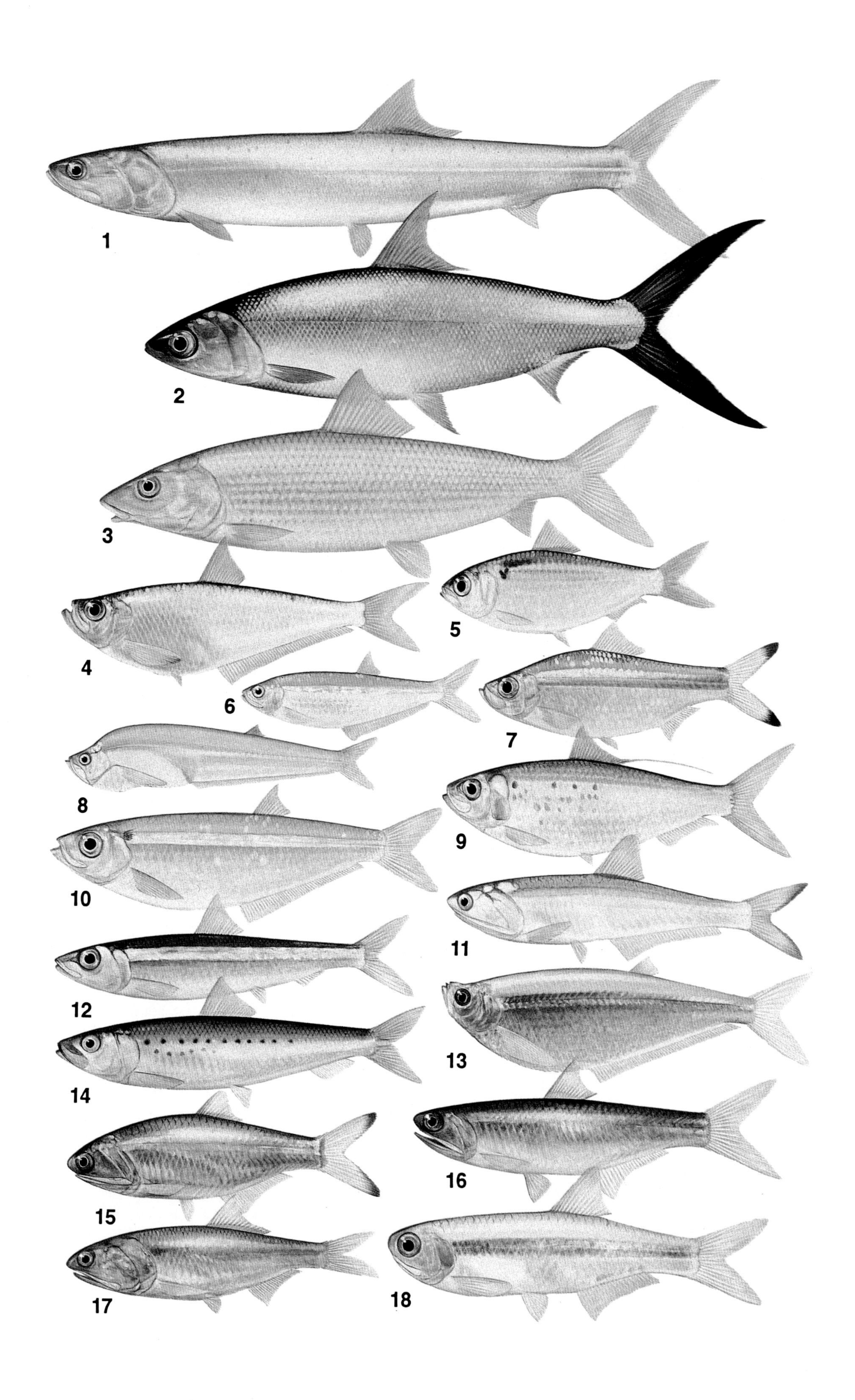
1
2
3
4
5
6
7
8
9
10
11
12
13
14
15
16
17
18

SEA CATFISHES (BAGRES, CONGOS, BARBUDOS, ALGUACILES, COMINATES)

FAMILY ARIIDAE

These fishes are easily recognized by their "whiskers" or barbels around the mouth (usually three pairs) and forked tail. Other features include an adipose fin, bony plates present on top of head and near origin of dorsal fin, and a stiff spine at front of dorsal and pectoral fin. The family occurs in all tropical and subtropical seas, and there are numerous freshwater (particularly in the Australia-New Guinea region) and estuarine dwellers as well. The eastern Pacific species are largely coastal fishes which frequent sand and mud bottoms of bays, harbours, and brackish river mouths; a few species range into deeper water and are seen in trawler catches. Freshly captured specimens should be handled with care as the sharp spines in front of the dorsal and pectoral fins are venomous. Fishermen often clip these off before untangling the fish from their nets. Their flesh is good eating and large numbers of catfishes are seen in fish markets throughout the region. Ariids feed on small fishes and a variety of benthic invertebrates, frequently shrimps, crabs, or molluscs. Worldwide there are approximately 20 genera containing an estimated 130 species. Patricia Kailola (Canberra, Australia) and Richard Cooke (Smithsonian Tropical Research Institute) supplied the information appearing in the species accounts below. However, there remains some confusion regarding the correct identity of several species, and the present generic allocation (particularly with regards to *Arius* and *Cathorops*) is somewhat unstable. Therefore, the classification presented here should be regarded as provisional and subject to change.

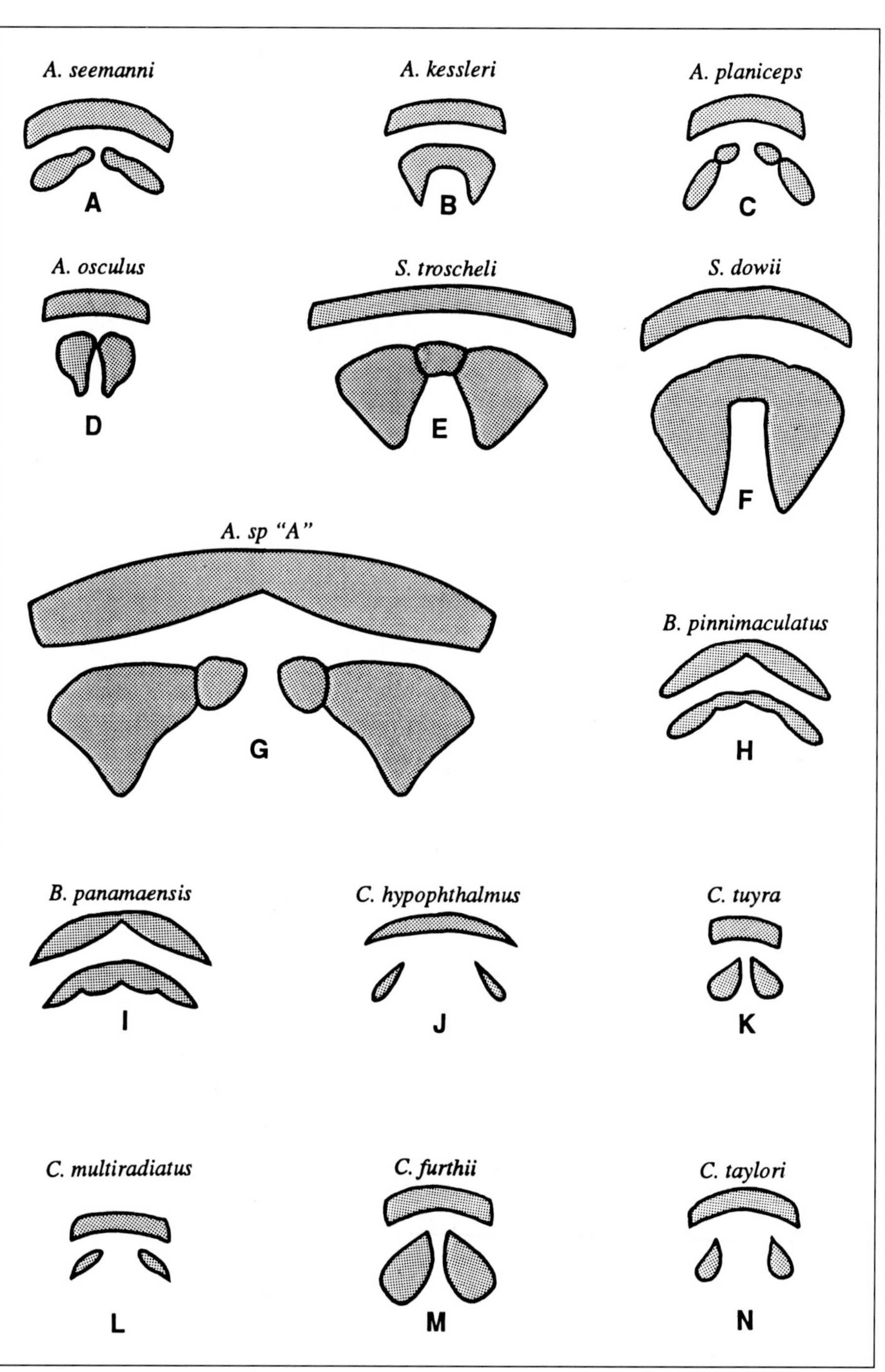

Diagrammatic drawing of premaxillary and palatine teeth patches of selected ariid catfishes. The shape and size of the patches is a useful feature for identification, although the shape is somewhat variable for a particular species and changes with increased growth. The median patch of palatine teeth that forms a bridge linking the large outer patches (for example, in B, E and F) may appear as a pair of small separate patches in young fish, which gradually become fused with increased growth. (A) *Arius seemanni*; (B) *Arius kessleri*; (C) *Arius planiceps*; (D) *Arius osculus*; (E) *Sciadeops troscheli*; (F) *Sciadeichthys dowii*; (G) *Arius* new species (not included below, but teeth patches very distinctive); (H) *Bagre pinnimaculatus*; (I) *Bagre panamensis*; (J) *Cathorops hypophthalmus*; (K) *Cathorops tuyra*; (L) *Cathorops multiradiatus*; (M) *Cathorops furthii*; and (N) *Cathorops taylori*.

BROADHEAD CATFISH
Arius dasycephalus Günther, 1864
(Plate V-9, p. 71)

Dorsal rays I,7; anal rays 19-23; pectoral rays I,10-11; gill rakers on first arch 4 + 5; no rakers on posterior surface of first arch; body robust; *eye 4.9-6.0 in head; smooth to rugose with 4 prominent ridges extending forward; 2 parallel ridges medially, and 2 outer ridges, each diverging towards eye; ridges with coarse to sharp granules or tubercles*; supraoccipital process triangular, slightly longer than wide; palatine teeth in 4 patches, the inner patches well separated at midline and continuous, with slightly larger and narrowly ovate outer patches; adipose fin base longer than dorsal basc; dark gray or bluish brown on back, sides gray to bluish; lower surface whitish; barbels black; fins dusky to blackish. Costa Rica and Panama; relatively common, occasionally in estuaries. Grows to 29 cm.

BLUE CATFISH
Arius guatemalensis Günther, 1864
(Plate V-11, p. 71)

Dorsal rays I,7; anal rays 17-20; pectoral rays I,10-11; pectoral spine very strong, with prominent serrations on inner margin; gill rakers on first arch 6-7 + 11-14; no rakers on posterior surface of first 2 arches; body moderately elongate; *eye small, about 4 in interorbital*; head shield rugose without ridges extending forward; *dorsomedian fontanel flat, scarcely apparent, ending well before base of supraoccipital process*; palatine teeth arranged in *4 patches*, the medial patches narrowly separated at midline and continuous with ovate and elongated outer patches; adipose fin base about three-fourths of dorsal fin base; bluish black or charcoal on back, greenish violet on sides, and silvery white on lower surface; fins dusky to blackish. Mexico to Panama; relatively common in coastal and brackish waters. Reaches 37 cm. *A. caerulescens* is a synonym.

KESSLER'S CATFISH
Arius kessleri Steindachner, 1876
(Plate V-10, p. 71)

Dorsal rays I,7; anal rays 17-20; pectoral rays I,10-11; gill rakers on first arch 3 + 8; no rakers on posterior surface of first 2 arches; body elongate; *snout projecting beyond broad (1.8-2.4 in head length) mouth; lips thick and fleshy; eye relatively small, 2.8-3.1 in internarial distance; dorsomedian fontanel shallow and tapered posteriorly, ending well short of supraoccipital; supraoccipital process triangular, its base subequal to its length; predorsal plate narrow and crescentric; palatine teeth in 4 patches, but inner patches usually joined across midline; inner margin of pectoral spine with fine serrae*; brown to blackish on back, silvery white below; fins dusky or dark bluish. Mexico (Sinaloa) to Ecuador; relatively common in coastal seas and brackish estuaries, but not entering fresh water. Grows to 47 cm. *A. insculptus* is a synonym. A similar, undescribed species (premaxillary and palatine teeth appear as figure E on page 66) is distinguished by a very thin, elongate supraoccipital process. It commonly occurs in fresh and brackish waters of Panama, and reaches 80 cm.

PANAMANIAN CATFISH
Arius lentiginosus (Eigenmann & Eigenmann, 1888)
(Plate V-15, p. 71)

Dorsal rays I,7; anal rays 23-28; pectoral rays I,10-11; gill rakers on first arch 2-3 + 4-5; no rakers on posterior surface of first 2 arches; body robust; *deep lateral cleft behind corner of mouth; lips thick; eye small, 8.0-9.5 in head*; head shield striate and granular, ridges extending forward towards eyes and snout; *supraoccipital process striate and granular, long and narrow (its width about 2 times in its length); palatine*

Arius platypogon (see description on p. 68)

teeth in 2 elongate strips, not joined across midline; adipose fin base longer than dorsal base; dark steel-blue on back, lighter bluish black on side and white on belly; fins dusky. Panama; infrequent in bays and estuaries. Attains 35 cm.

THICK-LIPPED CATFISH
Arius osculus Jordan & Gilbert, 1882
(Plate V-12, p. 71)

Dorsal rays I,7; anal rays 18; pectoral rays I,10-11; total gill rakers on first arch 11-13; body robust; *mouth small and narrow, 2.4-3.1 in head; lips thick*; head shield coarsely granular; *dorsomedian fontanel shallow and tapered posteriorly; palatine teeth in 4 patches, but the inner patches may be joined to the larger outer ones, thus appearing as only 2 large, somewhat triangular patches, separated at midline; adipose fin base about same length as dorsal fin base; mouth moderately small, snout rounded, lips thick*; gray to bronzy greenish on back and upper sides, whitish below; fins dusky. Costa Rica and Panama; common in river mouths. Reaches at least 35 cm.

FLATHEAD CATFISH
Arius planiceps Steindachner, 1875
(Plate V-13, p. 71)

Dorsal rays I,7; anal rays 17-19; pectoral rays I,10; gill rakers on first arch 8-10 (lower limb only); no rakers on posterior surface of first 2 gill arches; lips thick and crenate; *mouth subterminal, wide (1.8-2.4 in head); snout projecting slightly; eye large, 1.5-1.6 in distance between anterior nostrils; predorsal plate narrowly crescent-shaped; dorsomedian fontanel broad and flat, tapering posteriorly and extending to within 1 eye diameter of base of supraoccipital process; supraoccipital process triangular, its base subequal to its length; interorbital flat and smooth; palatine teeth arranged in 4, usually continuous patches, the inner patches smaller than outer ones*; adipose fin base equal to or longer than dorsal fin base; dark brown or grayish on back, white on ventral surface; fins dusky. Mexico (Sinaloa Province southwards) to Panama. Maximum size 58.5 cm.

SLENDER-SPINED CATFISH
Arius platypogon Günther, 1864
(Photograph on p. 67)

Dorsal rays I,7; anal rays 18-20; pectoral rays I,10-11; gill rakers on first arch 5-6 + 9-10; no rakers on posterior surface of first arch; body robust; mouth broad, lips thick; dorsal spine much thinner than in other eastern Pacific *Arius*; *eye large, 4.5-6.5 in head; supraoccipital triangular with strong keel; predorsal plate narrow and crescentic; dorsomedian fontanel extends posteriorly from interorbital as narrow deep groove to base of supraoccipital process; palatine teeth in 4, often continuous patches, the inner patches much smaller than triangular outer patches*; adipose fin base ½-⅔ base of dorsal fin; grayish to olive-brown on back, silvery gray on sides, and white below; fins dusky. Baja California to Panama; relatively common in coastal waters. To 45 cm.

SEEMANN'S CATFISH
Arius seemanni Günther, 1864
(Plate V-14, p. 71)

Dorsal rays I,7; anal rays 17-20; pectoral rays I,10-11; gill rakers on first arch 5-8 + 9-15; no rakers on posterior face of first arch; body moderately elongate; lips thick; *eye large, 1.6-2.8 in interorbital; head shield granular, the rough portion extending forward as broad, triangular patch on each side of flattened, smooth interorbital-postorbital; dorsomedian fontanel well-defined, a narrow, deep groove extending to base of supraoccipital process; supraoccipital process triangular with bluntly rounded apex; palatine teeth in 4 patches, inner patches often separated at midline and continuous with slightly larger, ovate outer patches; adipose fin base about two-thirds length of dorsal base*; dark blue to greenish brown on back; sides grayish or with greenish-bronzy sheen; white on lower surface; fins dusky. Mexico to Peru; common in coastal waters to at least 50 m depth, also entering estuaries. Reaches 40 cm. Often placed in the genus *Ariopsis*; *Arius jordani* is a synonym.

CHIHUIL CATFISH
Bagre panamensis (Gill, 1863)
(Plate V-1, p. 71)

Dorsal rays I,7; anal rays 25-30; pectoral rays I,12-13; gill rakers on first arch 5-7 + 12-14; no rakers on posterior face of first arch; body very robust; lips absent except at corner of mouth; *only 2 pairs of barbels; maxillary barbel broad, ribbon-like, reaching as far as pelvic fins*; head shield with striae, but without granules; palatine teeth in 4 patches, but fused to form narrow, continuous band; *pectoral spine with long, flattened filament reaching origin of anal fin*; lateral line bifurcate at tail base; dark blue on back, with brassy sheen; silvery on sides, and white below; fins dusky to pale. Baja California to Ecuador; commonly caught by trawlers. Attains 40 cm.

LONG-BARBLED CATFISH
Bagre pinnimaculatus (Steindachner, 1876)
(Plate V-2, p. 71)

Dorsal rays I,7; anal rays 27-32; pectoral rays I,13; gill rakers on first arch 1-2 + 3-4; no rakers on posterior face of first 2 arches; body moderately robust; lips thin; *only 2 pairs of barbels; maxillary barbel broad, ribbon-like, reaching to middle part of anal fin*; head shield mostly smooth; palatine teeth in 4 patches, but fused to form continuous band, except patches may be separate in young; *pectoral spine with long, flattened filament reaching to end of anal fin; dorsal spine with very elongate filament*; lateral line deflected upwards at caudal base; steel blue to nearly black on back, silvery white on side and white on belly; fins generally pale. Mexico to Ecuador; common in coastal seas and estuaries, also enters fresh water. Maximum size 70 cm.

CONGO CATFISH
Cathorops furthii (Steindachner, 1876)
(Plate V-5, p. 71)

Dorsal rays I,7; anal rays 19-23; pectoral rays I,10-11; gill rakers on first arch 5-7 + 10-12; rakers present on posterior surface of all gill arches; body moderately robust; *lips fleshy, but not thick; dorsomedian fontanel groove discontinuous, a short groove on snout, another posteriorly; eye small 3.0-4.0 in fleshy interorbital*; gill opening restricted, gill membranes forming low fold across isthmus or confluent with it medially; *premaxillary teeth in a continuous broad band; palatine teeth small and molariform in 2 large, oval patches*; dark blue or brownish on back, silvery white on sides and belly; fins dusky. Mexico to Peru; inhabits brackish estuaries and fresh water. Grows to at least 35 cm.

GLOOMY CATFISH
Cathorops hypophthalmus (Steindachner, 1876)
(Plate V-3, p. 71)

Dorsal rays I,7; anal rays 21-23; pectoral rays I,10-11; *gill rakers on first arch 10-11 + 27-30*; rakers present on posterior surface of all arches; body elongate; lips thin; *eye position relatively low, opposite corner of mouth; a small, narrow patch of conical teeth on each side of palate; supraoccipital process longer and narrower than in other* Cathorops; maxillary barbel reaches as far as dorsal fin; back and upper side bluish or brown, sometimes with purple sheen; silvery white below; fins dusky, dorsal and caudal

Cathorops tuyra

with darker margin. Costa Rica and Panama; inhabits brackish estuaries and fresh water. Reaches 38 cm.

MANY-RAYED CATFISH
Cathorops multiradiatus (Günther, 1864)
(Plate V-4, p. 71)

Dorsal rays I,7; anal rays 25-28; pectoral rays I,10; gill rakers on first arch 5-7 + 12-14; rakers on posterior surface of all arches; body elongate; *lips fleshy, but not thick; gill openings restricted, membranes connected to isthmus medially; premaxillary teeth in a continuous broad band; palatine teeth molariform in 2 well-separated, small patches (smaller and more widely separated than in other* Cathorops*); pectoral spine with many strong serrations on inner margin*; charcoal or bluish brown on back, silvery on sides, and white below; fins dusky. Costa Rica to Peru; inhabits brackish bays and river mouths, entering fresh water. Attains 34 cm.

TAYLOR'S CATFISH
Cathorops taylori (Hildebrand, 1926)
(Plate V-7, p. 71)

Dorsal rays I,7; anal rays 22-23; pectoral rays I,10; gill rakers on first arch 7 + 11-12; rakers present on posterior surface of all arches; body elongate; *lips fleshy, but not thick; dorsomedian fontanel a continuous groove; eye relatively large, 2.0-2.8 in fleshy interorbital*; gill openings moderately restricted, membranes forming fold across isthmus; *premaxillary teeth in continuous broad band; palatine teeth small and molariform in 2 small, well-separated, ovate patches*; dusky brownish on back, silvery white on sides and belly; fins dusky, pelvics often blackish on basal half. Guatemala to Panama; mainly inhabits freshwater streams, but sometimes found in brackish river mouths. To 36 cm.

TUYRA CATFISH
Cathorops tuyra (Meek & Hildebrand, 1923)

Dorsal rays I,7; anal rays 20; pectoral rays I,10-11; gill rakers on first arch 7 + 13-15; rakers on posterior surface of all arches; body moderately elongate; *lips fleshy, often thick*; gill opening restricted, gill membranes barely forming low fold across isthmus medially; *premaxillary teeth in broad band; palatine teeth molariform and very large (can easily be felt by putting finger in mouth), in a pair of well-separated, elongate oval patches*; gray on back and sides, white below; fins dusky bluish gray. Panama to Peru; inhabits brackish and fresh water. Maximum size to 40 cm.

FLAP-NOSED CATFISH
Sciadeichthys dowii (Gill, 1863)
(Plate V-6, p. 71)

Dorsal rays I,7; anal rays 17-21; pectoral rays I,10; gill rakers on first arch 7-9 + 15-17; rakers absent from posterior surface of first 2 arches; barbels very thick; *narrow skin flap connecting posterior nostril on each side of snout; predorsal plate enlarged, square or subpentagonal; palatine teeth conical, in 2 patches on each side, these coalesce with age, the smaller inner patches continuous with much larger, triangular outer patches; pectoral spine with moderately strong, curved serrae on inner margin*; brownish blue to bluish gray on back, grayish on sides, and white below. Panama to Ecuador; inhabits brackish and fresh water. To 90 cm.

CHILI CATFISH
Sciadeops troscheli (Gill, 1863)
(Plate V-8, p. 71)

Dorsal rays I,7; anal rays 17-18; pectoral rays I,10-11; gill rakers on first arch 4-7 + 7-10; body robust; *predorsal plate enlarged, square or subpentagonal; premaxillary teeth sharp and small, in a broad band; palatine teeth conical in 4 patches, 2 inner ones may be fused and joined with large outer patches; inner margin of pectoral spine finely serrated*; dark brown or bluish on back, silvery white on sides and belly; fins dusky. Mexico (Mazatlan) to Peru. Reaches 52 cm.

PLATE V

1 **CHIHUIL CATFISH** (*Bagre panamensis*)

2 **LONG-BARBLED CATFISH** (*Bagre pinnimaculatus*)

3 **GLOOMY CATFISH** (*Cathorops hypophthalmus*)

4 **MANY-RAYED CATFISH** (*Cathorops multiradiatus*)

5 **CONGO CATFISH** (*Cathorops furthii*)

6 **FLAP-NOSED CATFISH** (*Sciadeichthys dowii*)

7 **TAYLOR'S CATFISH** (*Cathorops taylori*)

8 **CHILI CATFISH** (*Sciadeops troscheli*)

9 **BROADHEAD CATFISH** (*Arius dasycephalus*)

10 **KESSLER'S CATFISH** (*Arius kessleri*)

11 **BLUE CATFISH** (*Arius guatemalensis*)

12 **THICK-LIPPED CATFISH** (*Arius osculus*)

13 **FLATHEAD CATFISH** (*Arius planiceps*)

14 **SEEMANN'S CATFISH** (*Arius seemanni*)

15 **PANAMANIAN CATFISH** (*Arius lentiginosus*)

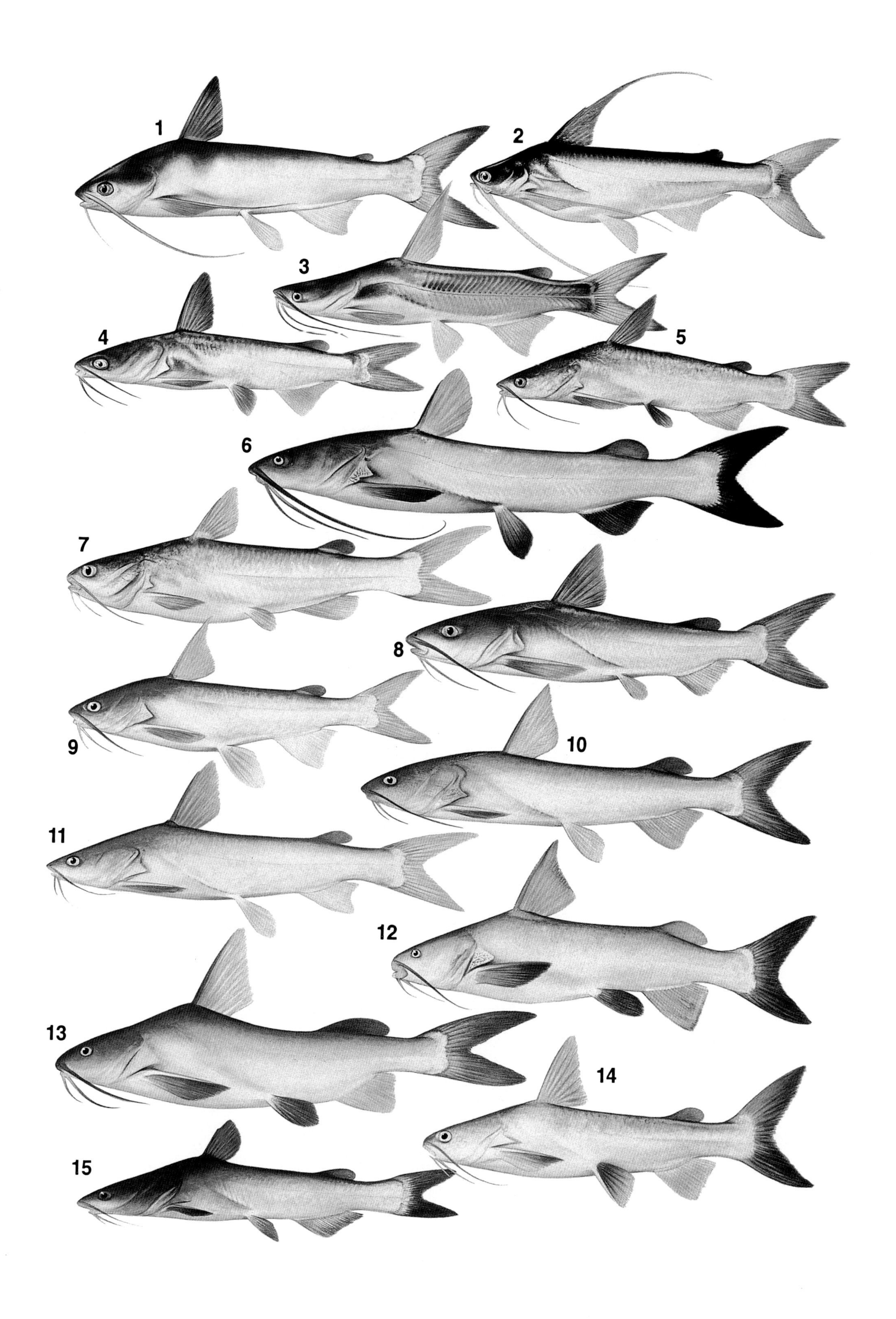
1
2
3
4
5
6
7
8
9
10
11
12
13
14
15

LIZARDFISHES (PECES IGUANA, LAGARTOS, GUARIPETES)

FAMILY SYNODONTIDAE

Lizardfishes are characterized by their slender body; reptile-like head; large mouth with numerous slender, sharp teeth; no fin spines; a single dorsal fin; a small adipose fin near the tail base; and relatively large pelvic fins with eight to nine rays. They are relatively sedentary, soft-bottom dwellers, although some Indo-West Pacific species inhabit coral reefs. They sometimes bury themselves in the substratum with only the eyes showing. This habit no doubt enables them to snare unsuspecting fish, which is the main dietary component. When prey is sighted they literally rocket off the bottom with amazing speed, and once a small fish is seized it has little chance of escape. The family contains five genera with about 40 species, most ocurring in the tropical Indo-Pacific. In our area, only four species are known; these were reviewed by Meek and Hildebrand (1923).

SHARPNOSE LIZARDFISH

Synodus evermanni Jordan & Bollman, 1889

Dorsal rays 10-11; *anal rays 10-11; lateral-line scales 43-50*; body very elongate, its depth 6.6-8.5 in standard length; snout long and sharply pointed; mouth large, *lower jaw ending in a fleshy knob*; premaxillary reaching well past eye; *dorsal fin origin closer to adipose fin than snout tip*; grayish brown on back, silvery whitish on sides and lower part of body, sometimes with about 8 diffuse dark blotches along lateral line; scales with narrow dark outlines; caudal fin dusky. Baja California to Peru; occurs on soft mud or sand bottoms in 10-60 m depth, frequently captured by shrimp trawlers. Reaches at least 28 cm.

Synodus evermanni ▲

Synodus lacertinus ▼

Synodus scituliceps ▼

REEF LIZARDFISH

Synodus lacertinus Gilbert, 1890

Dorsal rays 11-12; *anal rays 8-9; lateral-line scales 59-63*; body moderately elongate, its depth 5.9-6.5 in standard length; snout relatively short and not protruding; mouth large, *lower jaw without a fleshy knob at its tip*; premaxillary reaching well past eye; *dorsal fin origin slightly closer to tip of snout than to adipose fin*; mottled brown to reddish on upper half, white below; back with about 5 brown or red saddle-like markings, these usually extending onto side as irregular bars; lower side with 10-12 red to brown blotches; pectoral and pelvic fin with narrow dark bars. Gulf of California to Peru; a common inhabitant of sand-rubble areas near reefs in 2-20 m. Grows to at least 16 cm.

LANCE LIZARDFISH

Synodus scituliceps Jordan & Gilbert, 1881

Dorsal rays 10-11; *anal rays 11-14; lateral-line scales 57-66*; body very elongate, its depth 7.0-9.5 in standard length; snout relatively long and sharply pointed; mouth large, *lower jaw ending in a fleshy knob*; premaxillary reaching well past eye; *dorsal fin inserted about midway between anterior margin of eye and adipose fin*; grayish on back, gray on sides with silvery sheen, and white on ventral parts; scales with narrow blackish outline; caudal fin dusky. Gulf of California to Peru, including Galapagos; inhabits shallow muddy bottoms of bays. To 45 cm.

SAND LIZARDFISH
Synodus sechurae Hildebrand, 1946

Dorsal rays 11; *anal rays 12; lateral-line scales 56-59*; body moderately elongate, its depth 7.4-9.7 in standard length; snout moderately long, not sharply pointed; mouth large, *lower jaw without a fleshy knob at its tip*; premaxillary reaching well past eye; *dorsal fin origin slightly closer to tip of snout than to adipose fin*; mottled brown on upper half, white below. Gulf of California to Peru; occurs on open sandy bottoms, often considerable distances from reefs; depth range to at least 30 m. Attains 30 cm.

BROTULAS (CONGRIOS, CONGRIBADEJOS)

FAMILY OPHIDIIDAE

These fishes have an elongate, more or less eel-like body with the long-based dorsal and anal fins continuous with the caudal fin; the pelvic fins are either absent or consist of one or two slender filamentous rays; some species have barbels around the mouth. Unlike the similar-appearing cuskeels (family Bythitidae), brotulas are egg layers, and males therefore lack specialized copulatory organs. These reef-dwelling fishes are rarely seen because of their cryptic habits. They hide in caves and crevices during the day, and periodically emerge from cover at night to feed on crustaceans and fishes. The family occurs worldwide in both shallow and deep seas. It contains about 50 genera and 170 species. Only a few are found on reefs of the eastern Pacific.

CLARK'S BROTULA
Brotula clarkae Hubbs, 1944

Dorsal rays 108-112; anal rays 78-89; dorsal and anal fins continuous with caudal fin; *pectoral rays 27-28*; pelvic fins under middle of head, each consisting of a pair of filaments that are joined basally; 3 short barbels on each side of snout and 3 barbels on each side of lower jaw; dorsal and anal fins confluent, tail pointed; *generally reddish brown; median fins dusky dark brown to blackish; juveniles (under about 20 cm) with dark stripe on head behind eye, and numerous, large, dark brown spots on body.* Baja California to Peru; inhabits rocky reefs and adjacent sand bottoms to at least 60 m. Attains 82 cm.

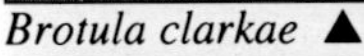
Brotula clarkae ▲

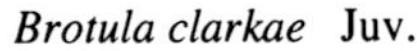
Brotula clarkae Juv.

Brotula ordwayi ▼

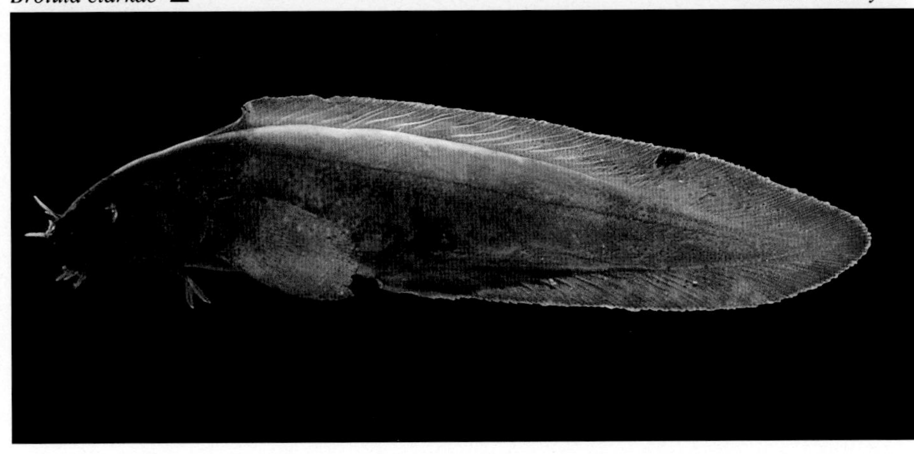

ORDWAY'S BROTULA
Brotula ordwayi Hildebrand & Barton, 1949

Dorsal rays 118-125; anal rays 86-94; dorsal and anal fins continuous with caudal fin; *pectoral rays 23*; pelvic fins under middle of head, each consisting of a pair of filaments that are joined basally; 3 short barbels on each side of snout, and 3 barbels on each side of lower jaw; dorsal and anal fins confluent, tail pointed; light gray-brown to dark gray with *blackish spots on head and anteriormost part of body; dorsal and anal fins dusky black with pronounced white margin*; barbels and pelvic filaments whitish. Galapagos Islands and Peru; inhabits caves and crevices of rocky reefs. Grows to at least 10 cm.

SPINESNOUT BROTULA
Lepophidium prorates (Jordan & Bollman, 1889)

Dorsal rays 124-133; anal rays 106-113; pectoral rays 21-24 (usually 22-23); total gill rakers on first arch 11-16 (usually 11-14); *a well-developed subdermal spine on snout*; body scales small and rounded in regular rows, somewhat covered with skin, also present on postorbital part of head and cheeks; *body elongate and somewhat compressed, the greatest depth about 7.8-8.9 in standard length*; origin of dorsal fin at level of middle of pectoral fins; pelvic fins inserted below eye, each consisting of pair of filamentous rays joined basally; dorsal and anal fins confluent, tail pointed; generally pale gray, grading to white ventrally; lateral-line pores whitish; margin of dorsal and anal fins black. Gulf of California to Peru; inhabits soft bottoms between 4-60 m depth. To 22 cm.

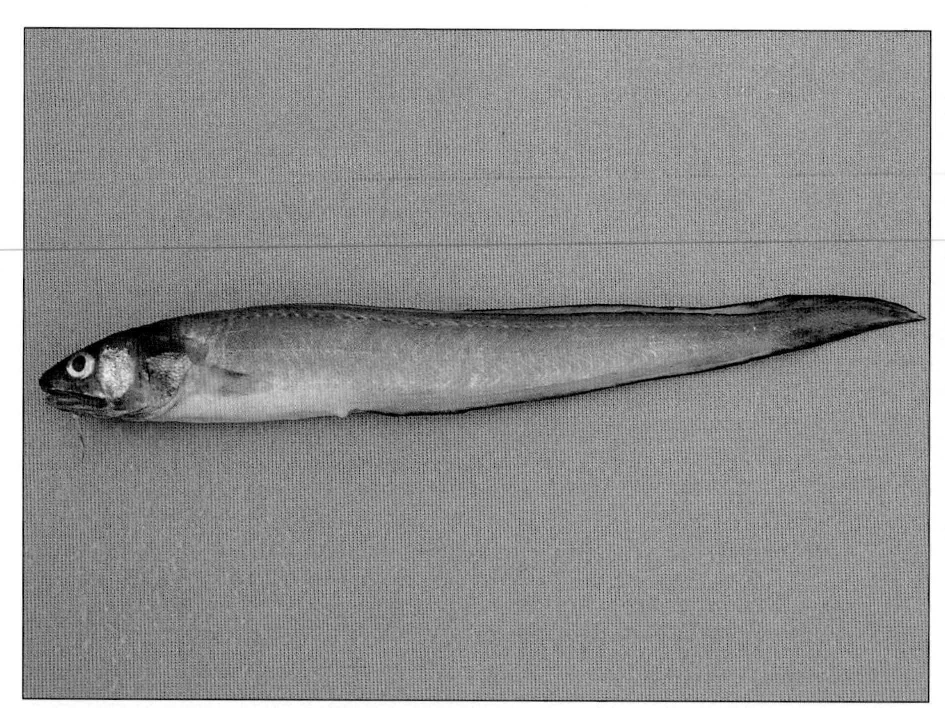

SHARK BROTULA
Ophidion galeoides (Gilbert, 1890)

Dorsal rays about 125; anal rays about 85-90; pectoral rays 21; dorsal and anal fins confluent, tail rounded; pelvic fins below eye, each consisting of 2 filamentous rays joined basally; head scaleless; body scales elongate or elliptical in shape and situated at oblique angles to each other; body elongate and compressed, the greatest depth 6.0-6.5 in standard length; *yellowish with golden sheen, white on breast and belly; pectoral fins yellow; dorsal and anal fins dusky blackish, a thin white margin on anal fin*. Gulf of California to Peru; inhabits flat, sandy bottoms; 1 collected by us in 20 m at Gulf of Chiriqui, Panama. To 13 cm.

GALAPAGOS BROTULA
Ophidion sp.

Dorsal rays about 150; anal rays about 122; pectoral rays 23-24; dorsal and anal fins confluent, tail rounded; pelvic fins below eye, each consisting of 2 filamentous rays joined basally; head scaleless; body scales cycloid, elongate or elliptical in shape, and situated at oblique angles to each other; body elongate and compressed, the greatest depth 6.5-7.0 in standard length; *generally light gray, somewhat silvery; outer edge of dorsal and anal fins blackish*. An undescribed species known only from the Galapagos; inhabits sand bottoms adjacent to reefs. Grows to at least 13 cm.

SPOTTED BROTULA
Ophidion sp.

Dorsal rays about 150-160; anal rays about 125-135; pectoral rays 21; dorsal and anal fins confluent, tail rounded; pelvic fins below eye, each consisting of 2 filamentous rays joined basally; head scaleless; body scales cycloid, elongate or elliptical in shape, and situated at oblique angles to each other; body elongate and compressed, the greatest depth 6.8-7.2 in standard length; *generally gray with large dark brown spots*; outer edge of front part of dorsal fin black; anal fin blackish. Panama, trawled in 30 m. Grows to at least 17 cm.

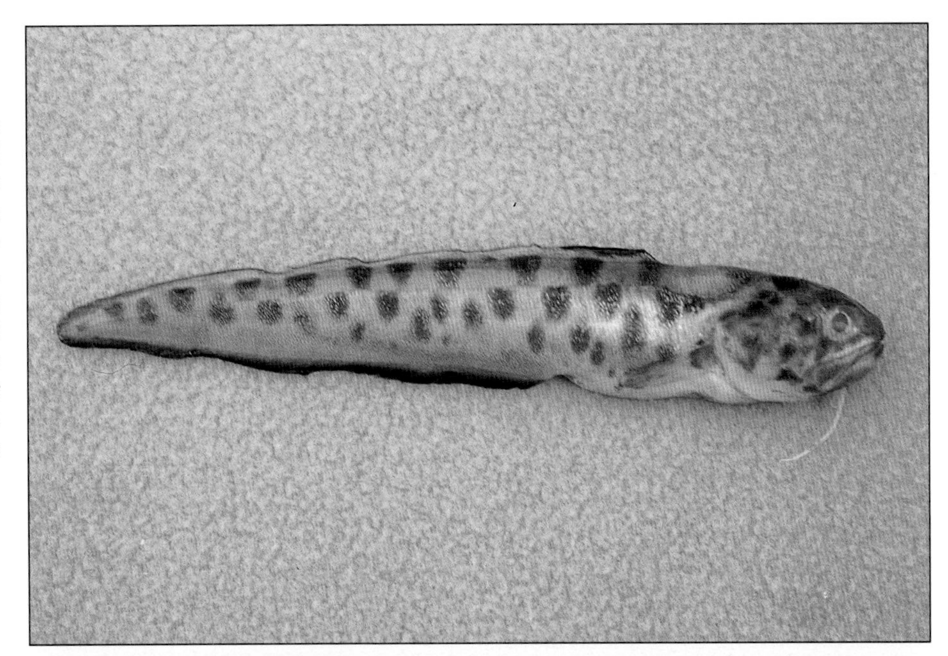

BARRED BROTULA
Otophidion indefatigable Jordan & Bollman, 1889

Dorsal rays about 110-120; anal rays about 85-90; *a stout subdermal spine present on snout*; scales small and rounded in regular rows, more or less imbricated; scales absent on head; body elongate and compressed, the greatest depth about 5.6-6.0 in standard length; origin of dorsal fin over middle of pectoral fins; pelvic fins inserted below eye, each consisting of 2 filamentous rays joined basally; dorsal and anal fins confluent, tail pointed; grayish to silvery white; *about 9-12 dusky dark bars on upper half of side*; margin of anterior dorsal and anal rays white. Panama to Galapagos; inhabits sandy bottoms; collected by us in 15 m at Perlas Islands, Panama. To at least 6 cm.

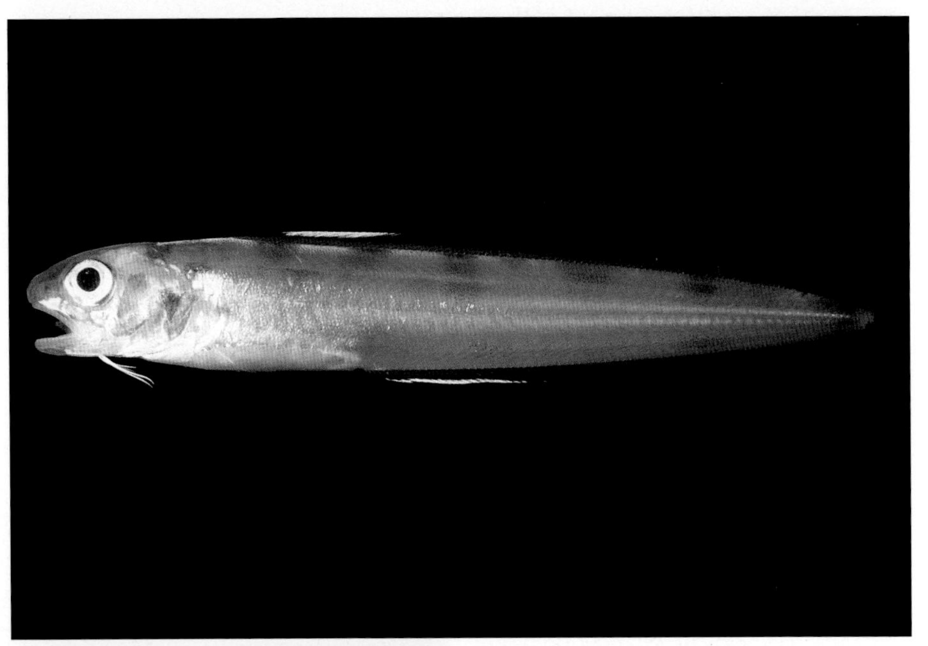

VELVETNOSE BROTULA
Petrotyx hopkinsi Heller & Snodgrass, 1903

Dorsal rays 82-88; anal rays 62-70; dorsal and anal fins confluent, tail pointed; snout covered with many small fleshy papillae; scales small and imbricate, about 130-140 in midlateral series; *rear margin of preopercle overgrown with skin; no spine on opercle*; dorsal fin origin above base of pectoral fin; *pelvic fins below rear part of head, each consisting of 2 rays united at the base*; generally brown; median fins dusky blackish. Gulf of California to Galapagos; inhabits rocky crevices. Attains 20 cm.

PEARLFISHES (PECES PERLA)
FAMILY CARAPIDAE

Pearlfishes are well known for their habit of living commensally in sea cucumbers, bivalves, starfishes, and tunicates. In addition, a few species are free-living. These fishes are very elongate, the body tapering to a slender, pointed tail; the dorsal and anal fins lack spines and are confluent around the tail. Other features include an unusually positioned anus (far forward on the throat), lack of pelvic fins (except in the four free-living species of *Pyramodon*), and pectoral fins that are either very small or absent. Carapids are sometimes called glass eels because of the transparent body exhibited by many species. There are two larval stages in the life cycle, the planktonic vexillifer, characterized by a streamer-like predorsal filament, and the benthic-dwelling tenuis stage. Most adult pearlfishes spend nearly full time inside their host and are seldom seen. Some species are parasitic, feeding on the gonads and gills of their host. Those that live in sea cucumbers are able to leave their host periodically, re-entering tail first through the cloacal opening. The family was recently reviewed by Markle and Olney (1990), who recognized 31 species in seven genera. We include a single representative (three species in our area).

PACIFIC PEARLFISH
Encheliophis dubius (Putman, 1974)

Body very elongate and compressed, tapering into long, slender tail; head 7.5-9.1 in total length; greatest body depth 10.0-14.5 in total length; dorsal fin very low and inconspicuous; anal fin somewhat taller, originating just anterior to pectoral fin base; pectoral fins pointed, about 1.4-2.0 in head length; *pelvic fins absent*; anus anterior to base of pectoral fins; *generally translucent white; lining of gut cavity silver*. Baja California to Colombia; also recorded from Hawaii; inhabits the body cavity of pearl shells (*Pinctada*), pen shells (*Pinna*), and cockles (*Laevicardium*). Reaches about 15 cm. Two other species of pearlfishes occur in our area: *Echiodon exsilium* Rosenblatt and *Encheliophis vermicularis* Müller. The latter species is distinguished by a lack of pectoral fins, and *E. exsilium* has a large fang at the front of the upper jaw and 1-2 similar fangs at the front of the lower jaw (only low, cardiform teeth present in other species). *Encheliophis* inhabits holothurians (sea cucumbers). The host of *E. exsilium* is unknown, although previous authors have suggested that members of *Echiodon* may be free-living.

CUSKEELS (BRÓTULAS)
FAMILY BYTHITIDAE

These are long, slender fishes with a long dorsal fin that is sometimes continuous with the caudal and anal fins; there are no fin spines; scales are usually present, although they may be embedded; there is usually a strong opercular spine; the eyes are small, and sometimes vestigial. These fishes bear their young alive; the males have an external intromittent organ. About 85 species are known, including some from fresh and brackish waters and others from deep-sea habitats. The reef species live deep in cracks and crevices. They are never seen unless flushed from their lairs with chemical ichthyocides (used by scientific collectors).

DEROY'S CUSKEEL
Ogilbia deroyi (Poll & Van Mol, 1966)

Dorsal rays 68-74; anal rays 50-53; pectoral rays 17-21 (usually 19-20); snout gently rounded; body covered with small imbricate scales; a sharp spine on opercle; body elongate and somewhat compressed, the greatest depth 4.6-6.3 in standard length; origin of dorsal fin at level of middle of pectoral fins; pelvic fins inserted below rear part of head, each consisting of a single filamentous ray; dorsal and anal fins separate from small caudal fin; generally golden brown, grading to whitish on ventral surface. Galapagos Islands; inhabits rocky crevices. Grows to 7 cm.

Encheliophis dubius ▼ *Ogilbia deroyi* ▲

Photo: Alex Kerstitch

GALAPAGOS CUSKEEL
Ogilbia galapagosensis (Poll & Leleup, 1965)

Dorsal rays 69-76; anal rays 54-62; pectoral rays 15-17 (usually 16-17); snout sloping obliquely, front of head somewhat pointed; body covered with small, imbricate scales; a sharp spine on opercle; body elongate and somewhat compressed, the greatest depth 4.7-5.8 in standard length; origin of dorsal fin at level of middle of pectoral fins; pelvic fins inserted below rear part of head, each consisting of a single filamentous ray; dorsal and anal fins separate from small caudal fin; generally brown, paler on lower half of head; a faint brown stripe from snout to rear of maxillary, and a faint stripe across head behind eye; median fins yellow. Galapagos Islands; inhabits rocky crevices. Attains at least 6 cm.

Ogilbia galapagosensis ▲

Ogilbia ventralis ▼

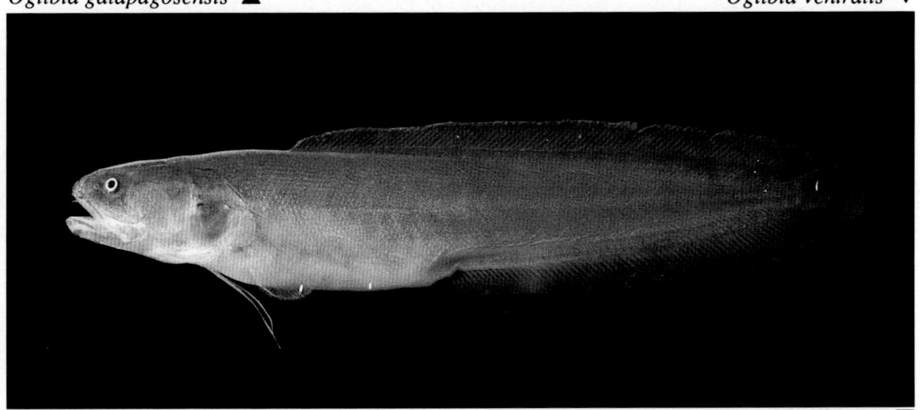

Ogilbia sp. ▼

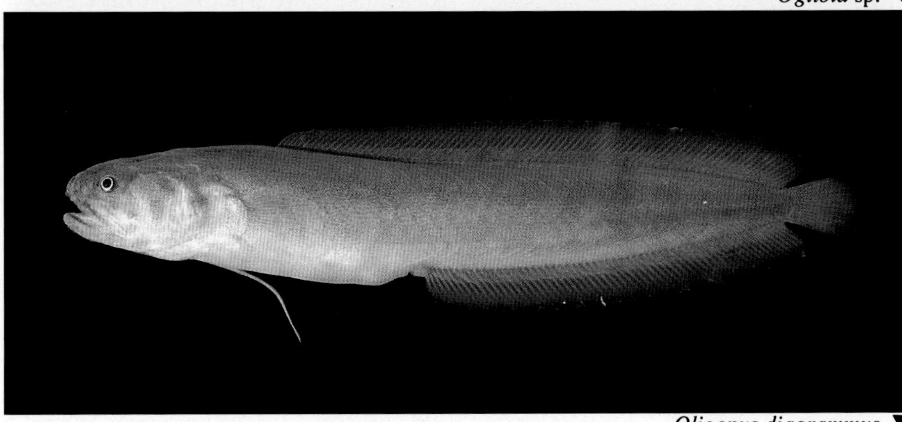

Oligopus diagrammus ▼

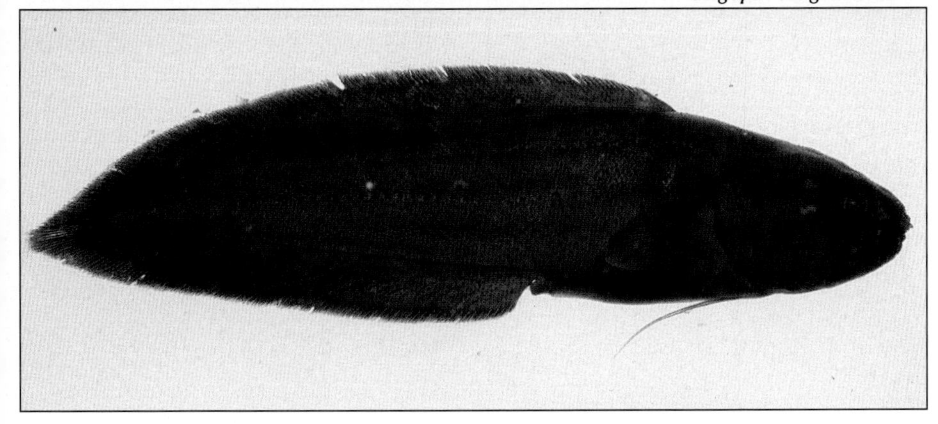

GULF CUSKEEL
Ogilbia ventralis (Gill, 1864)

Dorsal rays 78-85; anal rays 56-61; pectoral rays 21-24; vertebrae 12-13 + 30-32 = 43-44; snout rounded; body covered with tiny imbricate scales, about 110 in midlateral series; a sharp spine on opercle; body elongate and somewhat compresed, the greatest depth 5.0-5.7 in standard length; origin of dorsal fin at level of middle of pectoral fins; pelvic fins inserted below rear part of head, each consisting of a single filamentous ray; dorsal and anal fins separate from small caudal fin; brown, whitish on belly. Gulf of California to Panama; inhabits rocky crevices. Maximum size to 7 cm.

PANAMANIAN CUSKEEL
Ogilbia sp.

Dorsal rays 70-79; anal rays 53-60; pectoral rays 21-24; vertebrae 11-12 + 29-30 = 41; snout rounded; body covered with tiny imbricate scales, about 110 in midlateral series; a sharp spine on opercle; body elongate and somewhat compressed, the greatest depth 4.8-5.4 in standard length; origin of dorsal fin at level of middle of pectoral fins; pelvic fins inserted below rear part of head; yellowish brown, white on lower part of head and belly. Apparently an undescribed species, known thus far only from Panama; inhabits shallow reefs. Reaches at least 5.5 cm.

PURPLE CUSKEEL
Oligopus diagrammus (Heller & Snodgrass, 1903)

Dorsal rays 95-115; anal rays 76-91; pectoral rays 24-29; pelvic fins below rear part of head, each consisting of a single filamentous ray; dorsal and anal fins confluent, tail pointed; palatine teeth absent; body covered with small scales, about 100-115 in midlateral series, head partly scaleless; scattered papillae on head; origin of dorsal fin above middle part of pectoral fins; dark purplish to nearly black, including fins. Southern California to Panama and Galapagos; inhabits rocky crevices. Grows to about 6 cm.

TOADFISHES (PECES SAPO, PECES FRAILE, BRUJAS, PICADORES)
FAMILY BATRACHOIDIDAE

These fishes are characterized by a relatively large head and mouth, with the eyes positioned more dorsally than laterally. There are two or three stout dorsal spines followed by a separate, long second dorsal fin; the dorsal spines of *Daector* are hollow and connected to venom glands, but those of other species treated here are solid and not associated with venom glands; both second dorsal and anal fins without spines; head spines present on operculum and suboperculum (except latter spines absent in *Porichthys*); body with or without scales; no canine teeth (except in *Porichthys*); usually one, two or four lateral lines. Toadfishes are bottom dwellers that shelter under rocks, in crevices, or bury themselves in sand or mud. Some species migrate seasonally to spawn, and during this period are capable of producing curious "boat-whistle" sounds. Eggs are attached to rocks, or on the inner surface of pipes, tin cans or other submerged debris. Toadfishes occur in brackish coastal estuaries as well as to depths of at least 300 m. The diet consists of crabs, shrimps, molluscs (bivalves, gastropods, chitons, and octopus), echinoderms and fishes. Food is generally consumed whole; the stomach is capable of considerable expansion.

ESTUARY TOADFISH
Batrachoides boulengeri Gilbert & Starks, 1904

Dorsal rays III + 28-29; anal rays 26-27; body covered with small scales, head and breast scaleless; *a pair of strong divergent spines on opercle and additional pair on subopercle; a conspicuous pore at base of 16th-19th interradial membranes on inner side of pectoral fin*; 2 lateral lines present, the upper interrupted below posterior third of second dorsal fin and continued below posteriormost part of fin as 4-15 pores; *eye much smaller than in the Pacific Toadfish;* generally brown with diffuse dark saddles on back; belly whitish. Bay of Panama; inhabits bays and estuaries. Maximum size, 36 cm.

PACIFIC TOADFISH
Batrachoides pacifici (Günther, 1861)

Dorsal rays III + 25-27; anal rays 21-23; body covered with small scales, head and breast scaleless; *a pair of strong divergent spines on opercle, and additional pair on subopercle; a conspicuous pore at base of 10th-16th interradial membranes on inner side of pectoral fin*; 2 lateral lines present, the upper interrupted below middle portion of second dorsal fin and again below posteriormost dorsal rays; generally brown, darker on back; belly whitish; a dark brown oblique band sometimes evident on cheek. Panama to northern Peru; inhabits sand and mud bottoms adjacent to rocky areas; also occurs in tidepools. Reaches 29 cm.

BROWNBACK TOADFISH

Daector dowi (Jordan & Gilbert, 1887)

Dorsal rays II + 29-33 (usually 31); anal rays 28-30 (usually 29); pectoral rays 16-18 (usually 17); last dorsal and anal rays joined to caudal fin; head and body scaleless; *a sharp venomous spine on upper rear corner of gill cover*; dark brown on upper half of head and body, whitish below; dark upper portion with white vermiculations and prominent white stripe along lateral line; spinous dorsal fin black; outer edge of second dorsal and caudal fin blackish. Costa Rica to Peru; inhabits sand or mud bottoms of bays and estuaries. To 16 cm.

RETICULATED TOADFISH

Daector reticulata (Günther, 1864)

Dorsal rays II + 25-28 (usually 26); anal rays 24-27 (usually 25); pectoral rays 16-18 (usually 17); last dorsal and anal rays separate from caudal fin; head and body scaleless; *a sharp venomous spine on upper rear corner of gill cover; hair-like filaments on head and body less numerous than on* D. schmitti *(below), generally not present below infraorbital canal of head or below lateral line of body, and extending posteriorly on back no farther than second or third ray of second dorsal fin*; head and body covered with brown spots with interconnecting network of tan or whitish spaces between them; 3 relatively large, blackish saddles on upper back; anal fin with blackish submarginal stripe. Panama; inhabits sand and mud bottoms of bays, also occurs in tidepools. Grows to at least 27 cm.

Daector reticulata ▲ *Daector schmitti* ▼

SCHMITT'S TOADFISH

Daector schmitti Collette, 1968

Dorsal rays II + 22; anal rays 21; pectoral rays 15-16; last dorsal and anal rays separate from caudal fin; head and body scaleless; *a sharp venomous spine on upper rear corner of gill cover; hair-like filaments on head and body more numerous than on* D. reticulata *(above), generally extending below infraorbital canal of head and below lateral line on body, reaching posteriorly on back as far as eighth ray of second dorsal fin*; small specimens, about 8 cm, very similar in color to *D. reticulata*; larger specimens lose the interconnecting pale network and have pattern of small, dark spots in combination with the larger dark saddles on the back. Costa Rica and Panama; inhabits sand and mud bottoms in about 5-30 m. Reaches at least 13 cm.

SADDLED TOADFISH
Porichthys analis Hubbs & Schultz, 1939

Dorsal rays II + 34-41; anal rays 33-39; last dorsal and anal rays separate from caudal fin; head and body scaleless; a sharp, non-venomous spine on upper rear corner of gill cover; *4 lateral lines with conspicuous rows of photophores*; branchiostegal rows of photophores united in a broad V, without forward projection; tan or silvery whitish with 6-7 brown saddles dorsally. Gulf of California; inhabits sand bottoms. Grows to at least 28 cm.

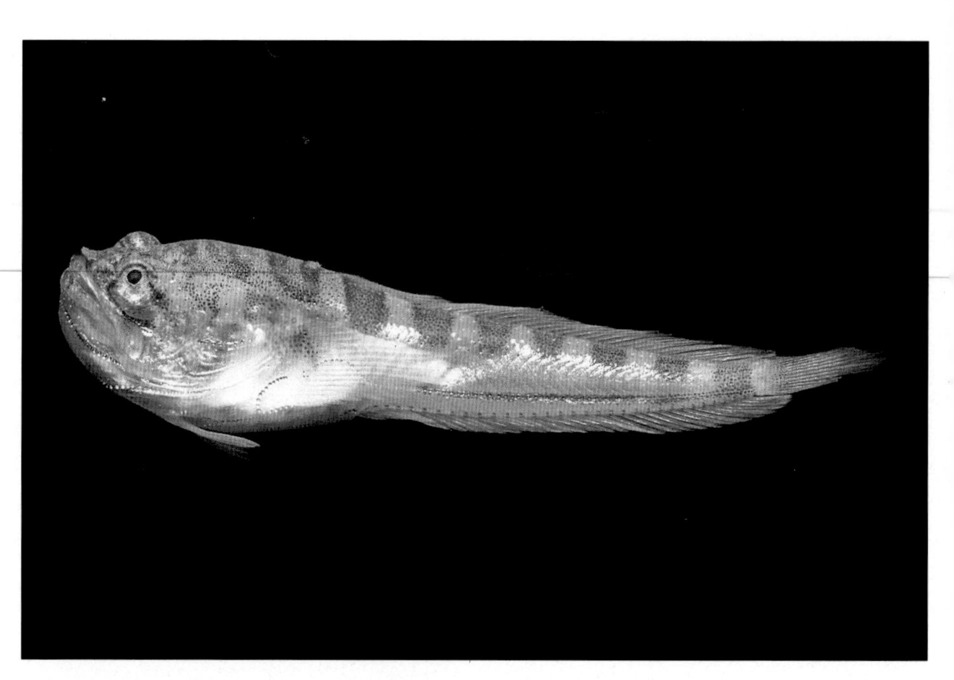

BRONZE TOADFISH
Porichthys margaritatus (Richardson, 1844)

Dorsal rays II + 32-37; anal rays 29-33; last dorsal and anal rays separate from caudal fin; head and body scaleless; a sharp, non-venomous spine on upper rear corner of gill cover; *4 lateral lines with conspicuous rows of photophores*; branchiostegal rows of photophores united in a broad V, without forward projection; bronze-colored, with 6-7 dark brown saddles, blotches, and spots on upper part of head and body. Gulf of Tehuantepec, Mexico to Colombia; inhabits sand or mud bottoms of trawling grounds to at least 50 m depth. To 17 cm. *P. greenei* Gilbert & Starks (Gulf of Tehuantepec to Panama) is similar in appearance, but the dorsal and anal fins are connected to the caudal fin.

MEXICAN TOADFISH
Porichthys myriaster Hubbs & Schultz, 1939

Dorsal rays II + 35-40; anal rays 33-38; last dorsal and anal rays separate from caudal fin; head and body scaleless; a sharp, non-venomous spine on upper rear corner of gill cover; *4 lateral lines with conspicuous rows of photophores*; branchiostegal rows of photophores with a U-shaped, forward-directed projection; tan or silvery whitish; *9-11 brown saddles on back of young, these disappearing with growth.* Southern California to outer coast of Baja California; inhabits sand bottoms close to shore. The largest species of *Porichthys*, reaching 43 cm.

ANGLERFISHES (BOCONES, RAPES)
FAMILY LOPHIIDAE

The lophiid anglerfishes are somewhat similar to their cousins, the frogfishes of the family Antennariidae. Both have a globular appearance with limb-like pectoral fins, a large mouth, first dorsal spine modified to form a fishing-pole apparatus (illicium) with a bait-like esca at its tip, and a gill opening situated in or near the "armpit" of the pectoral fins. However, the lophiid anglers are easily distinguished by their flattened head, which is armed with bony spines and four to six dorsal spines, the first three of which form isolated single units, and the remaining one to three are connected by a membrane, but are isolated from both the anterior three spines and posterior soft rays. Anglerfishes are "lie and wait" predators of small fishes and crustaceans. They inhabit soft sand or mud bottoms, generally below depths frequented by divers, with maximum depths of most species between 400-1 560 m. However, thc eastern Pacific species treated below is frequently caught by trawlers in depths as shallow as 14 m. The family occurs worldwide in tropical and temperate seas. Caruso (1981; 1983) recognized four genera and 25 species. Only species of *Lophiodes* are known from the eastern Pacific.

PACIFIC ANGLERFISH
Lophiodes caulinaris Garman, 1899

Dorsal rays I+I+I + III + 8; anal rays 8; pectoral rays 16-21 (usually 18-19); head broad and strongly depressed; pectoral fins limb-like; gill opening in "armpit" of pectoral fins; head with bony prickles; head and body with scattered skin flaps; *esca bearing pennant-like flap, long cirri, and usually with dark, stalked, eye-like appendages*; mottled brown; outer half of pectoral fins blackish with narrow white margin; *caudal fin dusky dark brown to blackish, with vertical row of white spots across middle of fin.* Gulf of California to Peru; depth range 14-310 m; frequently captured by trawlers. Reaches about 25-26 cm. The only other species of lophiid anglerfish occurring in the eastern Pacific, *L. spilurus* (Garman), is easily distinguished by the presence of minute dorsal spine cirri, a simple bulbous esca, thin loose skin, and lack of median row of white spots on the caudal fin. It ranges from Costa Rica to Peru in 119-475 m.

Lophiodes caulinaris ▲

Bloody Frogfish (*Antennarius sanguineus*), Isla Uva, Panama ▼

FROGFISHES (ZANAHORIAS, PECES PESCADORES, PECES ANTENA)
FAMILY ANTENNARIIDAE

These unusual fishes have globular, somewhat compressed bodies; limb-like pectoral fins with an "elbow" joint; a small, round gill opening; a very large, upward directed mouth; and a greatly modified first dorsal spine, termed the illicium, anteriorly on top of the head. The illicium forms a moveable "fishing rod" tipped with an enticing lure (esca). The rod is wiggled vigorously to attract fish prey that are swallowed whole. However, they do not always use their luring apparatus, and are able to slowly stalk fishes or crustaceans. Anglerfishes can engulf prey longer than themselves, as their abdomen can expand enormously. They are masters of camouflage. Their colors closely correspond to that of their surroundings, frequently bright-colored encrusting sponges. Gravid females expel a buoyant "raft" of up to 300 000 eggs, which remains afloat for several days until hatching. Frogfishes occur in all tropical and temperate seas. The family contains 13 genera and a total of 40 species, but only three are found in our area.

ROUGHJAW FROGFISH
Antennarius avalonis Jordan & Starks, 1907

Dorsal rays I+I+I + 12-14 (usually 13); anal rays 7-10 (usually 8); pectoral rays 11-14 (usually 13); illicium about equal to or slightly shorter than length of second dorsal spine; *second dorsal spine attached to head by membrane; esca an oval-shaped cluster of numerous short, vertically aligned appendages*; color highly variable: yellow, orange, red, brown, or black, always with pronounced light and dark mottling; *a large black spot surrounded by narrow orange ring on basal part of posterior dorsal fin.* Gulf of California to Peru; usually found on flat sand or mud bottoms to at least 200 m, usually below 20-30 m. Grows to 35 cm.

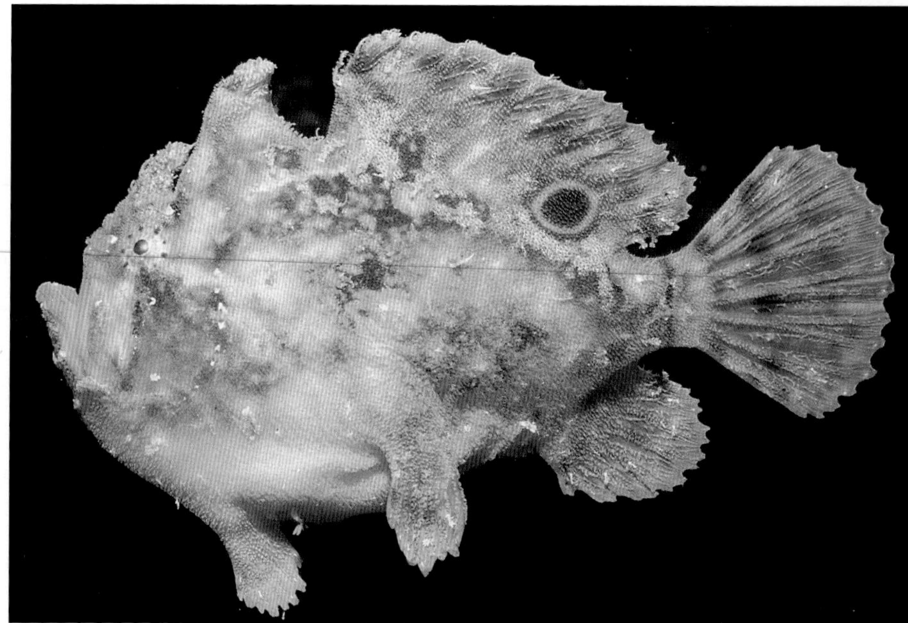

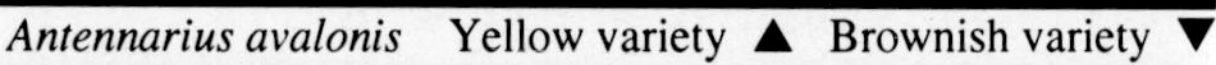

Antennarius avalonis Yellow variety ▲ Brownish variety ▼

BLOODY FROGFISH
Antennarius sanguineus Gill, 1863

Dorsal rays I+I+I + 12-14 (usually 13); anal rays 6-8 (usually 7); pectoral rays 10-12 (usually 11); illicium about equal to length of second dorsal spine; *second dorsal spine strongly curved and free, not connected to head by membrane; esca on elongate, tapering appendage with slender filaments and a cluster of darkly pigmented round swellings at base*; yellow or yellow-brown to reddish, with brown spotting and mottling; *belly with brown spots*; sometimes a weak ocellus or dark round spot below posterior part of dorsal fin. Gulf of California to Chile; the most common frogfish in the eastern Pacific, occurring in 1-40 m. Attains 10 cm.

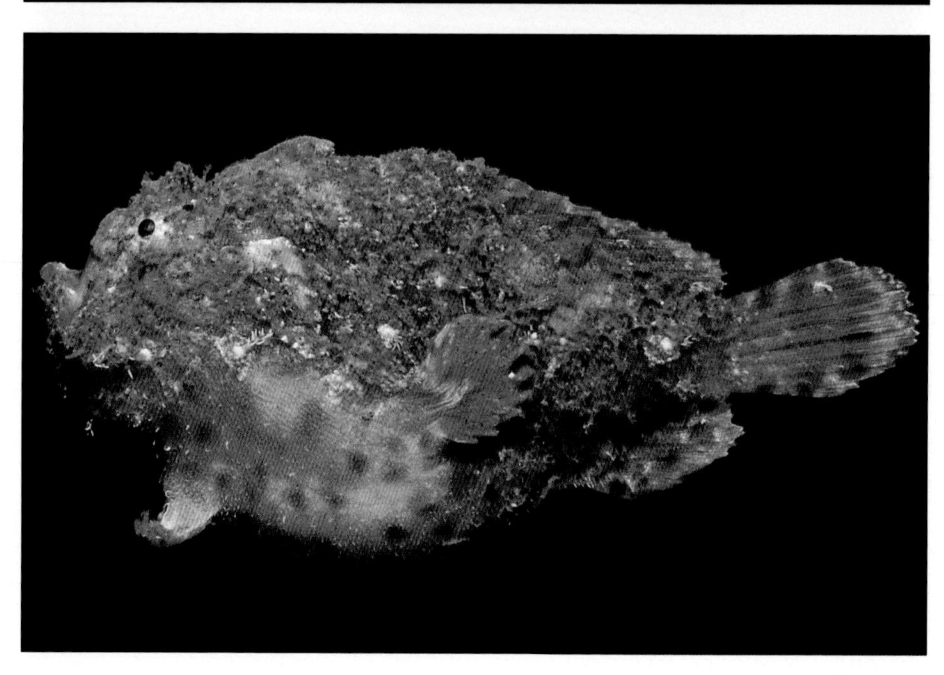

BANDTAIL FROGFISH
Antennatus strigatus (Gill, 1863)

Dorsal rays I+I+I + 11-13; anal rays 6-8; pectoral rays 9-11 (usually 10); illicium about equal in length to second dorsal spine; *second dorsal spine straight and free, not connected to head by membrane; esca a tiny lobed appendage, scarcely differentiated from illicium*; yellowish to pinkish red, with dark brown to blackish reticulations and marbling. Gulf of California to Colombia and Galapagos Islands; usually found in 1-15 m, but ranging to at least 40 m. Maximum length, 9 cm.

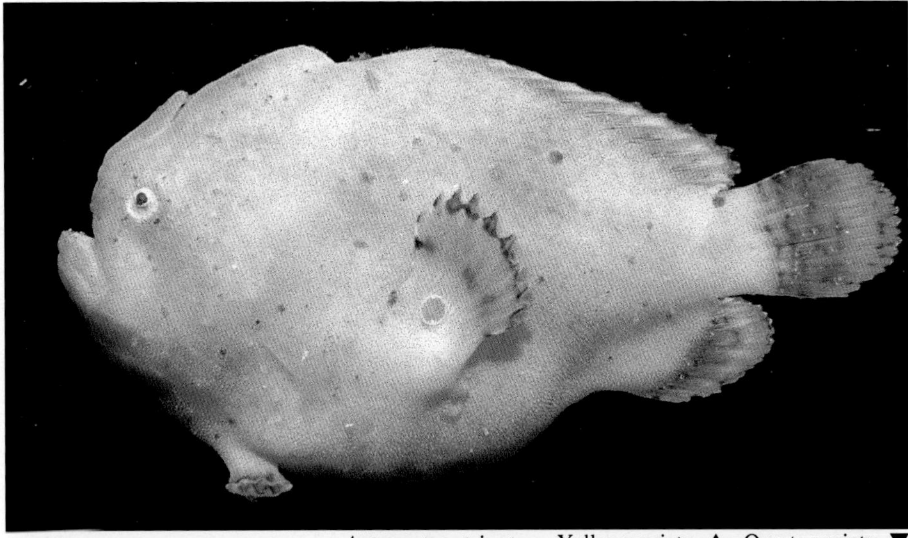

Antennatus strigatus Yellow variety ▲ Ornate variety ▼

BATFISHES (PECES MURCIÉLAGO)
FAMILY OGCOCEPHALIDAE

The batfishes are odd-shaped creatures that mainly inhabit the deep ocean floor, but a few species, including the two treated here, occur in relatively shallow depths. They belong to the family Ogcocephalidae, a group containing 57 species divided among nine genera. Batfishes have a strongly flattened, somewhat disc-like body, with large, arm-like pectoral fins and smaller pelvic fins. These fins are used to waddle across the bottom. They are sluggish swimmers, capable of only slow gliding movements. These predators of crustaceans, and occasional small fishes, rely on their camouflaged appearance and unusual luring behavior. On the snout there is a piston-like fishing rod apparatus that is thrust forward at the approach of potential prey. The end of the filamentous rod is equipped with a lumpy bait that is wiggled vigorously.

Galapagos Batfish (*Ogcocephalus darwini*), Isla Isabela, Galapagos

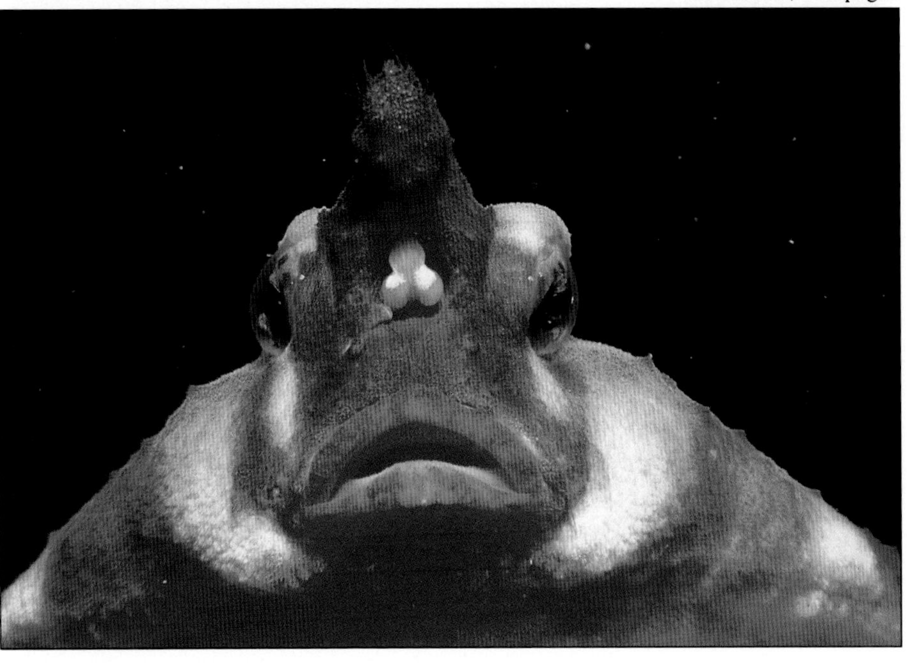

GALAPAGOS BATFISH
Ogcocephalus darwini Hubbs, 1952

Dorsal rays 4; anal rays 3; body flattened, disc-like; pectoral and pelvic fins forming limb-like appendages; *a pronounced bony horn on forehead*; light brown to grayish on back, white below; a dark brown stripe or series of blotches from top of head to caudal fin base on each side; snout and horn brownish; *lips bright red.* Galapagos Islands; inhabits sand bottoms, usually in 10-20 m, often encountered at Tagus Cove on Isla Isabela. Grows to about 20 cm.

SPOTTED BATFISH
Zalieutes elater (Jordan & Gilbert, 1881)

Dorsal rays 4; anal rays 4; body flattened, disc-like; pectoral and pelvic fins forming limb-like appendages; *numerous thorn-like spines on head and body* (except ventral surface); brown on dorsal surface of head and body, whitish on ventral surface; *a pair of black-rimmed orange ocelli (eye spots) straddling middle of back.* California to Peru; inhabits sand and mud bottoms. Grows to 13 cm.

Photograph: Alex Kerstitch

CLINGFISHES (CHUPAPIEDRAS, PEJESAPITOS, CHALACOS)
FAMILY GOBIESOCIDAE

These small fishes are distinguished by their thoracic sucking disc, which is a modification of the pelvic fin. The pattern of tiny papillae on the ventral surface of the pelvic disc (see accompanying diagram) is a useful feature in identifying genera. There is a single dorsal fin without spines, and the body is generally elongate. The pelvic disc is used to cling to rocks, weeds, and sessile invertebrates, and a few species can adhere to the surface of larger fishes. They feed on zooplankton, algae, and small benthic invertebrates. The family occurs worldwide and is represented by numerous species (many of them undescribed). There is need for a revision of this family, particularly the species inhabiting the Australian region, but the 1955 review by Briggs and several of his subsequent papers give comprehensive coverage of eastern Pacific species.

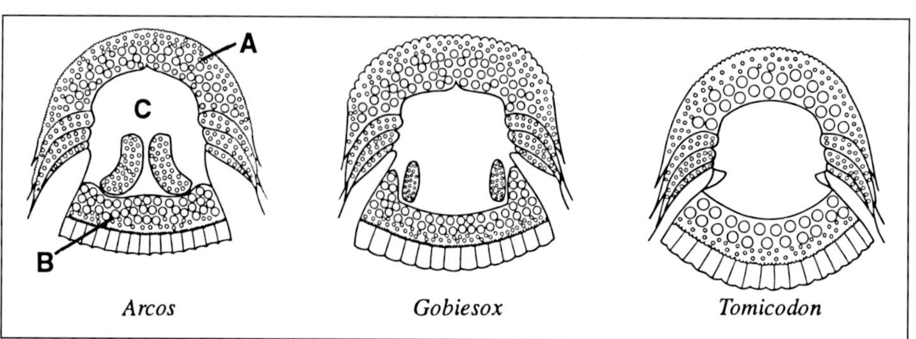

Papillae patterns on pelvic disc of eastern Pacific genera of clingfishes. Disc regions A, B, and C (as referred to in the text) are indicated on the drawing of *Arcos*.

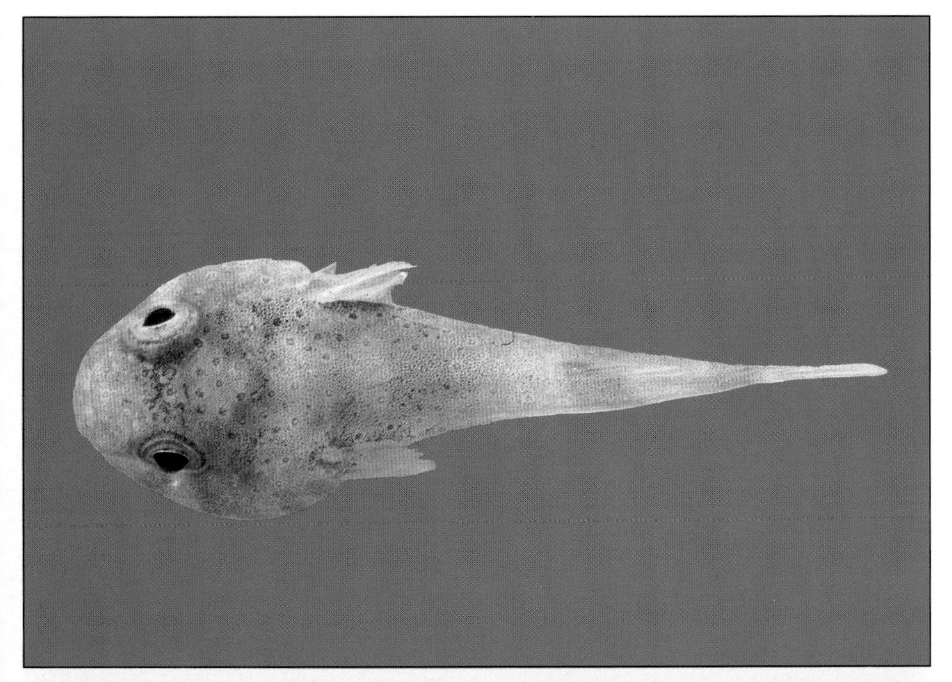

ELEGANT CLINGFISH
Arcos decoris Briggs, 1969

Dorsal rays 7-8; anal rays 8-9; *pectoral rays 22-24*; caudal rays 9-10; upper jaw with 2-3 pairs of rounded incisors at front, followed on each side by 2-3 canines and then a row of small conical teeth; lower jaw with 4 pairs of rounded incisors followed by 2-3 canines and then a row of small conical teeth; a well-developed, flattened, fleshy pad on pectoral base, with a free posterior margin; *length of disc 2.6-2.8 in standard length; postdorsal-caudal distance 0.8-0.9 in length of dorsal fin; 6-7 rows of flattened papillae across width of disc region A, 8-9 rows across disc region B, 3 longitudinal rows on each part of disc region C*; yellowish tan, with numerous small blue dots on dorsal portion of head and body; several darkish bars sometimes evident across back. Panama; inhabits rocky shores. To 2.8 cm.

ROCKWALL CLINGFISH
Arcos erythrops (Jordan & Gilbert, 1882)

Dorsal rays 6-7; anal rays 6; *pectoral rays 21-22*; caudal rays 7-8; upper jaw with 2-4 pairs of lightly compressed incisors at front, followed on each side by 2-4 weak canines; lower jaw with 4-5 pairs of incisors, followed on each side by 3-4 canines; a well-developed, flattened, fleshy pad on pectoral base, with a free posterior margin; *length of disc 2.3-2.4 in standard length; postdorsal-caudal distance 0.6-0.8 in length of dorsal fin; 9-14 rows of flattened papillae across width of disc region A, 11-14 rows across disc region B, 7-9 longitudinal rows across each part of disc region C*; greenish brown or reddish with intricate pattern of white, blue, and brown spots and intervening yellowish reticulum; a pair of broad, dark brown bars across rear portion of body. Mexico, including Gulf of California; inhabits surge zone along rocky shores. Attains 3.5 cm.

Arcos erythrops Green variety ▲ Red variety ▼

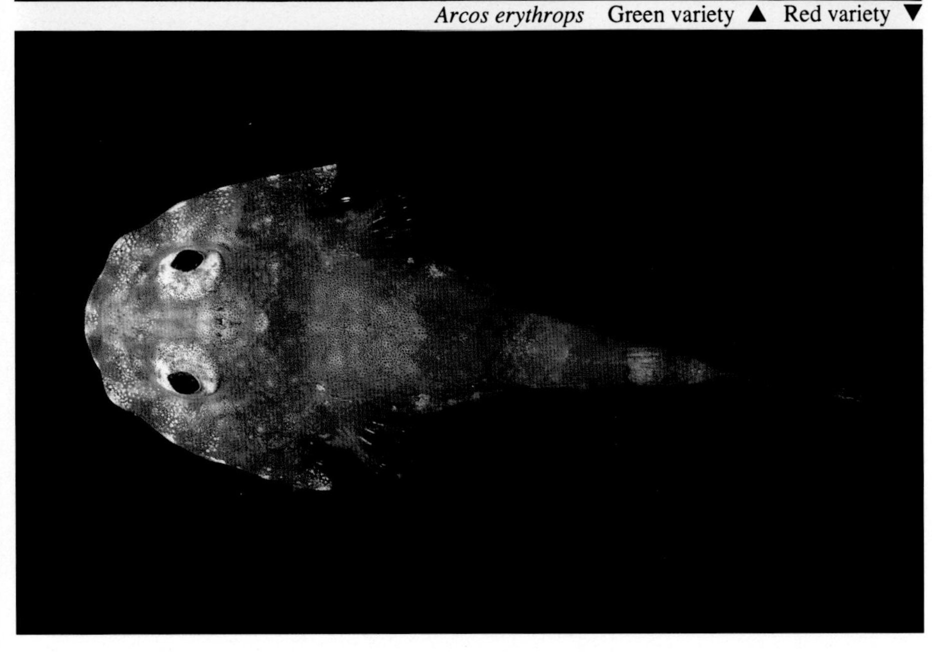

GALAPAGOS CLINGFISH
Arcos poecilophthalmus (Jenyns, 1842)

Dorsal rays 8; anal rays 7; *pectoral rays 23-24*; caudal rays 8-10; upper jaw with only compressed incisors, no canines; lower jaw with 2-3 pairs of incisors, followed on each side by 3-4 canines; a well-developed, flattened, fleshy pad on pectoral base with a free posterior margin; *length of disc 2.4-2.6 in standard length; postdorsal-caudal distance 0.9-1.4 in length of dorsal fin; 10-12 rows of flattened papillae across width of disc region A, 10-11 rows across disc region B, 5-7 longitudinal rows on each part of disc region C*; gray-brown with maze-like pattern of green bands; scattered white spots on back and sides, and whitish lines or bands on interorbital and cheek. Galapagos Islands; inhabits surge zone along rocky shores. Grows to at least 4 cm.

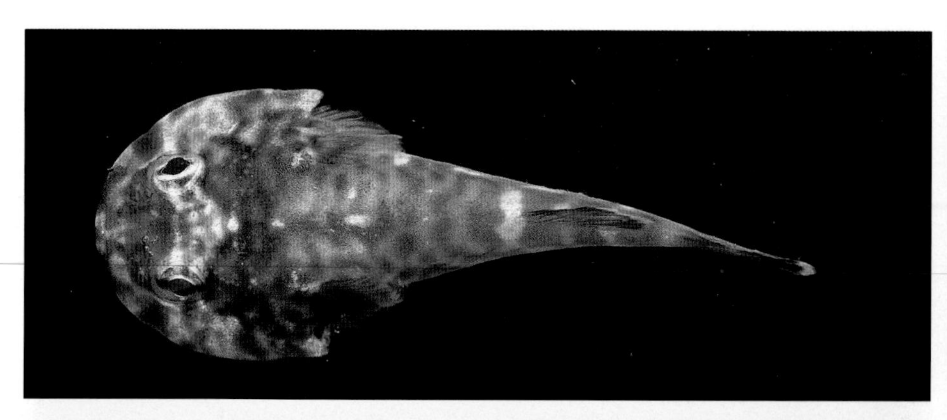

ROCK CLINGFISH
Arcos rhodospilus (Günther, 1864)

Dorsal rays 6-7; anal rays 5-7; *pectoral rays 20-22; upper jaw with incisor-like teeth only, no canines*; lower jaw with 4-6 pairs of incisors at front followed on each side by 2-4 canines; a conspicuous fleshy pad on pectoral fin base, with free posterior margin; *length of disc 2.7-2.9 in standard length; postdorsal-caudal distance 0.7-0.9 in length of dorsal fin; 5-8 rows of flattened papillae across width of disc region A, 7-9 rows across disc region B, 5-7 longitudinal rows across each part of disc region C*; brownish with reticulum of red and blue lines and spots on back and sides; also scattered, small white spots evident, these sometimes forming 4-5 bars across dorsal surface. Costa Rica to Ecuador; inhabits rocky shallows. Reaches about 3-5 cm.

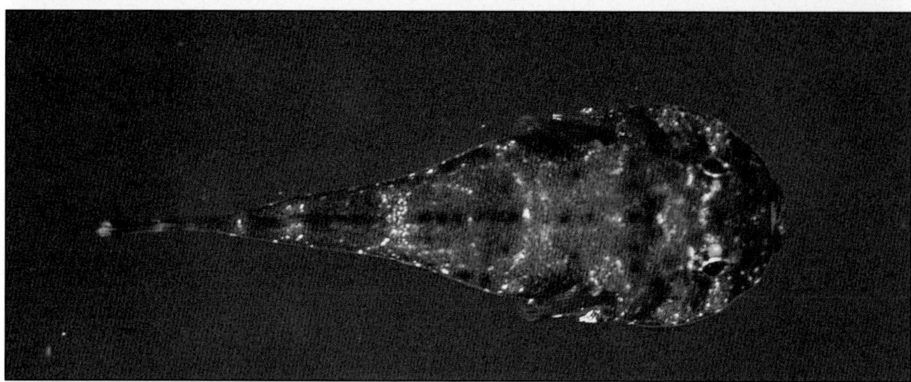

Arcos rhodospilus ▲

PANAMIC CLINGFISH
Gobiesox adustus Jordan & Gilbert, 1882

Dorsal rays 11; anal rays 9-10; pectoral rays 21-24; caudal rays 10-12; *well-developed, lobe-like papillae on head and along margin of upper lip*; upper jaw with patch of conical teeth at front, followed on each side by row of slightly recurved canines; lower jaw with 2 rows of incisors, those of outer row much larger, incisors followed on each side by a row of larger canines; *anus about midway between rear margin of disc and anal fin origin, well in front of dorsal fin origin*; 7-10 rows of flattened papillae across width of disc region A, 4-7 longitudinal rows on each part of disc region C; tan to whitish, with overall dorsal covering of light brown spots separated by fine reticulum of blue lines. Gulf of California to Ecuador; inhabits rocky reefs. Size to at least 4.5 cm.

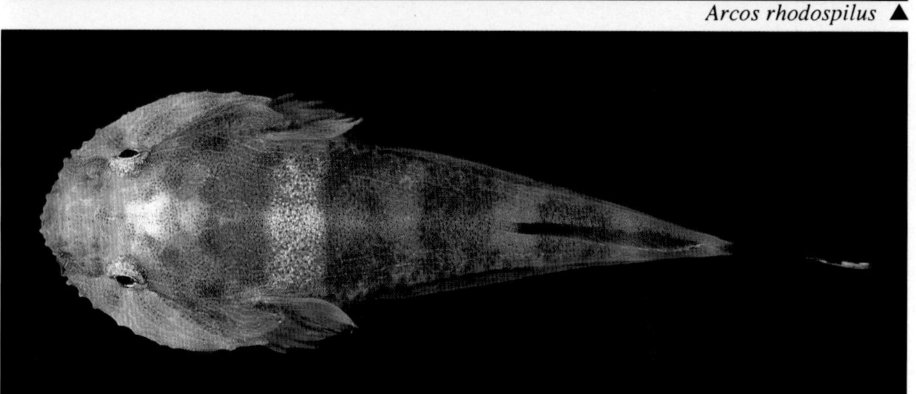

Gobiesox adustus ▲

Gobiesox papillifer ▼

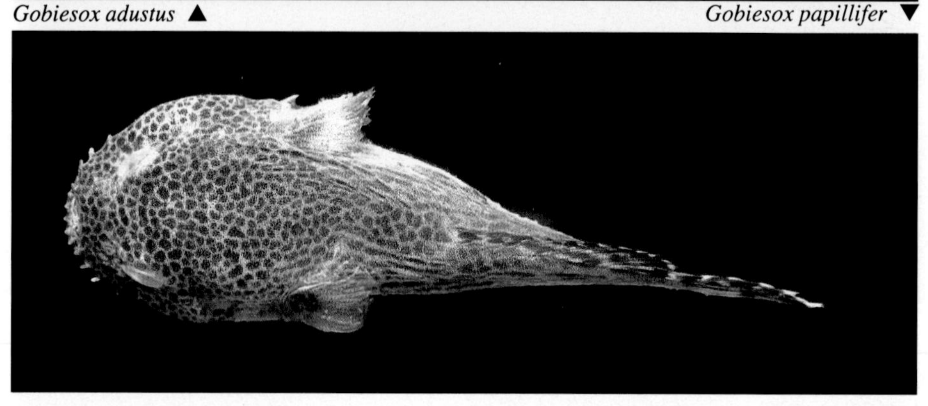

BEARDED CLINGFISH
Gobiesox papillifer Gilbert, 1890

Dorsal rays 14-15; anal rays 10-11; pectoral rays 23-25; caudal rays 11-12; *extensive development of papillae on head, some forming finger-like projections; usually 3 small papillae at center of upper lip*; patch of conical teeth at front of upper jaw, followed on each side by slightly recurved canines; lower jaw with 2 rows of incisors, those of outer row much larger, incisors followed on each side by a row of larger canines; *anus located close to anal fin origin*; 7-11 rows of flattened papillae across width of disc region A, 4-5 longitudinal rows on each part of disc region C; tan with *dense pattern of small, blackish, leopard-like spots* on dorsal surface, sometimes coalesced to form stripes. Southern California to Panama; inhabits rock outcrops, often in shallow sandy areas. To 4 cm.

WOODS' CLINGFISH
Gobiesox woodsi (Schultz, 1944)

Dorsal rays 12; anal rays 10; pectoral rays 21; caudal rays 12; *well-developed, lobe-like papillae on head and along margin of upper lip*; teeth of upper jaw conical, larger toward front; teeth of lower jaw in 2 rows, those of outer row much larger, 3 pairs of compressed incisors at front; *anus just in front of level of dorsal fin origin*; 8-9 rows of papillae across width of disc region A, 5-6 longitudinal rows on each part of disc region C; brown with reticulum of narrow whitish lines which may be somewhat thickened on sides to form longitudinal stripes; 4-5 broad, diffuse, brownish bars across back. Isla del Coco; inhabits rocky shores. To at least 4.8 cm.

Gobiesox woodsi ▲

Tomicodon bidens ▼

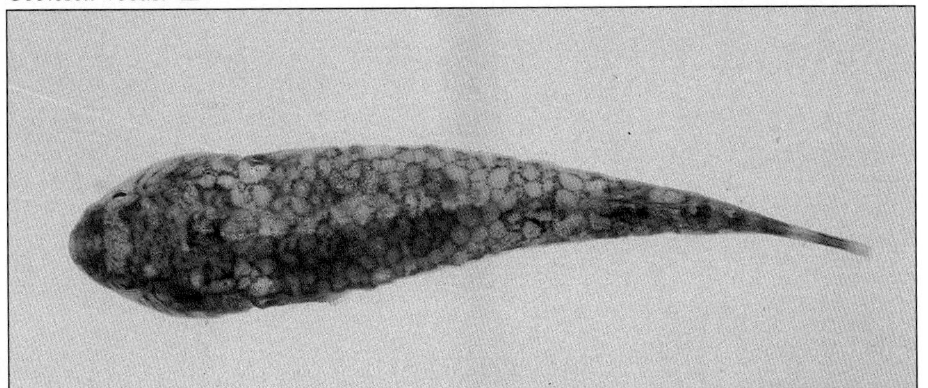

Tomicodon boehlkei ▼

Tomicodon eos ▼

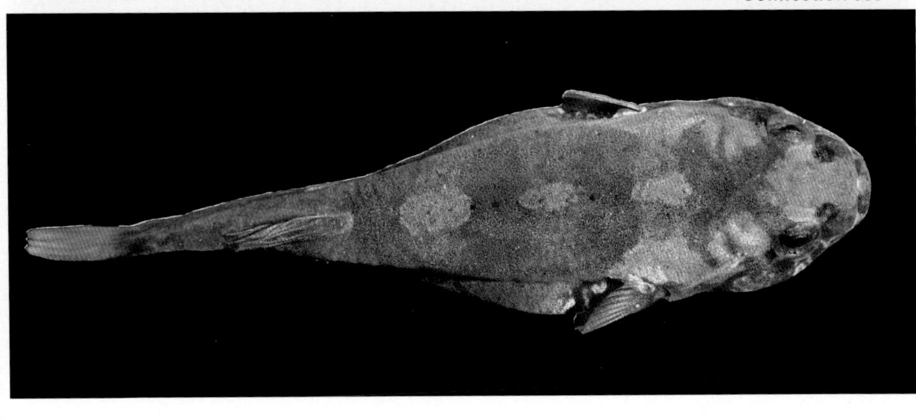

BIFID CLINGFISH
Tomicodon bidens Briggs, 1969

Dorsal rays 7-8; anal rays 6-7; *pectoral rays 21*; caudal rays 7-8; a single row of teeth in each jaw; *upper jaw with 3 pairs of bifid incisors,* followed on each side by 1 large canine; *lower jaw with pair of simple incisors at front, followed by 2 bifid incisors on each side, then a large canine; anus about halfway between anal fin origin and rear margin of disc*; a small, fleshy swelling on lower pectoral base, without a free posterior margin; 5-6 rows of flattened papillae across width of disc region A, 6-7 rows across disc region B, none on disc region C; overall olive, with reticulum of small, brown-edged whitish spots; sometimes a series of dark bars evident across dorsal surface. Panama; inhabits rocky shores. Reaches 2.8 cm.

CORTEZ CLINGFISH
Tomicodon boehlkei Briggs, 1955

Dorsal rays 6-8; anal rays 6-7; *pectoral rays 19-22*; caudal rays 8-9; a single row of teeth in each jaw; *upper jaw with 4 pairs of trifid incisors,* followed on each side by 1-2 well-developed canines; *lower jaw with 3 pairs of trifid incisors at front, followed by 1-2 well-developed canines; anus usually a little closer to anal fin origin than to rear margin of disc*; posterior nostril in front of anterior edge of eye; 5-7 rows of flattened papillae across width of disc region A, 6-8 rows across disc region B, none on disc region C; cream to greenish, with 4-5 irregular greenish brown bars across dorsal surface of body; close inspection reveals a maze-like pattern of green and tan to cream-colored bands. Gulf of California; inhabits large boulders and vertical outcrops, near surface of intertidal zone. To 6.5 cm.

ROSY CLINGFISH
Tomicodon eos (Jordan & Gilbert, 1882)

Dorsal rays 8-9; anal rays 7-8; *pectoral rays 21-23*; caudal rays 8-9; *upper jaw with 3-4 pairs of trifid incisors* at front, followed on each side by 1-2 small canines; *lower jaw with 3-4 pairs of trifid incisors at front, followed on each side by 2-5 small canines; anus a little closer to anal fin origin than to rear margin of disc; 3-5 rows of flattened papillae across width of disc region A, 4-5 rows across disc region B*, none on disc region C; rosy red to gray, with 4-5 diffuse dark bars across back; about 5 whitish blotches along dorsal midline, including large one on snout; fins yellowish. Mexico, including Gulf of California, inhabits rocky shallows close to shore. Maximum size to 4 cm.

SONORA CLINGFISH
Tomicodon humeralis (Gilbert, 1890)

Dorsal rays 8-9; anal rays 6-7; *pectoral rays 17-19*; caudal rays 9-10; *upper jaw with 5-7 pairs of trifid incisors* at front, followed on each side by 2-4 well-developed canines; *lower jaw with 4-5 pairs of trifid incisors, followed on each side by 3-4 well-developed canines; anus much closer to anal fin origin than to rear margin of disc*; 5-7 rows of flattened papillae across width of disc region A, 6-8 rows across disc region B, none on disc region C; greenish brown with narrow white bars on sides and across back; also scattered dark spots on back and a large dark blotch above each pectoral fin. Gulf of California; inhabits rocky intertidal zone; reported to be capable of surviving for 2 days out of water if kept moist. Attains 8.5 cm.

PETER'S CLINGFISH
Tomicodon petersi (Garman, 1875)

Dorsal rays 7-8; anal rays 6-7; *pectoral rays 19-21*; caudal rays 8-10; *upper jaw with 4-7 pairs of trifid incisors*, followed on each side by 1-2 well-developed canines; *lower jaw with 3-4 pairs of trifid incisors, followed on each side by 1-2 well-developed canines; anus usually closer to anal fin origin than to rear margin of disc*; 4-6 rows of flattened papillae across width of disc region A, 6-8 rows across disc region B, none on disc region C; white with brown reticulum, thickened in places to form a series of irregular brown bars across dorsal surface; a prominent dark brown spot above each pectoral fin. Mexico to Peru, including the Galapagos; inhabits rocky shores. To 3.5 cm.

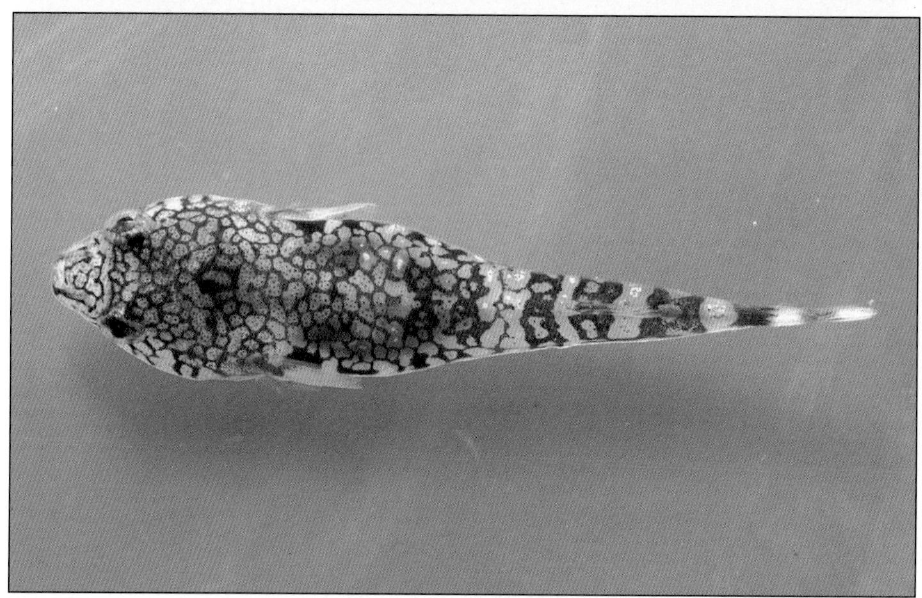

FLYINGFISHES (PECES VOLADORES)

FAMILY EXOCOETIDAE

Flyingfishes are aptly named for their habit of emerging quickly from the water and gliding for long distances (up to 200 to 300 m) with their outstretched pectoral fins (and enlarged pelvic fins in some species). These function as wings and are held rigid, without any flapping movements. When swimming, the pectorals are held flat against the body. Flyingfishes are primarily inhabitants of the open sea, but are often seen close to the outer edge of coral reefs, over deep water. They feed mostly on planktonic organisms. The eggs are relatively large, and have sticky filaments that attach to floating and benthic weeds. The family occurs in all warm seas and contains seven genera and about 60 species. We include only a few of the eastern Pacific representatives.

SPOTTED FLYINGFISH
Cypselurus callopterus (Günther, 1866)
(Plate VI-1, p. 93)

Dorsal rays 11-12; anal rays 8-10; midlateral scales about 42-46; snout much shorter than eye diameter; pectoral fins about 65-70 percent of standard length; pelvic fins relatively long, the tips reaching beyond middle of anal fin; origin of anal fin below middle of dorsal fin; dark blue to blackish on back, silvery below; *pectoral "wings" and pelvic fins yellowish with numerous black spots*. Gulf of California to Ecuador; pelagic in surface waters. Attains at least 22 cm.

TROPICAL TWO-WING FLYINGFISH
Exocoetus volitans (Linnaeus, 1758)
(Plate VI-3, p. 93)

Dorsal rays 13-15; anal rays 12-15; midlateral scales about 43-48; *snout shorter than eye diameter; pectoral fins 70 percent of standard length or longer; pelvic fins very short, their tips far short of reaching anal fin origin*; dark blue to blackish on back and upper sides, silvery white below; fins mainly pale or slightly dusky. Circumtropical distribution; in the eastern Pacific from Baja California to Chile and the Galapagos; pelagic in surface waters. Maximum size, 20 cm.

SHARPCHIN FLYINGFISH
Fodiator acutus (Valenciennes, 1846)
(Plate VI-2, p. 93)

Dorsal rays 9-10; anal rays 10-12; midlateral scales about 40-42; *snout longer than eye and pointed; lower jaw protruding; pectoral fins about 50 percent of standard length or shorter*; pelvic fins small, not reaching anal fin origin; irridescent blue on back and upper sides, silvery below; pectoral "wings" grayish; dorsal fin largely blackish. Tropical Atlantic and eastern Pacific; ranges from California to Peru; pelagic in surface waters. Grows to 15 cm.

HALFBEAKS (BALAOS, SALTADORES, MEDIO PICOS)
FAMILY HEMIRAMPHIDAE

Halfbeaks are elongate, slender fishes that characteristically have the lower jaw extended into a long beak (except in a few species); the upper jaw is short and triangular in shape. There are no spines in the fins; the dorsal and anal fins are posteriorly located with their bases about opposite one another; the pelvic fins are abdominal in position; the pectoral fins are high on the sides and of variable length; the caudal fin is forked or emarginate. The lateral line is positioned low on the sides of the body. Most species live near the sea's surface, frequently occurring in large schools. There are also a number of estuarine and freshwater species. They sometimes leap from the water and skitter across the surface. *Euleptorhamphus viridis* is capable of prolonged glides, similar to those of flying fishes. Halfbeaks exhibit diverse feeding habits; some are herbivores, feeding on floating seaweed, others are omnivores or carnivores, feeding on crustaceans and other small invertebrates. Their eggs are large, and have adhesive filaments with which they attach themselves to floating objects. Although bony, the flesh of halfbeaks is considered good eating in some regions. Halfbeaks occur circumglobally, mainly in tropical and temperate seas; the family contains 12 genera and about 80 species.

RIBBON HALFBEAK
Euleptorhamphus viridis (van Hasselt, 1823)
(Plate VI-5, p. 93)

Dorsal rays 20-25; anal rays 20-25; pectoral rays 8; total gill rakers on first arch 25-32; *body very long and slender, the greatest depth 16.1-23.5 in standard length; pectoral fins relatively long, 3.6-4.0 in standard length; lower jaw much elongated, its length 2.2-2.7 in standard length*; silvery on sides, bluish or greenish above; fins unpigmented. Tropical and temperate seas of the Indo-Pacific; in the eastern Pacific from Baja California to Ecuador. Grows to 40 cm.

JUMPING HALFBEAK
Hemiramphus saltator Gilbert & Starks, 1904
(Plate VI-7, p. 93)

Dorsal rays 12-15 (usually 13-14); anal rays 11-13 pectoral rays usually 11; gill rakers on lower limb of first arch 25-32; *lower jaw greatly prolonged with pale (red or orange in life) tip; pectoral fins relatively long, reaching beyond nasal opening when folded forward; caudal fin deeply forked; scales absent on snout; preorbital ridge absent*; bluish on back, silvery on side of head and body, and white ventrally; lower jaw blackish. Gulf of California to Peru and Galapagos. Grows to 35 cm.

COMMON HALFBEAK
Hyporhamphus unifasciatus (Ranzani, 1842)
(Plate VI-6, p. 93)

Dorsal rays 14-16; anal rays 15-17; pectoral rays 11-12 (usually 11); total gill rakers on first arch 30-37; *lower jaw greatly prolonged with reddish fleshy tip; pectoral fins relatively short, not reaching nasal opening when folded forward; caudal fin emarginate; scales present on snout; preorbital ridge well developed*; greenish on back, silvery white below. Tropical seas on both sides of the Americas; in the eastern Pacific ranging from California to Peru. Reaches 23 cm.

WINGED HALFBEAK
Oxyporhamphus micropterus (Valenciennes, 1846)
(Plate VI-4, p. 93)

Dorsal rays 13-15; anal rays 14-16; pectoral rays 12-13; total gill rakers on first arch 28-34; predorsal scales 28-33; *lower jaw projecting in juveniles, but short in adults, not forming long extension as in most halfbeaks; pectoral fins relatively long and*

School of Jumping Halfbeaks (*Hemiramphus saltator*), Isla Nalasco, Gulf of California

wing-like, but not reaching level of pelvic fins; dark blue to blackish on back and upper side, silvery below; dorsal, caudal, pectoral, and outer half of pelvic fins dusky blackish. Circumtropical distribution; in the eastern Pacific ranging from Baja California to Peru and the Galapagos; pelagic in surface waters. To 18 cm.

NEEDLEFISHES (AGUJAS, MARAOS)
FAMILY BELONIDAE

Needlefishes have very slender bodies and extremely elongate jaws with numerous needle-like teeth. The fins lack spines; the dorsal and anal fins are posterior in position, and the pelvics occur toward the end of the abdomen and contain six rays. The lateral line is low on the body, and the scales are small. Another characteristic feature is the green-colored bones. These fishes live at the surface and are protectively colored for this mode of life: green or blue on the back and silvery white on the sides and ventrally. When frightened (for example, by a light at night) they may leap from the water and skip at the surface. They have been known to injure people who lie in their path at this time, and fatalities have resulted. Their diet consists mainly of small pelagic fishes. Needlefishes are good eating, although bony. Their eggs are large, and have adhesive filaments with which they attach themselves to floating objects. The family is distributed worldwide in tropical and temperate seas; there are 10 genera and 32 species.

BARRED NEEDLEFISH
Ablennes hians (Valenciennes, 1846)
(Plate VI-9, p. 93)

Dorsal rays 22-25; anal rays 25-27; pectoral rays 13-15; predorsal scales 470-525; gill rakers absent; *head and body strongly compressed laterally*; dark bluish above, silvery white below, with several *black bars or vertically elongated blotches on sides*. Circumtropical. To at least 110 cm.

FLAT-TAILED NEEDLEFISH
Platybelone argalus pterura (Osburn & Nichols, 1916)
(Plate VI-8, p. 93)

Dorsal rays 11-14; anal rays 15-18; pectoral rays 11 or 12; predorsal scales 101-118; gill rakers 4 or 5 + 6 or 7; *caudal fin base greatly flattened, its width greater than the depth*; greenish to bluish on back, silvery on sides. Worldwide distribution in tropical and warm temperate seas; the subspecies *P. argalus pterura* is restricted to the eastern Pacific. Grows to about 45 cm.

YELLOWFIN NEEDLEFISH
Strongylura scapularis (Jordan & Gilbert, 1881)
(Plate VI-10, p. 93)

Dorsal rays 13-15; anal rays 15-17; pectoral rays 9-11 (usually 10); predorsal scales about 177-226; body elongate, rounded in cross-section; upper and lower jaws greatly elongated, with sharp teeth; *caudal peduncle without lateral keels and not strongly depressed*; brownish or olive on back, silvery on sides with golden tinge; *pectoral fins and iris of eye yellow*; dorsal and anal fins slightly yellow. Costa Rica to Peru; coastal seas, entering estuaries and river mouths. Grows to at least 38 cm.

Flat-Tailed Needlefish (*Platybelone argalus pterura*), Islas Secas, Panama

PACIFIC NEEDLEFISH
Tylosurus acus pacificus (Steindachner, 1875)
(Plate VI-11, p. 93)

Dorsal rays 21-22; anal rays 19-20; pectoral rays 14 or 15; predorsal scales 280-310; lower lobe of caudal fin much longer than upper lobe; *prominent, raised black keel on side of caudal peduncle*; dark bluish green above, silvery below; a dark blue stripe on middle of sides. Circumtropical. To 130 cm.

CROCODILE NEEDLEFISH
Tylosurus crocodilus fodiator Jordan & Gilbert, 1881
(Plate VI-12, p. 93)

Dorsal rays 18-22 (usually 20); anal rays 17-18 (usually 18); pectoral rays 14 or 15; predorsal scales 326-361; lower lobe of caudal fin much longer than upper lobe; *prominent, raised black keel on side of caudal peduncle*; dark bluish green above, silvery on sides. Circumtropical. To 130 cm.

LIVEBEARERS (SARDINAS VIVÍPARAS)
FAMILY POECILIIDAE

This family includes many popular aquarium fishes, including the world-famous Guppy (*Poecilia reticulata*), swordtails, platies, and mollies. They are inhabitants of fresh and brackish waters of low-lying areas of North and South America. As the common name suggests, the members of this family (except *Tomerus gracilus* of northeastern South America) give birth to live young. Males are characterized by the presence of a gonopodium, a specialized organ which facilitates internal fertilization. It is an elongated structure composed primarily from the third, fourth, and fifth anal rays. The family contains about 22 genera with approximately 150 species. A representative species, common in brackish estuaries, is treated here.

ELONGATE TOOTHCARP
Poeciliopsis elongata (Günther, 1866)

Dorsal rays 9; anal rays 8; midlateral scales 30-32; snout greatly depressed; jaw teeth very small, nearly indistinguishable on posterior part of upper jaw; *anal fin of males modified to form gonopodium*; olive on back, silvery on sides, and yellowish on belly; often with 10-12 faint darkish crossbars; fins plain. Costa Rica and Panama; forms large schools among mangroves in estuaries and brackish river mouths; also abundant in salt pans. Grows to 11 cm.

SILVERSIDES (PEJERREYES, CAUQUES)

FAMILY ATHERINIDAE

Silversides are small, schooling fishes, usually found inshore and sometimes over reefs. There are also a number of freshwater and estuarine species. They have two dorsal fins, the first consisting of a few slender spines; the pelvic fins, which have one spine and five rays, are usually abdominal in position; there is no lateral line. Many atherinid fishes have a broad silvery stripe on the side, and for this reason are called silversides. They are heavily preyed upon by other fishes such as jacks. Although usually classified in the older literature with mullets and barracudas, recent studies suggest that they are allied to the halfbeaks (Hemiramphidae). The California Grunion (*Leuresthes tenuis*) is one of the best-known members of this family, due to its habit of undergoing mass spawning runs on sandy beaches. The fish are easily caught by hand at this time, which is correlated with the phases of the moon. The spawning runs usually occur a day or two after each full or new moon during the spring and summer.

PANAMANIAN SILVERSIDE
Atherinella panamensis Steindachner, 1875
(Plate VI-14, p. 93)

Dorsal rays III-I,7; anal rays I,21; *midlateral scales 37-38*; body relatively slender, its depth about 4.5 in standard length; upper profile of body very flat from snout tip to tail base; *abdomen strongly compressed, forming keel-like structure*; scales strongly ctenoid; *pectoral fins greatly elongated, somewhat falcate, extending well beyond tips of pelvic fin, nearly to level of anal fin origin*; olive on back and upper sides, silvery white below, with narrow, silvery midlateral stripe. Panama; forms aggregations in shallow coastal areas. To 14 cm.

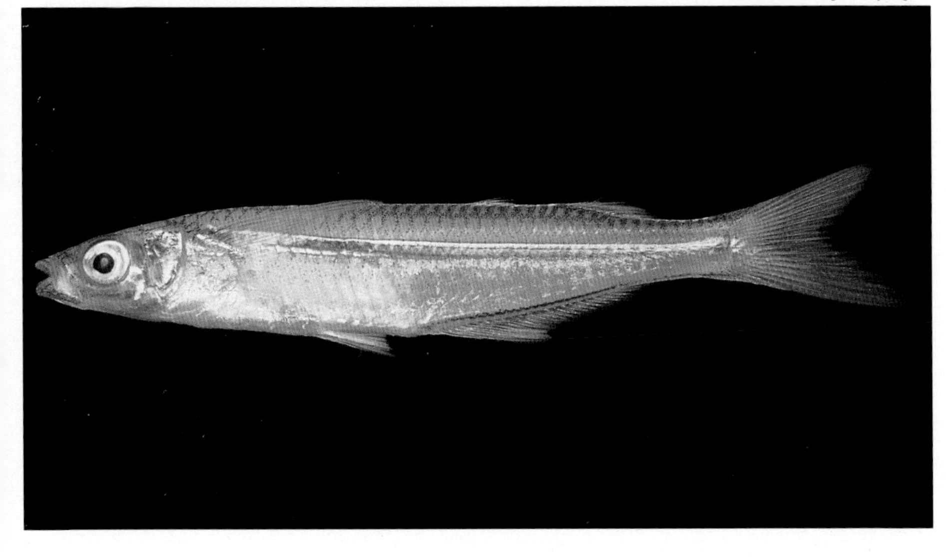

Melaniris pachylepis

SMOOTH SILVERSIDE
Eurystole eriarcha (Jordan & Gilbert, 1881)
(Plate VI-15, p. 93)

Dorsal rays IV+I,11-12; *anal rays I,27; midlateral scales about 48-49*; body relatively slender, its depth about 5.0 in standard length; abdomen rounded in cross-section, not compressed into a keel; *scales smooth; pectoral fins short, reaching to about middle of pelvic fins*; translucent pale olive with narrow dark scale outines and broad silver midlateral band. Baja California to Colombia; forms aggregations in shallow bays and near coastal reefs. Reaches at least 8 cm.

LAGOON SILVERSIDE
Melaniris pachylepis (Günther, 1864)

Dorsal rays III-V + 7-9; anal rays I,20-23; *midlateral scales 43-47*; abdomen rounded in cross-section, not compressed into a keel; *scales of body crenate; pectoral fins longer than head, reaching well past base of pelvic fins, frequently near their tips*; dull greenish or olive on back, sides silvery white with well-pronounced silver midlateral stripe. Panama to Peru; commonly forms aggregations in shallow bays and on coastal reefs. Grows to 16 cm.

MULLET SILVERSIDE
Mugilops cyanellus Meek & Hildebrand, 1923
(Plate VI-13, p. 93)

Dorsal rays VIII-IX + 15-17; anal rays I,18-22; *midlateral scales 43-47*; body relatively deep, 3.7-4.0 in standard length; abdomen rounded in cross-section, not compressed into a keel; *scales crenate on back, slightly crenate to smooth on sides; pectoral fin short, reaching slightly beyond base of pelvic fins*; dark blue on back and upper sides, silvery on lower half of sides and belly. Costa Rica and Panama; forms aggregations in the vicinity of reefs. Attains at least 10 cm.

PLATE VI

1 **SPOTTED FLYINGFISH** (*Cypselurus callopterus*)

2 **SHARPCHIN FLYINGFISH** (*Fodiator acutus*)

3 **TROPICAL TWO-WING FLYINGFISH** (*Exocoetus volitans*)

4 **WINGED HALFBEAK** (*Oxyporhamphus micropterus*)

5 **RIBBON HALFBEAK** (*Euleptorhamphus viridis*)

6 **COMMON HALFBEAK** (*Hyporhamphus unifasciatus*)

7 **JUMPING HALFBEAK** (*Hemirhamphus saltator*)

8 **FLAT-TAILED NEEDLEFISH** (*Platybelone argalus pterura*)

9 **BARRED NEEDLEFISH** (*Ablennes hians*)

10 **YELLOWFIN NEEDLEFISH** (*Strongylurus scapularis*)

11 **PACIFIC NEEDLEFISH** (*Tylosurus acus pacificus*)

12 **CROCODILE NEEDLEFISH** (*Tylosurus crocodilus fodiator*)

13 **MULLET SILVERSIDE** (*Mugilops cyanellus*)

14 **PANAMANIAN SILVERSIDE** (*Atherinella panamensis*)

15 **SMOOTH SILVERSIDE** (*Eurystole eriarcha*)

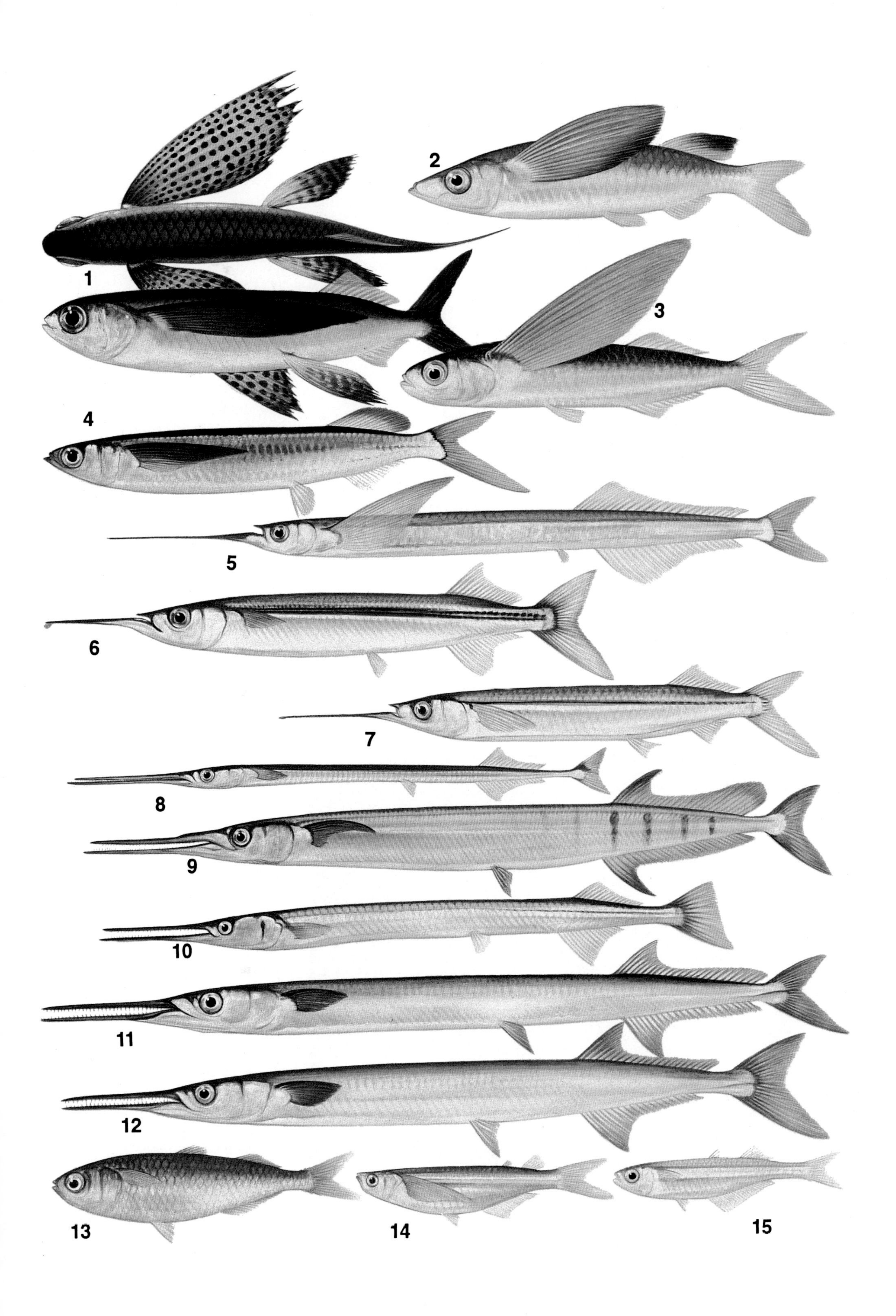
1
2
3
4
5
6
7
8
9
10
11
12
13
14
15

FLASHLIGHT FISHES (PECES LINTERNA)
FAMILY ANOMALOPIDAE

The flashlight or lanterneye fishes are relatives of the squirrelfishes (Holocentridae). Their most outstanding feature is the presence beneath the eye of a light organ that has a rotational or shutter mechanism that allows its light emission to be controlled. The light organ is composed of a series of tubes that contain luminous bacteria. During night forays the fishes turn on the lights intermittently, presumably to attract photopositive planktonic food and also to establish and maintain contact with conspecifics. Other characteristic features include a spinous dorsal fin, containing two to six spines, that is continuous with the soft portion of the fin or is strongly notched, an anal fin with two to three spines, and short subocular shelf. This small family contains four genera and six species. The group was reviewed by McCosker and Rosenblatt (1987), but since then an additional genus and species was described from the Society Islands. Only a single genus and species is known from our area. They are small fishes, mostly under 10 cm, although *Anomalops* from the Indo-West Pacific may reach 30 cm. They are nocturnally active, emerging from shelter at dusk to feed. Daytime retreats include deep crevices and caves.

FLASHLIGHT FISH
Phthanophaneron harveyi (Rosenblatt & Montgomery, 1976)

Dorsal rays IV-I,14-15; anal rays II,10; pectoral rays 16; pelvic rays I,6; scales very small and strongly ctenoid, about 80 in longitudinal series on upper side; a midventral row of about 12-13 enlarged, keeled scales on belly; bones of head and shoulder girdle strongly sculptured with numerous spine-bearing ridges; *a prominent, pearly white, luminous organ under eye*; overall blackish brown. Known only from a few specimens captured in the Gulf of California; the illustrated specimen was captured in 35 m on a steep sandy slope with profuse growth of black coral at Isla San Pedro Martir. Grows to at least 8 cm.

Phthanophaneron harveyi ▲

Myripristis berndti ▼

SOLDIERFISHES AND SQUIRRELFISHES (PECES ARDILLA, SOLDADOS, CANDILES)
FAMILY HOLOCENTRIDAE

Squirrelfishes and soldierfishes occur in all tropical seas, but most inhabit the Indo-Pacific region. The family contains eight genera with about 65 species; only four species are present in our area. Distinguishing characteristics include rough scales, prominent fin spines, a large eye, and usually a red coloration. They possess a moderately large mouth, but the teeth are small. The caudal fin is forked. Species of *Sargocentron* possess a sharp venomous spine at the lower corner of each cheek, and can inflict a painful wound if handled carelessly. During the day, holocentrids shelter in caves and under ledges, but emerge from their retreats at dusk and spend the night feeding in the open. *Myripristis* consume larger planktonic items such as crab larvae, whereas the species of *Sargocentron* feed mainly on benthic crabs and shrimps. Although most of the species are small, the flesh is considered good eating. Members of *Sargocentron* were formerly classified in the genus *Holocentrus* (now restricted to the Atlantic) and later in *Adioryx* (a synonym).

BIGSCALE SOLDIERFISH
Myripristis berndti (Jordan & Evermann, 1903)

Dorsal rays X,I,13-15; anal rays IV,11-13; lateral-line scales 28-31; lower jaw of adults prominently projecting when mouth closed; vomerine teeth in a subtriangular patch; *lower half of inner pectoral axil with small scales; no enlarged spine at lower rear corner of cheek*; silvery pink to pale yellow, with red scale edges; *edge of gill cover (opercular membrane) black; upper pectoral axil black*; spinous dorsal fin yellow on outer half. Widespread in the tropical Indo-Pacific from East Africa to the Americas; in the eastern Pacific it ranges from the Gulf of California to the Galapagos; common at offshore island areas. Attains 30 cm.

PANAMIC SOLDIERFISH
Myripristis leiognathus Valenciennes, 1846

Dorsal rays X,I,13-15; anal rays IV,11-13; lateral-line scales 34-40; gill rakers on first arch 9-10 + 19-24, total gill rakers 28-34; *inner pectoral axil lacking scales; no enlarged spine at lower rear corner of cheek;* overall red with darker scale margins, vague stripes on side; *edge of gill cover and pectoral axil lacking black pigment.* Baja California to Ecuador and offshore islands including Revillagigedos, Cocos, and Galapagos. Reaches 18 cm. Two additional species of *Myripristis* (not illustrated) also occur in the eastern Pacific: *M. clarionensis*, known from the Revillagigedos and Clipperton Island, has 3½ scale rows between the lateral line and dorsal origin (2½ in all other *Myripristis*); *M. gildi* Greenfield from Clipperton Island is similar to *M. leiognathus*, but has 34-47 total rakers, and the inner pectoral axil and opercular membrane are black.

ROUGHSCALE SOLDIERFISH
Plectrypops lima (Valenciennes, 1831)

Dorsal rays XII,14-16; anal rays IV,10-12; lateral-line scales 39-42; *scales above lateral line to base of middle dorsal spines 4; scales very rough to the touch; no enlarged spine at lower rear corner of cheek*; dorsal spines relatively short, the longest about 3.0 in head length; uniform bright red. Widespread in the tropical Indo-Pacific; known in the eastern Pacific from offshore islands. A seldom seen fish that hides in caves and crevices during the day. Grows to 16 cm.

TINSEL SQUIRRELFISH
Sargocentron suborbitalis (Gill, 1864)

Dorsal rays XI,13-14; anal rays IV,9; lateral-line scales 36-38; third anal spine very long, more than half of body depth; *a stout spine at lower rear corner of cheek, about equal in length to pupil width*; silvery with rosy to violet hue; dark scale margins especially pronounced on upper half of body. Gulf of California to Ecuador, including the Galapagos and other offshore islands; inhabits rocky caverns and coral crevices. Maximum size to about 25 cm.

CORNETFISHES (PECES CORNETA)

FAMILY FISTULARIIDAE

These extremely elongate fishes have a depressed body; a very long tubular snout with a short, oblique mouth at the end; minute teeth; no fin spines; a single dorsal fin posteriorly on the body directly over the anal fin; and a forked caudal fin with a long median filament. They feed by sucking in small invertebrates and fishes. Four species are known from tropical seas, one of which is common on reefs in our area.

REEF CORNETFISH
Fistularia commersonii Rüppell, 1835

Dorsal rays 15-17; anal rays 14-16; pectoral rays 15 (rarely 13-14); greenish dorsally, shading to silvery white below, with 2 blue stripes or rows of blue spots on back. Circumtropical distribution, including Baja California to the Galapagos; a free-swimming fish often seen over reefs and seaweed beds; it can quickly assume a dark-barred pattern when close to the bottom. Reaches 150 cm. A second species of cornetfish, *F. corneta*, also occurs in the eastern Pacific, but it inhabits deeper water (usually on open bottoms below 30 m). It differs from *commersonii* in having more dorsal, anal, and pectoral rays (usually 18-20, 17-18, and 16-18 respectively).

TRUMPETFISHES (PECES TROMPETA)

FAMILY AULOSTOMIDAE

See discussion of the single Indo-Pacific species of the family below.

TRUMPETFISH
Aulostomus chinensis (Linnaeus, 1758)

Dorsal rays VII-XII + 24-27; anal rays 26-29; pectoral rays 17; pelvic rays 6; *snout elongate and tubular; chin barbel present; series of short dorsal spines on back;* color variable, usually brown or greenish, with diffuse pale stripes and connecting bars on side; sometimes entirely yellow. Widespread in Indo-Pacific region; in eastern Pacific mainly at offshore islands; feeds on small fishes. Reaches 80 cm.

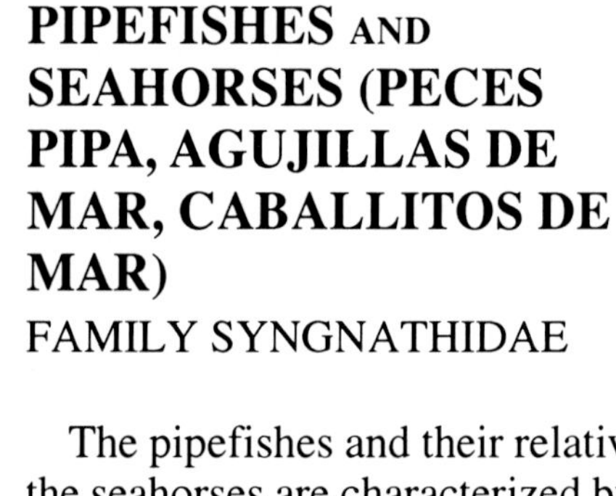

PIPEFISHES AND SEAHORSES (PECES PIPA, AGUJILLAS DE MAR, CABALLITOS DE MAR)

FAMILY SYNGNATHIDAE

The pipefishes and their relatives the seahorses are characterized by a long, slender body that is composed of a series of ring-like, bony segments. They have a very small gill opening, no spines in the fins, no pelvic fins, a single dorsal fin, and a very small anal fin (dorsal, anal, and/or pectoral fins absent on a few species). Most species are seldom seen because of their habit of remaining in crevices. Perhaps the greatest peculiarity displayed by this family is their habit of male egg incubation. The female deposits her eggs on the ventral surface of the male, usually in a pouch or on a specially vascularized surface. The "pregnant" male then carries these until hatching occurs. Worldwide the family contains about 200 species of pipefishes belonging to more than 50 genera and approximately 30 species of seahorses in the single genus *Hippocampus*. The pipefishes of the Indo-Pacific region were comprehensively reviewed by Dawson (1985).

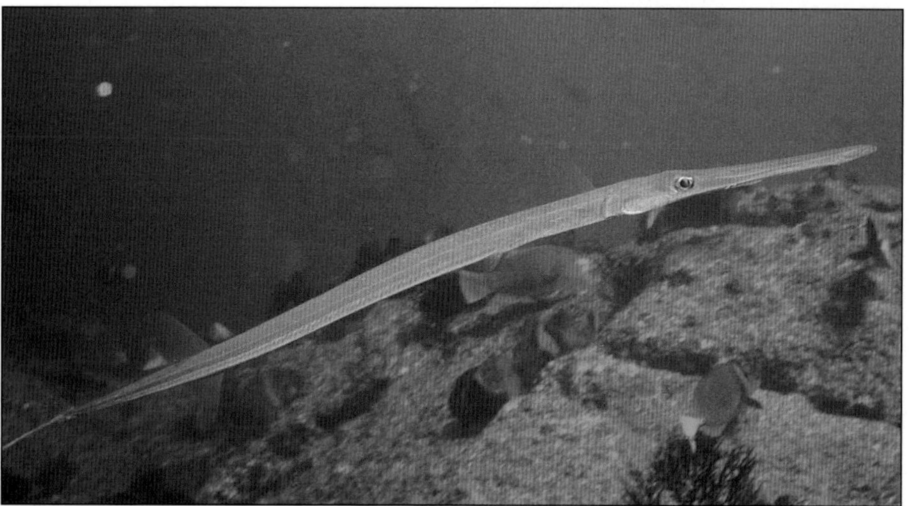

Fistularia commersonii ▲

Aulostomus chinensis ▼

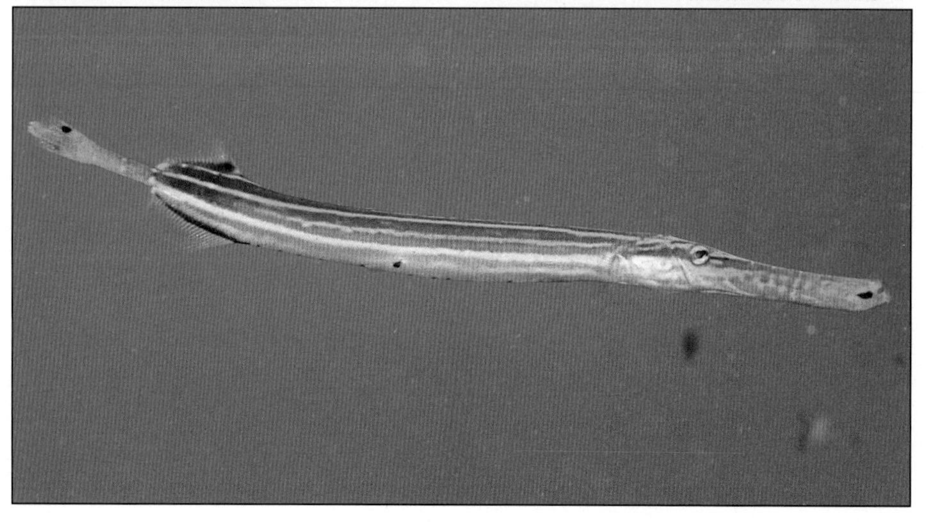

SUBNOSE PIPEFISH
Cosmocampus arctus (Jenkins & Evermann, 1889)

Dorsal rays 18-20; anal rays 2-4; pectoral rays 10-12 (usually 11); *body rings 14-17 (usually 15) + 36-41*; superior trunk and tail ridges discontinuous; lateral trunk ridge straight, ending near anal ring; inferior trunk and tail ridges continuous; head length 8.1-13.2 in standard length; snout length 2.3-4.0 in head length; dermal flaps frequently present on eye and elsewhere on body; male brood pouch under tail; reddish to dark gray with narrow whitish bars and/or pale saddles on back; usually a whitish stripe on head behind eye. Gulf of California to Peru, including the Galapagos; inhabits rock or coral reefs from shallow depths to at least 20 m; often found among red algae. Attains 13 cm. Dawson (1985) recognized 3 subspecies: *C. arctus arctus* from California to Mazatlan, Mexico; *C. arctus coccineus* (Herald) from Central Mexico to Peru, including Galapagos; and *C. arctus heraldi* Fritzsche from San Felix Island in the Juan Fernandez Islands, Chile.

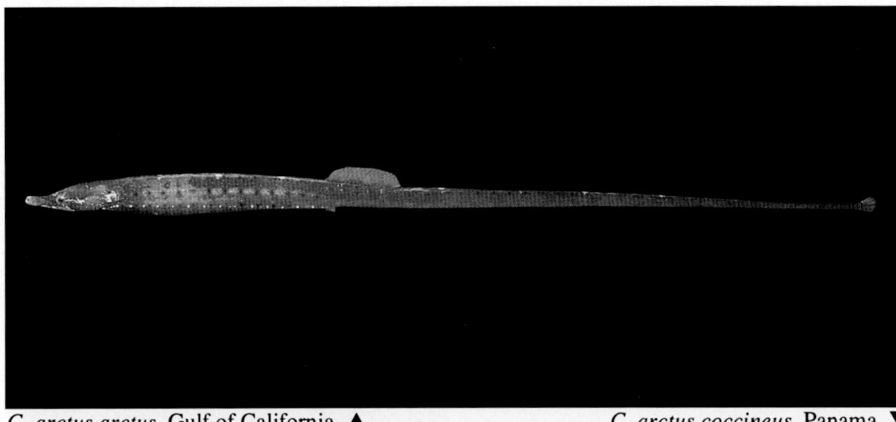

C. arctus arctus, Gulf of California ▲ *C. arctus coccineus*, Panama ▼

FANTAIL PIPEFISH
Doryrhamphus excisus Kaup, 1856

Dorsal rays 21-29; anal rays 4; pectoral rays 19-23; *body rings 16-19 + 13-17*; superior trunk and tail ridges discontinuous; inferior trunk ridge ending on anal ring; lateral trunk ridge confluent with inferior tail ridge; head length 3.9-4.9 in standard length; snout length 2.0-2.4 in head length; male brood pouch under trunk; generally reddish to red-brown, with a blue stripe on upper part of body and blackish stripe on side of snout, continued behind eye; *caudal fin rounded with narrow white margin and mainly blackish central portion containing 1 or more medial orange-red spots*. Widespread in the tropical Indo-Pacific from East Africa to the Americas; the eastern Pacific distribution extends from Baja California to Ecuador, including the Galapagos and other offshore islands; inhabits rock and coral reefs to 45 m. Reaches maximum size of 7 cm. Dawson (1985) recognized a separate subspecies, *D. excisus paulus* Fritzsche from the Revillagigedos Islands, based on its fewer (usually 16 vs. 17-19) trunk rings and smaller size (males sometimes brooding at only 24 mm standard length).

Doryrhamphus excisus ▲ *Syngnathus auliscus* ▼

Photograph: Alex Kerstitch

BARRED PIPEFISH
Syngnathus auliscus (Swain, 1882)

Dorsal rays 26-33; pectoral rays 10-14; *body rings 14-16 + 34-39*; superior trunk and tail ridges discontinuous; inferior trunk and tail ridges continuous; lateral trunk ridge ends near anal ring; head length 7.1-10.1 in standard length; snout length 2.0-2.8 in head length; male brood pouch under tail; generally greenish brown, frequently with dark midlateral stripe and/or series of dark bars on side. Southern California to Peru; usually among vegetation in bays and estuaries, occasionally among floating *Sargassum*. Reaches maximum size of 19 cm.

PACIFIC SEAHORSE
Hippocampus ingens Girard, 1858

Dorsal rays 19-21; anal rays 4; pectoral rays 15-17; body rings 11-12 + 38-40; *tail prehensile; snout elongate; coronet (crown-like structure on top of head) moderately high, becoming somewhat lower in large males; tubercles generally well developed, but becoming obscure in large males*; color variable according to its surroundings (weed, rock, coral, etc.); generally various shades of red, yellow, tan, brown, gray, black, or green; often with small blackish and white spots or whitish cross-bands. Southern California to Peru, including the Galapagos; inhabits weed beds and sea-whips on patch reefs. Grows to about 28 cm.

SCORPIONFISHES (DIABLOS, PUÑALES, PECES ESCORPIÓN)
FAMILY SCORPAENIDAE

Scorpionfishes obtain their name from the venomous fin spines possessed by many of the species. Other characteristics include a bony ridge (suborbital stay) across the cheek; the head is relatively large and spiny; the cheek margin has 3-5 spines; there is one or two spines on the opercle, and other spines are scattered on the head. There is a single dorsal fin that is usually strongly notched at the rear of the spinous part and consists of VII-XVIII spines and 4-14 soft rays; the anal fin has II-IV spines and 5-14 soft rays. Scorpionfishes are bottom-living predators that occur in a variety of depths from shallow tide pools to the oceanic abyss. Most reef species are secretive, dwelling in caves and crevices. They remain mostly stationary during daylight, but are nocturnally active, feeding mostly on crustaceans and fishes. Scorpaenids often exhibit variegated color patterns that blend well with their surroundings, thus enabling them to remain undetected by their prey. The dorsal, anal, and pelvic spines are all venomous. Poison is produced by glandular tissue in longitudinal grooves on each side of the spine or (in stonefishes of the Indo-West Pacific) a large oblong gland protruding from the side of the spine that is connected to the spine tip by a venom duct. Wounds from the spines vary from bee-sting intensity to unbelievable agony in the Indo-West Pacific genera *Synanceia* (stonefishes) and *Pterois*. Immersion of the injured limb in very hot water helps alleviate the pain, but treatment by a physician may be required. Scorpionfishes are found in all temperate and tropical seas and are commercially important in some areas. Most of the estimated 350 species and 70 genera occur in the Indo-Pacific; only a few are found on reefs of the tropical eastern Pacific, but numerous species occur in adjacent temperate seas, particularly along the California coast.

Spotted Scorpionfish (*Scorpaena plumieri mystes*), Isla Santiago, Galapagos

ROSY SCORPONFISH
Pontinus sp.

Dorsal rays XII,9-10; anal rays III,5; pectoral rays 16-18, all rays unbranched; *occipital pit absent; preorbital with 2 spines over upper jaw; suborbital ridge with 3-4 spines; overall reddish, pinkish to white on breast and belly; irregular oblique brown bars on side*; a broken brown stripe along lateral line; fins reddish, with brown spotting on soft dorsal, caudal and pectorals. Baja California to northern Peru (S. Poss, pers. comm.). Our collections contain a single 10 cm specimen; probably reaches at least 15-20 cm.

DARKBLOTCH SCORPIONFISH
Scorpaena histrio Jenyns, 1843

Dorsal rays XII,10; anal rays III,5; pectoral rays 19-20, the lower rays unbranched and thickened; *head less broad than deep*; occipital pit present behind interorbital; no pit below eye; *suborbital stay of adults with only a single suborbital spine*; numerous skin flaps on head and body; color highly variable, generally strongly mottled brown to reddish; *the most consistent diagnostic mark is a blackish to dark brown blotch on lateral line above middle of pectoral fin, often with red or orange skin flaps within this blotch.* Mazatlan, Mexico, to Peru; also known from Cabo San Lucas, Baja California and the Galapagos; inhabits crevices and ledges of weed-covered rocky reefs in 5-160 m. Reaches at least 24 cm. *S. pannosa* Cramer is a synonym.

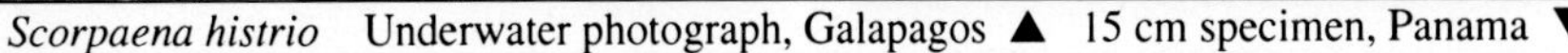

Scorpaena histrio Underwater photograph, Galapagos ▲ 15 cm specimen, Panama ▼

Scorpaena plumieri mystes Underwater photograph, Panama ▲ 15 cm specimen, Panama ▼

SPOTTED SCORPIONFISH
Scorpaena plumieri mystes Jordan & Starks, 1895

Dorsal rays XII,9-10; anal rays III,5-6; pectoral rays 18-21, the lower rays unbranched and thickened; *head as deep as broad; a deep occipital pit present behind interorbital; a smaller pit below and in front of eye; suborbital stay with 3-4 spines*; numerous skin flaps on head and body; color highly variable, usually a drab mottled mixture of gray, brown, red, green, and black; juveniles with pronounced dark bar posteriorly at level of soft dorsal fin; axil of pectoral fin usually dark with white spots, frequently arranged in rows; specimens from the Revillagigedos and Galapagos have a more uniform axillary coloration and may represent a separate subspecies. Baja California to Peru, including the Galapagos and other offshore islands; inhabits weed-covered reefs and open sand-rubble areas. To 34 cm. The nominal subspecies *S. plumieri plumieri* occurs in western Atlantic.

RED SCORPIONFISH
Scorpaena russula Jordan & Bollman, 1889

Dorsal rays XII,10; anal rays III,5; pectoral rays 21, the lower 14 unbranched and thickened; *body relatively slender, 30-33 percent of standard length; snout relatively short, 7-11 percent of standard length; interorbital width 4 percent of standard length; occipital pit behind interorbital weakly developed, only a slight depression; no pit below eye*; suborbital stay bearing 3 small spines; supraocular cirrus disappearing in adults; *skin flaps absent on body*; generally mottled red and brown; pectoral fins with blackish tip. Gulf of California to northern Peru; inhabits sand-rubble bottoms to 60 m depth. Our largest specimen is only 5 cm, but reported to at least 12 cm.

SONORA SCORPIONFISH
Scorpaena sonorae Jenkins and Evermann, 1888

Dorsal rays XII,10; anal rays III,5; pectoral rays 21, the lower 14 unbranched and thickened; body relatively slender, 30-33 percent of standard length; snout relatively short, 7-11 percent of standard length; occipital pit behind interorbital weakly developed, only a slight depression; no pit below eye; suborbital stay bearing 3 small spines; supraocular cirrus disappearing in adults; skin flaps absent on body; *very similar to* S. russula, *but has a rounded dark spot on soft dorsal fin (rather than irregular rows of small spots), and has slightly wider interorbital (5-6 percent versus about 4 percent of standard length)*; generally mottled red and brown with irregular dark bars on sides, ventral parts whitish. Gulf of California; inhabits sand-rubble bottoms to at least 30 m. To about 18 cm.

Photograph: Alex Kerstitch

RAINBOW SCORPIONFISH
Scorpaenodes xyris (Jordan & Gilbert, 1882)

Dorsal rays XIII,10; anal rays III,5; pectoral rays 18, the lower 10 unbranched and thickened; no occipital or suborbital pits; suborbital stay with 3 spines; skin flaps absent on scales of body; *palatine teeth absent*; blotchy or mottled brown to red, often with 3-5 irregular dark bars on side; *a white bar below eye extending across upper jaw; a prominent blackish spot on lower part of gill cover*. Southern California to Peru including the Galapagos; inhabits reef crevices. Reaches 8-9 cm.

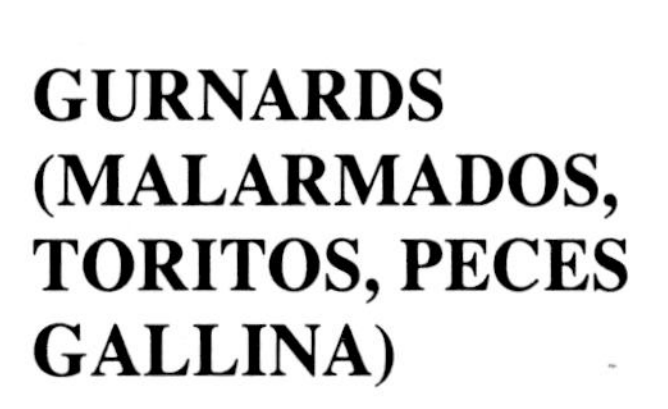

GURNARDS (MALARMADOS, TORITOS, PECES GALLINA)
FAMILY TRIGLIDAE

These scorpionfish relatives are characterized by a squarish, bony head that is frequently armed with spines, two separate dorsal fins, and two to three enlarged, free rays on the lower pectoral fin. The latter are used for detecting food in sand and rubble habitats. The body is either scaled or covered with elongate plates or spine-bearing plates. These fishes are capable of producing sounds with special muscles attached to their large air bladder. Triglids literally "walk" over the bottom using their pectoral "feelers". They feed mainly on molluscs and small crustaceans. The family inhabits all tropical and temperate seas from rocky shallows to considerable depths. There are two subfamilies, the unarmed or scaled Triglinae (10 genera and about 70 species, including the eastern Pacific species treated here), and the armored searobins in Peristidinae (four genera and 17 species). Some authors consider these as separate families. The New World genus *Bellator* Jordan and Evermann, containing seven species (three from the eastern Pacific) was revised by Miller and Richards (1991a). The same authors presented a nominal list of eastern Pacific and Atlantic *Prionotus*, in which 21 species were considered valid. However, they erroneously omitted *P. albirostris* (Richards, pers. comm.).

GULF GURNARD

Bellator gymnostethus (Gilbert, 1891)

Dorsal rays usually XI + 11; anal rays 11; pored lateral-line scales 48-52; *rostral plate prolonged, forming divergent lobes on each side of snout tip; breast and belly entirely scaleless; second dorsal spine tallest*; pectoral fins relatively short, 23-34 percent of standard length; mottled brown or reddish, with faint brown bars on sides; 1-2 ocellated black spots between fourth and fifth dorsal spines. Baja California to Peru; inhabits sand-rubble bottoms near reefs. Reaches at least 10 cm.

Photograph: Alex Kerstitch

SHOVELNOSE GURNARD

Bellator xenisma (Jordan & Bollman, 1889)

Dorsal rays X or XI + 11 (rarely 10); anal rays 10; pored lateral-line scales 35-40; *rostral plate prolonged, forming a "scalloped" shovel-shaped extension on each side of snout tip; breast and belly with small scales; first dorsal spine tallest, about half of head length; a strong spine projecting laterally from cheek bone*; usually 3-4 small, barely detectable spines at rear of spinous dorsal; pectoral fins relatively short, 24-39 percent of standard length; generally mottled reddish brown with largely blackish or dark brown pectoral fins; a prominent ocellus between fourth and fifth dorsal spines. California to Peru; inhabits sand-mud bottoms of trawling grounds to 90 m. Reaches at least 12 cm. A third eastern Pacific species, *B. loxias*, also has the breast and belly scaled, but differs in lacking a strong spine on the cheek bone, by having 11 anal rays, and 49-55 pored lateral-line scales.

WHITESNOUT GURNARD

Prionotus albirostris Jordan & Bollman, 1889

Dorsal rays X + 12; anal rays 11; lateral-line scales 50-55; *center of radiation of bony striae on cheek without a spine*; no transverse groove across top of head behind eyes; *pectoral fins relatively long, reaching below last few dorsal rays; rear margin of jaw not reaching level of front of eye; about 5-6 well developed gill rakers on lower limb of first arch*; gray brown with darker mottling, white on ventral surface; dorsal fins somewhat dusky with brown mottling or spotting, outer edge of spinous dorsal with blackish spots; caudal fin with dark gray base and broad blackish or dark gray outer margin; *pectoral fins black on lower half, with black spotting at base and on upper half.* Gulf of California to Peru; inhabits trawling grounds to at least 60 m. Attains about 19 cm.

HORRIBLE GURNARD
Prionotus horrens Richardson, 1845

Dorsal rays X + 12; anal rays 10; lateral-line scales 93-105; *center of radiation of bony striae on cheek with a spine* (actually on middle of suborbital stay); no transverse groove across top of head behind eyes; *pectoral fins relatively short, barely reaching level of front of second dorsal fin*; vomerine teeth in a continuous band; generally gray-brown with dark blotch below rear part of second dorsal fin, and broad brown band below eye, with an extension behind mouth; *pectoral fin blackish with large whitish patch on outer half; caudal fin with large dark spots basally and blackish longitudinal streaks on outer half.* Baja California to Peru, common in the Gulf of Panama, where it is trawled in 20-50 m. Reaches 25 cm.

GALAPAGOS GURNARD
Prionotus miles Jenyns, 1842

Dorsal rays X + 12; anal rays 10-11; lateral-line scales about 75-80; *center of radiation of bony striae on cheek without a spine*; no transverse groove across top of head behind eyes; *pectoral fins relatively long, reaching to middle of anal fin*; snout emarginate, with prominent spines on each lobe; generally gray-brown; outer three-fourths of spinous dorsal fin black; *pectoral fin blackish with whitish margin; caudal fin with dark blotches.* Galapagos Islands; inhabits sand-rubble bottoms in the vicinity of rocky reefs. Reaches 25 cm.

BLACKFIN GURNARD
Prionotus stephanophrys Lockington, 1880

Dorsal rays X + 12; anal rays 11; lateral-line scales 50-55; center of radiation of bony striae on cheek without a spine; no transverse groove across top of head behind eyes; *pectoral fins relatively long, reaching below middle of second dorsal fin; rear margin of jaw reaching slightly beyond level of front of eye; about 8 well-developed gill rakers on lower limb of first arch*; gray-brown, white ventrally; *pectoral fins a strongly contrasted black; a black spot on outer part of dorsal fin between fourth and fifth spines*; second dorsal and caudal fin with blackish spots. Southern California to Peru; inhabits sand rubble bottoms to at least 90 m depth. Attains at least 15 cm. *P. quiescens* Jordan & Bollman is a synonym.

SNOOKS (ROBALOS, GUALAJOS)
FAMILY CENTROPOMIDAE

The family Centropomidae is composed of two subfamilies represented by the snooks (*Centropomus*) from the Americas and the barramundi (*Lates*) and its relatives from the Indo-West Pacific and African fresh waters. Each subfamily contains about 10 species. The New World snooks of the genus *Centropomus* are generally medium-sized, silvery, perch-like fishes, often with a dark lateral line. Other characteristic features include two separate dorsal fins (the first containing eight spines), three anal spines, caudal fin forked, mouth large and protractile, and margin of preopercle serrate. Snooks are very common in mangrove areas and exhibit a wide tolerance to salinity fluctuations. They readily enter rivers and may occur in pure fresh water. Snooks feed on a wide variety of invertebrates and fishes. They are fine angling fishes, but not good eating. However, they form an important portion of the commercial catch in the region. The eastern Pacific species of *Centropomus* were reviewed by Rivas (1986).

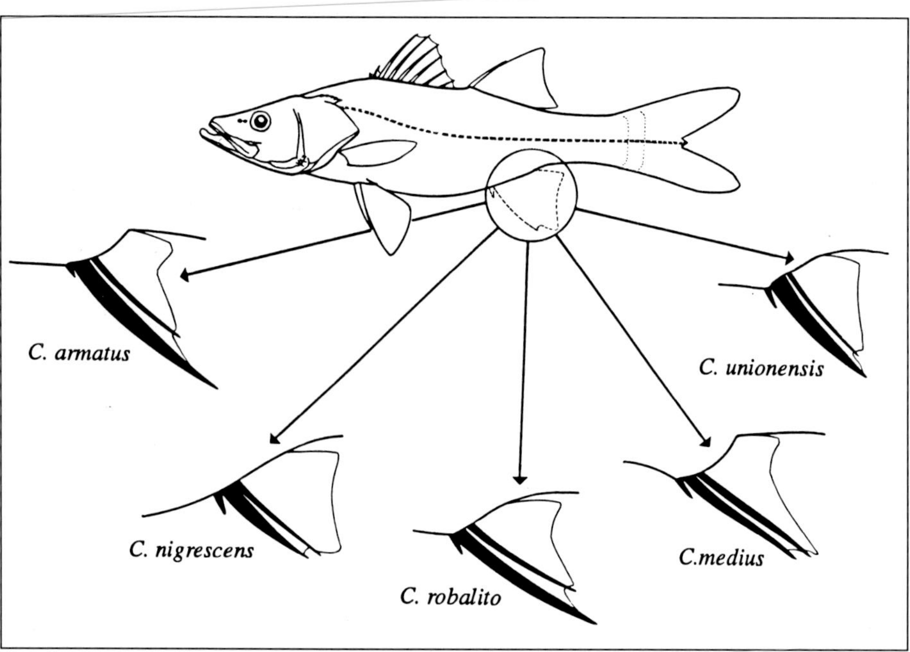

Relative anal-spine lengths of species of *Centropomus*.

LONGSPINE SNOOK
Centropomus armatus Gill, 1863

Dorsal rays VIII + I,10 (rarely 11); anal rays III,6 (rarely 7); pectoral rays 14-16 (usually 15); *total gill rakers on first arch including rudiments 20-25 (usually 21-24); lateral-line scales 49-57 (usually 51-54); scales around caudal peduncle 20-25 (usually 22-24); second anal spine very stout, much longer than third, reaching to or beyond level of caudal base when depressed, and clearly exceeding longest anal soft ray*; overall silvery; base of pectoral fin brown. Mazatlan, Mexico, to Ecuador; inhabits bays, estuaries, and lower reaches of freshwater streams. To at least 37 cm.

BIGEYE SNOOK
Centropomus medius Günther, 1864

Dorsal rays VIII + I,10 (rarely 9); *anal rays III,7 (rarely 8); pectoral rays* 13-15 *(usually 14)*; total gill rakers on first arch including rudiments 22-26 (usually 23-25); lateral-line scales 61-72 (usually 64-67); scales around caudal peduncle 18-21; *second and third anal spines relatively slender, about equal in length, not exceeding longest anal soft ray; pectoral fins much shorter than pelvic fins*; overall silvery with blackish lateral line. Baja California to Ecuador; inhabits bays, estuaries, and lower reaches of freshwater streams. Grows to 47 cm. Closely related to *C. pectinatus* Poey from the western Atlantic.

Centropomus nigrescens ▲

BLACK SNOOK
Centropomus nigrescens Günther, 1864

Dorsal rays VIII +I,10 (rarely 9 or 11); anal rays III,6; pectoral rays 14-16 (usually 15); total gill rakers on first arch including rudiments 19-23 (usually 20-22); *lateral-line scales 68-75 (usually 71-74)*; scales around caudal peduncle 25-30 (usually 27-29); *second and third anal spines relatively short for the genus, about equal in length, not exceeding longest anal soft ray*; generally silver with blackish lateral line. Gulf of California to Colombia; inhabits bays and estuaries; also descends rivers and streams. Attains 55 cm. *Similar to* C. viridis, *which has a similar geographical range, but differs in usually having 9 soft dorsal rays, 10-11 developed (excluding rudiments) rakers on the lower limb of the first gill arch (usually 8-9 in* nigrescens*), and the third dorsal spine is shorter than the fourth spine.*

Centropomus robalito ▲

Centropomus unionensis ▼

LITTLE SNOOK
Centropomus robalito Jordan & Gilbert, 1881

Dorsal rays VIII + I,10 (rarely 9 or 11); anal rays III,6 (rarely 7); pectoral rays 14-16 (usually 15); *total gill rakers on first arch including rudiments 26-31 (usually 27-30)*; lateral-line scales 47-55 (usually 50-54); scales around caudal peduncle 18-22 (usually 19-21); *second anal spine very stout, much longer than third, reaching to or beyond level of caudal base when depressed, and clearly exceeding longest anal soft ray*; overall silvery; a dark bar at base of pectoral fin; anal and pelvic fins yellow. Gulf of California to Peru; inhabits bays and estuaries; also ascends freshwater streams. Reaches at least 35 cm. Closely related to *C. ensiferus* Poey from the western Atlantic.

HUMPBACK SNOOK
Centropomus unionensis Bocourt, 1868

Dorsal rays VIII + I,9 (occasionally 10); anal rays III,6 (rarely 5); *pectoral rays 15-17 (usually 16)*; total gill rakers on first arch, including rudiments, 22-24; *lateral-line scales 46-52 (usually 48-51)*; scales around caudal peduncle 22-25 (usually 23-24); *second anal spine much stouter than third spine and slightly longer, also exceeding longest anal soft ray*; generally silvery with yellowish hue on head; pectoral, pelvic, and anal fins yellowish. El Salvador to Peru; inhabits bays, estuaries, and lower reaches of freshwater streams. Attains at least 36 cm. Similar in appearance to *C. armatus* and *C robalito*, but the lateral line is usually paler, and the second anal spine is shorter and less robust than in these two species.

WHITE SNOOK
Centropomus viridis Lockington, 1877

Dorsal rays VIII +I,8-10 (usually 9); anal rays III,6; pectoral rays 14-16 (usually 15); total gill rakers on first arch including rudiments 19-23 (usually 20-22); *lateral-line scales 67-75 (usually 69-73)*; scales around caudal peduncle 25-30 (usually 26-28); *second and third anal spines relatively short for the genus, about equal in length, not exceeding longest anal soft ray*; generally silver with blackish lateral line. Gulf of California to Peru and the Galapagos; inhabits bays, estuaries, and lower parts of freshwater streams. Attains 58 cm. *Similar to* C. nigrescens, *which has a similar geographical range, but differs in usually having 10 soft dorsal rays, 8-9 developed (excluding rudiments) rakers on the lower limb of the first gill arch (usually 10-11 in* viridis*), and the third dorsal spine is usually longer than the fourth spine.*

GROUPERS (MEROS, CABRILLAS, CHERNAS, SERRANOS)
FAMILY SERRANIDAE

Yellow phase of Leopard Grouper (*Mycteroperca rosacea*), Gulf of California. The other fish is the Greybar Grunt (*Haemulon sexfasciatum*)

This large family of reef fishes is represented circumglobally by 35 genera and nearly 400 species. Although most are bass-like in appearance, there is considerable diversity of form and habits. They range in size from the colossal (to 270 cm and over 400 kg) *Epinephelus lanceolatus* to tiny fishes of the genus *Plectranthias*, which are fully grown at 3 to 4 cm. Although both of these examples are from the Indo-West Pacific region, the contrast is appropriate for our area. Here the largest species is *Epinephelus itajara* (240 cm and 370 kg) and the smallest is *Serranus socorroensis* (8 cm). Because of their great diversity it is difficult to define the Serranidae in terms of external characteristics. Most species have three spines on the gill cover (operculum), and the posterior margin of the preopercle is almost always serrate or has small spines. The maxilla (posterior bone of the upper jaw) is fully exposed on the cheek when the mouth is closed; the mouth is large and there is more than one row of teeth on the jaws. The scales are small and usually ctenoid or secondarily cycloid. Groupers and their relatives comprise one of the major groups of predatory fishes found on reefs. They feed on a wide variety of fishes and invertebrates (frequently crustaceans). Most species live on or near the bottom, but members of the genera *Hemanthias* and *Paranthias* form large midwater aggregations and feed on plankton. Serranids are hermaphroditic, with adult females capable of transformation to the male sex. Individuals of some species have both male and female organs, which can function at the same time. Some groupers form aggregations at certain times of the year to engage in nocturnal spawning. The larger species are considered good eating and are frequently seen in markets throughout the region.

MUTTON HAMLET

Alphestes immaculatus Breder, 1936

Dorsal rays XI,18; anal rays III,9; pectoral rays 17; gill rakers on first arch 21-24; *a single strong, anteriorly directed spine on lower posterior corner of preopercle*; rusty brown with darker brown blotches forming irregular bars on side; also numerous dark brown and pale spots, and larger pale blotches on head and body. Gulf of California to Peru; a secretive nocturnal predator found hiding in rocky crevices or among weed during the day. Reaches 30 cm. Closely related and sometimes confused with *A. afer* from the western Atlantic.

MANYSPOTTED HAMLET

Alphestes multiguttatus (Günther, 1866)

Dorsal rays XI,18-20; anal rays III,9; pectoral rays 17-19; gill rakers on first arch 20-23; posterior margin of preopercle serrate; *a single strong, anteriorly directed spine on lower posterior corner of preopercle; generally brown with dense covering of dark brown spots, coalesced to form stripes on posterior part of body*; pectoral fins yellowish with 5-6 dark brown bars. Gulf of California to Panama and the Galapagos; inhabits tidepools, rocky reefs, and adjacent sand patches to at least 30 m depth. Grows to about 25 cm.

Alphestes multiguttatus Juv.

GREY THREADFIN BASS

Cratinus agassizii Steindachner, 1878
(Plate VII-4, p. 117)

Dorsal rays X,13; anal rays III,7; pectoral rays 19; midlateral scale rows about 80-90; gill rakers on first arch 19-20; *third to fifth (sometimes also sixth and seventh) dorsal spines prolonged and bearing filaments*, at least in adults; generally grayish to brownish, sometimes with 6-7 faint dusky bars on side. Galapagos and Peru; inhabits mangrove areas and shallow bays, occasionally seen on coral reefs. Grows to at least 60 cm.

LEATHER BASS

Dermatolepis dermatolepis Boulenger, 1895

Dorsal rays XI,18-20; anal rays III,9; pectoral rays 19-20; gill rakers on first arch 21-24; *body relatively deep (about 2.0-2.4 in standard length); forehead profile steep; alternating dark gray to blackish and white to light gray bars overlain with numerous white blotches*; fin margins narrowly yellow; juveniles with alternating light and dark bars. Baja California to Ecuador, including the Galapagos; inhabits rocky reefs; small juveniles may shelter among spines of the sea urchin *Centrostephanus coronatus*. Grows to 100 cm.

Dermatolepis dermatolepis ▼

Dermatolepis dermatolepis Juv.

PYGMY SANDPERCH
Diplectrum macropoma (Günther, 1864)
(Plate VII-2, p. 117)

Dorsal rays X,11-13; anal rays III,7; pectoral rays 16-18 (usually 17); pored lateral-line scales 44-51; midlateral scale rows 48-63; predorsal scales 10-18; *gill rakers on first arch, including rudiments 17-24 (usually 18-22)*; pale copper to light brown on upper half, white below; juveniles with pair of dark brown stripes on side, these faint and interrupted by pale pigment in adults; an oval blackish blotch at base of caudal fin; *pectoral and pelvic fins pale orange to reddish orange; upper lobe of caudal fin bright reddish orange.* Baja California to Peru, including the Galapagos; inhabits open mud or sand bottoms in 9-80 m, most frequently in 20-40 m. Maximum size, 17 cm. In addition to the 3 species of *Diplectrum* treated here, there are 6 other members of the genus in the eastern Pacific: *D. conceptione* (Valenciennes), *D. eumelum* Rosenblatt & Johnson, *D. euryplectrum* Jordan & Bollman, *D. labarum* Rosenblatt & Johnson, *D. maximum* Hildebrand, and *D. sciurus* Gilbert. Readers are referred to the work of Bortone (1977) for further information.

PACIFIC SANDPERCH
Diplectrum pacificum Meek & Hildebrand, 1925

Dorsal rays X,12-13; anal rays III,7-8; pectoral rays 16-18 (usually 17); pored lateral-line scales 47-51; midlateral scale rows 58-74; predorsal scales 11-19; *gill rakers on first arch, including rudiments 15-24 (usually 15-22)*; juveniles generally whitish with blackish midlateral stripe connected to intense black blotch on middle of caudal fin base; a second, more diffuse blackish stripe on upper back; *stripes of adults less apparent and replaced to certain extent by diffuse grayish bars.* Baja California to Peru; inhabits mud-sand bottoms, sometimes close to weed-covered rocky reefs in 0.1-90 m, but most frequently in 15-30 m. The most common shallow-water member of the genus. Grows to about 26 cm.

Diplectrum pacificum Juv.

BARSNOUT SANDPERCH
Diplectrum rostrum Bortone, 1974
(Plate VII-1, p. 117)

Dorsal rays X,12-13; anal rays III,7-8; pectoral rays 16-18 (usually 17); pored lateral-line scales 48-51; midlateral scale rows 60-70; predorsal scales 12-16; *gill rakers on first arch, including rudiments 22-25*; light tan to whitish, with 5-7 faint blackish bars on side; also an indistinct blackish midlateral stripe; a large blackish blotch at base of caudal fin; *each side of snout with brown bar between eye and upper lip.* Gulf of California to Colombia and the Galapagos; inhabits sand bottoms in 12-80 m. Reaches at least 19 cm.

GULF CONEY
Epinephelus acanthistius (Gilbert, 1892)
(Plate VII-11, p. 117)

Dorsal rays IX,17; anal rays III,9; gill rakers on lower limb of first arch 14-17; *membranes of dorsal spines deeply notched; larger specimens with elevated third and fourth dorsal spines*; maxilla scaleless; *generally rosy red* (sometimes with brownish tinge); fins grading to dark brownish; a thin, *dark brown band above upper jaw.* Gulf of California to Panama; inhabits rocky reefs and nearby sand patches, usually below 40 m and ranging to at least 90 m. Maximum size to at least 100 cm. An excellent food fish.

Diplectrum pacificum ▼ *Epinephelus acanthistius* ▲

SPOTTED CABRILLA
Epinephelus analogus Gill, 1864

Dorsal rays X,16-18; anal rays III,8; pectoral rays 19-20; gill rakers on first arch 26-28; posterior margin of preopercle serrate, the lower limb smooth; tan to grayish white, with *red-brown spots on head, body, and fins; 4 relatively broad, dark bars may also be evident on upper half of side.* Southern California to Peru; inhabits rocky reefs to at least 50 m depth; also occurs in shallow estuaries. A good-eating fish reported to reach 80 cm and 15 kg. Very similar to *E. adscensionis* of the western Atlantic.

JEWFISH
Epinephelus itajara (Lichtenstein, 1822)
(Plate VII-3, p. 117)

Dorsal rays XI,15-16; anal rays III,8; pectoral rays 19; gill rakers on first arch 22-24; fish under 100 cm greenish to tawny brown with oblique, irregular darker brown bars; larger fish *gray or greenish with pale blotches and smaller dark brown or blackish spots scattered over upper parts of head, body, and on pectoral fin.* Western Atlantic and eastern Pacific from the Gulf of California to Peru; inhabits rocky reefs. *The largest bony reef fish in the region,* growing to 240 cm and 320 kg.

Epinephelus labriformis ▲

FLAG CABRILLA
Epinephelus labriformis (Jenyns, 1843)

Dorsal rays XI,16-18; anal rays III,8; pectoral rays 18-19; gill rakers on first arch 23-26; *olive-green to reddish brown with scattered irregular white spots and blotches; a distinct black saddle on upper part of caudal peduncle.* Gulf of California to Peru including the Galapagos; inhabits rocky reefs from tidepool depths to at least 30 m. To about 50 cm.

Epinephelus panamensis ▼

STARSTUDDED GROUPER
Epinephelus niphobles Gilbert & Starks, 1897
(Plate VII-7, p. 117)

Dorsal rays XI,13-15; anal rays III,9; pectoral rays 17-21; gill rakers on first arch 23-26; *head and body chocolate brown to blackish with scattered white spots (faint or absent in larger adults); an intense black saddle on upper half of caudal peduncle; caudal fin abruptly pale.* Outer Baja California and Gulf of California to Panama; inhabits rocky reefs and adjacent sandy bottoms to at least 120 m. Reaches 75 cm. Often confused with and misidentified as *E. niveatus* from the western Atlantic.

PANAMA GRAYSBY
Epinephelus panamensis (Steindachner, 1876)

Dorsal rays IX,14; anal rays III,8; pectoral rays 17-18; *gill rakers on first arch 16-19*; caudal fin rounded; *body with alternating dark brown to blackish and pale gray bars*; head with oblique pale bands and network of blue and orange spots; *adults with large dark patch behind eye.* Gulf of California to Colombia, including the Galapagos and other offshore islands; perhaps the most abundant grouper in the region, inhabits rocky shores from shallows to at least 75 m. Grows to 30 cm.

ROSE THREADFIN BASS

Hemanthias peruanus (Steindachner, 1874)
(Plate VII-5, p. 117)

Dorsal rays IX or X,13-15; anal rays III,7-9; *pectoral rays 17-18 (rarely 16)*; pored lateral-line scales 52-55 (occasionally 56-59); total gill rakers on first arch, including rudiments, 31-33 (rarely 34); *third dorsal spine long and filamentous*, at least in adults; *caudal fin deeply forked; reddish pink with red spotting on back and upper part of head; sides pink with yellow blotches and spots*; fins pink to orange-red. Baja California to Peru; inhabits rocky reefs in 10-117 m. Grows to 44 cm. A second member of the genus, *H. signifer* (Garman) ranges from central California to Peru, occurring in 23-306 m. It differs from *H. peruanus* by having a sharp projecting spine on the lower edge of the urohyal bone at the front of the isthmus between lower edges of the gill covers, usually 19 pectoral rays, and the outermost rays of the caudal lobes are longest (middle rays of caudal lobes are longest in *H. peruanus*).

RAINBOW BASSLET

Liopropoma fasciatum Bussing, 1980
(Plate VII-9, p. 117)

Dorsal rays VIII,12; anal rays III,8; pectoral rays 15; total gill rakers on first arch, including rudiments, 17; pored lateral-line scales 46-47; greatest body depth 4.1 in standard length; *head and belly pinkish red; a dark brown, midlateral stripe with yellow stripe above and below; also a red to brownish stripe below base of dorsal fin with narrow yellow stripe above*. Southern California to Panama; inhabits rocky reefs in deep water (about 40-250 m). To 15 cm.

GULF GROUPER

Mycteroperca jordani (Jenkins & Evermann, 1889)
(Plate VII-8, p. 117)

Dorsal rays XI,15-17; anal rays III,10-11; *pectoral rays 17* (rarely 16); *total gill rakers on first arch 21-26; preopercle rounded without a distinct notch or lobe; caudal fin margin truncate or slightly concave, without projecting fin rays*; scales in longitudinal series above lateral line 90-120; *adults overall grayish brown; juveniles gray or gray-brown, with large dusky brown blotches* and diffuse dark streaks radiating from eye; median fins with narrow whitish margin in juveniles and adults. Southern California to Gulf of California; inhabits rocky reefs, usually below 30 m in summer, but in shallow depths during the remainder of the year. Grows to 150 cm and 90 kg.

BACALAO GROUPER

Mycteroperca olfax (Jenyns, 1840)

Dorsal rays XI,16-17; anal rays III,11; *pectoral rays 16-17; total gill rakers on first arch 24-29; first 3 dorsal spines elongated, forming a definite lobe, with the second spine longer than third*; scales in longitudinal series above lateral line 100-120; *generally pale gray with numerous, close-set brown spots; 10-12 narrow brown bars or partial brown bars sometimes apparent on upper side; occasional individuals are entirely bright yellow* (similar to yellow phase of *M. rosacea* from the Gulf of California). Galapagos Archipelago; inhabits rocky reefs in 5 to at least 50 m. Reaches at least 100 cm.

SAWTAIL GROUPER

Mycteroperca prionura Rosenblatt & Zahuranec, 1967
(Plate VII-6, p. 117)

Dorsal rays XI (rarely X),16-18; anal rays III,10-12 (usually 11); *pectoral rays 15-16 (usually 16); total gill rakers on first arch 34-38*; scales in longitudinal series above lateral line 90-110; *caudal fin with jagged rear margin; juveniles pale gray or whitish, with combination of small brown spots and larger oval-shaped brown blotches; adults similar, but small spots are much more numerous and larger blotches are generally faint*. Gulf of California; inhabits rocky reefs to at least 50 m. Reaches 100 cm.

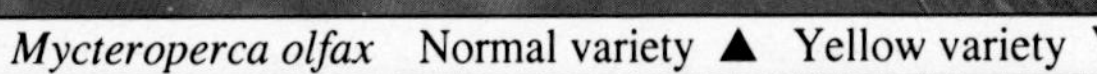

Mycteroperca olfax Normal variety ▲ Yellow variety ▼

Mycteroperca prionura ▼

LEOPARD GROUPER

Mycteroperca rosacea (Streets, 1877)

Dorsal rays XI,16-18 (usually 17); anal rays III,10-11; *pectoral rays 15-17 (usually 16); total gill rakers on first arch 38-43 (usually 39-40); caudal fin without jagged rear margin*; scales in longitudinal series above lateral line 110-130; small juveniles light tan to nearly whitish, with relatively widely-spaced, small brown spots; spotting becomes denser in adults, which have a greenish to gray-brown ground color; on the upper half there is often a series of irregular dark saddles with narrow whitish streaks between them; occasional (about 1 percent of population) individuals are overall golden yellow, sometimes with a few small patches of dark pigment. Gulf of California and outer coast of Baja California, north to Magdalena Bay; inhabits rocky reefs to at least 50 m. To 100 cm and 12.2 kg.

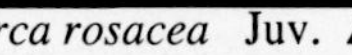

Mycteroperca rosacea Juv. ▲

BROOMTAIL GROUPER

Mycteroperca xenarcha Jordan, 1888

Dorsal rays XI,16-17; anal rays II,10-11; *pectoral rays 16-18 (usually 17); total gill rakers on first arch 29-33*; scales in longitudinal series above lateral line 110-120; *caudal fin with jagged rear margin; light brown with elongate dark brown blotches (often with light brown centers)*. Northern California to Peru; inhabits rocky reefs and mangrove estuaries. To about 120 cm and 45 kg.

PACIFIC CREOLEFISH
Paranthias colonus Hildebrand, 1946

Dorsal rays IX,19-21; anal rays III,9-11; pectoral rays 19-23; *gill rakers on first arch 37-44; caudal fin strongly lunate*; young often bright yellow with 5 small, dark spots on back; *adults greenish brown dorsally, reddish below, with 5 white or blue-white spots on back; fins reddish.* Gulf of California to Peru, including the Galapagos and other offshore islands; forms large midwater schools above rocky reefs. To 35 cm. *P. furcifer* (Valenciennes) from the western Atlantic is closely related and similar in appearance.

Juvs. ▲ Yellow juv. ▼

WHITE-SPOTTED SANDBASS
Paralabrax albomaculatus (Jenyns, 1840)

Dorsal rays X,14; anal rays III,7; pectoral rays 16-17; midlateral scale rows about 96; body depth 2.6 in standard length; *third dorsal spine greatly elevated, about 3 times longer than second*; light brown on upper two-thirds, white below; *about 6-10 large white spots on upper half of side,* and smaller, irregular, dark brown blotches in surrounding areas; caudal fin white basally with dark brown bar across middle portion and broad yellowish area posteriorly. Galapagos Islands; inhabits rocky reefs and nearby sand patches. Grows to at least 38 cm.

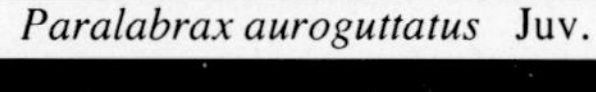

Paralabrax auroguttatus Juv. ▼

GOLD-SPOTTED SANDBASS
Paralabrax auroguttatus Walford, 1936
(Plate VII-12, p. 117)

Dorsal rays X,13-14; anal rays III,6-8; pectoral rays 16-17; midlateral scale rows about 90; body depth 3.0-3.2 in standard length; *third dorsal spine greatly elevated, about 3 times longer than second;* overall brownish, becoming white ventrally; *head and sides covered with numerous golden orange to brownish spots*; lining of gill cavity orange; juveniles white with 3 dark brown stripes and a lighter brown stripe or longitudinal row of spots between each dark stripe; the stripes break up into spots with increased growth. Southern California to the Gulf of California; inhabits rocky bottoms in 20-80 m. To at least 35 cm.

PARROT SANDBASS
Paralabrax loro Walford, 1936
(Plate VII-10, p. 117)

Dorsal rays X,13-14; anal rays III,6-8; pectoral rays 16-17; midlateral scale rows about 70; *third dorsal spine greatly elevated, about 3 times longer than second*; body depth 2.6-2.8 in standard length; *head with orange spots; body and fins (except pectorals) covered with numerous brown spots; a row of 5-6 large brown blotches along lower side*. Gulf of California to Panama; inhabits rocky reefs and adjacent sand patches. Grows to at least 30 cm.

SPOTTED SANDBASS
Paralabrax maculatofasciatus
(Steindachner, 1868)

Dorsal rays X,13-14; anal rays III,6-8; pectoral rays 16-17; midlateral scale rows about 90; body depth 3.0-3.3 in standard length; *third dorsal spine greatly elevated, about 3 times longer than second; generally white with numerous black, brown, and orange spots that more or less coalesce to form a dark midlateral stripe and 6-7 dark bars on side of body*. Central California to the Gulf of California; inhabits low-profile, often weed-covered reefs adjacent to sandy bottoms in 1-60 m. Attains 56 cm.

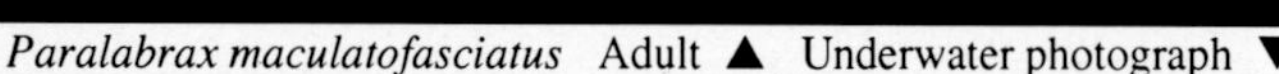

Paralabrax maculatofasciatus Adult ▲ Underwater photograph ▼

BANDED SERRANO
Serranus psittacinus Valenciennes, 1855

Dorsal rays X (rarely IX),11-13; anal rays III,7; pectoral rays 17; lateral-line scales 44-50; body depth 3.2-3.4 in standard length; brown on upper half, white below; *about 8 dark brown, vertically elongate blotches on upper side, these often coalesced to form a broad stripe, continued along lower half of body as row of more or less isolated, large brown spots*; a large black spot or saddle preceded by white bar or saddle on caudal peduncle. Gulf of California to Peru, including the Galapagos; inhabits small patch reefs or rocky outcrops in sand-rubble areas in 2-60 m. To 11 cm. Usually referred to as *S. fasciatus* by previous authors.

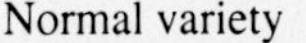
Serranus psittacinus Normal variety ▲ Dusky variety ▼

SOCORRO SERRANO
Serranus socorroensis Allen & Robertson, 1992

Dorsal rays X,12-13; anal rays III,7-8; pectoral rays 16; lateral-line scales 49-51; body depth 3.6-3.8 in standard length; *a small, isolated patch of scales on interorbital region; mainly red on upper half of head and body, white below; 4-5 pearl-white patches on back, and 6-7 reddish, bar-like extensions on lower sides*; 4 brownish spots on dorsal fin and adajcent part of back, and 2 smaller brown spots on upper surface of caudal peduncle. Known thus far only from Socorro Island in the Revillagigedo Group; inhabits sand-rubble areas. Maximum size, 8 cm.

Serranus socorroensis ▼

COCOS SERRANO
Serranus sp.

Dorsal rays X,12; anal rays III,7; pectoral rays 16; lateral-line scales 47-50; body depth 3.6-3.7 in standard length; *head bluish brown with pair of red-orange stripes on snout and behind eye; about 8 bars on side of body, these red on upper half and bright yellow below*; bars may coalesce to form more or less solid red area, in which case the yellow portions have the appearance of a row of vertically elongate, rectangular blotches; a whitish bar across caudal peduncle; base of caudal fin red. Known only from Isla del Coco; inhabits sand-rubble fringe of reefs. Grows to at least 8.5 cm.

Serranus sp. ▼

PLATE VII

1 **BARSNOUT SANDPERCH** (*Diplectrum rostrum*)

2 **PYGMY SANDPERCH** (*Diplectrum macropoma*)

3 **JEWFISH** (*Epinephelus itajara*)

4 **GREY THREADFIN BASS** (*Cratinus agassizii*)

5 **ROSE THREADFIN BASS** (*Hemanthias peruanus*)

6 **SAWTAIL GROUPER** (*Mycteroperca prionura*)

7 **STARSTUDDED GROUPER** (*Epinephelus niphobles*)

8 **GULF GROUPER** (*Mycteroperca jordani*)

9 **RAINBOW BASSLET** (*Liopropoma fasciatum*)

10 **PARROT SANDBASS** (*Paralabrax loro*)

11 **GULF CONEY** (*Epinephelus acanthistius*)

12 **GOLD-SPOTTED SANDBASS** (*Paralabrax auroguttatus*)

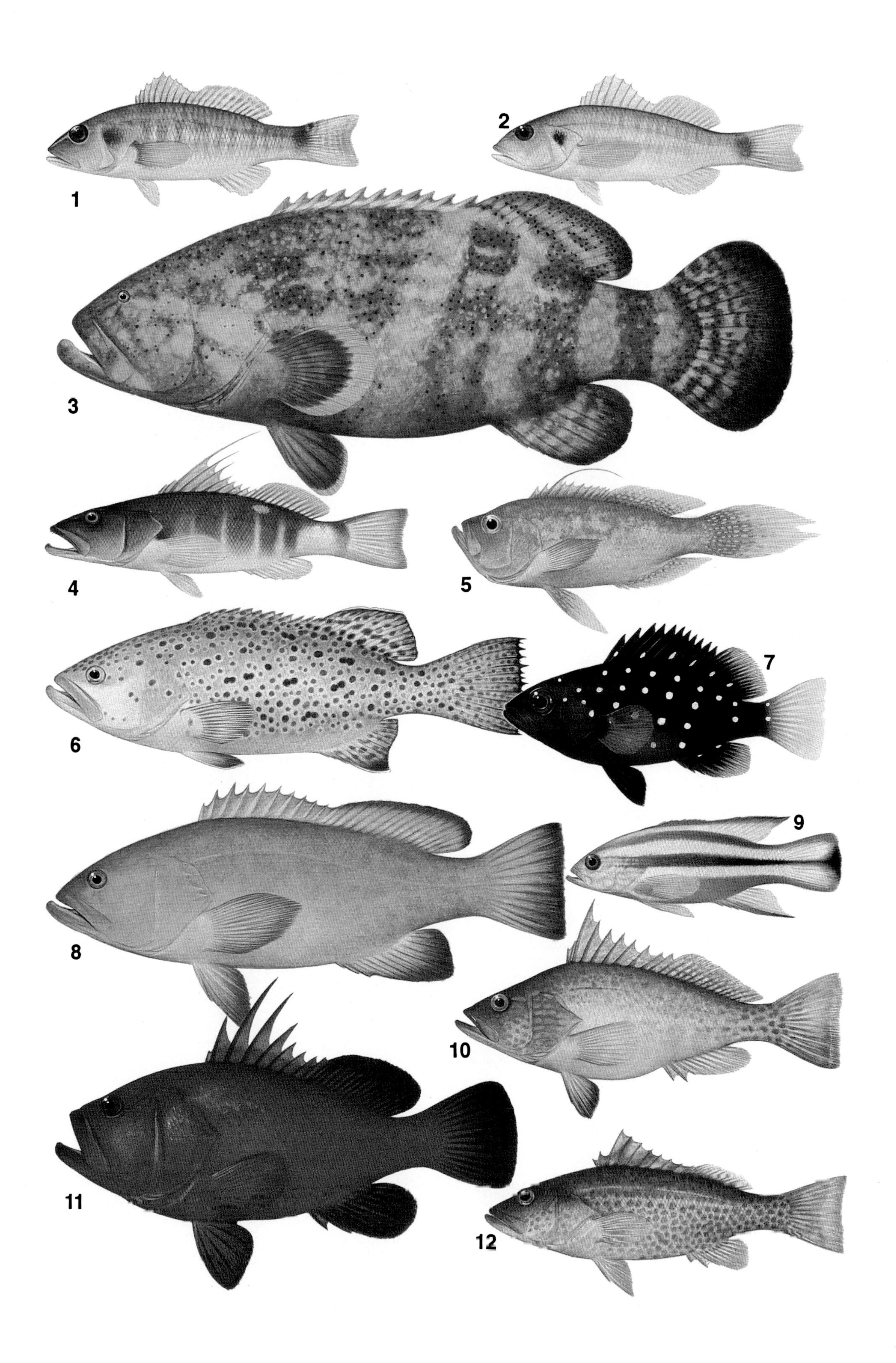
1
2
3
4
5
6
7
8
9
10
11
12

SOAPFISHES (JABONEROS, JABONCILLOS)
FAMILY GRAMMISTIDAE

Soapfishes are small (usually less than 20 cm), grouper-like fishes that inhabit reef caves and crevices. The young are sometimes common in estuaries and mangrove areas. The common name is derived from the soapy slime that covers the external body surface. This bitter-tasting substance, known as grammistin, apparently serves as an effective deterrent against potential predators. The substance is present in members of the subfamily Grammistinae, which includes *Rypticus*, but not *Pseudogramma* (subfamily Pseudogrammatinae). Diagnostic features include a dorsal fin with II-IX spines and 12-27 soft rays; an anal fin with 0-III spines and 8-17 soft rays; lower jaw projecting and frequently a chin appendage present; opercle with 2-3 distinct spines; innermost pelvic ray attached to the belly by a membrane; and small scales (about 75-140 in lateral series) that are either mainly ctenoid and exposed or cycloid and often embedded. Their diet consists mainly of small fishes and benthic crustaceans such as crabs and shrimps. Worldwide there are 25 species in 10 genera; four are found in our area. The genus *Rypticus* was revised by McCarthy (1979).

PACIFIC REEF BASS
Pseudogramma thaumasium (Gilbert, 1900)

Dorsal rays VII (rarely VIII),20-24; anal rays III,17-20; midlateral scales 48-55; lateral line interrupted below posterior part of soft dorsal fin, reappearing lower down on side; *preopercular margin with a single broad spine; opercle without visible spine*; generally brown, with darker scale centers often giving the appearance of fine, dark, longitudinal lines on sides; head pale brown or tannish, with 2-3 brown bands behind and below eye; *a prominent black spot on opercle*. Gulf of California to Panama; a secretive dweller of rocky crevices. Attains 10 cm.

MOTTLED SOAPFISH
Rypticus bicolor Valenciennes, 1846

Dorsal rays III (rarely II),23-26 (usually 24); anal rays 16-18 (usually 17); pectoral rays 14-17 (usually 16); total gill rakers on first arch 7-9 (usually 8); preopercle and opercle each with 2-3 spines; *a prominently projecting lower jaw with fleshy protuberance; pores along ventral surface of lower jaw and margin of preopercle numerous, small, and occurring in patches* (in specimens in excess of 65 mm SL); generally brown; paler on head and anteriodorsal part of body; body with numerous tan flecks or spots; small (under about 3 cm) juveniles with 3 narrow stripes behind eye extending posteriorly on side of body; 4-5 cm specimens with ocelli-type spots on side. Baja California to Peru, including the Galapagos and other offshore islands; inhabits rocky reefs, sheltering in caves and crevices, but also encountered in the open. To 12 cm. Closely related to *R. saponaceus* (Bloch & Schneider) from the western Atlantic. In addition, a similar species, *R. courtenayi* McCarthy (1979), is known from the Revillagigedo Islands.

Juvs. of *R. bicolor* (top) and *R. nigripinnis* ▲

Rypticus bicolor ▲

TWICE-SPOTTED SOAPFISH
Rypticus nigripinnis Gill, 1861

Dorsal rays II (rarely III),24-28 (usually 26); anal rays 14-18 (usually 16); pectoral rays 15-17 (usually 15); total gill rakers on first arch 7-10 (usually 8); preopercle and opercle each with 2-3 spines; *a slightly projecting lower jaw without fleshy protuberance; juvenile and adult with 4 large, oval pores along preopercular margin and 4 pores on each side of ventral surface of lower jaw* (except in specimens less than 65 mm SL, posteriormost pore may be divided into a few small pores); generally brown with numerous yellow-tan spots and blotches; specimens under 2-3 cm are brown dorsally and yellowish white below, with a pair of pale stripes behind eye, the lowermost extending to caudal fin base; the dorsal and anal fins of juveniles have prominent yellow bands basally and the dark caudal fin is edged with yellow; larger juveniles and subadults usually have at least some of the pale blotches and spots with dark centers. Gulf of California to Peru. To 19 cm.

Rypticus nigripinnis Juv., 50 mm SL

R. nigripinnis Adult ▼ Juv., 17 mm SL ▲

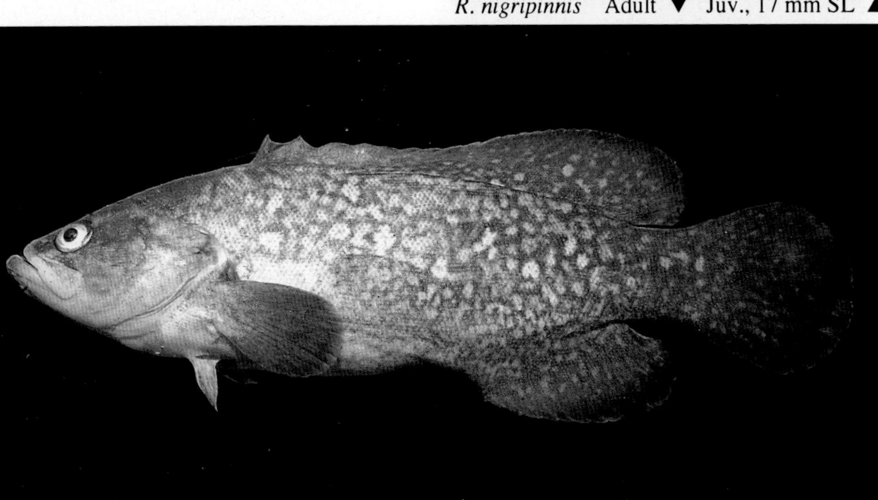

FLAGTAILS (SARDINAS DE MAR FALSAS, COLAS DE BANDERA)
FAMILY KUHLIIDAE

These are small perch-like fishes that are usually silvery in color, and occur in either marine shallows or fresh water. The family contains two genera, *Parakuhlia* of West Africa, with a single species, and *Kuhlia* of the Indo-Pacific, with about six species. They are characterized by a single dorsal fin that is deeply notched; a well-developed scaly sheath at the base of the dorsal and anal fins; III anal spines; fine teeth in the jaws, arranged in bands; fine teeth also present on roof of mouth; caudal fin moderately forked and usually marked with black stripes or spots (thus the name flagtail). They are nocturnal predators that feed mainly on planktonic crustaceans. Juveniles and young adults frequently inhabit tidepools or form schools close to shore. There is a single species present in the eastern Pacific.

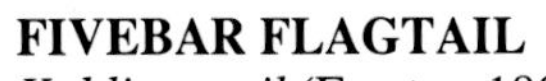

FIVEBAR FLAGTAIL
Kuhlia mugil (Forster, 1801)

Dorsal rays X,9-11; anal rays III,9-11; pectoral rays 14; lateral-line scales 48-56; preorbital and preopercular margins serrate; operculum with 2 spines; a single dorsal fin with notch between spinous and soft portions; dorsal and anal fins with well-developed scaly sheath; overall silvery; *caudal fin with 5 black bands*. Wide-ranging in the tropical Indo-Pacific, from East Africa to the eastern Pacific; ranges from Cape San Lucas to Colombia and the Galapagos; sometimes seen in large aggregations in areas affected by surge; the young inhabit tide pools. Reaches 20 cm.

BIGEYES (CATALANUS, CATALUFAS, SEMAFOROS)

FAMILY PRIACANTHIDAE

Bigeyes are somewhat similar in appearance to the squirrelfishes and soldierfishes (Holocentridae), but are distinguished by smaller scales, a larger, upturned mouth, and lack of spines on the head. They also exhibit similar behavior, being essentially nocturnal and spending the daylight hours in caves. At night they emerge to feed on larger zooplankton such as crab and shrimp larvae, fish larvae, small squids and octopuses, and larval polychaete worms. The family occurs in all warm seas and contains four genera with 18 species, all but three of which inhabit the Indo-West Pacific region. Many of the species occur in shallow reef habitats, but others range to depths as great as 500 m. The family was reviewed by Starnes (1988).

Photograph: John Randall

LONG-FINNED BIGEYE
Cookeolus japonicus (Cuvier, 1829)

Dorsal rays X,12-14 (usually 13); anal rays III,12-14 (usually 13); pectoral rays 17-19 (usually 18); *lateral-line scales 60-73; scale rows between lateral line and dorsal fin origin 16-20*; gill rakers on first arch 23-27; *pelvic fins elongate, greatly exceeding head length*; crimson to scarlet red, including fins and iris of eye; pelvic fin membranes dusky to blackish. Circumglobal in tropical seas; in the eastern Pacific it ranges from Baja California to Peru; generally found in deep water (60-400 m). Attains 55 cm.

GLASSEYE
Heteropriacanthus cruentatus (Lacepède, 1801)

Dorsal rays X,12-13 (usually 13); anal rays III,13-14 (usually 14); pectoral rays 18-19 (usually 18); *lateral-line scales 78-96; scale rows between lateral line and dorsal fin origin 9-12*; gill rakers on first arch 21-25; pelvic fins relatively short, usually less than head length; pink to red, including fins and iris of eye; *caudal and soft parts of dorsal and anal fins with elliptical dark specks*; can also assume pattern that is overall silvery with red and pink blotches. Circumglobal in tropical seas; in the eastern Pacific it ranges from Baja California to the Galapagos, but is mainly restricted to offshore islands; inhabits shallow reefs, usually in less than 20 m. Reaches 30 cm.

HAWAIIAN BIGEYE
Priacanthus alalaua Jordan & Evermann, 1904

Dorsal rays X,12-14 (usually 13); anal rays III,13-14 (usually 14); pectoral rays 17-19 (usually 18); *lateral-line scales 65-79; scale rows between lateral line and dorsal fin origin 10-13*; gill rakers on first arch 19-23; pelvic fins relatively short, usually less than head length; overall crimson, including fins and iris of eye; membranes of pelvic fin black; no dark spotting on fins. Hawaiian Islands and tropical eastern Pacific; in the latter region it is known from Baja California and the Revillagigedos Islands, but is probably more widespread; inhabits rocky bottoms in 10-50 m, but occurs to at least 275 m in the Hawaiian Islands. To 30 cm.

Photograph: L. Stockland

POPEYE CATALUFA
Pristigenys serrula (Gilbert, 1891)

Dorsal rays X,10-11 (usually 10); anal rays III,10; pectoral rays 17-18; *lateral-line scales 40-45*; scale rows between lateral line and dorsal fin origin 11-13; gill rakers on first arch 23-25; pelvic fins relatively short, usually about equal or less than head length; dark red, including fins and iris of eye; outer part of pelvic fins blackish; young specimens with several faint pale bars on sides. Central California to Chile, including Galapagos and other offshore islands; inhabits rocky bottoms from 3-4 m to at least 100 m depth. Grows to 30 cm.

CARDINALFISHES (CARDENALES)

FAMILY APOGONIDAE

These are small (usually under 10 cm), laterally compressed, often brightly colored fishes. They have two separate dorsal fins (the first with six to eight spines), two anal spines, large eyes, a moderately large oblique mouth, and preopercle with a ridge preceding the margin. Apogonidae is one of the largest coral-reef fish families. They are represented in all tropical and warm temperate seas, but the majority are distributed in the Indo-Pacific region. Worldwide there are an estimated 250 species in 21 genera; only six species, all in the genus *Apogon*, are known from the tropical eastern Pacific. The common name is derived from the red coloration displayed by many of the species. Cardinalfishes occur in a diversity of habitats, but individual species are restricted to relatively narrow ecological zones. Most prefer caves and crevices of rock or coral reefs. At dusk they emerge from these retreats for nocturnal feeding. Most species are seen as solitary individuals, in pairs, or in small aggregations. However, some eastern Pacific species form dense aggregations near the reef's surface. Cardinalfishes are one of few marine fish families in which oral brooding is found. Courtship is often accompanied by flicking movements of the dorsal and pelvic fins. The prospective mates may also engage in chasing bouts and non-injurious nipping. During spawning the female releases a large, gelatinous egg mass containing up to several hundred ova. The mass is summarily fertilized by the male, who then engulfs it with his mouth. Egg brooding males are easily distinguished by the swollen throat region, and the eggs can clearly be seen when the mouth is partially opened. Incubation lasts several days, during which time the male is unable to feed. Its main activity consists of periodically juggling the position of the egg mass. The larvae, measuring 2-4 mm, are planktonic for up to several weeks before settlement and subsequent transformation to the juvenile stage. Diet is variable according to species, but most consume some form of zooplankton (often copepods), and small benthic invertebrates such as crabs and shrimps are also eaten.

PLAIN CARDINALFISH
Apogon atricaudus Jordan & McGregor, 1898

Dorsal rays VI + I,9; anal rays II,8; pectoral rays 11-12; lateral-line scales 24-25; gill rakers on first arch, including rudiments, 5 + 13; caudal fin truncate or only slightly emarginate; *yellow-orange on head and belly; remainder of body lavender or bluish red; middle portion of first dorsal fin with blackish streak.* Revillagigedos Islands and southern tip of Baja California; inhabits rocky reefs. To 8 cm.

BLACKTIP CARDINALFISH
Apogon atradorsatus Heller & Snodgrass, 1903

Dorsal rays VI + I,9; anal rays II,8; pectoral rays 12; lateral-line scales 24-26; gill rakers on first arch, including rudiments, 5 + 13-14; caudal fin moderately emarginate; generally dusky brownish red, except yellow-orange on belly region and lower half of head; *outer half of second dorsal fin with promiment black blotch; caudal lobes with black tips.* Galapagos and Malpelo Island; inhabits rocky reefs. Grows to at least 7 cm.

TAILSPOT CARDINALFISH
Apogon dovii Günther, 1861

Dorsal rays VI + I,9; anal rays II,8; pectoral rays 11-12; lateral-line scales 23-27; gill rakers on first arch, including rudiments, 5 + 13-14; caudal fin moderately emarginate; generally red with *large black spot at base of caudal fin; a broad, dusky stripe on snout and continued behind eye to edge of gill cover*; also a narrow, dusky stripe sometimes present on anterior half of lateral line; a pair of silvery white stripes on eye. Gulf of California to Peru; inhabits rock and coral reefs. Maximum size to about 10 cm.

GUADALUPE CARDINALFISH
Apogon guadalupensis (Osburn & Nichols, 1916)

Dorsal rays VI + I,9; anal rays II,8; pectoral rays 11-12; lateral-line scales 25; gill rakers on first arch, including rudiments, 5 + 13 = 18; caudal fin truncate or slightly emarginate; head, breast, and belly yellow; remainder of body, including median fins pale lavender; *very similar to* A. atricaudus, *but lacks blackish streak in middle part of first dorsal fin.* Southern California to tip of Baja California; also Islas Revillagigedo; inhabits rocky reefs. Reaches 7 cm.

PINK CARDINALFISH
Apogon pacifici Herre, 1935

Dorsal rays VI + I,9; anal rays II,8; pectoral rays 11-12; lateral-line scales 24-26; gill rakers on first arch including rudiments 4-5 + 11-12; caudal fin moderately emarginate; generally red or pinkish red, with yellowish wash on head, breast, and belly; an abbreviated, relatively *narrow black bar on back below second dorsal fin; black stripe on snout; pair of golden stripes on eye*. Gulf of California to Peru; inhabits rock and coral reefs. Grows to 10 cm. *A. parri* Breder is a synonym which is frequently used for this species.

BARSPOT CARDINALFISH
Apogon retrosella (Gill, 1863)

Dorsal rays VI + I,9-10; anal rays II,7; pectoral rays 12; lateral-line scales 24-27; gill rakers on first arch including rudiments 4-5 + 11-13; caudal fin moderately emarginate; generally red or reddish pink; *a dark brown or blackish bar below second dorsal fin, and blackish spot at middle of caudal fin base*; dark stripe on snout, continued behind eye. Gulf of California to southern Mexico; inhabits rocky reefs. Reaches length of 10 cm.

TILEFISHES (CABEZONES, MATAJUELOS, PECES BLANQUILLO)

FAMILY MALACANTHIDAE

Tilefishes generally occur in sandy areas, frequently near reefs. They are characterized by a long-based dorsal fin with 22-84 elements (spines plus segmented rays), and there is often a single, sharp opercular spine. The teeth are mainly small and villiform, and teeth are lacking on the vomer and palatines. The postlarval stages have elongate, serrated spines on the head, and because of their relatively large size (to 8-10 cm), were once thought to be separate species. Some authors recognize two closely related families, Malacanthidae and Branchiostegidae, but we follow Nelson (1984) in uniting them. The subfamily Malacanthinae, containing *Malacanthus* and *Hoplolatilus* (an Indo-West Pacific genus not found in our area) is characterized by a very slender body (depth usually about 20 percent of standard length), a dorsal fin with 22-64 total fin-ray elements, and an anal fin with 14-56 total elements. They generally occur in depths less than 50 m and live either in burrows or mounds which they construct. Members of the subfamily Latilinae, including *Caulolatilus*, have a ridge of skin in front of the dorsal fin, and have a deeper body (depth usually about 27 percent of standard length), with a relatively steep snout profile. They generally have fewer dorsal and anal fin-ray elements (22-36 and 14-28 respectively). Many of the species dwell in depths below 50 m. Tilefishes feed on a wide variety of small fish and invertebrate items. *Caulolatilus* consumes eels, anchovies, lanternfishes, shrimps, crabs, polychaetes, brittle stars, urchins, molluscs, ascidians, and bryozoans. The family contains five genera and about 35 species; it is distributed in tropical and temperate seas. The family was revised by Dooley (1978).

PACIFIC GOLDEN-EYED TILEFISH
Caulolatilus affinis Gill, 1865

Dorsal rays VII-IX (usually VIII),22-25; anal rays I or II,21-24; pectoral rays 19; *pored lateral-line scales 80-91; body depth as percentage of standard length 23-32 (mean 29); snout profile relatively steep; caudal fin truncate*; olive-green to bluish gray, with silvery sides; a broad yellow stripe in front of eye; a large dark area above pectoral fin axil. Gulf of California to Peru; inhabits rock and sand bottoms in 30-91 m. Grows to 50 cm.

OCEAN WHITEFISH
Caulolatilus princeps (Jenyns, 1842)

Dorsal rays VII-X (*usually IX*),24-27; anal rays II,22-26; pectoral rays 19; *pored lateral-line scales 99-115; body depth as percentage of standard length 23-28 (mean 25); snout profile generally less abrupt than in* C. affinis*; caudal fin deeply lunate in adults*; generally light bluish gray; caudal fin yellowish. British Colombia to Peru, including the Galapagos; inhabits rock and sand bottoms in 8-150 m. Reaches 32 cm. A third eastern Pacific species of *Caulolatilus*, *C. hubbsi* Dooley, is known from the Galapagos. It is similar to *C. princeps*, but has a slightly deeper body, thick fleshy lips, and a truncate rather than emarginate caudal fin.

FLAGTAIL BLANQUILLO
Malacanthus brevirostris Guichenot, 1848

Dorsal rays I-IV,52-60; anal rays I,46-53; pectoral rays 15-17; lateral-line scales 146-181; *body long and slender*; caudal fin truncate; light gray, yellowish above eye; *2 black stripes on caudal fin*. Widespread in the tropical Indo-Pacific from East Africa and Red Sea to the Americas (Costa Rica to Colombia and Galapagos); inhabit open sand-rubble bottoms; usually seen in pairs; live in a burrow of their own construction, often under a rock. Grows to 30 cm.

SUCKERFISHES OR REMORAS (RÉMORAS, PEGADORES)
FAMILY ECHENEIDAE

These peculiar fishes are easily distinguished by the sucking-disc on top of the head, which represents a modification of the spinous dorsal fin. The body is slender and somewhat rounded in cross-section; the second dorsal and anal fins are similar in shape and opposite one another. These seagoing hitchhikers use their suction apparatus to attach themselves to large marine animals including sharks, rays, bony fishes, turtles, whales, and dolphins. Occasionally they latch onto boats or even unsuspecting divers. They feed on scraps that result from the feeding activities of their hosts; in addition, they sometimes eat parasitic crustaceans that have attached to the hosts. Some species are restricted to specific hosts. The family contains four genera and eight species, most of which occur worldwide. We present accounts of two representative species. The group was comprehensively covered by Robins et al. (1986).

SLENDER SUCKERFISH
Echeneis naucrates Linnaeus, 1758
(Plate VIII-1, p. 131)

Dorsal rays 34-42; anal rays 32-38; pectoral rays 21-24; distinctive flat head with sucking disc of 21-27 transverse laminae; caudal fin slightly rounded; lower jaw projecting; gray with *a white-edged black stripe on side, from tip of lower jaw through eye to caudal fin*, this stripe broadest anteriorly on body. Circumtropical; often seen on or near reefs. Attaches to a wide variety of hosts, but frequently encountered free-living. Reaches 100 cm.

REMORA
Remora remora (Linnaeus, 1758)
(Plate VIII-2, p. 131)

Dorsal rays 22-26; anal rays 22-24; pectoral rays 23-27; gill rakers on lower limb of first arch (including rudiments) 12-14; distinctive flat head with sucking disc of 15-19 transverse laminae; caudal fin slightly emarginate or truncate; lower jaw projecting; *uniformly dark grayish brown*. Circumtropical; commonly attached to sharks, sometimes in gill chamber. Grows to 62 cm.

JACKS (JURELES, PAMPANOS, COJINUAS, COCINEROS)
FAMILY CARANGIDAE

Jacks are well represented in all tropical and subtropical seas. The family contains about 25 genera and approximately 140 species. These fishes are generally silvery in color, and exhibit a wide size range – from the scads that attain about 30 cm, to the Giant Trevally (*Caranx ignobilis*) of the Indo-West Pacific, which grows to 170 cm and may weigh more than 35 kg. Jacks are powerful midwater swimmers characterized by a streamlined shape, laterally compressed body, slender tail base, and a strongly forked caudal fin. Most species have the posterior scales of the lateral line modified into spiny, plate-like structures known as scutes. The dorsal and anal fins are generally low, but often have elongated rays on their anterior portions. The first dorsal fin is composed of III-IX spines and the second of one spine and 18-37 soft rays. There are usually III anal spines, with the first two detached from the rest of the anal fin, and usually 15-31 soft rays. One or more of the dorsal and anal spines are often obsolete or embedded in some species. Carangids are pelagic spawners that release large numbers of tiny, buoyant eggs. Judging from the widespread distribution of most species, the larvae may lead a pelagic existence for extended periods. The juveniles of several species, including the Bigeye Crevalle Jack (*Caranx sexfasciatus*), are sometimes encountered in brackish estuaries or in fresh water. Most of the jacks are highly esteemed as food fishes, and are targeted both by sport anglers and commercial fishermen. They frequently occur in large schools that roam considerable distances. Although not strictly reef fishes, jacks are common along the edge of reefs, sometimes adjacent to steep slopes. Voracious predators, they feed on a variety of fishes. Some species, such as the Golden Jack (*Gnathanodon speciosus*), also eat molluscs and crustaceans, and the scads (genus *Decapterus*) eat mainly planktonic invertebrates. Dr William F. Smith-Vaniz provided the information that appears in the following species accounts.

AFRICAN POMPANO (FLECHUDO)
Alectis ciliaris (Bloch, 1788)
(Plate VIII-3, p. 131)

Dorsal rays VII (embedded and not apparent in adults) + I,18-22; anal rays II + I,18-20; gill rakers on first arch (excluding rudiments) 4-6 + 12-17; lateral line with a pronounced, moderately long arch anteriorly; straight part of lateral line with 6-11 scutes; body superficially scaleless, scales actually minute and embedded where present; silvery with light, metallic, bluish tinge dorsally; a small, dark, diffuse spot on upper end of gill opening; juveniles with 5 chevron-shaped bars on body; *readily distinguished by its highly compressed body, steep angular forehead, and juveniles with extremely long and filamentous anterior dorsal and anal rays*. Worldwide in tropical seas; adults solitary in coastal seas, young usually pelagic and drifting. Maximum size to 130 cm.

ISLAND JACK (JUREL ISLEÑO)
Carangoides orthogrammus Jordan & Gilbert, 1881
(Plate VIII-8, p. 131)

Dorsal rays VIII+I,29-32; anal rays II+I,24-36; gill rakers on first arch (excluding rudiments) 8-10 + 20-23; lateral line anteriorly with very slight arch, the curved part slightly longer than or equal to straight part; scales on straight part of lateral line 21-34, followed by 19-31 small scutes; generally silvery, brassy to greenish blue on back, and whitish on ventral parts; adults with several relatively large, elliptical yellow spots on side; *distinguished by its combination of spots, relatively short dorsal and anal fin lobes, and presence of isolated scaleless areas on breast and base of pectoral fin.*

Steel Pompano (*Trachinotus stilbe*), Isla Marchena, Galapagos

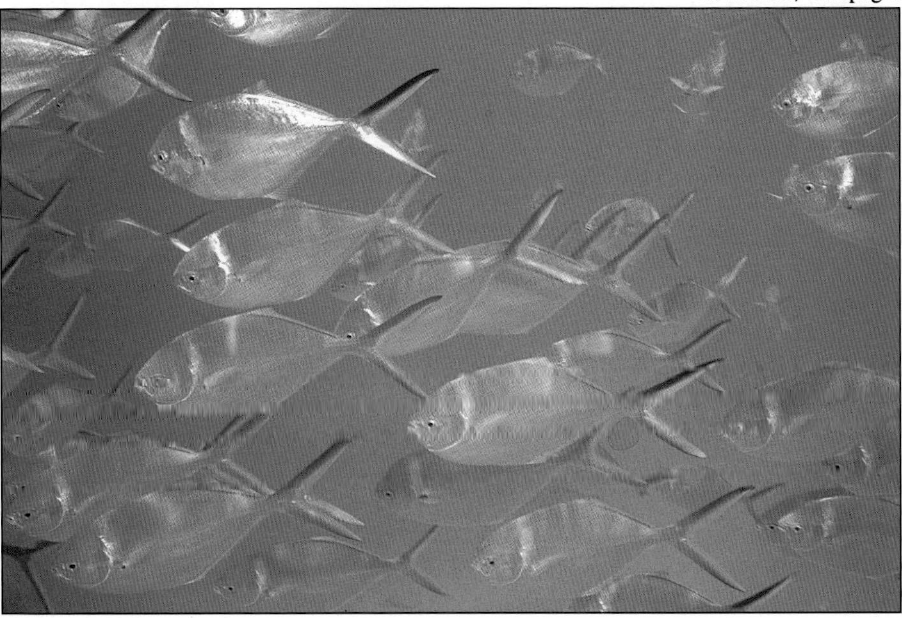

Widely distributed in the Indo-Pacific region, from East Africa to the Americas. To 70 cm. *C. jordani* Nichols is a synonym.

THREADFIN JACK
Carangoides otrynter Jordan & Gilbert, 1883
(Plate VIII-11, p. 131)

Dorsal rays VIII+I,18-19; anal rays II+I,16-17; lateral line anteriorly with moderate arch, the curved part about equal to straight part; generally silvery or silvery gray, without distinguishing marks; *distinguished by elongate dorsal and anal lobes, and large, continuous scaleless area on breast and pectoral fin base.* Baja California to Panama, including offshore islands. Grows to 60 cm. *C. dorsalis* Gill a synonym.

GREEN JACK
Caranx caballus Günther, 1869
(Plate VIII-7, p. 131)

Dorsal rays VIII+I,22-25; anal rays II+I,18-21; gill rakers on first arch (excluding rudiments) 13-15 + 28-30; lateral line with a pronounced but short anterior arch; scales on straight part of lateral line 0-7, followed by 45-56 strong scutes; breast completely scaled; light olive to dark bluish green on back, silvery gray to golden below; juveniles with about 7 dark bars on body. A common, schooling species, *distinguished by its relatively slender body and black spot on gill cover.* Southern California to Peru, including the Galapagos. Grows to at least 70 cm.

PACIFIC CREVALLE JACK
Caranx caninus Günther, 1869
(Plate VIII-4, p. 131)

Dorsal rays VIII+I,19-21; anal rays II+I,16-17; gill rakers on first arch (excluding rudiments) 3-6 + 15-17; breast naked except for a small median patch of scales in front of pelvic fins; lateral line with a pronounced, moderately long anterior arch; 35-42 strong scutes; greenish blue or bluish black on back, silvery white to yellowish or golden on lower side; *distinguished by its moderately deep body, black spot on gill cover, and yellowish caudal and anal fins.* Eastern Pacific Ocean, from Southern California to Ecuador; common in estuaries. Attains 75 cm.

BLACK JACK
Caranx lugubris Poey, 1860
(Plate VIII-5, p. 131)

Dorsal rays VIII+I,20-22; anal rays II+I,16-19; gill rakers on first arch (excluding rudiments) 6-8 + 17-22; lateral line with pronounced anterior arch; straight part of lateral line with 26-32 strong scutes; breast completely scaled; *body and fins mostly uniform gray to black*; small dark spot on upper corner of gill opening; *scutes often black.* Circumtropical distribution; commonly seen on reefs surrounding offshore islands. Grows to 80 cm.

Caranx sexfasciatus

BLUE CREVALLE
Caranx melampygus Cuvier, 1835
(Plate VIII-9, p. 131)

Dorsal rays VIII+I,21-24; anal rays II+I,17-20; gill rakers on first arch (excluding rudiments) 5-9 + 17-20; lateral line with pronounced anterior arch; straight part of lateral line with 27-42 strong scutes; breast completely scaled; *head and dorsal half of body brassy, suffused with blue, and covered with small blue-black spots* (forming at about 20 cm length and increasing in number with size); *second dorsal, anal, and caudal fins electric blue*; pectoral fins of juveniles yellow. Indo-Pacific region from East Africa to the Americas; solitary or in small schools near reefs. Maximum size to 100 cm.

BIGEYE CREVALLE JACK
Caranx sexfasciatus Quoy & Gaimard, 1824
(Plate VIII-6, p. 131)

Dorsal rays VIII+I,19-22; anal rays II+I,14-17; gill rakers on first arch (excluding rudiments) 6-8 + 15-19; lateral line with pronounced, but relatively long arch anteriorly; straight part of lateral line with 27-36 strong scutes; breast completely scaled; iridescent blue-green on back, shading to silvery white below; a small blackish spot near upper end of gill opening; lateral line scutes yellowish to black. *Easily identified in the field by its large eye and white tip on the dorsal fin lobe.* Widely distributed in the tropical Indo-Pacific, from East Africa to the Americas; a nocturnal species, usually seen milling in stationary aggregations during the day. Reaches 78 cm.

COCINERO
Caranx vinctus Jordan & Gilbert, 1882
(Plate VIII-13, p. 131)

Dorsal rays VIII+I,22-24; anal rays II+I,18-21; gill rakers on first arch (excluding rudiments) 11-12 + 28-30; lateral line with a pronounced but short anterior arch; scales on straight part of lateral line 0-4, followed by 46-53 strong scutes; body dusky bluish above, silvery with golden or greenish reflections on side; *8-9 incomplete darkish bars on side, and a distinct black blotch on edge of gill cover.* Baja California to Peru. Maximum length to 35 cm.

PACIFIC BUMPER
Chloroscombrus orqueta Jordan & Gilbert, 1882
(Plate VIII-10, p. 131)

Dorsal rays VIII+I,25-28; anal rays II+I,25-28; lateral line with a pronounced, short anterior arch; 6-12 very weak scutes; body and head dark metallic blue above, silvery on sides and belly; a dark spot on upper edge of gill opening; *distinguished by its highly compressed, ovate shape, with the ventral profile more convex than the dorsal profile; also has a black saddle marking on the upper part of the tail base.* Southern California to Peru; forms schools in shallow coastal seas; also enters estuaries. To 30 cm.

MACKEREL SCAD
Decapterus macarellus Cuvier, 1833
(Plate VIII-12, p. 131)

Dorsal rays VIII+I,31-37 (including detached finlet); anal rays II+I,27-31 (including detached finlet); gill rakers on first arch (including rudiments) 10-13 + 34-41; lateral line with long, low arch anteriorly; *scales in straight part of lateral line 18-36,* followed by 24-39 scutes; *total scales and scutes in lateral line (excluding caudal scales) 110-138; posterior end of upper jaw straight on upper part; scales on top of head usually extend forward to at least anterior margin of pupil*; body long, slender; bluish green on back, silvery below; a small black blotch on margin of gill cover near upper edge; caudal fin yellow-green. Worldwide in tropical seas. Grows to 32 cm. *D. pinnulatus* (Eydoux & Souleyet, 1841) a synonym.

SHORTFIN SCAD
Decapterus macrosoma Bleeker, 1851
(Plate VIII-14, p. 131)

Dorsal rays VIII+I,33-39 (including detached finlet); anal rays II+I,27-31 (including detached finlet); gill rakers on first arch (including rudiments) 10-12+34-38; lateral line with long, low arch anteriorly; scales in straight part of lateral line 14-29, followed by 24-40 scutes; total scales and scutes in lateral line (excluding caudal scales) 110-126; *posterior end of upper jaw concave on upper part; scales on top of head usually do not extend forward of posterior margin of pupil*; body long and slender; metallic blue on back, silvery below; a small black blotch on upper margin of gill cover; caudal fin translucent to dusky. Widely distributed in the tropical Indo-Pacific, from East Africa to the Americas (Gulf of California to Peru). Attains 32 cm. *D. afuerae* Hildebrand (1946) is a synonym.

RAINBOW RUNNER
Elagatis bipinnulata (Quoy & Gaimard, 1825)
(Plate IX-1, p. 133)

Dorsal rays VI+I,25-28 + 2; anal rays I+I,18-20 + 2; lateral line with slight arch anteriorly; no scutes; breast scaled; end of upper jaw terminating distinctly before eye (to below front margin of eye in young); grooves present dorsally and ventrally on tail base; dark olive-green to blue above, white below; 2 narrow light blue or bluish white stripes along sides, with a broader olive or yellowish stripe between them; fins with an olive or yellow tint; *easily recognized by the torpedo-shaped body, 2-rayed dorsal and anal finlets, and distinctive color pattern.* Circumtropical; a pelagic species, usually found near the surface, sometimes far offshore. Grows to 120 cm, but common to 80 cm.

GOLDEN JACK
Gnathanodon speciosus (Forsskål, 1775)
(Plate IX-3, p. 133)

Dorsal rays VII+I,18-20; anal rays II+I,15-17; gill rakers on first arch (excluding rudiments) 19-22 + 27-30; *adults without teeth* (young with a few feeble teeth in lower jaw); *lips noticeably thick and fleshy*; lateral line with moderate arch anteriorly; straight part of lateral line with 17-24 scales followed by 17-26 scutes; breast completely scaled; silvery gray on sides, yellowish on belly; frequently with scattered black patches or spots on sides; *juveniles and young adults silvery to yellow with 7-11 black bars on side*; all fins yellow; tips of caudal fin lobes black. Tropical Indo-Pacific from East Africa to the Americas (Gulf of California to Colombia); a bottom feeder that uses its protractile mouth to root for crustaceans, molluscs, and small fishes. Young display "piloting" behavior with sharks and other large fishes. To 110 cm.

YELLOWFIN JACK
Hemicaranx leucurus (Günther, 1864)
(Plate IX-2, p. 133)

Dorsal rays VII or VIII+I,25-28; anal rays II+I,20-25; gill rakers on first arch (excluding rudiments) 7-10 + 20-22; lateral line with pronounced, short arch anteriorly; scutes in straight part of lateral line 50-58; breast completely scaled; deep blue on upper half; lower sides and belly silvery; a jet black blotch at base of pectoral fin; *fins yellow or orange-yellow, the caudal narrowly edged with black; juveniles have 6-9 dark bars on side.* Mexico to Ecuador; common inshore, often found in schools. Attains 30 cm. *H. atrimanus* (Jordan & Gilbert, 1862) is a synonym.

BLACKFIN SCAD
Hemicaranx zelotes Gilbert, 1898
(Plate IX-4, p. 133)

Dorsal rays VII or VIII+I,25-31; anal rays II+I,22-25; gill rakers on first arch (excluding rudiments) 7-10 + 18-23; lateral line with pronounced, short arch; scutes in straight part of lateral line 47-55; breast completely scaled; dark olive to deep blue above, dusky silvery below; a large, jet black blotch on base of pectoral fins; *none of fins yellow*; juveniles with 4-6 dark bars on side. Mexico to Ecuador; common inshore, often found in schools. Maximum size to 30 cm. *H. sechurae* (Hildebrand, 1946) a synonym.

PILOTFISH
Naucrates ductor (Linnaeus, 1758)
(Plate IX-11, p. 133)

Dorsal rays IV or V (some spines often minute or embedded in adults) + I,25-29; anal rays II+I,15-17; gill rakers (including rudiments) 6-7 + 15-20; upper jaw very narrow and extending to about below anterior margin of eye; *no scutes; caudal fin base with a well-developed, lateral, fleshy keel on each side, and with dorsal and ventral groove; silvery with 6-7 black bars; white tips on upper and lower caudal fin lobes.* Circumtropical; usually in company with sharks, rays, turtles, or large fishes; juveniles found in floating weed or with jellyfishes. Reaches 70 cm.

LONGJAW LEATHERJACKET
Oligoplites altus (Günther, 1868)
(Plate IX-5, p. 133)

Dorsal rays IV or V+I,20-21; anal rays II+I,19-20; posterior 11-15 dorsal and anal soft rays forming semidetached finlets; gill rakers on first arch (excluding rudiments) 2-5 + 8-13; lateral line slightly arched over pectoral fin, but mainly straight; scales needle-like and embedded, but visible; no scutes; *upper jaw terminating behind rear margin of eye, its length 58-70 percent of head length; upper jaw with single band of villiform teeth*; overall silvery, greenish on upper back; pectoral and caudal fins yellow; belly sometimes yellow. Gulf of California to Ecuador; occurs in schools along sandy beaches and in muddy estuaries. Grows to 50 cm; common to 30 cm. *O. mundus* Jordan & Starks (1898) is a synonym.

SHORTJAW LEATHERJACKET
Oligoplites refulgens Gilbert & Starks, 1904
(Plate IX-9, p. 133)

Dorsal rays IV or V+I,19-21; anal rays II+I,19-22; posterior 11-15 dorsal and anal soft rays forming semidetached finlets; gill rakers on first arch (excluding rudiments) 6-8 + 19-22; lateral line slightly arched over pectoral fin, but mainly straight; scales needle-like and embedded, but visible; no scutes; *upper jaw terminating anterior to rear margin of pupil, its length 41-46 percent of head length; upper jaw teeth in 2 distinct rows*; overall silvery, caudal fin yellow. Panama to Ecuador; occurs in schools along sandy beaches and in estuaries. Attains 35 cm; common to 20 cm.

YELLOWTAIL LEATHERJACKET
Oligoplites saurus (Bloch & Schneider, 1801)
(Plate IX-7, p. 133)

Dorsal rays V (rarely IV or VI)+I,19-21; anal rays II+I,19-21; gill rakers on first arch

Elagatis bipinnulata

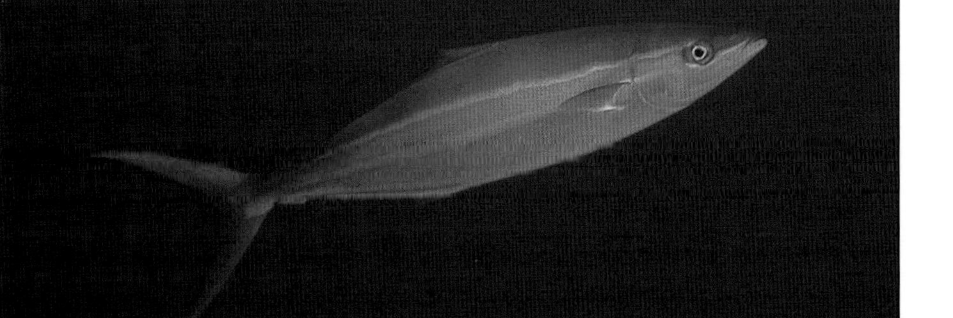

Seriola lalandi

gill rakers on first arch (excluding rudiments) 7-10 + 15-20; lateral line with slight arch over pectoral region; no scutes; *caudal fin base with a slight fleshy keel on each side, and with dorsal and ventral grooves; blue on upper back; sides and belly silvery to white; a narrow bronze stripe along middle of body, becoming yellow posteriorly; most fins, including caudal, yellowish.* Circumglobal in temperate and subtropical waters; in our area from central Mexico northwards, and south of Peru; forms large offshore schools. Maximum size to at least 150 cm and 50 kg.

(excluding rudiments) 5-7 + 14-28; lateral line slightly arched over pectoral fin, but mainly straight; scales needle-like and embedded, but visible; no scutes; *upper jaw terminating at or slightly beyond rear margin of pupil, its length 52-57 percent of head length; upper jaw teeth in 2 distinct rows*; overall silvery, greenish to bluish on uppermost part of back; caudal fin yellow. Gulf of California to Panama; occurs in schools along sandy beaches and in muddy estuaries. Grows to 35 cm; common to 25 cm.

PURSE-EYED SCAD
Selar crumenophthalmus (Bloch, 1793)
(Plate IX-12, p. 133)

Dorsal rays VIII+I,24-27; anal rays II+I,21-23; gill rakers on first arch 9-12 + 27-37; anterior part of lateral line with long, low arch; straight part of lateral line with 0-11 scales and 29-42 scutes; *a deep furrow in lower margin of gill opening with a large papilla immediately above it and a smaller papilla near upper edge; adipose eyelid covering eye, except for broad oval slit centered on pupil*; metallic blue to bluish green above, shading to white below; yellow stripe sometimes present from opercle margin to upper part of caudal fin base. Worldwide in tropical and subtropical seas; forms inshore schools. To 30 cm.

PACIFIC LOOKDOWN
Selene brevoortii (Gill, 1863)
(Plate IX-6, p. 133)

Dorsal rays VIII+I,20-24; anal rays II+I,17-18; dorsal and anal fin spines become embedded with age; gill rakers on lower limb of first arch (excluding rudiments) 29-34; body superficially scaleless, scales very small and embedded; scutes in lateral line very weak; *body pentagonal and strongly compressed*; head very deep, with angular nape and very steep forehead-snout profile; *dorsal and anal fin lobes prolonged in adults*; juveniles have anterior dorsal spines extremely elongated and filamentous; pelvic fins elongate in small juveniles, becoming very short in adults. *Similar to* S. oerstedii, *but has more dorsal rays and lower arch gill rakers*; overall silvery in color. Gulf of California to Peru. Reaches 30 cm.

HAIRFIN LOOKDOWN
Selene oerstedii Lutken, 1880
(Plate IX-8, p. 133)

Dorsal rays VIII+I,17-20; anal rays II+I,17-20; dorsal and anal fin spines become embedded with age; gill rakers on lower limb of first arch (excluding rudiments) 15-16; body superficially scaleless, scales very small and embedded; scutes in lateral line very weak; *body pentagonal and strongly compressed*; head very deep, with angular nape and very steep forehead-snout profile; *dorsal and anal fin lobes prolonged in adults*, juveniles have prolonged dorsal spines, but not filamentous as in *S. brevoorti*; pelvic fins elongate in small juveniles, becoming very short in adults; overall silvery in color. *Similar to* S. brevoorti, *but has fewer dorsal rays and lower arch gill rakers.* Baja California to Peru; juveniles common in estuaries. Attains 30 cm.

PACIFIC MOONFISH
Selene peruviana (Guichenot, 1866)
(Plate IX-10, p. 133)

Dorsal rays VIII+I,20-24; anal rays II+I,16-19; dorsal and anal fin spines become embedded with age; gill rakers on first arch (excluding rudiments) 7-9 + 30-34; body superficially scaleless, scales very small and embedded; scutes in lateral line very weak; *body somewhat rectangular and strongly compressed*; head moderately deep with angular nape and steep, slightly concave forehead-snout profile; *dorsal and anal fin lobes short in adults*, juveniles with anterior dorsal spines elongated; overall silvery; pectoral and caudal fins yellowish. Southern California to Peru. Maximum size to 30 cm.

YELLOWTAIL AMBERJACK
Seriola lalandi Valenciennes, 1833
(Plate X-2, p. 135)

Dorsal rays VII+I,30-35; anal rays II (reduced or embedded in large fish) + I,19-22;

FORTUNE JACK
Seriola peruana Steindachner, 1876
(Plate X-3, p. 135)

Dorsal rays VII+I,31-35; anal rays II (reduced or embedded in large fish) + I,20-22; gill rakers on first arch (excluding rudiments) 9-11 + 21-23; lateral line with slight arch over pectoral region; no scutes; *grooves present on dorsal and ventral surface of caudal fin base, but no lateral fleshy keel*; olivaceous to bluish on back; sides silvery to slightly bronzy; fins somewhat dusky; *no bars or stripe on head or body.* Southern Mexico to Peru, including the Galapagos. Reaches at least 60 cm; common to 40 cm.

ALMACO JACK
Seriola rivoliana Valenciennes, 1833
(Plate X-4, p. 135)

Dorsal rays VII+I,27-33; anal rays II (reduced or embedded in large fish) + I,18-22; total gill rakers on first arch (excluding rudiments) 22-26; lateral line with slight arch over pectoral region; no scutes; *grooves present on dorsal and ventral surface of caudal fin base, but no lateral fleshy keel*; bluish to greenish on upper back, silvery below; *distinguished from other Seriola in the region by its deeper body, oblique dark band from snout to front of dorsal fin, and relatively tall dorsal and anal fin lobes*; juveniles (up to 20 cm) with 6 dark bars on side. Circumtropical, in the eastern Pacific from Southern California to Peru, including the Galapagos. Attains at least 100 cm; common to 60 cm.

BLACKBLOTCH POMPANO
Trachinotus kennedyi Steindachner, 1875
(Plate X-7, p. 135)

Dorsal rays VI+I,17-19; anal rays II+I,16-17; gill rakers on first arch (excluding rudiments) 5-8 + 9-12; lateral line slightly arched over pectoral region; no scutes; color generally silver, sometimes golden or bronzy tinge, especially on freshly dead specimens; *distinguished from similar species by the very blunt, broadly rounded snout and large black blotch on the inner pectoral fin base.* Baja California to Ecuador; common in estuaries. To 75 cm; common to 40 cm.

PALOMA POMPANO

Trachinotus paitensis Cuvier, 1832
(Plate X-8, p. 135)

Dorsal rays VI+I,24-27; anal rays II+I,21-25; gill rakers on first arch (excluding rudiments) 2-6 + 8-10; lateral line nearly straight; no scutes; *similar to the Blackblotch Pompano, but snout not as rounded, body more slender, lateral line straight, and lacks black mark on inner pectoral fin base;* generally silvery, greenish or bluish on upper back, white on belly. Baja California to Chile. Maximum size to at least 45 cm; common to 30 cm.

GAFFTOPSAIL POMPANO

Trachinotus rhodopus Gill, 1863
(Plate X-6, p. 135)

Dorsal rays VI+I,19-21; anal rays II+I,18-20; second dorsal and anal fin lobes very elongate in adults; gill rakers on first arch (excluding rudiments) 8-11 + 13-15; lateral line slightly arched anteriorly; no scutes; *easily distinguished by its pattern of 4-6 yellowish to brown bars on a silvery background and the prolonged fin lobes,* also has yellow to reddish fins. Southern California to Peru; often common around rocky offshore islets, where it is seen close to the surface. Grows to at least 40 cm.

STEEL POMPANO

Trachinotus stilbe Jordan & McGregor, 1898
(Plate X-9, p. 135)

Dorsal rays V+I,25-26; anal rays II+I,24-25; gill rakers on first arch (excluding rudiments) 13-18 + 20-32; lateral line nearly straight; no scutes; sides and belly silvery white; distinguished from other members of the genus by its relatively slender shape, low dorsal and anal lobes, tiny pelvic fins, and *dark margins on the caudal fin; live individuals possess a distinctive white bar across the rear part of the head.* Generally rare, except around certain offshore areas such as the Revillagigedos and Galapagos; known only from the tropical eastern Pacific; forms large schools, which churn up the surface while feeding. Attains a maximum length of at least 40 cm; common to 30 cm.

JACK MACKEREL

Trachurus symmetricus (Ayres, 1855)
(Plate X-11, p. 135)

Dorsal rays VIII+I,31-33; anal rays II+I,26-30; gill rakers on first arch (excluding rudiments) 14-18 + 38-43; lateral line moderately arched anteriorly; *scales in curved part as well as straight part of lateral line enlarged and scute-like; body long and slender, slightly compressed;* metallic blue to olive-green on back; lower two-thirds paler, usually whitish to silvery; no distinctive markings except black spot on upper edge of gill opening. Alaska to the Gulf of California. Reaches about 80 cm; common to 55 cm. A similar species, *T. murphyi* Nichols, occurs at the Galapagos, Peru, and Chile.

WHITEMOUTH JACK

Uraspis helvola (Forster, 1801)
(Plate X-5, p. 135)

Dorsal rays VIII+I,25-30 (posterior spines of first dorsal fin may become embedded with growth); anal rays II (embedded in all but very young fish) + I,19-22; gill rakers on first arch (excluding rudiments) 5-8 + 13-17; lateral line with pronounced arch anteriorly; straight part of lateral line with 23-40 scutes; isolated scaleless patches present on breast and base of pectoral fin; *tongue, roof and floor of mouth white, remainder of mouth blue-black; body dusky with 6 broad, dark bars with narrow pale spaces between them.* Circumtropical and temperate seas; known in the eastern Pacific from southern California to Costa Rica; typically occurs in small schools. Grows to at least 50 cm.

ROOSTERFISH
FAMILY NEMATISTIIDAE

See discussion of the single species of the family below.

ROOSTERFISH

Nematistius pectoralis Gill, 1862
(Plate X-10, p. 135)

Dorsal rays VIII+I, 25; anal rays II, 15; pectoral rays 16; scales small and in irregular rows, about 130 in lateral series; *spines of first dorsal fin, except first one, forming long filaments (except in juveniles); anterior rays of second dorsal and anal fins moderately elevated; caudal fin strongly forked; pectoral fins very long and falcate; bluish gray with silvery reflections; a broad, blackish bar across rear part of head, and a pair of obliquely curved black bands on side.* Gulf of California to Peru; inhabits sandy shores, occasionally in estuaries; an angling favorite that is sometimes caught by trolling close to beaches. Grows to at least 100 cm and 35 kg.

DOLPHINFISHES (DORADOS, LLAMPUGAS)
FAMILY CORYPHAENIDAE

Dolphinfishes are oceanic dwellers of moderately large size, characterized by an elongate, compressed body; small cycloid scales; a long dorsal fin extending nearly the entire distance between the nape and caudal fin; no finlets or scutes; and a deeply forked caudal fin. Adult males develop a bony crest on the forehead, and the anterior snout profile becomes vertical. The family contains only two species, both of which have a worldwide distribution in tropical and subtropical seas. They are pelagic in habit, generally living near the surface, often congregating around floating objects or sometimes following ships. Dolphinfishes are swift predators that feed on fishes (especially flyingfishes) and squid. They are much sought after by anglers, who catch them by trolling a lure near the surface. The flesh is very tasty, and in some places is marketed under the Hawaiian name Mahimahi.

COMMON DOLPHINFISH

Coryphaena hippurus Linnaeus, 1758
(Plate X-1, p. 135)

Dorsal rays 58-67; anal rays 25-30; *greatest body depth in adults less than 25 percent of standard length; an oval patch of teeth on tongue; pectoral fin more than half of head length;* back brilliant metallic blue-green, shading to golden yellow ventrally, with scattered iridescent blue-green spots; dorsal fin deep blue-green; caudal, anal, and pelvic fins mainly yellow; small juveniles golden with about 12 dark bars on side. Brilliant colors fade to silvery gray with black spots and dark fins soon after death. Circumtropical. Attains 200 cm and 39.6 kg. The similar *C. equiselis* is distinguished by a broad, square tooth patch on the tongue, body depth more than 25 percent of standard length, and pectoral fin equal to about half of head length.

PLATE VIII

1 **SLENDER SUCKERFISH** (*Echeneis naucrates*)

2 **REMORA** (*Remora remora*)

3 **AFRICAN POMPANO** (*Alectis ciliaris*)

4 **PACIFIC CREVALLE JACK** (*Caranx caninus*)

5 **BLACK JACK** (*Caranx lugubris*)

6 **BIGEYE CREVALLE JACK** (*Caranx sexfasciatus*)

7 **GREEN JACK** (*Caranx caballus*)

8 **ISLAND JACK** (*Carangoides orthogrammus*)

9 **BLUE CREVALLE** (*Caranx melampygus*)

10 **PACIFIC BUMPER** (*Chloroscombrus orqueta*)

11 **THREADFIN JACK** (*Carangoides otrynter*)

12 **MACKEREL SCAD** (*Decapterus macarellus*)

13 **COCINERO** (*Caranx vinctus*)

14 **SHORTFIN SCAD** (*Decapterus macrosoma*)

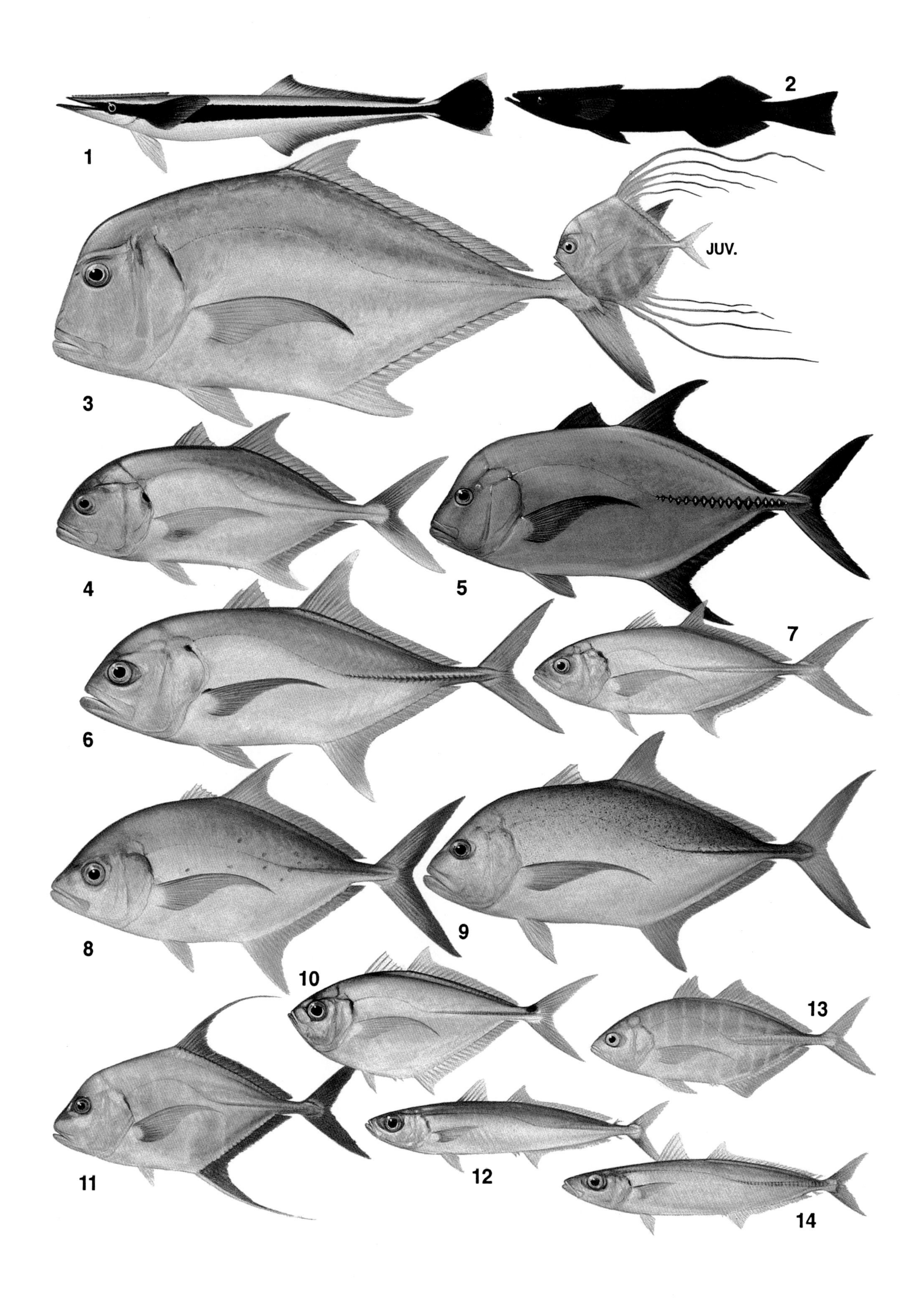
1
2
JUV.
3
4
5
6
7
8
9
10
11
12
13
14

PLATE IX

1 **RAINBOW RUNNER** (*Elagatis bipinnulata*)

2 **YELLOWFIN JACK** (*Hemicaranx leucurus*)

3 **GOLDEN JACK** (*Gnathanodon speciosus*)

4 **BLACKFIN SCAD** (*Hemicaranx zelotes*)

5 **LONGJAW LEATHERJACKET** (*Oligoplites altus*)

6 **PACIFIC LOOKDOWN** (*Selene brevoortii*)

7 **YELLOWTAIL LEATHERJACKET** (*Oligoplites saurus*)

8 **HAIRFIN LOOKDOWN** (*Selene oerstedii*)

9 **SHORTJAW LEATHERJACKET** (*Oligoplites refulgens*)

10 **PACIFIC MOONFISH** (*Selene peruviana*)

11 **PILOTFISH** (*Naucrates ductor*)

12 **PURSE-EYED SCAD** (*Selar crumenophthalmus*)

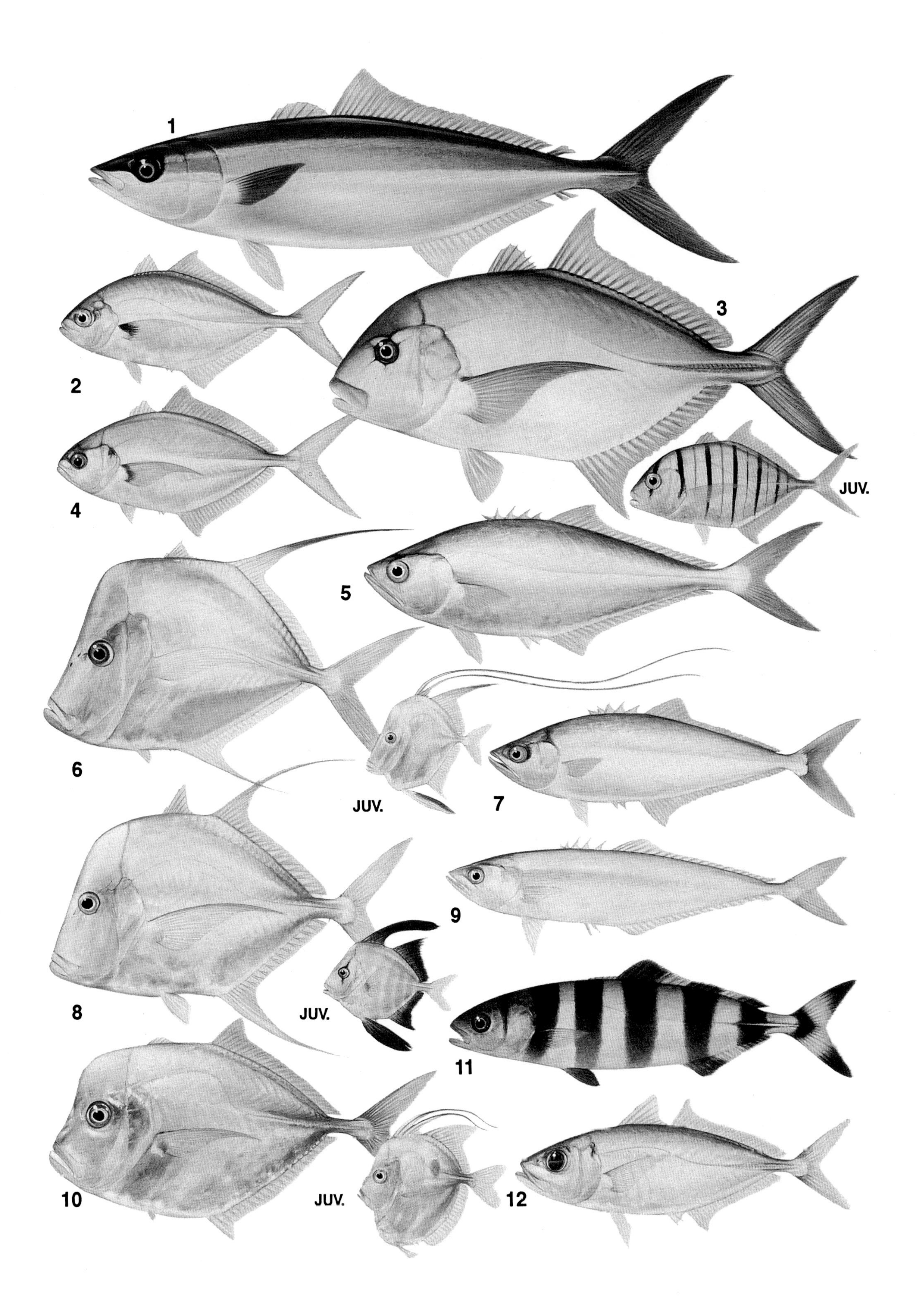
1
2
3
JUV.
4
5
6
JUV.
7
8
9
JUV.
11
10
JUV.
12

PLATE X

1 **COMMON DOLPHINFISH** (*Coryphaena hippurus*)

2 **YELLOWTAIL AMBERJACK** (*Seriola lalandi*)

3 **FORTUNE JACK** (*Seriola peruana*)

4 **ALMACO JACK** (*Seriola rivoliana*)

5 **WHITEMOUTH JACK** (*Uraspis helvola*)

6 **GAFFTOPSAIL POMPANO** (*Trachinotus rhodopus*)

7 **BLACKBLOTCH POMPANO** (*Trachinotus kennedyi*)

8 **PALOMA POMPANO** (*Trachinotus paitensis*)

9 **STEEL POMPANO** (*Trachinotus stilbe*)

10 **ROOSTERFISH** (*Nematistius pectoralis*)

11 **JACK MACKEREL** (*Trachurus symmetricus*)

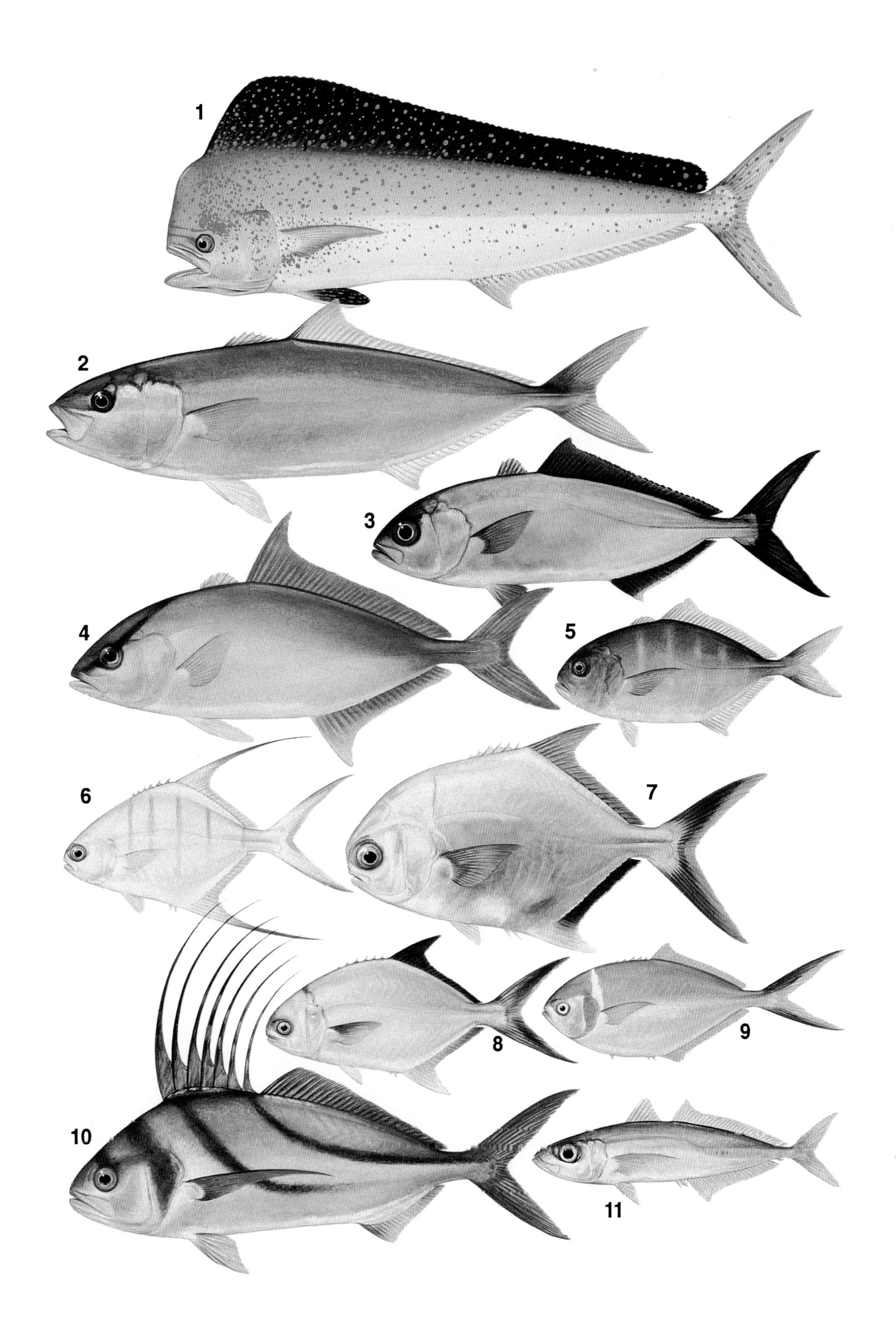
1
2
3
4
5
6
7
8
9
10
11

SNAPPERS (PARGOS, RABIRUBIAS, HUACHINANGOS)

FAMILY LUTJANIDAE

Snappers are small to medium-sized fishes with an ovate to elongate, moderately compressed body. Other features include a single dorsal fin that may be notched in the middle or is sometimes deeply incised between the spines; dorsal spines usually X, the soft rays 8-18; anal fin with III spines and 7-11 soft rays; caudal fin truncate to deeply forked; pelvic axillary scale process usually well-developed; jaws usually with more or less distinct canines; vomer and palatines usually with small conical teeth; scales ctenoid; cheek and operculum scaly; snout, preorbital and lower jaw scaleless. The approximately 100 members of the family are distributed in all tropical seas, but the majority inhabit the Indo-Pacific region. The genus *Lutjanus* is the largest containing 64 species, including nine eastern Pacific representatives. Most snappers dwell in shallow to intermediate depths (to 100 m) in the vicinity of reefs, although there are some species largely confined to depths between 100 and 500 m. Snappers are active predators, feeding mainly at night on a variety of items, but fishes are dominant in the diet of most species. Other common foods include crabs, shrimps, various other crustaceans, gastropods, cephalopods, and planktonic organisms, particularly urochordates. The maximum lifespan of snappers has been estimated between four and 21 years, based on studies of growth rings on bony structures such as otoliths and vertebrae. In general, the larger snappers have longer lifespans, perhaps in the range of 15 to 20 years. Snappers are considered good eating, and are frequently offered at markets and restaurants.

Blue and Gold Snappers (*Lutjanus viridis*), Isla Marchena, Galapagos

BARRED PARGO
Hoplopagrus guntheri Gill, 1862

Dorsal rays X,14; anal rays III,9; pectoral rays 16 or 17; scales medium-sized, about 45-49 in lateral line; gill rakers on lower limb of first arch (including rudiments) 11-15; scale rows on upper back parallel to lateral line; *teeth on lateral part of jaws molar-like; vomer with several large molars; anterior nostril long and tubular, posterior nostril in deep groove*; head and back dark brown, sides with about 8 brown bars, belly often reddish; juveniles have closely paired bars on the side and possess a blackish spot at the base of the posteriormost dorsal rays. Baja California (including Gulf of California) to Panama; a nocturnal predator of crustaceans and small fishes, it shelters on rocky reefs during the day; common in 3-10 m, but ranges to at least 30 m. Grows to a maximum length of about 80 cm.

MULLET SNAPPER
Lutjanus aratus (Günther, 1864)

Dorsal rays XI or XII,12; anal rays III,7 or 8; pectoral rays 15; gill rakers on lower limb of first arch (including rudiments) 11 or 12, total rakers 16 or 17; *scale rows on upper back parallel to lateral line*; teeth conical to caniniform, those at front of jaws generally enlarged and fang-like; *vomerine tooth patch V-shaped or crescentic, without a medial posterior extension*; gray to greenish, darker on back; center of each scale yellowish white, forming *alternate dark and light stripes on side*; fish from deeper water sometimes mainly reddish. Baja California to Ecuador, including offshore islands such as the Galapagos Archipelago; juveniles found in coastal embayments and estuaries; adults range into clear, deep waters and sometimes form large aggregations in midwater over rocky bottoms. Reaches about 100 cm.

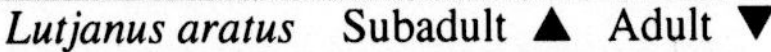

Lutjanus aratus Subadult ▲ Adult ▼

YELLOW SNAPPER
Lutjanus argentiventris (Peters, 1869)

Dorsal rays X,14; anal rays III,8; pectoral rays 16 or 17; gill rakers on lower limb of first arch (including rudiments) 12 or 13; *scale rows on upper back parallel to lateral line*; teeth conical to caniniform, those at front of jaws generally enlarged and fang-like; *vomerine tooth patch triangular or crescentic, with an elongate medial posterior extension; rosy red anteriorly, becoming bright yellow to orange over most of body*; fins mainly yellow or orange; a blue horizontal streak below eye. Southern California to Peru, including offshore islands; occurs over rocky bottoms; usually found close to shelter; juveniles and fish under 30 cm common in estuaries; a nocturnal predator of fishes, shrimps, crabs, and molluscs. Maximum size to about 60 cm.

COLORADO SNAPPER
Lutjanus colorado Jordan & Gilbert, 1882

Dorsal rays X,13 or 14; anal rays III,8 (occasionally 7); pectoral rays 16 or 17; gill rakers on lower limb of first arch (including rudiments) 11 or 12; *scale rows on upper back parallel to lateral line*; teeth conical to caniniform, those at front of jaws generally enlarged and fang-like; *vomerine tooth patch crescentic, without a medial posterior extension; body and fins mostly red; pink or whitish on belly*; small specimens reported to have dusky bars on side. Southern California to Panama; occurs on offshore rocky reefs to at least 50 m; young and adults often found inshore, sometimes in shallow estuaries. Reported to reach 90 cm.

SPOTTED ROSE SNAPPER
Lutjanus guttatus (Steindachner, 1869)

Dorsal rays X,12 or 13; anal rays III,8; pectoral rays 17; gill rakers on lower limb of first arch (excluding rudiments) 14; *scale rows on upper back rising obliquely above lateral line*; teeth conical to caniniform, those at front of jaws generally enlarged and fang-like; *vomerine tooth patch crescentic to triangular, with a relatively short posterior extension*; pale crimson or pinkish to yellowish, with silvery sheen, with narrow golden-green to brownish oblique stripes; *a distinctive black spot or smudge on back below middle portion of dorsal fin*. Gulf of California to Peru; occurs inshore in sandy bays, also in deeper trawling grounds. Reported to reach 80 cm; common to 40 cm.

GOLDEN SNAPPER
Lutjanus inermis (Peters, 1869)

Dorsal rays X,13; anal rays III,11; pectoral fins short and somewhat rounded, with 17 rays; gill rakers on lower limb of first arch 14 or 15 (only 10 are distinct); *scale rows on upper back rising obliquely above lateral line*; teeth small and conical; enlarged canines of jaws relatively small and slender, 2 in upper jaw and 3 or 4 on each side of lower jaw; *vomerine tooth patch triangular, with a short medial posterior extension; generally reddish with narrow brown stripes on side, those above lateral line obliquely directed; peduncle and caudal fin yellowish; a white to yellow blotch below base of posteriormost dorsal rays*. Mexico to Panama; locally common on rocky reefs, sometimes forming large, diurnal aggregations; juveniles are effective mimics of the damselfish *Chromis atrilobata*, and are often seen swimming with them. Grows to at least 35 cm. Often placed in a separate genus, *Rabiruba*.

Lutjanus inermis Resting adult ▲ Normal coloration ▼

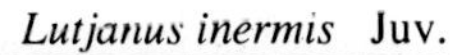

Lutjanus inermis Juv.

Lutjanus jordani Normal variety ▲ Dark variety ▼

JORDAN'S SNAPPER
Lutjanus jordani (Gilbert, 1897)

Dorsal rays X,14; anal rays III,9; pectoral rays 16 or 17; gill rakers on lower limb of first arch 12 (only 7 are well formed and distinct); *scale rows on upper back parallel to lateral line*; teeth conical to caniniform, those at front of jaws generally enlarged and fang-like; *vomerine tooth patch diamond-shaped; overall grayish or dark olive to reddish, with silvery sheen on lower sides*; freshly captured specimens often exhibit silvery-white spots on center of scales forming longitudinal rows, but these soon fade and fish becomes mainly dark brown to blackish. Southern Mexico to Peru, including offshore islands; inhabits inshore areas, including shallow mangrove-lined embayments; also occurs on deeper offshore trawling grounds. Reaches about 55 cm.

DOG SNAPPER
Lutjanus novemfasciatus Gill, 1862

Dorsal rays X,13 or 14; anal rays III,8; pectoral rays 26 or 17; gill rakers on lower limb of first arch 12 (including 5 very low rudiments); *scale rows on upper back rising obliquely above lateral line*; teeth conical to caniniform, those at front of jaws generally enlarged and fang-like; *vomerine tooth patch crescentic, without a medial posterior extension*; dark olive-brown to copper red on back and sides, becoming silvery white on lower sides; *juveniles and adults with 8-9 dusky, brownish bars on upper half, these sometimes obscure in large fish*; inhabits rocky reefs, usually in or near shelter during the day; common in estuaries; adults may ascend rivers as far as 20 km; feeds largely on crustaceans and fishes at night; depth range about 4-35 m. Attains about 100 cm and 45 kg.

PACIFIC RED SNAPPER
Lutjanus peru (Nichols & Murphy, 1922)

Dorsal rays X,13 or 14; anal rays III,8; pectoral rays 16 or 17; *scale rows on upper back rising obliquely above lateral line*; teeth conical to caniniform, those at front of jaws generally enlarged and fang-like; *vomerine tooth patch roughly diamond-shaped; large adults develop a groove from front of eye to nostrils; mainly red to pink*, with silvery sheen; fins reddish; generally occurs in deeper offshore waters to at least 90 m; sometimes seen in markets. Maximum total length about 90 cm; common to 50 cm.

BLUE AND GOLD SNAPPER
Lutjanus viridis (Valenciennes, 1845)

Dorsal rays X,14 or 15; anal rays III,8; pectoral rays 16 or 17; *scale rows on upper back rising obliquely above lateral line*; teeth conical to caniniform, those at front of jaws generally enlarged and fang-like; *vomerine tooth patch V-shaped without a medial posterior extension; a deep notch in lower edge of cheek margin (preopercle); bright yellow with 5 black-edged bluish-white stripes*. Baja California (including Gulf of California) to Ecuador, including Galapagos Archipelago and other offshore islands; sometimes forming large diurnal aggregations on rocky or coral reefs. Maximum size to 30 cm.

TRIPLETAILS (DORMILONES)
FAMILY LOBOTIDAE

See discussion of the single species of the family below.

TRIPLETAIL
Lobotes surinamensis (Bloch, 1790)

Dorsal rays XI-XII,15-16; anal rays III,11-12; pectoral rays 17; lateral-line scales 43-45; gill rakers on first arch 6-7 + 13-15; body oval to oblong, its depth 2.0-2.5 in standard length; a single dorsal fin with stout spines and elevated posterior section; the common name is derived from the rounded lobes of the dorsal, anal, and caudal fins; scales ctenoid, covering body and head except preorbital region and jaws; jaws with outer row of short, close-set canines and an inner band of smaller teeth; no teeth on roof of mouth; preopercle serrate, the serrae decreasing in size and increasing in number with age; dark brown or greenish yellow on back, silvery gray on sides; pectoral fins pale yellow; other fins dusky; juveniles generally dark brownish, but may be mottled. Tropical and subtropical seas around the world; occurs in bays, brackish estuaries, fresh water, or sometimes well out to sea around floating objects; juveniles sometimes in floating *Sargassum* weed; they may lay on their side at the surface and mimic a drifting dark leaf. Grows to 100 cm and at least 15 kg.

Lobotes surinamensis ▲

Diapterus aureolus ▼

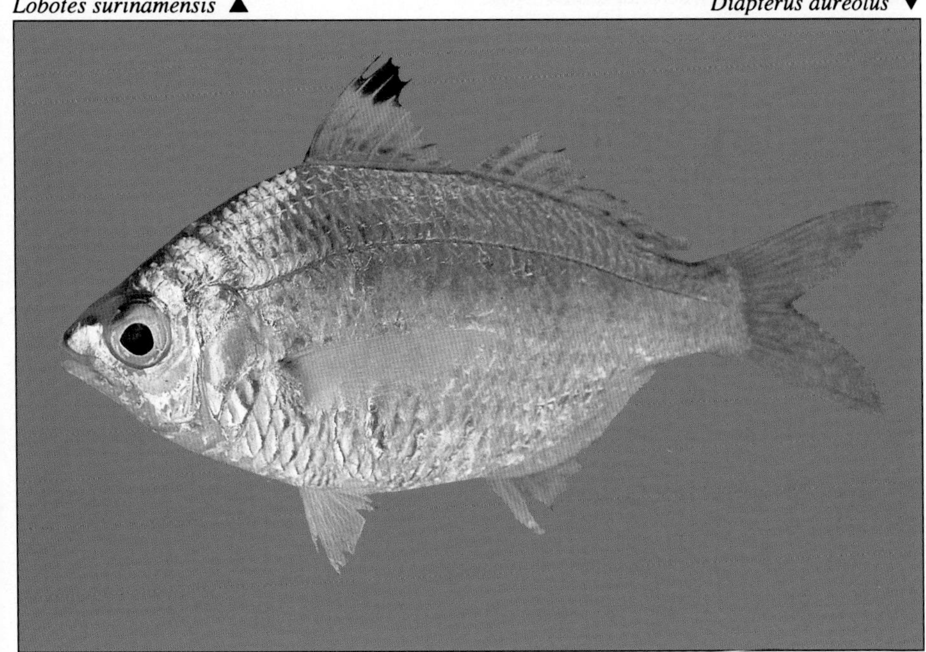

MOJARRAS (PERICHES)
FAMILY GERREIDAE

These are silvery fishes with moderately deep to somewhat slender, compressed bodies and a distinctive concave ventral profile to the head; the mouth is very protractile, extending downward when protruded; the teeth of the jaws are tiny and brush-like, and none exists on the roof of the mouth; there is a single dorsal fin, elevated anteriorly, and the dorsal and anal fins fold into a scaly sheath at their base; the scales are large and finely ctenoid. These fishes occur on sand or mud bottoms, frequently in brackish environments. A few species are regularly seen on the sandy margin of reefs. Gerreids are predators of buried organisms, which they capture by plunging their protrusible mouth into the sediment, afterwards ejecting the sand through the gill openings. Food items include polychaete worms and small crustaceans.

HIGHLINE MOJARRA
Diapterus aureolus (Jordan & Gilbert, 1882)

Dorsal rays XI,10; anal rays III,8; lateral-line scales 35-36; margin of preorbital smooth; *margin of preopercle serrate; body depth 2.2-2.3 in standard length; dorsal fins notched deeply to base and entirely separate; lateral line position very high on side, well above longitudinal axis of body*; scales on anterior part of body enlarged, becoming much smaller on caudal peduncle; silvery white, with black spot at tip of dorsal fin. Costa Rica to Peru; inhabits sand-mud bottoms of shallow coastal seas. Grows to at least 15 cm.

PERUVIAN MOJARRA
Diapterus peruvianus (Valenciennes, 1830)

Dorsal rays IX,10; anal rays III,8; lateral-line scales 37-40; *margin of preorbital smooth; margin of preopercle serrate; body depth 1.9-2.0 in standard length*; lips not thickened; pectoral fins relatively long, reaching slightly beyond anal fins; overall silvery; pelvic and anal fins yellowish. Gulf of California to Peru; inhabits sandy bays and estuaries, including brackish lower parts of streams. To 30 cm.

BLACKSPOT MOJARRA
Eucinostomus currani Yanez-Aranciba, 1978

Dorsal rays IX,10; anal rays III,7; lateral-line scales 44-48; margins of preorbital and preopercle smooth; body depth 2.4-2.5 in standard length; overall silvery white; *a prominent black spot with whitish lower border on apex of spinous portion of dorsal fin*. California to Peru; inhabits sand and mud bottoms of bays and estuaries; also penetrates freshwater streams at least 20 km inland. Reaches at least 20 cm. *E. entomelas* Yanez-Aranciba is a similar species with an overlapping distribution, but lacks the black spot on the dorsal fin. It is characterized by a black blotch on the uppermost part of the inner surface of the gill cover.

DOW'S MOJARRA
Eucinostomus dowii (Gill, 1863)

Dorsal rays IX,10; anal rays III,7; lateral-line scales 44-46; margins of preorbital and preopercle smooth; body depth 2.5-3.3 in standard length; *second anal spine shorter than the third, but much more robust; overall silvery white, without distinguishing dark spot or blotch on dorsal fin or inner surface of opercle*. Panama to Peru, including the Galapagos; inhabits sand or mud bottoms of shallow coastal waters, including estuaries. Grows to 15 cm.

SLENDER MOJARRA
Eucinostomus gracilis (Gill, 1862)

Dorsal rays IX,10; anal rays III,7; lateral-line scales 44-48; margins of preorbital and preopercle smooth; *body depth 2.8-3.3 in standard length; anal spines slender, second anal spine shorter than the third and only slightly more robust*; silvery white, without distinguishing marks, although juveniles may have 3-5 diagonal dark bars on side. Baja California to Peru; inhabits sand-mud bottoms of shallow coastal waters, including mangrove estuaries. To about 20 cm.

SHORT-FINNED MOJARRA
Eugerres brevimanus (Günther, 1864)

Dorsal rays IX,9-10; anal rays III,8; lateral-line scales 39-40; *margins of preorbital and preopercle serrate; body depth 2.2-2.5 in standard length; second anal spine much enlarged, thick and robust; lips noticeably swollen; pectoral fins short, scarcely reaching level of anus*; silvery white, with narrow dark lines corresponding with longitudinal scale rows; pectoral, pelvic, and anal fins yellowish. Northern Mexico to Panama; inhabits sand, mud, and rubble bottoms of coastal waters to at least 30 m. Maximum size about 28 cm.

YELLOWFIN MOJARRA
Gerres cinereus (Walbaum, 1792)

Dorsal rays IX,10; anal rays III,7; lateral-line scales 39-44; *margins of preorbital and preopercle smooth; body depth 2.3-2.7 in standard length; silvery white, with about 8 faint dark bars on side; pelvic and anal fins yellowish.* Western Atlantic and eastern Pacific from Baja California to Peru; inhabits sand bottoms adjacent to reefs; also enters brackish estuaries. Maximum size about 15 cm.

GRUNTS (BURROS, RONCADORES, CURRACAS, BOCAYATES)

FAMILY HAEMULIDAE

The grunts are a family of small to medium-sized fishes that occur worldwide in tropical and temperate seas. The group contains about 17 or 18 genera and approximately 120 species. A total of 29 species in nine genera occur in the tropical eastern Pacific. They generally resemble snappers (Lutjanidae), but differ in having a smaller mouth, and the teeth in the jaws are conical and small with none developed as canines. There is a single dorsal fin, except in *Xenocys* (two dorsals), and in *Xenichthys* and *Xenistius* the dorsal is very deeply notched. Dorsal and anal fin ray counts are highly variable; the number of dorsal spines, which range from IX to XIV, is often useful in identifying a particular genus. The common name is derived from their habit of emitting a grunting noise that results from grinding the upper and lower pharyngeal teeth. This sound is greatly amplified by the gas bladder. Grunts are primarily nocturnal. During the day they shelter in the reef, then disperse for feeding at dusk. Many of the eastern Pacific species form huge daytime resting aggregations next to the reef. There are also several species that inhabit estuaries and the lower reaches of streams. The diet of grunts consists of a wide variety of benthic invertebrates, although the young may feed predominately on plankton. The flesh is generally considered to be good eating. Most of the species have distinctive markings and are readily recognized, except species of *Pomadasys* are often similar in appearance and their correct identity may present a challenge. The family was formerly known as Pomadasyidae.

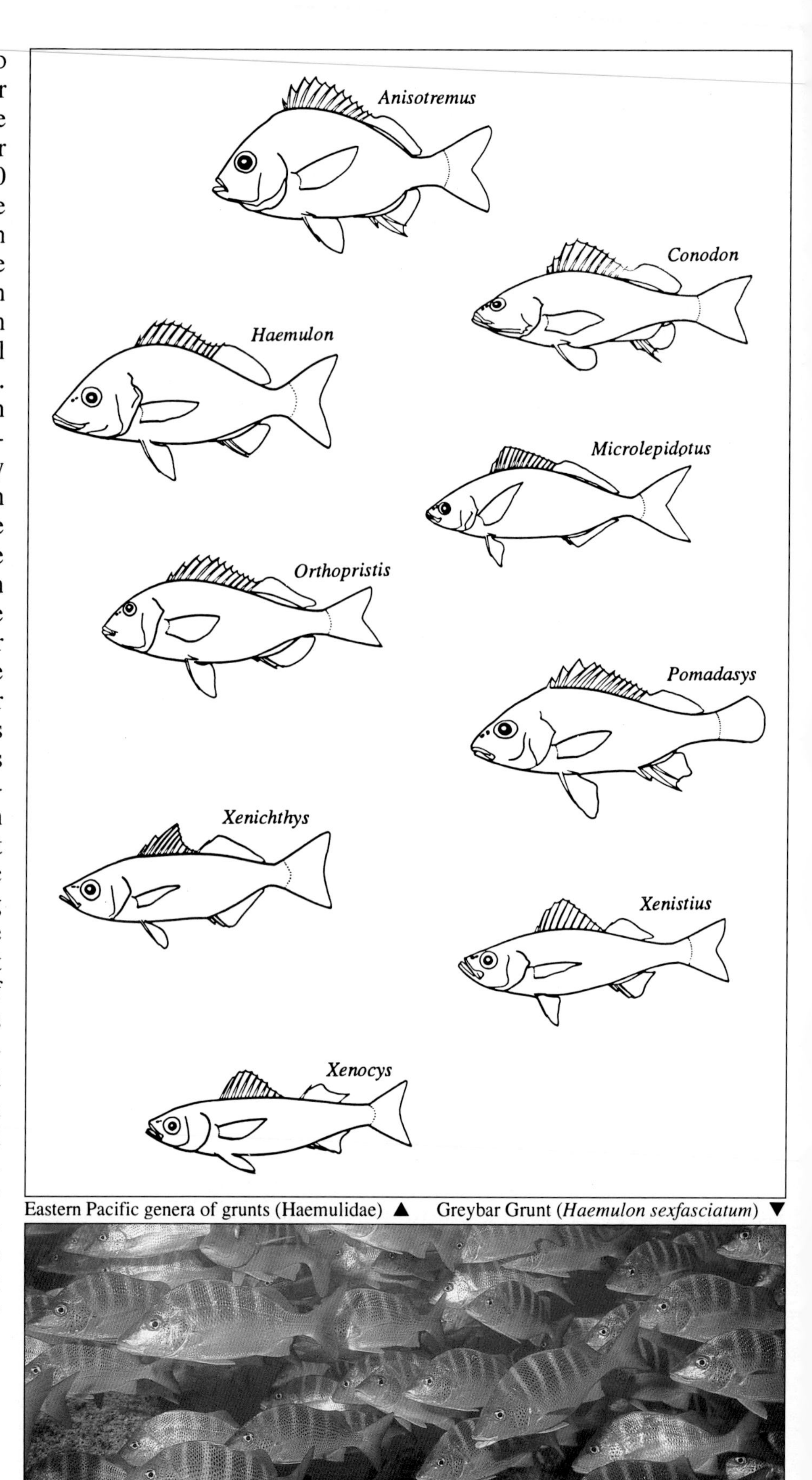

Eastern Pacific genera of grunts (Haemulidae) ▲ Greybar Grunt (*Haemulon sexfasciatum*) ▼

SILVERGREY GRUNT
Anisotremus caesius (Jordan & Gilbert, 1881)

Dorsal rays XII,15-16; anal rays III,9-10; lateral-line scales 50-54; body depth 2.0-2.2 in standard length; longitudinal scale rows of back parallel to lateral line; pectoral fins relatively long, reaching to or beyond anal fin origin; lips thick; second anal spine noticeably enlarged; *most of head and body golden or yellowish, becoming white on lower sides; a broad blackish bar immediately behind head*; fins mainly yellowish. Mexico (Mazatlan) to Panama; occurs on rocky reefs in 3-12 m; sometimes seen with *A. taeniatus*, but usually not as common as that species. Grows to at least 25 cm. Also known as Longfin Grunt.

SARGO
Anisotremus davidsoni (Steindachner, 1875)

Dorsal rays XI-XII,14-16; anal rays III,9-11; lateral-line scales 50-52; body depth 2.2-2.4 in standard length; scale rows of back ascending obliquely; lips thick; second anal spine not noticeably enlarged; overall silvery with yellowish tinge; *a prominent blackish bar on side at level of outer part of pectoral fin*; young specimens have 2-3 blackish stripes on the side and lack a spot at the caudal fin base. Central California to the Gulf of California; found on rocky, weed-covered reefs, sometimes forming schools. Attains 58 cm, but common to 30 cm.

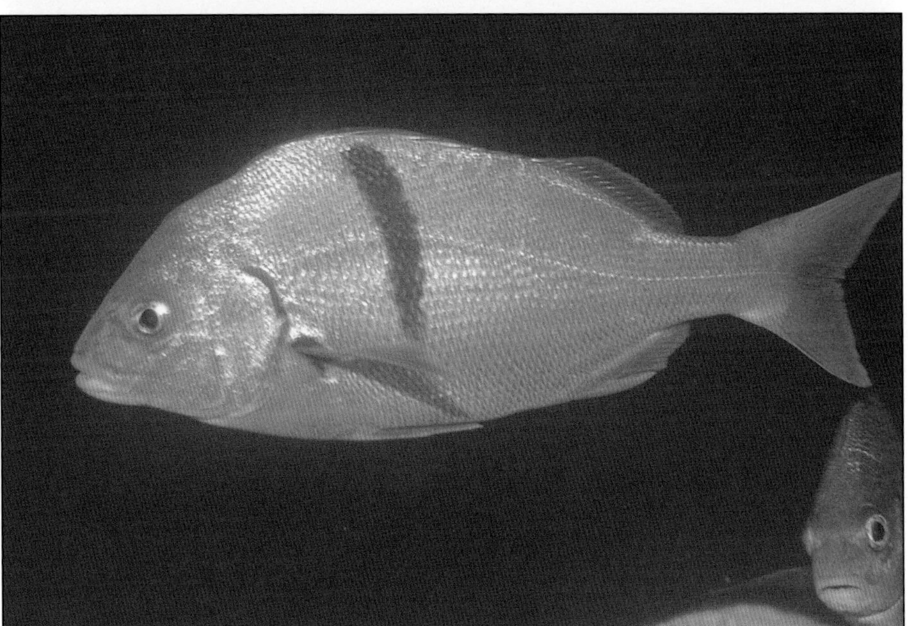

BLACKBARRED GRUNT
Anisotremus dovii (Günther, 1864)

Dorsal rays XI,13-16; anal rays III,9-10; lateral-line scales 47-51; body depth 2.0-2.2 in standard length; longitudinal scale rows of back parallel to lateral line; lips thick; *caudal fin strongly emarginate; 8-9 rows of scales between lateral line and base of first dorsal spine; whitish to dusky pale gray, with 4-5 blackish bars on body and head*; soft dorsal fin and soft part of anal fin dusky blackish. Gulf of California to Peru; inhabits trawling grounds in 10-40 m. Reaches 40 cm.

BURRITO GRUNT
Anisotremus interruptus (Gill, 1862)

Dorsal rays XII,16-17; anal rays III,8-9; lateral-line scales 46-50; body depth 2.0-2.4 in standard length; scale rows of back ascending obliquely above lateral line; lips thick; second anal spine noticeably enlarged; *overall silvery, sometimes with yellowish sheen; scale centers dusky blackish, thus imparting a spotted pattern on sides*; fins yellowish to greenish brown. Baja California to Peru, including Galapagos; common on inshore rocky reefs. Maximum length 46 cm.

PACIFIC GRUNT
Anisotremus pacifici (Günther, 1864)

Dorsal rays XI,14; anal rays III,9-10; body depth 2.0-2.5 in standard length; longitudinal scale rows of back parallel to lateral line; lips thick; *caudal fin slightly emarginate, nearly truncate; 6-7 rows of scales between lateral line and base of first dorsal spine; whitish to dusky pale gray with 4-5 blackish bars on body and head* (bars quickly disappear after death); soft dorsal fin and soft part of anal fin dusky blackish. Costa Rica to Peru; inshore waters on sandy or muddy bottoms; also enters estuaries and rivers. To at least 25 cm.

PERUVIAN GRUNT
Anisotremus scapularis (Tschudi, 1845)

Dorsal rays XII,14-17; anal rays III,12-13; body depth 2.1-2.5 in standard length; scale rows of back ascending obliquely; lips thick; caudal fin truncate or slightly emarginate; *anterior rays of dorsal and anal fin much taller than posterior rays, giving these fins a somewhat triangular outline; dark charcoal gray to brownish on upper half, white below; a silvery white streak along lateral line (rapidly fades after death)*. Galapagos Islands and Peru; forms large schools on rocky reefs in about 5-20 m. Grows to 40 cm.

PANAMIC PORKFISH
Anisotremus taeniatus Gill, 1861

Dorsal rays XII,16-17; anal rays III,9-10; lateral-line scales 49-57; body depth 1.9-2.4 in standard length; scale rows of back ascending obliquely; lips thick; 10-11 scale rows between lateral line and base of first dorsal spine; *bright yellow with dark-edged blue stripes on side; a pair of blackish bars anteriorly, 1 through eye and another immediately behind head*; juveniles with yellow head and anterodorsal region of body, a dark brown midlateral stripe, and large black spot on base of caudal fin. Baja California to Ecuador; generally seen in small aggregations on rocky reefs, but not particularly common. To 30 cm.

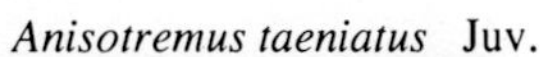

Anisotremus taeniatus Juv.

SERRATED GRUNT
Conodon serrifer Jordan & Gilbert, 1882

Dorsal rays XII,12; anal rays III,7-8; lateral-line scales 50; body depth 2.8-3.4 in standard length; *preopercle strongly serrate, including 1-2 enlarged spines at angle, serrae on lower margin directed forward*; brownish on back, silvery blue-gray on sides, and white below; 6-7 blackish bars on upper side usually evident; soft dorsal, anal, and pelvic fins yellow. Gulf of California to Ecuador; frequents trawling grounds, occasionally encountered in fish markets. Grows to about 30 cm.

Conodon serrifer ▲

Haemulon flaviguttatum ▼

CORTEZ GRUNT
Haemulon flaviguttatum Gill, 1863

Dorsal rays X-XII,15-18 (usually XII,17); anal rays III,10-11; lateral-line scales 50-53; gill rakers on first arch 26-31 (usually 27-28); body depth 2.7-3.3 in standard length; overall silvery gray; *small white spots on side arranged in oblique rows; fins yellow, except caudal mainly dusky blackish (with yellowish outer margin)*; juveniles with narrow, dark midlateral stripe and second stripe above, slanting from snout to upper surface of caudal peduncle, and also a black spot at base of caudal fin. Baja California to Panama; occurs near shore on rocky reefs; often seen in schools. Reaches 41 cm. Also known as Yellow-Spotted Grunt.

SPOTTAIL GRUNT
Haemulon maculicauda (Gill, 1863)

Dorsal rays XII-XIV,15-18 (usually XIII,15-16); anal rays III,8-11 (usually 10); lateral-line scales 50-52; gill rakers on first arch 22-29 (usually 23-26); body depth 2.8-3.2 in standard length; *scale rows of side parallel to longitudinal axis of body (obliquely oriented in other species of* Haemulon*); scale centers white, forming narrow white stripes on side; basal half of caudal fin blackish*; juveniles with dark midlateral stripe, and spot at base of caudal fin. Baja California to Panama; forms large aggregations on coral or rocky reefs. To 23 cm.

Adult ▲ School of adults ▼

MOJARRA GRUNT
Haemulon scudderi Gill, 1863

Dorsal rays XI-XII,14-17; anal rays III,7-8; lateral-line scales 46-50; gill rakers on first arch 15-21 (usually 17-19); body depth 2.4-2.7 in standard length; *overall silvery to dusky silver-gray; sides covered with small blackish spots, 1 per scale; no dark spot at base of caudal fin; fins whitish to dusky blackish*; juveniles with dark midlateral stripe converging with black spot at base of caudal fin, and second stripe from upper part of eye to dorsal surface of caudal peduncle. Baja California to Ecuador, including Galapagos; often in large aggregations which hover near rocky reefs during the day. Maximum size 35 cm.

Haemulon scudderi Subadult ▲ Juv. ▼

GREYBAR GRUNT
Haemulon sexfasciatum Gill, 1863

Dorsal rays XI-XII,16-18 (usually XII,16-17); anal rays III,9-10; lateral-line scales 49-52; gill rakers on first arch 18-22 (usually 18-20); body depth 2.5-2.8 in standard length; *about 5-6 prominent blackish bars on upper half to two-thirds of side*; spaces between bars white to yellow, remainder of head and body silvery white; juveniles with dark midlateral stripe converging with black spot on caudal fin base, also a pair of dark stripes above: 1 on midline of snout to dorsal fin origin and another from upper margin of eye to dorsal surface of caudal peduncle; these stripes gradually fade as the bars begin to appear. Baja California to Panama; very abundant in some parts of the Gulf of California, forming huge daytime aggregations. Grows to 48 cm.

Haemulon sexfasciatum ▲ *Haemulon steindachneri* ▼

LATIN GRUNT
Haemulon steindachneri (Jordan & Gilbert, 1882)

Dorsal rays XI-XII,15-17 (usually XII,16); anal rays III,8-10 (usually 9); lateral-line scales 49-53; gill rakers on first arch 19-25 (usually 23); body depth 2.5-2.8 in standard length; overall silvery to silvery gray; *sides with narrow, oblique, brownish bars corresponding with scale rows; a prominent black spot at base of caudal fin*; fins yellowish; juveniles with brownish orange midlateral stripe, with pearly white upper and lower border converging with black spot at base of caudal fin, also 2-3 diffuse brownish orange stripes on back. Western Atlantic and eastern Pacific; ranges from Baja California to Peru; forms schools in sandy areas adjacent to reefs. Attains 30 cm.

Haemulon steindachneri Juv.

SHORTFIN GRUNT

Microlepidotus brevipinnis
(Steindachner, 1869)

Dorsal rays XIII,16-17; anal rays III,12-13; lateral-line scales 60-62; *gill rakers on lower limb of first arch 16; body slender, depth 2.6-2.7 in standard length; dorsal and anal fins covered with small scales; silvery gray, with brown spot on each scale of back giving impression of dark, oblique bands*; dorsal and anal fins dusky yellowish; pelvic and pectoral fins yellowish; caudal fin brown. Gulf of California to Peru; forms schools in sandy areas, usually in 5-30 m. Reaches 40 cm. This species has been placed in the genus *Orthopristis* by most previous authors.

Microlepidotus brevipinnis Subadult

WAVYLINE GRUNT

Microlepidotus inornatus Gill, 1862

Dorsal rays XIV,15; anal rays III,12; lateral-line scales about 80; body slender, depth 3.1-3.3 in standard length; gill rakers on lower limb of first arch 17; *similar in general appearance to species of* Orthopristis, *but has 14 instead of 12 or 13 dorsal spines*; dorsal and anal fins mainly scaleless; overall silvery gray with *7-9 narrow orange stripes on side; some of stripes above lateral line broken and forming wavy pattern.* Northern and Central Mexico including Baja California (Magdalena Bay southwards) and Gulf of California; usually seen in sandy areas near the edge of reefs. To 45 cm.

Microlepidotus inornatus ▲ *Orthopristis chalceus* ▼

BRASSY GRUNT

Orthopristis chalceus (Günther, 1864)

Dorsal rays XII (rarely XIII),14-15; anal rays III,10-11; lateral-line scales 55-60; *gill rakers on lower limb of first arch 12-14; body depth 2.2-2.4 in standard length; dorsal and anal fins scaleless; silvery gray, with oblique orange bands corresponding with scale rows on upper half of side*; white on ventral parts; fins whitish to dusky blackish (caudal), except pectorals slightly yellow. Gulf of California to Peru, including the Galapagos; usually found over sand or rubble bottoms to at least 35 m. Grows to 35 cm.

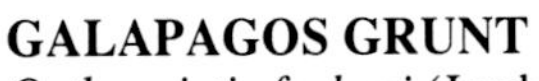

GALAPAGOS GRUNT

Orthopristis forbesi (Jordan & Starks, 1897)

Dorsal rays XII,15; anal rays III,11; *lateral-line scales 80-85; gill rakers on lower limb of first arch 13-14; body slender, depth 2.8-3.0 in standard length; dorsal and anal fins scaleless; silvery gray, sometimes with narrow orange stripes corresponding with scale rows on sides*; opercular membrane blackish. Galapagos Islands; forms schools in sandy areas near rocky reefs. Attains at least 25 cm.

Orthopristis forbesi Subadult

Orthopristis forbesi Adults ▲ Blotched coloration ▼

BRONZE-STRIPED GRUNT

Orthopristis reddingi Jordan & Richardson, 1895

Dorsal rays XII,15; anal rays III,9-10; lateral-line scales 52-53; *gill rakers on lower limb of first arch 12-15; body slender, depth 2.9-3.1 in standard length; dorsal and anal fins scaleless; overall silvery, often with orange or bronze stripes or oblique bands on side corresponding with scale rows*; fins whitish, except caudal dusky. Gulf of California; occurs in schools over sand or rubble bottoms. Reaches a total length of at least 30 cm.

AXIL GRUNT
Pomadasys axillaris (Steindachner, 1869)

Dorsal rays XII,14-15; anal rays III,8; lateral-line scales 49-54; *gill rakers on lower limb of first arch 14-17*; body depth 2.7-3.3 in standard length; *pectoral fins reaching level of anus; second anal spine longer than eye diameter*; generally silvery with large blackish blotch just behind upper edge of gill opening; *a triangular black spot in axil of pectoral fin*; narrow brown stripes corresponding with scale rows often present on sides; 3-4 bars often evident on side. Mexico (Mazatlan) to Panama. Reaches 30 cm.

FRESHWATER GRUNT
Pomadasys bayanus (Jordan & Evermann, 1898)

Dorsal rays XII,12; anal rays III,7; scales in longitudinal series just above lateral line 60-70; body depth 3.2-3.3 in standard length; *pectoral fins relatively short, not reaching tips of depressed pelvic fins; eye 3.5-5.4 in head length; second anal spine very long and stout, its length about 2 times eye diameter or more*; dark olive on back, silvery on sides; fins dusky. Gulf of California to Ecuador; tolerates salinities between 20-30 ppt, but penetrates far up rivers, even fast-flowing ones. Grows to 36 cm.

Pomadasys bayanus ▲

BRANICK'S GRUNT
Pomadasys branickii (Steindachner, 1879)

Dorsal rays XIII,11-12; anal rays III,7; scales in longitudinal series just above lateral line 47-50; body depth 2.8-3.2 in standard length; *pectoral fins relatively long, reaching tips of depressed pelvic fins or beyond; eye relatively large, 2.6-3.2 in head length; second anal spine very long and stout, its length about 2 times eye diameter or more*; generally silvery with yellowish brown opercular membrane and yellow iris. Mazatlan, Mexico to Peru; inhabits muddy bays and estuaries; also enters river mouths. Grows to at least 25 cm.

Pomadasys branickii ▲

Pomadasys elongatus ▼

ELONGATE GRUNT
Pomadasys elongatus (Steindachner, 1879)

Dorsal rays XII,14-15; anal rays III,7-8; lateral-line scales 50-54; *gill rakers on lower limb of first arch 10-14*; body depth 2.9-3.3 in standard length; *pectoral fin reaching beyond tip of pelvics, but not reaching anus; second anal spine longer than eye diameter; length of depressed anal spines reaching only to base of penultimate anal ray*; generally silver; *upper edge of gill opening narrowly blackish*; diffuse dark stripes sometimes evident on side. Mexico (Mazatlan) to Panama; frequents coastal seas on sandy or muddy bottoms. Attains 30 cm.

WHITE GRUNT
Pomadasys leuciscus (Günther, 1864)

Dorsal rays XII,14; anal rays III,7-8; lateral-line scales 50-53; gill rakers on lower limb of first arch 9-14; body depth 2.9-3.2 in standard length; *predorsal or nape profile strongly elevated, and ventral profile of head rather straight; pectoral fin reaching tips of pelvic or slightly beyond, but not reaching anus; second anal spine longer than eye diameter, length of depressed anal spines reaching to base of last anal ray or beyond*; generally silver; *upper edge of gill opening narrowly blackish*; diffuse dark stripes sometimes evident on side. Gulf of California to Peru; inhabits coastal seas on sandy or muddy bottoms. Reaches 37 cm.

BIGSPINE GRUNT
Pomadasys macracanthus (Günther, 1864)

Dorsal rays XII (rarely XI),13; anal rays III,7-8; lateral-line scales 48-50; gill rakers on lower limb of first arch 10-13; *body depth 2.5-2.8 in standard length; pectoral fins very long, reaching to level of anus; second anal spine very long and robust; caudal fin emarginate to nearly truncate; overall silvery with 4-5 blackish bars on side (sometimes indistinct in large fish).* Gulf of California to Panama; coastal seas on sandy or muddy bottoms, entering brackish estuaries and river mouths. Maximum size 37 cm.

SILVER GRUNT
Pomadasys nitidus (Steindachner, 1869)

Dorsal rays XII,14-15; anal rays III,7-9; lateral-line scales 50-53; *gill rakers on lower limb of first arch 14-16*; body depth 2.8-3.3 in standard length; *dorsal and ventral profile of slope of head about equal; pectoral fins generally reaching level between tip of pelvic fins and anus; second anal spine very short, about equal to or less than eye diameter; third anal spine about same length as second spine, reaching to base of penultimate soft anal ray when depressed*; generally silver; a black blotch at beginning of lateral line; diffuse dark stripes sometimes evident on side. Gulf of California to Peru; coastal seas on sandy or muddy bottoms. Attains 30 cm.

PANAMANIAN GRUNT
Pomadasys panamensis (Steindachner, 1875)

Dorsal rays XII,12-13; anal rays III,7-8; lateral-line scales 48-52; gill rakers on lower limb of first arch 12-14; body depth 2.5-2.7 in standard length; *pectoral fins very long, reaching to level of anal fin*; second anal spine enlarged, but not reaching tips of longest soft rays when depressed; *silvery with prominent blackish blotch behind upper edge of gill cover; fins yellowish.* Gulf of California to Peru; generally captured by shrimp trawlers in 10-50 m. Grows to at least 30 cm.

WHITE SALEMA
Xenichthys agassizi Steindachner, 1875

Dorsal rays XI,I,17-18; anal rays III,17; lateral-line scales 56-60; gill rakers on lower limb of first arch 17; body elongate, depth 3.2-3.5 in standard length; *dorsal fin deeply notched, but 2 portions joined at base*; pectoral fins long and somewhat falcate, reaching to level of anal fin origin; base of spinous part of dorsal fin shorter than base of soft part; eye large, about 2.7 in head length; lower jaw projecting; margin of preopercle finely serrated; *silvery bluish without distinguishing marks.* Galapagos Islands; forms schools in rocky shallows. To 20 cm.

LONGFINNED SALEMA
Xenichthys xanti Gill, 1863

Dorsal rays XI,I,17-18; anal rays III,17-18; lateral-line scales 55-62; gill rakers on lower limb of first arch 14-18; body elongate, depth 3.2-3.5 in standard length; *dorsal fin deeply notched, but 2 portions joined at base*; pectoral fins short, not reaching level of anus; base of spinous part of dorsal fin shorter than base of soft part; eye 3.0-3.3 in head; lower jaw projecting; *silvery with about 6 dark stripes on upper two-thirds of side*, those on middle of side brown, becoming blackish on back; a prominent black spot at base of caudal fin. Baja California to Peru; generally occurs in mixed schools with other grunters. Attains 25 cm.

CALIFORNIA SALEMA
Xenistius californiensis (Steindachner, 1875)

Dorsal rays IX-XI,I-II,12-14; anal rays III,11-12; lateral-line scales 50-55; gill rakers on lower limb of first arch 15-20; body elongate, depth 3.2-3.5 in standard length; *dorsal fin deeply notched, but 2 portions joined at base; base of spinous part of dorsal fin longer than base of soft part; silvery with 6-8 orange-brown stripes on side-* only 3-4 stripes on upper half of body evident in small specimens. Southern California to Gulf of California; forms large schools around piers and on weed-covered rocky reefs. Reaches 30 cm.

BLACK-STRIPED SALEMA
Xenocys jessiae Jordan & Bollman, 1889

Dorsal rays X+I,13-14; anal rays III,10-11; lateral-line scales 50-52; gill rakers on lower limb of first arch 23; body elongate, depth 3.5-4.0 in standard length; *spinous and soft portions of dorsal fin completely separate; silvery with about 7 black stripes on side.* Galapagos Islands; forms huge daytime aggregations, containing hundreds of individuals, along rocky shores. Attains 30 cm.

PORGIES (SARGOS, BESUGOS, PALMAS, MOJARRONES)
FAMILY SPARIDAE

Porgies are snapper-like fishes, with an oblong to ovate compressed body; dorsal fin with X-XIII spines and 8-15 soft rays; anal fin with III spines and 8-14 soft rays; margin of preopercle smooth; scales usually weakly ctenoid; cheeks and opercles usually scaly; teeth conical or incisiform, molars present in some species; no teeth on roof of mouth; and caudal fin forked to emarginate. They occur in all temperate and tropical seas; approximately 100 species are known. Southern Africa has a particularly rich sparid fauna with over one-third of the world's species found there. Porgies frequent a variety of habitats including brackish estuaries, bays, coastal reefs, and deeper waters of the continental shelf. However, only a few species are seen around reefs in the tropical Indo-Pacific. Some porgies are hermaphroditic, undergoing sex reversal from male to female. They feed on a variety of plants and animals, although benthic invertebrates such as molluscs, crabs, and urchins form most of the diet of many species.

GALAPAGOS SEABREAM

Archosargus pourtalesii (Steindachner, 1881)

Dorsal rays XIII,9-10; anal rays III,9-10; lateral-line scales 48; jaws with broad incisors at front and molariform teeth on sides; snout and suborbital scaleless; pectoral fins long, reaching level of second anal spine; caudal fin forked; light bluish on back, silvery white below; *a prominent black blotch on upper side above pectoral fin, and 7 golden stripes on side*. Galapagos Islands; seen infrequently in shallow bays. Grows to at least 26 cm.

PACIFIC PORGY

Calamus brachysomus (Lockington, 1880)

Dorsal rays XII-XIII,11-13 (usually XII,12); anal rays III,10-11 (usually 10); pectoral rays 13-16 (usually 15); lateral-line scales 45-52; total gill rakers on first arch 9-12; *slope of head profile moderately steep; pectoral fins moderate, 2.3-4.1 in standard length; body relatively deep, 1.9-2.6 in standard length*; jaw teeth in bands, outer row teeth enlarged at front of jaws and somewhat caniniform; molariform teeth on lateral part of jaws; pectoral fins long, reaching level of anal spines; caudal fin forked; overall silvery; lips and chin white; *axil of pectoral fin blackish*. Baja California to Peru; usually seen on flat sandy bottoms in clear water. To 44 cm.

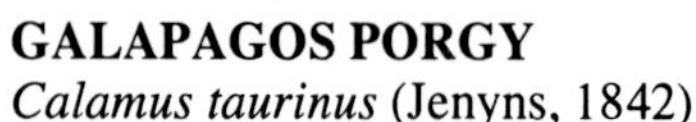

GALAPAGOS PORGY

Calamus taurinus (Jenyns, 1842)

Dorsal rays X-XIII,11-13 (usually XII,12); anal rays III,10-11 (usually 10); pectoral rays 14-16 (usually 15); lateral-line scales 46-50; total gill rakers on first arch 9-12; *slope of head profile gentle and rather smoothly rounded; pectoral fins short, 2.6-4.0 in standard length; body depth 2.2-2.4 in standard length*; jaw teeth in bands, outer row teeth enlarged at front of jaws and somewhat caniniform; molariform teeth on lateral part of jaws; caudal fin forked; generally silver gray with irregular blackish spots or blotches on body; rear edge of gill cover blackish; *a black bar on upper pectoral fin base, and axil of fin blackish*. Galapagos Islands; inhabits sand-rubble bottoms adjacent to reefs in 10-40 m. Attains 40 cm.

CROAKERS OR DRUMMERS (CORVINAS, COCOCHAS, CAJERAS, BOTELLONES, BERRUGATOS)

FAMILY SCIAENIDAE

This is one of the largest families in the region. Approximately 80 species have been recorded from our area and its surrounding waters. We provide coverage of 50 of the most common inshore species. Sciaenids are found worldwide, mainly in tropical and subtropical seas, but a number of species also inhabit fresh waters (particularly in South America). They are usually elongate to ovate fishes, with a long-based dorsal fin that is deeply notched (or completely separated in a few species) between the spinous and soft portions. Other general features include a relatively blunt snout in most species; conspicuous pores on the snout and lower jaw; a single barbel or patch of small barbels on the chin of some species; a bony flap above the gill opening; large, cavernous sensory canals on the head; unusully large otoliths; and a swim bladder that frequently has many branches. These fishes derive their common name from the croaking or drumming sounds that are produced by specialized muscles on the body wall connected to the swim bladder. The bladder serves as a resonating chamber that amplifies the sound. Sciaenids are generally bottom dwellers that inhabit sandy or muddy areas, frequently off beaches or in sheltered bays, estuaries, and river mouths. Other species occur offshore on the continental shelf (usually in less than 50 m depth), and are an important component of trawl fisheries. The annual landing of sciaenids is very substantial, both from trawlers and from gill-net fishermen. Indeed, the various species of corvina are a common element of fish markets throughout the region. They are generally carnivorous, feeding on a variety of small fishes and benthic invertebrates. Worldwide there are an estimated 80 genera and about 300 species. Our species accounts are based on material provided by L. Chao (currently reviewing the eastern Pacific species for a United Nations FAO publication), and published works by Mc Phail (1958) and Araya (1984).

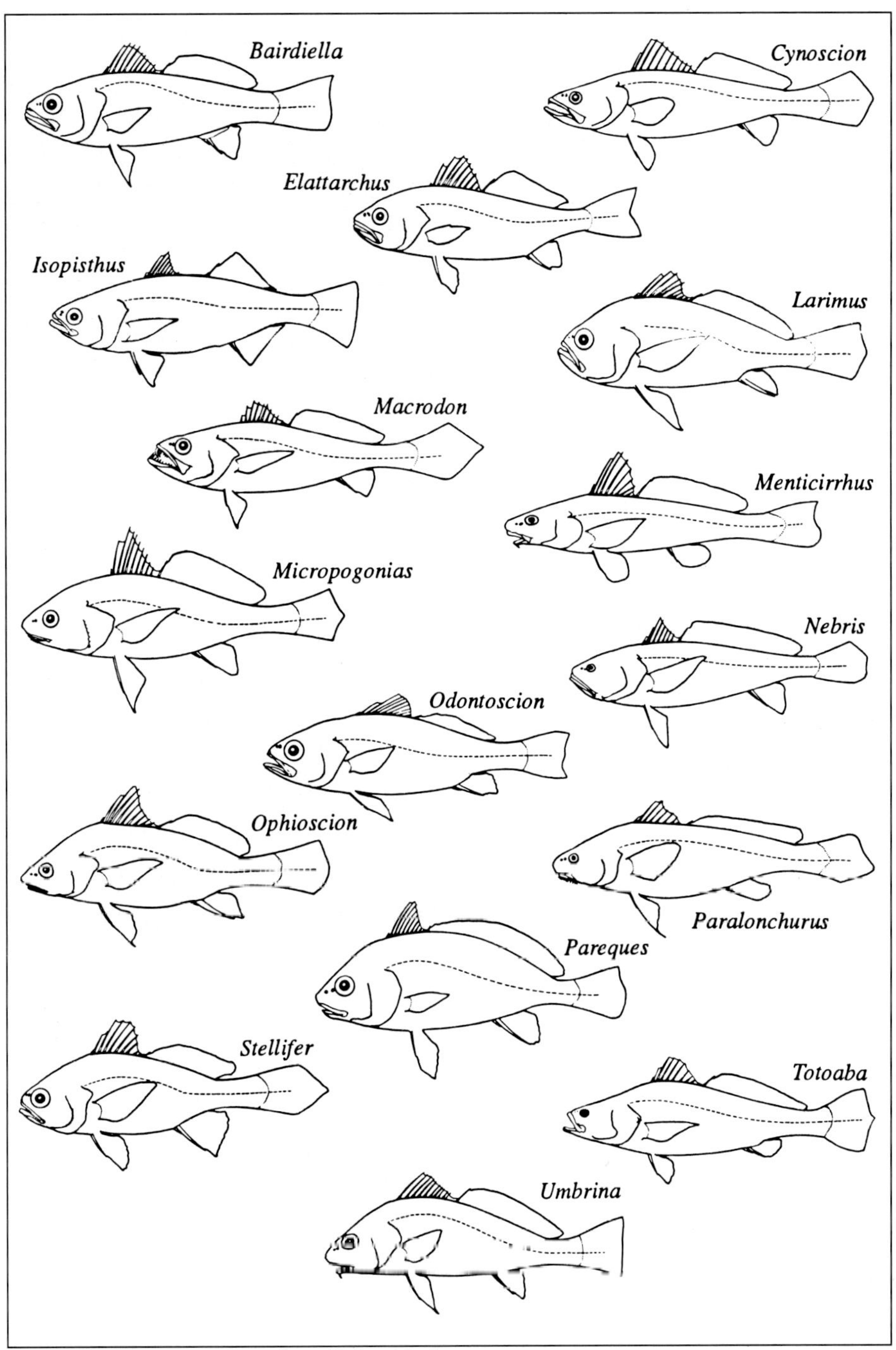

Eastern Pacific genera of croakers (Sciaenidae)

ARMED CROAKER
Bairdiella armata Gill, 1863

Dorsal rays X+I, 21-23; anal rays II,8; pored lateral-line scales 46-49; gill rakers on lower limb of first arch 14-16, total rakers 22-24; mouth sub-terminal; no canines in lower jaw; 5 mental pores on chin, *no barbel; preopercular margin with 8-12 fine spines and 2 sharp spines at angle, the lowermost directed downwards or slightly forward; second anal spine very strong, longer than base of anal fin*; caudal fin slightly rounded; scales ctenoid, except cycloid around eyes and on fins; mainly silvery, with fins (especially pectoral and pelvic fins) slightly yellowish; faint wavy stripes or dots sometimes apparent along horizontal scale rows of body. Gulf of California to Panama; inhabits inshore shallows, entering estuaries and coastal rivers; feeds mainly on crustaceans. Reaches maximum length of about 25-30 cm.

Bairdiella armata ▲ *Bairdiella ensifera* ▼

SWORDSPINE CROAKER
Bairdiella ensifera Jordan & Gilbert, 1862

Dorsal rays X+I,22-24; anal rays II,8-9; pored lateral-line scales 47-55; gill rakers on first arch 16-18, total rakers 24-26; mouth terminal, oblique; no canines in lower jaw; snout short, less than eye diameter; 5 mental pores on chin, *no barbel; preopercular margin strongly serrated with 8-12 spines, including 2-3 strong spines at angle, the lowermost directed downwards and forward; second anal spine extremely strong, reaching caudal fin base when depressed*; caudal fin double emarginate to nearly truncate; scales ctenoid, except cycloid below eyes and on fins; silvery, with dark spot on axil of pectoral fins; pectoral, pelvic, anal, and caudal fins yellowish; *tip of lower jaw with black lining*. Mexico to Gulf of Guayaquil, Peru; inhabits coastal seas, estuaries, and river mouths; sometimes found in pure fresh water; feeds on crustacean larvae, shrimps, and fishes. Grows to about 30 cm.

ROMEO CROAKER
Bairdiella icistia (Jordan & Gilbert, 1881)
(Plate XI-I, p. 171)

Dorsal rays X+I or II,25-29; anal rays II,8; pored lateral-line scales 49-54; gill rakers on lower limb of first arch 15-18, total rakers 25-27; mouth terminal, oblique; no canines in lower jaw; snout blunt, length slightly greater than eye diameter; 5 mental pores on chin, *no barbel; preopercular margin with 8-9 spines, including 2-4 strong spines at angle, the lowermost directed downwards and forward; second anal spine about as long as first soft anal ray*; caudal fin truncate; scales ctenoid, except cycloid around eyes and on fins; silvery with yellowish to clear fins; a dark spot at base of pectoral fins; *upper lip and tip of lower jaw dusky*. Gulf of California to Chiapas, Mexico; inhabits inshore areas, including estuaries and river mouths; feeds on crustacean larvae, shrimps and fishes. To 30 cm.

VACUOCUA
Corvula macrops Steindachner, 1876
(Plate XI-5, p. 171)

Dorsal rays X or XI+I or II,24-27; anal rays II,9-11; pored lateral-line scales 45-54; gill rakers on lower limb of first arch 14-17, total rakers 22-26; mouth subterminal, oblique; snout blunt, shorter than eye diameter; 5 mental pores on chin, the median pair sharing a single opening, *no barbel; margin of preopercle smooth to slightly denticulate; second anal spine strong, about three-fourths length of first soft ray*, slightly longer than anal fin base; caudal fin truncate to slightly rounded; scales ctenoid, except cycloid on head and fins; mainly silvery, with faint stripes or dots along scale rows above lateral line; a dark spot in pectoral fin axil; *a dark lining behind front teeth of lower jaw*. Gulf of California (southern part) to Ecuador and Galapagos. Inhabits rocky coastal areas. Maximum size to about 25 cm.

WHITE CORVINA
Cynoscion albus (Günther, 1864)

Dorsal rays IX or X+I,19-22; anal rays II,8-9; pored lateral-line scales about 54-58; *gill rakers long and slender, 8-9 on lower limb of first arch*; mouth oblique, lower jaw projecting beyond upper jaw; *no pores or barbel on chin; teeth in multiple rows with 1-2 large canines at tip of upper jaw*; margin of preopercle without serrations; *pectoral fins relatively long, reaching halfway to anus*; rear margin of caudal fin lanceolate, the middle rays much longer than the others; *lateral line arched anteriorly and straight well before level of anus*; mainly silvery, bluish gray above, with black opercular lining; fins clear or whitish. Southern Mexico to Ecuador; inhabits coastal waters, including estuaries and rivers. Reaches 70 cm. Also known as Whitefin Weakfish.

GULF CORVINA
Cynoscion orthonopterus Jordan & Gilbert, 1881
(Plate XI-2, p. 171)

Dorsal rays IX or X+I,23-27; anal rays II,10-11; pored lateral-line scales 54-66; gill rakers long and well developed, 13-16 on lower limb, 22-27 total rakers on first arch; mouth oblique, lower jaw projecting beyond upper jaw; *no pores or barbel on chin; teeth in multiple rows, with 1-2 large canines at tip of upper jaw; margin of preopercle with serrations; pectoral fins reaching slightly beyond tips of depressed pelvic fins; soft dorsal fin with a scaly sheath covering at least lower half of fin*; caudal fin squarish or slightly concave; *lateral line arched anteriorly and becoming straight near level of soft dorsal fin origin*; silvery, blue-gray above, with dark dots on scales; tip of chin dusky; lining of opercle dusky; fins yellowish. Gulf of California; inhabits coastal seas. Reaches at least 65 cm. Also known as Gulf Weakfish.

SHORTFIN CORVINA
Cynoscion parvipinnis Ayres, 1862
(Plate XI-7, p. 171)

Dorsal rays VIII to X+I,21-24; anal rays II,10-11; pored lateral-line scales 65-75; gill rakers on lower limb of first arch 7-9, total rakers 10-12; mouth oblique, lower jaw projecting beyond upper jaw; *no pores or barbel on chin; margin of preopercle without serrations; teeth in multiple rows, with 1-2 large canines at tip of upper jaw*; upper jaw extends beyond hind margin of eye; *pectoral fins short, extending well short of tips of depressed pelvic fins; caudal fin slightly concave; soft dorsal fin without a scaly sheath, or with narrow sheath at base of fin; lateral line arched anteriorly*; silvery, blue-gray dorsally; pectoral fin axil dusky; inside of mouth yellow-orange. Southern California and Gulf of California to Mazatlan, Mexico; inhabits inshore areas, usually off sandy beaches. Grows to 60 cm.

Cynoscion phoxocephalus

CACHEMA WEAKFISH
Cynoscion phoxocephalus Jordan & Gilbert, 1882

Dorsal rays IX or X-I,20-22; anal rays II,9-10; pored lateral-line scales 60-70, total rakers 9-11; mouth oblique, lower jaw projecting beyond upper jaw; *no pores or barbel on chin; teeth in multiple rows, with pair of large canines at tip of upper jaw; margin of preopercle without serrations*; upper jaw not reaching level of rear margin of eye; pectoral and pelvic fins short; caudal fin lanceolate, the middle rays much longer than the others; *soft dorsal fin without a scaly sheath; lateral line arched anteriorly*; silvery, bluish gray on upper back; dorsal and caudal fins dusky, other fins whitish; axil of pectoral fin blackish. El Salvador to Peru; inhabits coastal waters; juveniles enter river mouths and shallow estuaries. To 60 cm.

STRIPED WEAKFISH
Cynoscion reticulatus (Günther, 1864)

Dorsal rays X+I,25-29; anal rays II,9; pored lateral-line scales 56-61; gill rakers on lower limb of first arch 6-8, total rakers 8-11; mouth oblique, lower jaw projecting beyond upper jaw; *at least 2, sometimes several tiny pores on tip of chin, no barbel; teeth in 1-3 rows, a pair of large canines at tip of upper jaw; margin of preopercle without serrations; pectoral fins reach beyond tips of depressed pelvic fins*; caudal fin truncate, rhombic, or pointed in young; *soft dorsal fin without a scaly sheath; generally silver, with brownish wavy streaks on back and sides*; pale along lateral line; axil of pectoral fin dark; margin of dorsal fin dusky, other fins yellowish. Gulf of California to Colombia; inhabits coastal waters including deeper parts of estuaries. Attains maximum length of 90 cm.

SCALEFIN WEAKFISH
Cynoscion squamipinnis (Günther, 1869)
(Plate XI-4, p. 171)

Dorsal rays VII or VIII+I,21-23; anal rays II,9; pored lateral-line scales 51-61; gill rakers on lower limb of first arch 9-10, *total rakers 12-16*; mouth oblique, lower jaw projecting beyond upper jaw; *no pores or barbel on chin; teeth in 1-2 series, a pair of large canines at tip of upper jaw; margin of preopercle without serrations; pectoral fins reach beyond tips of depressed pelvic fins; caudal fin rhombic, the middle rays longest; soft dorsal and anal fin completely or mostly covered with scaly sheath*; silvery, grayish on upper back; tip of lower jaw dusky; inside of mouth yellow-orange; pectoral axil dark; dorsal and anal fins dusky; tip of caudal fin blackish. Gulf of California to northern Peru; inhabits coastal seas; common in shrimp trawls, but rare in estuaries. Grows to a maximum size of 64 cm.

STOLZMANN'S WEAKFISH
Cynoscion stolzmanni (Steindachner, 1879)
(Plate XI-10, p. 171)

Dorsal rays X+I,19-21; anal rays II,8-9; pored lateral-line scales 60-61; gill rakers on lower limb of first arch 8-9, total rakers 10-12; mouth oblique, lower jaw projecting beyond upper jaw; *no pores or barbel on chin; teeth in 1-2 series, a pair of large canines at tip of upper jaw; margin of preopercle without serrations; pectoral fins short, extending well short of tips of depressed pelvic fins; caudal fin double emarginate, pointed in young; soft dorsal fin without a scaly sheath; lateral line decurved anteriorly*; silver, steel blue on upper back; pectoral axil dark; soft dorsal fin with dusky margin, other fins yellowish. Mexico to Peru; inhabits coastal waters, including estuaries. Attains at least 90 cm.

ORANGEMOUTH CORVINA
Cynoscion xanthulus (Jordan & Gilbert, 1881)
(Plate XI-8, p. 171)

Dorsal rays IX+I,19-20; anal rays II,8-9; pored lateral-line scales 58-68; gill rakers short and thick; gill rakers on lower limb of first arch 8-10, total rakers 12-14; head elongate, compressed; snout pointed; mouth oblique, lower jaw projecting beyond upper jaw; *no pores or barbel on chin; teeth in 1-2 series, a pair of moderately large canines at tip of upper jaw; margin of preopercle without serrations; pectoral fins short, not reaching beyond tips of depressed pelvic fins; caudal fin double emarginate, pointed in young; soft dorsal fin without a scaly sheath; lateral line arched anteriorly, becoming straight at level of anus*; fourth dorsal spine the longest; silver, bluish on upper back; inside of mouth bright yellow-orange; pectoral axil blackish; caudal fin yellow. Northern Gulf of California to Acapulco, Mexico; inhabits coastal waters. Grows to 90 cm.

BLUESTREAK DRUM
Elattarchus archidium (Jordan & Gilbert, 1882)

Dorsal rays X+I or II,24-28; anal rays II,8-9; pored lateral-line scales 47-56; *gill rakers on lower limb of first arch 13-18, total rakers 19-25*; snout blunt, mouth terminal and oblique; *5 mental pores on chin, no barbel; teeth in 2-3 rows in upper jaw, the outer row enlarged and canine-like, a pair of large canines at tip of lower jaw; margin of preopercle with 5-7 spines at angle, the lowermost directed downwards or slightly forward; pectoral fins long, reaching beyond tips of depressed pelvic fins; second anal spine strong, longer than base of anal fin; caudal truncate*; silver, bluish gray on upper back; dark, oblique stripes often present on sides (following scale rows); tip of spinous dorsal fin dusky, pectoral and anal fins clear to yellowish; *young with pair of black stripes on upper side, and black spot at base of caudal fin.* Gulf of California to Peru; inhabits shallow coastal waters, often off sandy beaches or in sheltered bays, but seldom in estuaries; the young form large aggregations on or near rocky reefs. To 25 cm.

Elattarchus archidium Juvs. ▲ Juv. ▼

SILVER WEAKFISH

Isopisthus remifer (Jordan & Gilbert, 1881)

Dorsal rays VII to IX+I or II,21-23; *anal rays II, 17-19*; pectoral rays 17-18; pored lateral-line scales 52-57; gill rakers on lower limb of first arch 7-10, total rakers 11-15; mouth oblique, lower jaw projecting beyond upper jaw; *no pores or barbel on chin; a pair of large canines at tip of upper jaw; margin of preopercle smooth or ciliated; spinous dorsal fin well separated from soft dorsal fin; pectoral fins long, reaching beyond tips of depressed pelvic fins; caudal fin double truncate to truncate; soft dorsal fin and anal fins with well-developed scaly sheath*; silver, bluish gray on upper back; tip of lower jaw and opercular lining black; pectoral axil dark; fins pale to yellowish. Baja California to northern Peru; inhabits coastal waters and lower parts of estuaries. Size to 35 cm.

STEEPLINED DRUM

Larimus acclivis Jordan and Bristol, 1898

Dorsal rays X+I,27-30; anal rays II,6; pored lateral-line scales 46-52; gill rakers on lower limb of first arch 13-16, total rakers 18-22; mouth strongly oblique, lower jaw projecting and directed upward; *eye large, 3.6-3.7 in head length; 4 minute pores on tip of chin, no barbel; teeth small, in a single series; margin of preopercle without serrations; pectoral fins reaching beyond tips of depressed pelvic fins to about level of anus; second anal spine strong, longer than first soft ray; caudal fin double truncate, pointed in young; soft dorsal and anal fins with scaly sheath on basal half*; silver, grayish on upper back; *sides with distinct blackish stripes following scale rows*; fins yellowish, especially pectoral, pelvic, anal, and caudal fins. Gulf of California and west coast of Baja California to northern Peru; inhabits coastal waters. Attains a length of 26 cm.

Larimus acclivis ▲

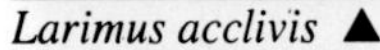

Larimus argenteus ▼

SILVER DRUM

Larimus argenteus (Gill, 1863)

Dorsal rays X+I,27-29; anal rays II,6; pored lateral-line scales 46-55; gill rakers on lower limb of first arch 15-18, total rakers 23-26; *mouth very strongly oblique (almost vertical), lower jaw projecting; nape slightly concave; eye moderate, 3.9-5.0 in head length; bony ridge on top of head spongy to touch; 2 minute pores on tip of chin, no barbel; teeth small, in single row; margin of preopercle without serrations; pectoral and pelvic fin tips reaching to level of anus; second anal spine strong, slightly shorter than first soft ray; caudal fin double truncate, pointed in young; soft dorsal and anal fins with scaly sheath on basal half*; silvery, grayish blue on upper back; sometimes with faint dark stripes on sides; axil of pectoral fin dark; fins clear to slightly yellowish. Gulf of California to northern Peru; inhabits coastal waters; common in estuaries. To 35 cm.

SHINING DRUM

Larimus effulgens Gilbert, 1898
(Plate XI-9, p. 171)

Dorsal rays X+I,28-30; anal rays II,6; pored lateral-line scales 45-49; gill rakers on lower limb of first arch 19-21, total rakers 26-31; mouth strongly oblique, lower jaw projecting; *eye 4.2-4.8 in head length; 2 minute pores on tip of chin, no barbel; teeth small, in 1-2 rows; margin of preopercle without serrations; pectoral fins reaching beyond tips of depressed pelvic fins and beyond level of anus; second anal spine strong, slightly shorter than first soft ray; caudal fin double truncate, pointed in young; soft dorsal and anal fins with scaly sheath on basal one-third*; silver, grayish on upper back; axil of pectoral fin blackish; pectoral, pelvic, anal, and caudal fins frequently yellow. Gulf of California to northern Peru; inhabits coastal waters. Maximum size 28 cm.

PACIFIC DRUM

Larimus pacificus Jordan & Bollman, 1889
(Plate XI-6, p. 171)

Dorsal rays X+I,26-28; anal rays II,6; pored lateral-line scales 50; gill rakers on lower limb of first arch 20-21, total rakers 31-33; *mouth not as oblique as most other members of the genus, much more horizontal; lower jaw projecting slightly; eye 3.6-4.0 in head length; no barbel on chin; teeth small, in 1-2 series; margin of preopercle without serrations; pectoral fins relatively long, reaching to about level of anus; pelvic fins shorter, reaching about halfway to anal fin origin; second anal spine strong, but shorter than soft rays; caudal fin double emarginate to truncate; soft dorsal and anal fins with scaly sheath on basal half*; silver, grayish on upper back; sides sometimes with darkish stripes following scale rows; axil of pectoral fin dark; fins pale to dusky. Gulf of California to Colombia; inhabits coastal waters. Reaches length of at least 15 cm.

DOGTOOTH WEAKFISH

Macrodon mordax Gilbert & Stark, 1904

Dorsal rays IX or X+I,27-30; anal rays I or II,9-10 ; pored lateral-line scales 45-55; gill rakers on lower limb of first arch 7-8, total rakers 9-11; *mouth large and oblique, lower jaw projecting*; eye moderate, 5.6-6.5 in head length; minute pores often present on tip of chin, no barbel; *prominent canine teeth in both upper and lower jaws*; margin of preopercle without serrations; pectoral fins long, reaching beyond tips of depressed pelvic fins; dorsal and anal spines weak; caudal fin double emarginate, pointed in young; soft dorsal and anal fins with scaly sheath covering almost entire fin; silvery, bluish or greenish grayish on upper back; axil of pectoral fin dark; fins whitish to slightly dusky. Panama to Colombia; inhabits coastal waters, including bays and estuaries. Grows to 43 cm.

PACIFIC KINGCROAKER

Menticirrhus elongatus (Günther, 1864)
(Plate XI-11, p. 171)

Dorsal rays X+I,22-24; *anal rays I,7; pectoral rays 17-20 (usually 19 or fewer)*; pored lateral-line scales 49-60; gill rakers on lower limb of first arch 0-6, total rakers 7-10; mouth small, inferior to upper jaw which encloses it; *a short, rigid barbel on chin*; teeth villiform, no canines; outer row teeth of upper jaw scarcely enlarged if at all; *margin of preopercle weakly serrated*; pectoral fins reaching beyond tips of depressed pelvic fins; anal fin with a single, weak spine; margin of caudal fin S-shaped with pointed upper lobe; soft dorsal and anal fins with low, basal scaly sheath; scales of head and body ctenoid, except smaller cycloid scales on breast; silvery, grayish blue on upper back; dorsal and anal fins dusky. Baja California to southern Peru; inhabits coastal waters, frequently off sandy beaches. Attains a length of 50 cm. Another species, *M. paitensis* (not illustrated), has a similar distribution and general apperance. It is also characterized by 7 anal rays, but differs from *M. elongatus* in having 21-23 pectoral rays.

HIGHFIN KINGCROAKER

Menticirrhus nasus (Günther, 1864)
(Plate XI-12, p. 171)

Dorsal rays X+I,21-23; *anal rays I,8; pectoral rays 21-22*; pored lateral-line scales 49-56; gill rakers on lower limb of first arch 0-6, total rakers 4-9; mouth small, inferior to upper jaw which encloses it; *a short, rigid barbel on chin*; teeth villiform, no canines; outer row teeth of upper jaw slightly enlarged; *margin of preopercle weakly serrated*; pectoral fins reaching beyond tips of depressed pelvic fins; anal fin with a single weak spine; margin of caudal fin S-shaped with elongated upper lobe; soft dorsal and anal fins without basal scaly sheath; grayish, lighter below; faint dark streaks following scale rows on back, often with numerous dark spots on sides; spinous dorsal fin and pectorals blackish; remaining fins dusky to blackish. Baja California to northern Peru; inhabits coastal waters, including bays and the lower reaches of estuaries. To 50 cm.

Macrodon mordax ▲ *Menticirrhus panamensis* (see description on p. 163) ▼

PANAMA KINGCROAKER

Menticirrhus panamensis (Steindachner, 1875)
(Photograph on p. 162)

Dorsal rays IX-X+I,18-22; *anal rays I,8-9; pectoral rays 20-23*; pored lateral-line scales 52-55; gill rakers on lower limb of first arch 0-5, total rakers 3-8; mouth small, inferior to upper jaw which encloses it; *a short, rigid barbel on chin; teeth villiform, no canines; outer row teeth of upper jaw distinctly enlarged; margin of preopercle weakly serrated*; pectoral fins reaching beyond tips of depressed pelvic fins; anal fin with a single weak spine; margin of caudal fin S-shaped; soft dorsal and anal fins without basal scaly sheath; dusky silver, grayish brown on upper back; head brownish; fins dusky, except *pectorals jet black* and pelvics whitish. Baja California to Chile and Peru; inhabits coastal waters, including bays and off sandy beaches. Grows to 55 cm.

Micropogonias altipinnis ▲ *Odontoscion eurymesops* ▼

HIGHFIN CORVINA

Micropogonias altipinnis (Günther, 1864)

Dorsal rays X+I,19-23; anal rays II,7-8; pectoral rays 17-19; pored lateral-line scales 48-52; gill rakers on lower limb of first arch 15-16, total rakers 24-25; mouth small, inferior to upper jaw; lips thin; preorbital very broad; *eye relatively small, its width much less than snout length (eye 2.0 or more in snout); chin with 4-5 pairs of small barbels*; teeth villiform, no canines; outer row teeth of upper jaw slightly enlarged; *margin of preopercle strongly serrated, 2 spines at the angle; first dorsal fin tall, reaching beyond origin of second dorsal fin when depressed*; margin of caudal fin double truncate, upper lobe slightly concave; soft dorsal fin with low, basal scaly sheath; silvery with slight golden hue; scale rows above lateral line with faint brownish longitudinal streaks; also several oblique bars running upward and backward, crossing arched portion of lateral line; fins yellowish, dorsal and caudal fins somewhat dusky. Gulf of California to Peru; inhabits coastal waters. Attains 66 cm.

VERRUGATO

Micropogonias ectenes Jordan & Gilbert, 1881
(Plate XII-1, p. 173)

Dorsal rays X+I,23-26; anal rays II,7-8; pectoral rays 16-18; pored lateral-line scales 51-54; gill rakers on lower limb of first arch 13-16, total gill rakers 22-24; mouth small, inferior to upper jaw; lips thin; preorbital very broad; *eye width much less than snout length (eye about 1.7-1.8 in snout); lower jaw with series of minute barbels on each side*; teeth villiform, no canines; outer row teeth of upper jaw slightly enlarged; *margin of preopercle strongly serrated; first dorsal fin barely reaching or short of origin of second dorsal fin when depressed*; margin of caudal fin slightly double emarginate; soft dorsal fin with low, basal scaly sheath; silvery gray with brassy hue; upper sides with dark, oblique streaks extending upward and backward along scale rows; fins yellowish, spinous dorsal fin with blackish outer margin. Gulf of California; inhabits coastal waters. Grows to about 40 cm. Also known as Slender Croaker.

BIGEYE CROAKER

Micropogonias megalops Gilbert,1890
(Plate XII-4, p. 173)

Dorsal rays X+I,27-29; anal rays II,7-8; pectoral rays 17-18; pored lateral-line scales 49-53; gill rakers on lower limb of first arch 15-17, total rakers 23-27; mouth small, inferior to upper jaw; lips thin; preorbital very broad; *eye relatively large, its width 4.0-4.7 in head length; chin with 3 pairs of small barbels*; teeth villiform, no canines; outer row teeth of upper jaw slightly enlarged; *margin of preopercle strongly serrated, 2 spines at angle; first dorsal fin just short of origin of second dorsal fin when depressed*; margin of caudal fin S-shaped; soft dorsal fin with low, basal scaly sheath; silvery with slight golden hue; scale rows above lateral line with faint, brownish, oblique streaks; fins yellowish. Gulf of California to vicinity of Acapulco, Mexico; inhabits coastal waters. Reaches at least 40 cm.

PACIFIC SMALLEYE CROAKER

Nebris occidentalis Vallant, 1897
(Plate XII-3, p. 173)

Dorsal rays VI or VII+I,28-31; *anal rays II,11-13*; pored lateral-line scales 45-59, 100-120 scales in lateral series above lateral line; gill rakers on lower limb of first arch 14-15, total rakers on first arch 20-23; *head broad, skull bones very cavernous, spongy to touch; mouth large, oblique, the lower jaw projecting; no barbels on chin; eye very small, 8.0-10.0 in head length*; teeth small, villiform, in a narrow ridge on jaws; margin of preopercle without serrations; margin of caudal fin rounded to double truncate; soft dorsal and anal fins with well-developed scaly sheath; silvery, grayish on upper back; sides with brownish spotting; fins dusky, except pelvics and anal yellowish; young with broad, vertical, dark bars on sides. Southern Mexico to Peru; inhabits coastal waters. Attains at least 60 cm.

GALAPAGOS CROAKER

Odontoscion eurymesops Heller & Snodgrass, 1903

Dorsal rays XI+II or III,24-27; anal rays II,8-9; pectoral rays 13-15; pored lateral-line scales 47-52; gill rakers on lower limb of first arch 16-18, total rakers on first arch 24-26; mouth large, oblique, the upper jaw slightly projecting; *no barbels on chin; eye large, 3.9-4.4 in head length; a row of somewhat enlarged sharp teeth in each jaw; margin of preopercle without serrations*; margin of caudal fin truncate or slightly rounded; soft dorsal and anal fins with well-developed scaly sheath; brownish with silvery reflections; *sides with narrow, longitudinal brown stripes following scale rows, these becoming oblique above lateral line; fins dusky to dark brown*; juvenile probably similar to that of *O. xanthops* (see below). Galapagos Islands; inhabits rocky reefs. To about 25 cm.

YELLOWEYE CROAKER
Odontoscion xanthops Gilbert,1898

Dorsal rays XI-XII+I,25-27; anal rays II,8-9; pectoral rays 14-16; pored lateral-line scales 48-50; gill rakers on lower limb of first arch 14-16, total rakers on first arch 21-23; mouth large, oblique, the lower jaw slightly projecting; *no barbels on chin; eye large, 3.3-3.9 in head length; usually enlarged canine teeth at tip of lower jaw; upper jaw with an outer row of slightly enlarged teeth; margin of preopercle smooth; second anal spine long, nearly equal to length of first soft anal ray*; margin of caudal fin truncate or slightly rounded; soft dorsal and anal fins with well-developed scaly sheath; brownish with silvery reflections; *sides with narrow, longitudinal brown stripes following scale rows, these becoming oblique above lateral line; fins dark brown*; juveniles silvery, with distinct, midlateral brown stripe and narrower stripes following scale rows; fins slightly dusky, except pelvics and anal dark brown. Gulf of California to Panama; inhabits rocky areas and coral reefs. To about 30 cm.

BLINKARD CROAKER
Ophioscion imiceps (Jordan & Gilbert,1881)
(Plate XII-2, p. 173)

Dorsal rays XI+I,25-26; anal rays II,8; pectoral rays 16-17; pored lateral-line scales 47-49; gill rakers on lower limb of first arch 13-15, total rakers on first arch 19-22; *snout bulbous, projecting well in front of upper jaw*; mouth horizontal or only slightly oblique; no barbels on chin; eye 4.2-4.5 in head length; teeth in villiform bands, those in outer row of upper jaw slightly enlarged; *margin of preopercle with 8-10 spines, strong at angle*; margin of caudal fin double truncate, middle rays the longest; soft dorsal and anal fins with scaly sheath; dull brownish with silvery reflections; fins dusky brown. Southern Mexico to Colombia; inhabits sand bottoms of bays. To about 25 cm.

TUZA CROAKER
Ophioscion scierus (Jordan & Gilbert,1881)

Dorsal rays X+I,24-26; anal rays II,7; pectoral rays 18-20; pored lateral-line scales 48-52; gill rakers on lower limb of first arch 9-13, total rakers on first arch 18-23; *snout scarcely projecting in front of upper jaw*; mouth horizontal or only slightly oblique; no barbels on chin; eye 4.5-5.5 in head length; *margin of preopercle with about 8-10 spines; second dorsal spine strong, about two-thirds length of third spine*; margin of caudal fin double truncate, middle rays the longest; pectoral fins relatively long, reaching past tips of depressed pelvic fins, but not reaching level of anus; soft dorsal and anal fins with scaly sheath; silvery, with dusky scale margins imparting overall duskiness; fins dusky brownish, except pelvics sometimes whitish. Gulf of California to Peru; inhabits coastal waters, including bays and river mouths. Grows to 35 cm.

Odontoscion xanthops Adult ▼ Juv. ▲

Ophioscion scierus ▼

SNUBNOSE CROAKER

Ophioscion simulus Gilbert,1898
(Plate XII-5, p. 173)

Dorsal rays XI+I,26; anal rays II,7; pectoral rays 17-18; about 50-51 pored scales in lateral line; gill rakers on lower limb of first arch 14-15, total rakers 21-22; *snout bulbous, mouth oblique*; no barbels on chin; *margin of preopercle with 14-16 spines or serrae, increasing in size ventrally*; second dorsal spine thick and elongated; second anal spine very strong, about as long as first soft anal ray; anus well in front of anal fin origin; margin of caudal fin rounded to S-shaped, middle rays the longest; pectoral fins reaching to about level of depressed pelvic fin tips or slightly beyond; soft dorsal and anal fins with scaly sheath; silvery gray, whitish below, without dark stripes or streaks, although scale margins may be dusky giving overall speckled impression; fins dusky brown. Costa Rica and Panama; inhabits sand bottoms of bays. Reaches at least 20 cm.

COCODIA CROAKER

Ophioscion strabo Gilbert,1896

Dorsal rays XI+I,22-24; anal rays II,7; pectoral rays 17-19; pored lateral-line scales 47-50; gill rakers on lower limb of first arch 11-13, total rakers on first arch 17-21; *snout bulbous, projecting slightly in front of upper jaw*; mouth horizontal or only slightly oblique; no barbels on chin; eye 4.1-4.3 in head length; *margin of preopercle with 5-7 spines or serrae*; second dorsal spine strong; second anal spine stout, nearly equal in length to first soft anal ray; margin of caudal fin double truncate, middle rays the longest; pectoral fins relatively short, barely reaching tips of depressed pelvic fins; soft dorsal and anal fins with scaly sheath; dull brownish with silvery reflections; fins dusky brown, except pelvics and anal nearly black. Gulf of California to Colombia; inhabits sand bottoms of bays, entering estuaries and river mouths. Attains at least 15 cm.

Ophioscion strabo ▲

Ophioscion typicus ▼

Ophioscion vermicularis ▼

POINT-NOSED CROAKER

Ophioscion typicus Gill,1863

Dorsal rays X+I,21-23; anal rays II,7; pectoral rays 16-18; pored lateral-line scales 46-52; gill rakers on lower limb of first arch 14-17; *snout bulbous, projecting well in front of upper jaw*; mouth horizontal or only slightly oblique; no barbels on chin; eye 3.8-4.9 in head length; teeth in villiform bands, those in outer row of upper jaw enlarged; *margin of preopercle with about 8-10 spines or serrae, increasing in size ventrally*; second dorsal and second anal spines very stout; margin of caudal fin double truncate or lanceolate, middle rays the longest; pectoral fins relatively short, falling well short of tips of depressed pelvic fins; soft dorsal and anal fins with scaly sheath; silvery with dusky scale margins, grayish on upper back; fins dusky except pelvics and anal spines whitish. Costa Rica to Colombia; inhabits sand bottoms of bays, abundant in estuaries. To about 30 cm.

WORM-LINED CROAKER

Ophioscion vermicularis (Günther, 1869)

Dorsal rays X+I,26-28; anal rays II,8; pectoral rays 19-20; pored lateral-line scales 48-50; *snout bulbous*; mouth moderately oblique; no barbels on chin; eye 4.9-5.3 in head length; *margin of preopercle with about 15-17 fine serrae, those at angle not greatly enlarged*; margin of caudal S-shaped, upper lobe pointed; pectoral fins relatively long, usually reaching slightly past tips of depressed pelvic fins; soft dorsal and anal fins with scaly sheath; dusky gray with silvery reflections; *prominent, oblique, blackish stripes following scale rows on body*; fins dusky brown to blackish. Panama to northern Peru; inhabits coastal waters. To at least 26 cm.

SUCO CROAKER
Paralonchurus dumerilii (Bocourt, 1869)

Dorsal rays IX+I,23-26; anal rays II,7; pectoral rays 17-18; pored lateral-line scales 49-52; gill rakers on lower limb of first arch 6-9, total rakers on first arch 13-16; snout bulbous, projecting beyond upper jaw; mouth inferior, moderately oblique; *lower jaw with 11-13 pairs of small barbels; eye relatively large, 4.7-5.3 in head length; teeth in villiform bands, none notably enlarged; margin of preopercle without serrations or spines*; caudal fin S-shaped, the longest rays below middle of fin; pectoral fins equal to or shorter than head; soft dorsal and anal fins with basal scaly sheath; *silvery white with 5-6 prominent black bars extending onto lower side*; fins dusky, although pelvics and anal mainly whitish; El Salvador to Peru; inhabits coastal waters. Maximum size about 45 cm.

CORVALO
Paralonchurus goodei Gilbert, 1898
(Plate XII-8, p. 173)

Dorsal rays X+I,27-28; anal rays II,7; pectoral rays 19; pored lateral-line scales 48-51; gill rakers on lower limb of first arch 6-7, total rakers on first arch 11-12; snout bulbous, projecting beyond upper jaw; mouth inferior, moderately oblique; *lower jaw with more than 30 pairs of small barbels; eye relatively small, 6.0-6.6 in head length; teeth in villiform bands, none notably enlarged; margin of preopercle without serrations or spines*; caudal fin S-shaped, the longest rays below middle of fin; pectoral fins about equal to head; soft dorsal and anal fins with basal scaly sheath; *dusky brownish with several faint blackish bars; body white ventrally; pectoral fins black*, other fins dusky; Gulf of California to Peru; inhabits coastal waters. Maximum size about 35 cm.

PETER'S CROAKER
Paralonchurus petersi Bocourt, 1869
(Plate XII-7, p. 173)

Dorsal rays X+I,30-31; anal rays II,7; pectoral rays 20-21; pored lateral-line scales 48-51; gill rakers on lower limb of first arch 10-11, total rakers on first arch 15-16; snout bulbous, projecting beyond upper jaw; mouth inferior, moderately oblique; *lower jaw with 22-24 pairs of small barbels; eye very small, about 9.0-10.0 in head length; teeth in villiform bands, outer row of upper jaw notably enlarged; margin of preopercle with weak serrations*; caudal fin more or less lanceolate, the longest rays below middle of fin; *pectoral fins very large, longer than head, reaching to origin of anal fin or beyond*; soft dorsal and anal fins with basal scaly sheath; *dusky blackish without bars*, whitish on ventral parts; fins dusky blackish, except *pectorals jet black*; El Salvador to Colombia; inhabits coastal waters. Grows to at least 35 cm.

BEARDED BANDED CROAKER
Paralonchurus rathbuni (Jordan & Bollman, 1889)

Dorsal rays X+I,28-29; anal rays II,8-9; pectoral rays 15-17; pored lateral-line scales 53-55; gill rakers on first arch 8-10, total rakers on first arch 12-15; snout bulbous, projecting beyond upper jaw; mouth inferior, slightly oblique; *lower jaw with 9-10 pairs of small barbels; eye relatively large, 4.3-5.0 in head length; teeth in villiform bands, those in outer row of upper jaw slightly enlarged; margin of preopercle weakly serrated*; caudal fin double truncate, the middle rays longest; pectoral fins short, 1.5-1.8 in head; soft dorsal and anal fins with basal scaly sheath; *silvery white with 6-7 prominent, dusky, blackish bars extending onto lower side, the anteriormost bar (at level of pectoral fin base) about twice width of other bars*; fins dusky, except pelvics and anal whitish; Panama to Peru; inhabits coastal waters. Attains at least 30 cm.

Pareques perissa Juv. ▲ Tiny juv. ▼

Pareques viola Large juv. ▲ Small juv. ▼

GALAPAGOS ROCK CROAKER
Pareques perissa (Heller & Snodgrass, 1903)
(Plate XII-6, p. 173)

Dorsal rays IX+I,32-33; anal rays II,8; pectoral rays 16-18; pored lateral-line scales 50-53; gill rakers on lower limb of first arch 6-7, total rakers on first arch 16-18; snout blunt, projecting slightly beyond upper jaw; mouth inferior, more or less horizontal; lower jaw or chin without barbels; eye relatively large, 3.0-4.0 in head length; teeth in villiform band, those in outer row of upper jaw distinctly enlarged; margin of preopercle more or less smooth or with very weak serrations; caudal fin asymmetrically rounded; second dorsal and anal fins covered with a thick scale sheath; *overall charcoal, including fins; juveniles silvery white with distinctive black stripes on sides*; Galapagos Islands; inhabits rocky reefs. To about 30 cm.

ROCK CROAKER
Pareques viola (Gilbert, 1904)

Dorsal rays IX+I,40-41; anal rays II,7-8; pectoral rays 15-17; pored lateral-line scales 50-56; gill rakers on lower limb of first arch 6, total rakers on first arch about 17; snout blunt, projecting slightly beyond upper jaw; mouth inferior, more or less horizontal; lower jaw without barbels; eye relatively large, 3.6-4.1 in head length; teeth in villiform band, those in outer row of upper jaw enlarged and canine-like; margin of preopercle without serrations or spines, but membranous edge finely ciliated; caudal fin rounded, median rays longest; second dorsal and anal fins covered with a thick scale sheath; *overall charcoal, including fins; often a purplish sheen on head and body; mouth whitish; small juveniles (known in the aquarium trade as "high-hats" because of the tall spinous dorsal fin) are white with distinctive black stripes and black bars on the spinous dorsal and pelvic fins; the stripes increase in number in larger juveniles, but gradually disappear in subadults*; Panama to Peru; inhabits rocky coastal reefs; adults are secretive, emerging from crevices at dusk to feed during the night; juveniles form small aggregations close to shelter. Maximum size to 25 cm. A very similar species, *P. fuscovittatus*, occurs in the Gulf of California and ranges south along the Mexican coast to at least Puerto Vallarta.

SHORTNOSE STARDRUM
Stellifer chrysoleuca (Günther, 1867)
(Plate XII-9, p. 173)

Dorsal rays X+I,20-23; anal rays II,8-9; pectoral rays 18-19; pored lateral-line scales 48-51; *gill rakers on lower limb of first arch 13-17 (usually 15-16), total rakers on first arch 21-26*; snout blunt, projecting slightly beyond upper jaw; mouth inferior, slightly oblique; chin without barbels; eye 4.5-5.2 in head length; teeth in villiform bands, those in outer row of upper jaw enlarged; *margin of preopercle with 9-14 spiny serrations, spine at angle and those below it enlarged and usually directed downward*; margin of caudal fin double truncate to rounded; silvery with dusky scale margins and faint, dusky, blackish stripes following scale rows; fins dusky blackish, except pectorals slightly yellowish and pelvics white. Mexico to Peru; inhabits coastal waters. Reaches about 32 cm.

Pareques viola ▼

HOLLOW STARDRUM
Stellifer ericymba (Jordan & Gilbert, 1882)
(Plate XII-10, p. 173)

Dorsal rays X or XI+II or III,20-23; anal rays II,8; pectoral rays 17-20; pored lateral-line scales 41-46; *gill rakers on lower limb of first arch 18-21, total rakers on first arch 28-36; head cavernous and spongy to touch*; snout blunt, not projecting beyond upper jaw; mouth terminal, oblique; lower jaw without barbels, but tip of chin with a bony knob; eye 4.4-4.8 in head length; teeth in villiform bands, those in outer row of upper jaw enlarged; *margin of preopercle with 4-10 radiating spines or serrae, those at angle usually enlarged but not directed downward*; margin of caudal fin lanceolate with the middle rays elongated; silvery, bluish gray on upper back; dusky spotting and faint blackish stripes following scale rows; head dusky; spinous dorsal fin and remaining fins slightly dusky. Central Mexico to Peru; inhabits coastal waters. To at least 16 cm.

WHITE STARDRUM
Stellifer furthii (Steindachner, 1875)

Dorsal rays IX-XI+I or II,22-26; anal rays II,8-9; pectoral rays 18-20; pored lateral-line scales 47-51; *gill rakers on lower limb of first arch 18-24 (usually 21-22), total rakers on first arch 31-39; head cavernous, but firm to touch*; snout blunt, projecting slightly beyond upper jaw; mouth inferior, slightly oblique; lower jaw or chin without barbels; eye 3.7-4.5 in head length; teeth in villiform bands, those in outer row of upper jaw enlarged; *margin of preopercle with 2 strong spines at angle , the lower 1 directed downward; pelvic fins with filamentous tips*; margin of caudal fin double truncate or rounded; silvery, light gray on upper back; ventral parts white, sometimes with pinkish tinge; fins pale yellowish or with pinkish tinge; roof of mouth black. Honduras to Peru; inhabits coastal waters. Reaches about 20 cm.

Stellifer furthii ▲

Stellifer illecebrosus ▼

SILVER STARDRUM
Stellifer illecebrosus (Gilbert, 1904)

Dorsal rays X or XI+II or III,18-22; anal rays II,11-12 (rarely 10); pectoral rays 17-19; pored lateral-line scales 47-51; *gill rakers on lower limb of first arch 12-14, total rakers on first arch 17-21*; snout blunt, projecting slightly beyond upper jaw; mouth inferior, nearly horizontal; lower jaw or chin without barbels; eye 4.1-5.0 in head length; teeth in villiform bands, those in outer row of upper jaw enlarged; *margin of preopercle with 9-12 spines or serrae; pelvic fins with filamentous tips*; margin of caudal fin double truncate, the middle rays elongated; silvery, bluish gray on upper back; dorsal fin dusky; caudal fin dusky with yellow tinge; pectorals, pelvics, and anal yellowish. Mexico to Peru; inhabits coastal waters. Grows to at least 25 cm.

YAWNING STARDRUM
Stellifer oscitans (Jordan & Gilbert, 1882)
(Photograph on p. 169)

Dorsal rays IX or X+II or III,21-25; anal rays II,8-9 (usually 8); pectoral rays 18-21; pored lateral-line scales 47-52; *gill rakers on lower limb of first arch 27-30, total rakers on first arch 45-52*; snout blunt, barely or not projecting beyond upper jaw; mouth terminal, strongly oblique; lower jaw or chin without barbels; eye 3.8-4.9 in head length; teeth in villiform bands, those in outer row of upper jaw enlarged; *margin of preopercle with 2 strong spines, the lower 1 directed downward and backward*; margin of caudal fin double truncate or somewhat rounded, the middle rays elongated; silvery, grayish above; fins dusky gray, except pectorals yellow. Honduras to Peru; inhabits coastal waters, abundant in estuaries. Reaches 25 cm.

SOFTHEAD STARDRUM

Stellifer zestocarus (Gilbert, 1898)

Dorsal rays IX or X+II or III,17-20; anal rays II,10 (rarely 9 or 11); pectoral rays 17-19; pored lateral-line scales 48-50; *gill rakers on lower limb of first arch 18-20, total rakers on first arch 29-30; top of head extremely cavernous and spongy to touch*; snout very short and blunt, not projecting beyond upper jaw; mouth terminal, oblique; lower jaw or chin without barbels; *eye very large, 3.3-3.4 in head length*; teeth in narrow bands, none notably enlarged; *margin of preopercle with a single strong spine at angle*; margin of caudal fin double truncate, the middle rays elongated; silvery or silvery gray; fins grayish to whitish. Costa Rica to Ecuador; inhabits coastal waters. Grows to at least 20 cm.

TOTUAVA

Totoaba macdonaldi (Gilbert, 1890)
(Plate XI-3, p. 171)

Dorsal rays X+I,23-26; anal rays II,7-8; pectoral rays 15-17; pored lateral-line scales 50-56; *gill rakers on lower limb of first arch 10-16, total rakers on first arch 15-20*; snout sharp, lower jaw projecting; mouth large, oblique; lower jaw or chin without barbels; *teeth in narrow bands, somewhat enlarged and conical in outer row of upper jaw, including a few sharp teeth at tip of jaw*; inner row teeth of lower jaw slightly larger than those of outer row; *margin of preopercle very finely serrate*; margin of caudal fin double truncate to double emarginate, the middle rays elongated; dusky silvery; fins blackish to slightly dusky. Gulf of California; formerly very abundant, but stock now seriously depleted due to overfishing. A much sought food fish that grows to 190 cm and nearly 100 kg. Now protected by law.

GALAPAGOS DRUM

Umbrina galapagorum Steindachner, 1878
(Plate XII-12, p. 173)

Dorsal rays X+I,26-30; anal rays II,6-7; pectoral rays 16-19; pored lateral-line scales 50-53; gill rakers on lower limb of first arch 11-14, total rakers on first arch 14-16; snout blunt, projecting beyond upper jaw; mouth inferior, more or less horizontal; *tip of chin with a single, thick barbel*; teeth in villiform bands, those in outer row of upper jaw somewhat enlarged; *margin of preopercle very finely serrate*; margin of caudal fin slightly emarginate, the upper rays elongated; silvery gray, yellowish below; faint, oblique, dark lines following scale rows; fins pale to slightly dusky. Galapagos Islands; inhabits sandy bays. To at least 45 cm.

Stellifer oscitans (see description on p. 168) ▼ *Stellifer zestocarus* ▲

YELLOWFIN DRUM

Umbrina roncador (Jordan & Gilbert, 1881)
(Plate XII-11, p. 173)

Dorsal rays X+I,24-30; anal rays II,7 (rarely 6); pectoral rays 16-19; pored lateral-line scales 51-54; gill rakers on lower limb of first arch 10-14, total rakers on first arch 15-22; snout blunt, projecting beyond upper jaw; mouth inferior, more or less horizontal; *tip of chin with a single, thick barbel*; teeth in villiform bands, those in outer row of upper jaw slightly enlarged; *margin of preopercle very finely serrate; second anal spine stout*, 2.2-2.7 in head length; margin of caudal fin slightly emarginate; bright silvery, bluish above; sides with oblique, undulating brown or olive lines following scale rows; fins yellow; *inner side of gill cover blackish*. Point Conception, California, south to Gulf of California; inhabits coastal areas, including sandy bays and harbors, to a depth of about 45 m. Attains 38 cm.

POLLA DRUM

Umbrina xanti Gill, 1862
(Plate XII-13, p. 173)

Dorsal rays X+I,26-30; anal rays II,6-7; pectoral rays 16-19; pored lateral-line scales 45-51; gill rakers on lower limb of first arch 12-13, total rakers on first arch 16-21; snout blunt, projecting beyond upper jaw; mouth inferior, more or less horizontal; *tip of chin with a single, thick barbel*; teeth in villiform bands, those in outer row of upper jaw slightly enlarged; *margin of preopercle very finely serrate; second anal spine slender*, 2.0-3.3 in head length; margin of caudal fin truncate to emarginate; silvery, greenish brown on upper back; dark lines following scale rows on back and sides; dorsal and caudal fins dusky, pectorals, pelvics and anal fin yellow; *inner lining of gill cover intensely black*. Southern California to Peru; inhabits coastal waters to a depth of about 35 m. Grows to about 35 cm.

PLATE XI

1 **ROMEO CROAKER** (*Bairdiella icistia*)

2 **GULF CORVINA** (*Cynoscion orthonopterus*)

3 **TOTUAVA** (*Totoaba macdonaldi*)

4 **SCALEFIN WEAKFISH** (*Cynoscion squamipinnis*)

5 **VACUOCUA** (*Corvula macrops*)

6 **PACIFIC DRUM** (*Larimus pacificus*)

7 **SHORTFIN CORVINA** (*Cynoscion parvipinnis*)

8 **ORANGEMOUTH CORVINA** (*Cynoscion xanthulus*)

9 **SHINING DRUM** (*Larimus effulgens*)

10 **STOLZMANN'S WEAKFISH** (*Cynoscion stolzmanni*)

11 **PACIFIC KINGCROAKER** (*Menticirrhus elongatus*)

12 **HIGHFIN KINGCROAKER** (*Menticirrhus nasus*)

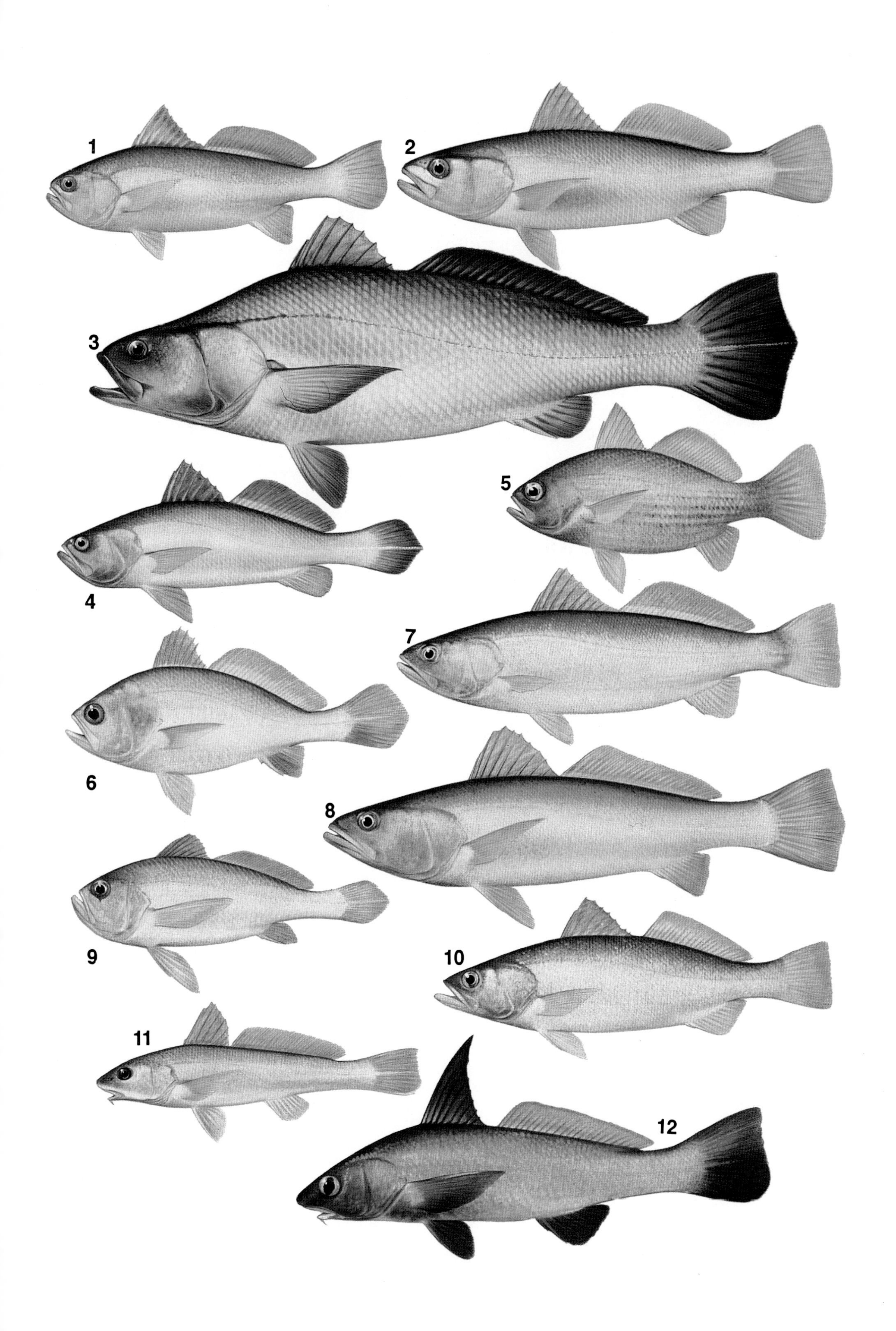
1
2
3
4
5
6
7
8
9
10
11
12

PLATE XII

1 **VERRUGATO** (*Micropogonias ectenes*)

2 **BLINKARD CROAKER** (*Ophioscion imiceps*)

3 **PACIFIC SMALLEYE CROAKER** (*Nebris occidentalis*)

4 **BIGEYE CROAKER** (*Micropogonias megalops*)

5 **SNUBNOSE CROAKER** (*Ophioscion simulus*)

6 **GALAPAGOS ROCK CROAKER** (*Pareques perissa*)

7 **PETER'S CROAKER** (*Paralonchurus petersi*)

8 **CORVALO** (*Paralonchurus goodei*)

9 **SHORTNOSE STARDRUM** (*Stellifer chrysoleuca*)

10 **HOLLOW STARDRUM** (*Stellifer ericymba*)

11 **YELLOWFIN DRUM** (*Umbrina roncador*)

12 **GALAPAGOS DRUM** (*Umbrina galapagorum*)

13 **POLLA DRUM** (*Umbrina xanti*)

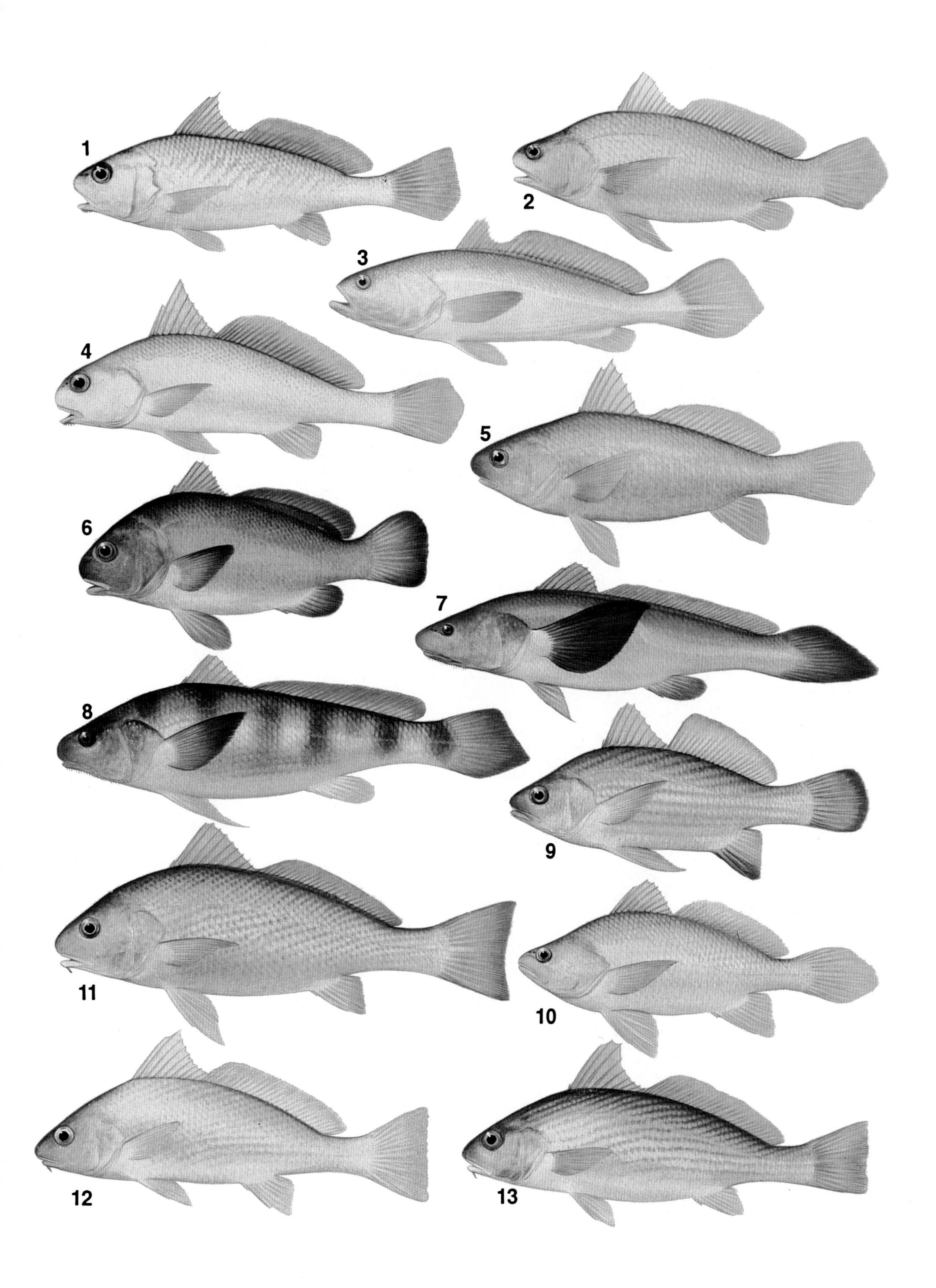
1
2
3
4
5
6
7
8
9
11
10
12
13

GOATFISHES (SALMONETES, TRILLAS, PECES CHIVO)

FAMILY MULLIDAE

Goatfishes occur in all warm temperate and tropical seas, usually in the vicinity of reefs. Worldwide there are six genera with about 55 species; most inhabit the Indo-West Pacific. Only two are found in our area. One of their most distinctive features is the presence of a pair of long chin barbels, which possess chemosensory organs and are used for detecting food. The barbels are also used by males to attract females during courtship. When they are not being used, the barbels are tucked tightly under the chin. Other characteristics include a moderately elongate body, a pair of well-separated dorsal fins, relatively small villiform or conical teeth, and a forked caudal fin. Goatfishes feed on small fishes and a variety of mainly sand- and weed-dwelling organisms including worms, shrimps, crabs, molluscs, and echinoderms. Most species are considered good eating.

MEXICAN GOATFISH

Mulloidichthys dentatus (Gill, 1863)

Dorsal rays VIII+I,8 (first dorsal spine minute, often embedded); anal rays I-II,6; pectoral rays 15-17; lateral-line scales 34-38; body relatively slender, the depth about 4.0 in standard length; yellow to greenish yellow on back and top of head, whitish on lower parts; *a broad, bright yellow midlateral stripe with bluish stripe immediately above; caudal fin bright yellow*. Gulf of California to Peru and the Galapagos; inhabits coral and rocky reefs; often seen over adjacent sand and rubble bottoms. Reaches 30 cm. Closely related to *M. martinicus* of the western Atlantic.

RED GOATFISH

Pseudupeneus grandisquamis (Gill, 1863)

Dorsal rays VIII+I,8 (first dorsal spine minute, often embedded); anal rays II,6; pectoral rays 13-16; lateral-line scales 28-32; body relatively deep, the depth 3.0-3.6 in standard length; *brownish red to violet on back, reddish or pink on sides, whitish below; a blackish blotch on lateral line below spinous dorsal fin; pearly spots and streaks on head, and narrow pearly stripes or rows of spots on upper side*. Gulf of California to Chile; usually found away from reefs on sand or mud bottoms; frequently caught by trawlers. To about 28 cm.

CHUBS (CHOPAS, BABUNCOS, GALLINAZOS)

FAMILY KYPHOSIDAE

The chubs, sometimes called rudderfishes or drummers, are moderately deep-bodied, compressed fishes with a small head and a small terminal mouth; the teeth are incisiform, and the maxilla slips partially under the preorbital bone when the mouth is closed. The scales are small and ctenoid, covering most of the head and soft portions of the median fins. The dorsal fin is continuous, and the caudal fin is emarginate to forked; the paired fins are relatively short, with the origin of the pelvics behind the pectoral base. These fishes are omnivorous, but feed mainly on benthic algae; their digestive tract is very long, as would be expected from their plant-feeding habits. They sometimes occur in huge aggregations that swarm over the reef's surface. The family is found circumglobally and contains about 45 species in 17 genera. It is divisible into two subfamilies, the Kyphosinae and the Girellinae (often regarded as a separate family). The latter group is mainly subtropical or temperate in distribution.

Cortez Chub (*Kyphosus elegans*), Perlas Islands, Panama

DUSKY CHUB
Girella freminvillei Jordan & Fesler, 1893

Dorsal rays XIII,15; anal rays III,12; scales in midlateral series about 50-55; jaw teeth freely movable; upper lip enlarged; overall bluish gray with blackish margin on gill cover; *a diagnostic whitish patch just in front of each eye*. Galapagos Islands; usually seen in grazing aggregations in 3-20 m. Maximum size to 45 cm.

GULF OPALEYE
Girella simplicidens Osburn & Nichols, 1916

Dorsal rays XIV,14; anal rays III,12; scales in midlateral series about 50; jaw teeth freely movable; dark brownish gray, usually slightly lighter on head and breast; *3 or 4 diffuse whitish spots on upper back below dorsal fin base*. Gulf of California; abundant in the northern Gulf, but scarce in the southern portion, often seen in large schools that graze on algae. To 46 cm.

ZEBRAPERCH
Hermosilla azurea Jenkins & Evermann, 1889

Dorsal rays XI,11; anal rays III,10; scales in oblique series below lateral line about 55; jaw teeth fixed, not freely movable; head scaleless; *sides with about 10 grayish brown bars with light gray spaces between them; a black spot on upper edge of gill cover and similar spot below pectoral fin base*; ventral portion of head and body whitish. California (Monterey southwards), Baja California, and Gulf of California; often seen in schools on shallow, weed-covered reefs; it is most abundant in the northern and central parts of the Gulf of California. Grows to 43 cm.

STRIPED SEA CHUB
Kyphosus analogus (Gill, 1863)

Dorsal rays XI,14; anal rays III,13-14; scales in oblique series just below lateral line 70-80; jaw teeth fixed, not freely movable; *anal fin very low and uniform in height; bluish gray with narrow brassy stripes between each scale row on sides*; small juveniles dark gray with elongated silvery white spots on sides. Southern California to Peru, including the Galapagos and other offshore islands; usually seen in schools at depths between 5-20 m. Grows to about 38 cm. Misidentified as *K. elegans* by Thomson et al. (1979).

Difference in dorsal and anal fin shapes of *Kyphosus analogus* (top) and *K. elegans*

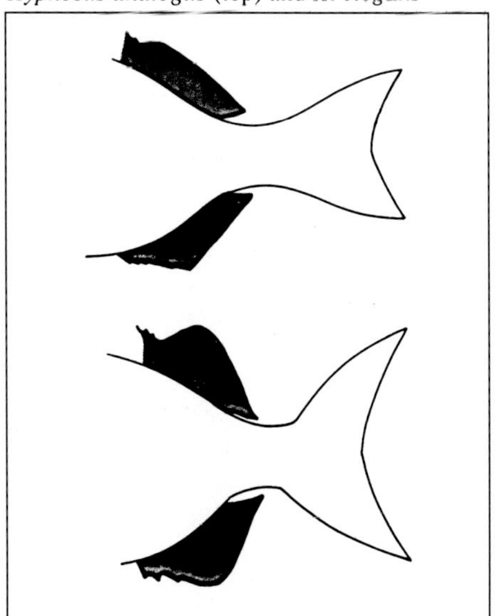

Kyphosus analogus ▲

CORTEZ CHUB
Kyphosus elegans (Peters, 1869)

Dorsal rays XI,13; anal rays III,12; scales in oblique series just below lateral line 59-62; jaw teeth fixed, not freely movable; *anterior rays of anal fin elevated; silvery gray with faint brownish stripes between each scale row on sides*; can quickly assume pattern of numerous eye-sized white spots. Gulf of California to Panama; often occurs in large schools in 3-12 m. Attains 38 cm.

Kyphosus elegans Spotted coloration ▲

RAINBOW CHUB
Sectator ocyurus (Jordan & Gilbert, 1881)

Dorsal rays XI,13; anal rays III,14; scales in midlateral series about 75-80; body relatively elongate; *caudal fin deeply forked*; silvery olive on back, white on lower parts; *a broad, bright blue midlateral stripe with yellow stripe immediately below, and blue band along base of dorsal fin; caudal fin bright yellow*. Society Islands to the tropical eastern Pacific; in our region from the tip of Baja California to the Galapagos; most common around offshore islands; also frequently seen near floating logs far out to sea. Reaches at least 60 cm.

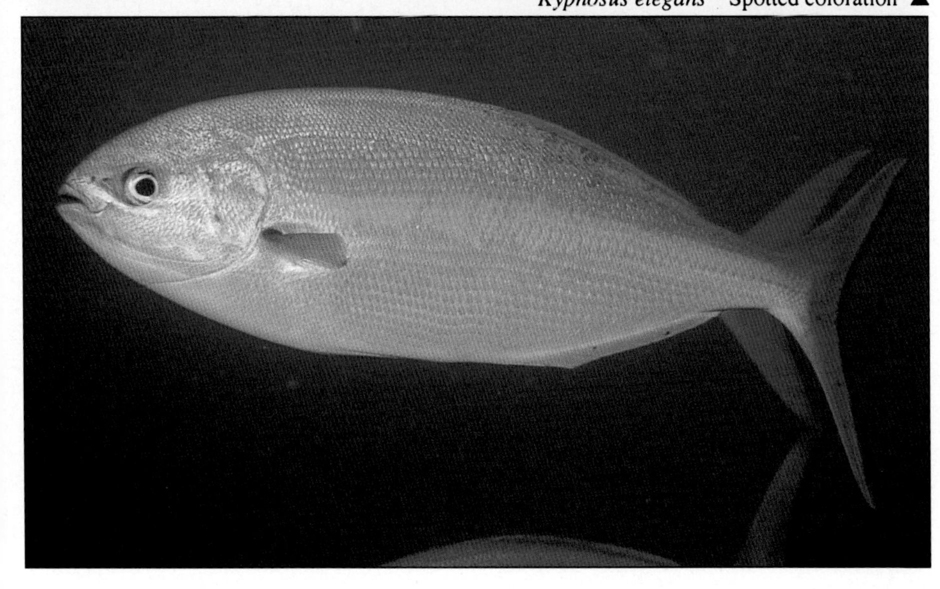

SPADEFISHES (CURACAS, PAGUALAS, PELUQUEROS)

FAMILY EPHIPPIDIDAE

Spadefishes are characterized by their deep, laterally compressed body, small terminal mouth with a band of brush-like teeth in the jaws, and a dorsal fin consisting of distinct spinous and soft-rayed portions (at least in eastern Pacific species). Young spadefishes frequently display a pattern of black bands or bars. The group consists of five genera and 17 species, and is divisible into three subfamilies. The subfamilies Drepaninae and Platacinae (containing *Platax*, popular aquarium fishes) are mainly Indo-West Pacific and are absent in our area. The subfamily Ephippidinae contains local representatives, as well as those from the western Pacific, Indian, and Atlantic oceans. Food items include algae and a variety of benthic and planktonic invertebrates including sponges, zoantharians, polychaete worms, gorgonians, and tunicates. The flesh of spadefishes is highly edible.

PACIFIC SPADEFISH

Chaetodipterus zonatus (Girard, 1858)

Dorsal rays VIII+I,18-23, *dorsal spines connected by membrane, the third spine tallest*; anal rays II-III,16-20; pectoral rays 17-18; lateral-line scales 46-49; body somewhat orbiculate and laterally compressed; dorsal and anal spines very low in comparison with soft parts of these fins; a band of brush-like teeth in jaws, the outer row the largest; silvery white to grayish, with *6 black bars on head and side*; fins dusky blackish, except pectoral and caudal yellowish; juveniles coppery brown, but dark bars evident. Southern California to Peru; inhabits bays and inlets with sand or rubble bottoms; also captured by trawlers on open sandy bottoms. To 65 cm.

PANAMA SPADEFISH

Parapsettus panamensis Steindachner, 1875

Dorsal rays IX,28; *spinous dorsal composed of short free spines, scarcely rising above surface of scales, the last dorsal spine tallest*; anal rays III,24; pectoral rays 18; body somewhat orbiculate and laterally compressed; dorsal and anal spines very low in comparison with soft parts of these fins; a band of brush-like teeth in jaws, the outer row the largest; *pelvic fins very small*; silvery gray with dusky blackish fins; pectorals and caudal fin with yellowish wash. Gulf of California to Peru; inhabits sand and mud bottoms. Grows to 40 cm.

BUTTERFLYFISHES (MARIPOSAS, MUNECAS)
FAMILY CHAETODONTIDAE

Butterflyfishes are renowned for their striking color patterns, delicate shapes, and graceful swimming movements. They have deep, compressed bodies, and small protractile mouths with brush-like teeth in the jaws. The scales are ctenoid and cover the head, body, and median fins. There is a single dorsal fin with VI to XVI stout spines and no notch between the spinous and soft portions; the membranes of the anterior spines are deeply incised; the anal fin has III-V stout spines; there is a scaly axillary process at the upper base of the pelvic fins; and the caudal fin varies in shape from rounded to slightly emarginate. These fishes have a distinctive late postlarval stage called a tholichthys, which has large bony plates on the head and anterior body. The family contains 116 species, which occur mainly in tropical seas around coral reefs; only four are known to regularly occur in the tropical eastern Pacific, but several "vagrants" from the western Pacific (*Chaetodon auriga, C. kleini,* and *C. lunula*) are infrequently encountered at the Galapagos. Most of the species dwell in depths of less than 20 m, but some are restricted to deeper sections of the reef, to at least 200 m. Butterflyfishes are active during daylight hours, and seek shelter close to the reef's surface at dusk. They often assume a drab nocturnal color pattern. Most species are restricted to a relatively small area of the reef, perhaps an isolated patch reef or part of a more extensive reef system. They travel extensively throughout their home range, foraging for food. Many species feed on live coral polyps; others consume a mixed diet consisting of small benthic invertebrates and algae. A few species feed in midwater on zooplankton. Young butterflyfishes are highly prized as aquarium fishes. Most species grow to a maximum length that is less than 30 cm.

Barberfish (*Johnrandallia nigrirostris*), Isla Marchena, Galapagos

SCYTHEMARKED BUTTERFLYFISH
Chaetodon falcifer Hubbs & Rechnitzer, 1958

Dorsal rays XII,19-20; anal rays III,14-16; pectoral rays 13-14; lateral-line scales 45-56; *membranes between anterior dorsal spines deeply incised; second anal spine very elongate, usually longer than pelvic fin length; snout produced, somewhat tubular; generally yellow with prominent, black, scythe-like marking on side*; also a black diagonal band from snout to dorsal fin origin. Southern California (Santa Catalina Island and La Jolla), southern tip of Baja California, Guadalupe Island, West San Benito Island, and Galapagos Islands; usually seen in deep water (30-150 m), but in the Galapagos it is sometimes encountered adjacent to steep slopes in only 10-12 m. Usually seen alone or in small groups. Grows to 17 cm.

THREEBANDED BUTTERFLYFISH
Chaetodon humeralis Günther, 1860

Dorsal rays XIII,18-20; anal rays III,15-17; pectoral rays 15-16; lateral-line scales 34-39; membranes between dorsal spines not deeply incised; dorsal spines increasing in length to fourth or fifth spine then decreasing in length; second anal spine much shorter than pelvic fin length; *generally white or silvery white, with prominent black bars at level of pectoral fin and just in front of caudal fin base*; also a black band through eye to nape; dorsal, anal, and caudal fins with black bands. Gulf of California to Peru, including offshore islands; a common species, seen individually or in small groups in 3-55 m depths; most common in 5-12 m. Attains 15 cm.

LONGNOSE BUTTERFLYFISH
Forcipiger flavissimus Jordan & McGregor, 1898

Dorsal rays XII,22-24; anal rays III,17 or 18; pectoral rays 15; lateral-line scales 74-80; greatest body depth 1.9-2.4 in standard length; *snout extremely long and attenuate, its length 1.6-2.1 in body depth; mouth with a distinct gape (hence forceps-like)*; overall bright yellow; upper half of head and nape black, white below; a black spot on anal fin just below base of caudal fin. East Africa to the Americas; ranges from Mexico to the Galapagos; inhabits rock and coral reefs; feeds on hydroids, small crustaceans, tubed feet of echinoderms, pedicillaria of sea urchins, and polychaete tentacles. To 22 cm.

BARBERFISH
Johnrandallia nigrirostris (Gill, 1863)

Dorsal rays XII,24-25; anal rays III,18-20; pectoral rays 16; lateral-line scales 52-63; membranes between anterior dorsal spines moderately incised; dorsal spines increasing in length to third or fourth spine, then decreasing in length; second anal spine much shorter than pelvic fin length; *yellow, with black bands on snout, forehead, and along base of dorsal fin; also a black area around eye, and edge of gill cover black.* Baja California to Panama, also the Galapagos and other offshore islands; usually seen in small to very large aggregations in 6-40 m. Feeds on algae, molluscs, and crustaceans; also "cleans" crustacean ectoparasites from other fishes. Maximum size 20 cm.

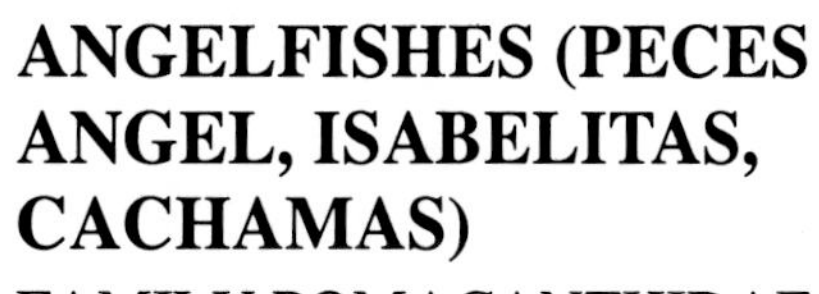

ANGELFISHES (PECES ANGEL, ISABELITAS, CACHAMAS)
FAMILY POMACANTHIDAE

Angelfishes are close relatives of the butterflyfishes, and until recently were considered to belong in the same family. They share a number of characteristics such as deep compressed bodies, ctenoid scales which extend out onto the median fins, a single, unnotched dorsal fin, and a small mouth with brush-like teeth. They differ from butterflyfishes, however, in having a long spine at the corner of the preopercle (also smaller spines on the preopercle, opercle, and preorbital); they lack a scaly axillary process at the base of the pelvic fins; the scales are more strongly ctenoid and have distinct ridges on the exposed part, and adults may have auxiliary scales; and the postlarvae lack bony plates on the head and anterior body. Most species are inhabitants of tropical Indo-Pacific seas, being found mainly in the vicinity of coral reefs. They occur both solitarily and in aggregations. Many species inhabit shallow water, from only a few meters to 10-15 m depth. Others are restricted to deep water. Angelfishes are favorite aquarium pets, well known for their brilliant array of color patterns. Species of *Holacanthus* and *Pomacanthus* exhibit dramatic changes from the juvenile to adult stage. Most angelfishes are dependent on the presence of shelter in the form of boulders, caves, and coral crevices. Typically, they are somewhat territorial, and spend daylight hours near the bottom in search of food. The diet varies according to species: some feed almost exclusively on algae and detritus; others, such as the species of *Holacanthus* and *Pomacanthus*, prefer mainly sponges supplemented by a variety of benthic invertebrates. Divers are sometimes startled by the powerful drumming or thumping sound which is produced by large adult angels in the genus *Pomacanthus*. The family has a circumtropical distribution and is represented by 76 species in nine genera; only four species are known from the eastern Pacific.

Clarion Angelfish (*Holacanthus clarionensis*), Los Frailes, Gulf of California

Holacanthus clarionensis Adult ▼ Juv. ▲

CLARION ANGELFISH
Holacanthus clarionensis Gilbert, 1890

Dorsal rays XIV,17-19; anal rays III,18-19; pectoral rays 17-18; lateral line weakly developed, about 50 scales in longitudinal series; a stout spine on lower edge of cheek; most of body brownish, head dark brown; *a broad, bright orange area just behind head; caudal fin orange*; small juveniles brownish orange with narrow blue bars on side and pair of similar bars on head; bars gradually become narrower and eventually disappear with increased growth. Southern tip of Baja California, Revillagigedo Islands, and Clipperton Island; occurs solitarily or in aggregations in 10-30 m; generally rare at Baja California (Cape San Lucas region), but common at offshore islands. Reaches 20 cm.

Holacanthus clarionensis Small juv.

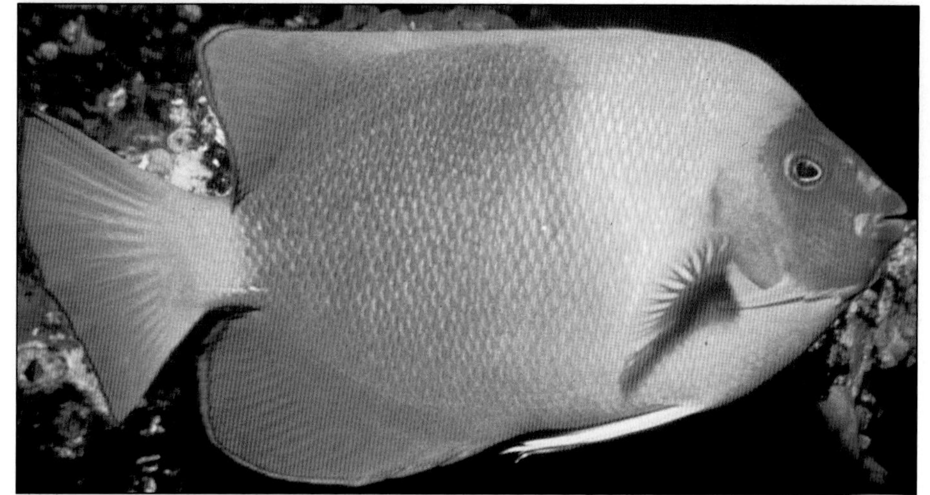

Holacanthus limbaughi ▼

CLIPPERTON ANGELFISH
Holacanthus limbaughi Baldwin, 1963

Dorsal rays XIV,17-18; anal rays III,17-18; pectoral rays 17-18; lateral line weakly developed, 45-48 scales in longitudinal series; a stout spine on lower edge of cheek; *generally dark blue-gray with white caudal fin; a small (slightly larger than eye) white spot on upper side above tip of pectoral fin*; margins of dorsal, anal, and pelvic fins blue; juveniles dark gray with about 7 narrow blue bars, including 2 on head. Known only from Clipperton Island; common in 5-10 m, but also ranging to at least 30 m. Grows to at least 24 cm.

Holacanthus passer ▼

KING ANGELFISH
Holacanthus passer Valenciennes, 1846

Dorsal rays XIV,18-20; anal rays III,17-19; pectoral rays 18-20; lateral line weakly developed, about 50 scales in longitudinal series; a stout spine on lower edge of cheek; *velvety blue-gray with blue scale centers; a narrow white bar on side at level of rear edge of pectoral fin; caudal fin yellow*; juveniles orangish anteriorly, brown posteriorly, 5-6 narrow blue bars on side, and pair of similar bars on head surrounding broad brown bar through eye. Gulf of California to Ecuador, including the Galapagos; relatively common on rocky reefs or coralline areas in 3-12 m, but ranging to at least 80 m; juveniles sometimes found in tide pools. Maximum size about 25 cm.

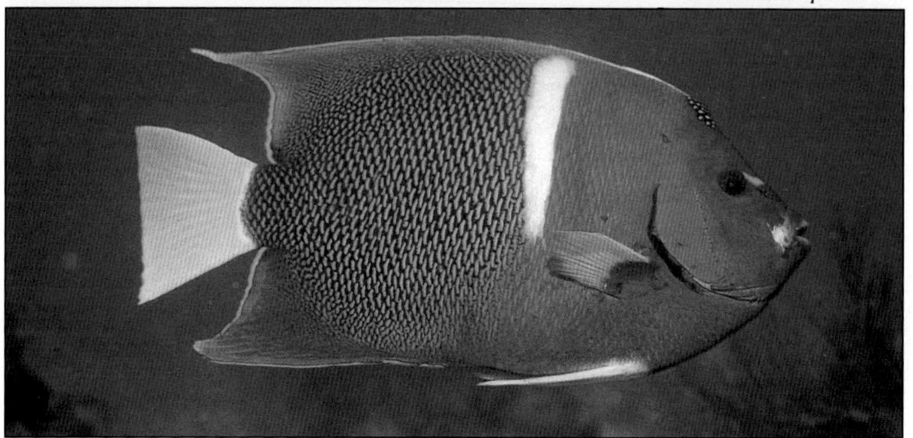

Holacanthus passer Juv.

CORTEZ ANGELFISH
Pomacanthus zonipectus (Gill, 1862)

Dorsal rays XI,24-25; anal rays III,20-22; pectoral rays 19-20; a stout spine on lower edge of cheek; anterior part of head grayish, central part of body and adjacent portion of dorsal fin light gray, rear part of body and adjacent dorsal and anal fins charcoal or blackish; *a broad band of yellow, just behind margin of cheek, and second, more narrow yellow bar bordered with black just behind pectoral fin base*; caudal fin pale yellowish; juvenile black with 6 curved yellow bars on head, body, and tail, with narrow blue bars between them. Baja California and Gulf of California to Peru, including the Galapagos and other offshore islands; usually seen solitarily or in pairs on rocky reefs in 6-15 m, but ranging to at least 35 m. Juveniles "clean" parasites from the surface of other fishes. To 46 cm.

Pomacanthus zonipectus Juv.

KNIFEJAWS (PECES LORO, PERICOS)
FAMILY OPLEGNATHIDAE

This small, unusual family contains six species (all in the genus *Oplegnathus*) that occur in coastal seas of the Galapagos, Peru, South Africa, southern Australia, and Japan. The main distinguishing feature is the parrot-like beak formed from the fused jaw teeth. A similar dentition is found in the parrotfishes (Scaridae). This apparatus is adapted for crushing barnacle shells, molluscs, and sea urchins.

Oplegnathus insignis

PACIFIC BEAKFISH
Oplegnathus insignis (Kner, 1867)

Dorsal rays XI,17; anal rays III,12-13; pectoral rays 16-17; gill rakers on first arch 5 + 12; scales very small, about 115-130 in midlateral series; lateral line strongly arched; mouth small; *teeth fused, forming continuous beak-like structure with median suture in each jaw*; body depth 1.6-2.1 in standard length; head 2.8-3.1 in standard length; margin of preopercle finely serrate; *black with profuse white spotting, the spots frequently joined to form irregular maze pattern*; lower jaw, breast, and belly white; small juveniles pale with 6 dark bars. Galapagos, Peru, and Chile; inhabits rocky reefs. Grows to 60 cm.

SURFPERCHES
FAMILY EMBIOTOCIDAE

Surfperches are small (up to about 45 cm), ovate, laterally compressed fishes that occur mainly along the Pacific coast of North America. Other features include a small mouth, cycloid scales, forked caudal fin, a single continuous dorsal fin of VI-XI spines and 9-28 soft rays, and anal fin with III spines and 15-35 soft rays. The most notable characteristic of this family is their rather unusual mode of reproduction. Unlike other perciform fishes, they have internal fertilization and bear their young alive. Breeding generally occurs in summer, but the gestation period is lengthy, and birth of the young (up to 80, depending on species) does not occur until fall or the following spring. The young are virtually mature at birth. The family contains about 23 species in 13 genera; all but two (from Japan and Korea) inhabit North America. By far the most species occur along the coast of Southern and Central California; a single freshwater species is also found in California. Surfperches inhabit surge areas off sandy beaches, but are also found in kelp beds, tide pools, and sheltered bays and harbors. Food items include molluscs, crustaceans, and other small invertebrates. Some species are ectoparasite "cleaners" of other fishes. The family was reviewed by Tarp (1952). We include a single representative, which occurs in the Gulf of California.

PINK SURFPERCH
Zalembius rosaceus (Jordan & Gilbert, 1880)

Dorsal rays X,18; anal rays III,20; lateral-line scales about 50; body depth 2.5, head length 3.3-3.4, both in standard length; middle dorsal spines longer than soft dorsal rays; upper caudal fin lobe of males prolonged into filament; rose pink with pair of brown to blackish spots on back below rear half of dorsal fin. Northern California to central Baja California; also an isolated population in central Gulf of California around Guaymas; inhabits rocky reefs and open trawl grounds in 9-230 m. Grows to 20 cm.

Zalembius rosaceus

Photograph: Alex Kerstitch

DAMSELFISHES (CASTAÑUELAS, PETACAS, PECES CACHO, PINTAÑOS)

FAMILY POMACENTRIDAE

The damselfish family Pomacentridae occurs worldwide in tropical and temperate seas. Approximately three-fourths of the 321 known species are found in the Indo-West Pacific region, where they are common inhabitants of coral reefs. In the eastern Pacific, 22 species have been reported, all of which are endemic to this region; 17 are found in the tropical area covered here. They are elongate to ovoid, compressed fishes with a single continuous dorsal fin of XII-XIV spines and 10-17 soft rays (up to XVII spines and 21 soft rays in other regions); the base of the spinous portion is longer than the soft portion; the anal fin has II spines and 10-14 soft rays (up to 16 in other areas); the caudal fin varies from slightly emarginate to forked or lunate. The scales are moderately large and ctenoid; the head is largely scaled, as are the basal parts of the median fins. Damselfishes are one of the most abundant groups of reef fishes. Most inhabit the tropics, but a number of species live in cooler temperate waters. They display remarkable diversity of habitat preference, feeding habits, and behavior. Coloration is highly variable, ranging from drab hues of brown, gray, and black to brilliant combinations of orange, yellow, and neon-blue. A number of species have juvenile stages characterized by a yellow body with bright blue stripes crossing the upper head and back. Most damselfishes are territorial, particularly algal-feeding species such as *Stegastes*. They zealously defend their small plot against all intruders, regardless of size. Damsels exhibit a highly stereotyped mode of reproduction in which one or both partners clear a nest site on the bottom and engage in courtship displays of rapid swimming and fin extension. Males generally guard the eggs, which are attached to the bottom by adhesive strands. The eggs hatch within about two to seven days, and the fragile larvae rise to the surface. They are transported by ocean currents for periods which vary between about 10 to 50 days, depending on the species. Eventually the young fish settle to the bottom, and their largely transparent bodies quickly assume the juvenile coloration. The growth rate of juveniles generally ranges from about 5 to 15 mm per month and gradually tapers off as maturity approaches. There is very little reliable data concerning their longevity, but it appears they are capable of living to at least an age of 10 years. Damselfishes feed on a wide variety of plant and animal material. Generally, the drab-colored species feed mainly on algae, whereas many of the brightly patterned species, and also members of the genus *Chromis*, obtain their nourishment from current-borne plankton. Past authors have used the generic names *Pomacentrus* and *Eupomacentrus* for the eastern Pacific species now placed in *Stegastes*. The family was recently reviewed by Allen (1991).

PANAMIC NIGHTSERGEANT

Abudefduf concolor (Gill, 1863)

Dorsal rays XIII,12-13 (usually 13); anal rays II,10; pectoral rays 18-20; lateral-line scales 20-21; gill rakers on first arch 19-21; *suborbital attached to preopercle (cheek), its lower border completely obscured by scales*; greatest body depth 53.8-59.2 percent of standard length; *gray with 4-5 relatively narrow whitish bars*; most of body scales with narrow dark outline; wedge-shaped black mark on pectoral fin base; lips whitish. El Salvador to Peru, including the Galapagos; found on rocky inshore reefs exposed to wave action, in 1-5 m depths. Maximum size to 18 cm.

Abudufdef concolor

MEXICAN NIGHTSERGEANT
Abudefduf declivifrons (Gill, 1862)

Dorsal rays XIII,12-13 (usually 13); anal rays II,10; pectoral rays 19; lateral-line scales 20-21; gill rakers on first arch 19-23; *suborbital usually separated from preopercle and clearly demarcated by free lower edge, but sometimes partly attached to cheek and its lower border obscured by scales*; greatest body depth of adults 59.2-62.6 percent of standard length; gray with 4-5 narrow, whitish bars; most of body scales with narrow dark outline; narrow blackish bar of wedge-shaped mark on pectoral fin base. Gulf of California to Acapulco, Mexico. Inhabits rocky reefs exposed to surge, in 1-5 m depths. To 18 cm. Formerly considered a synonym of *A. concolor*, but a recent study by Lessios, Allen, Wellington and Birmingham revealed that it is a valid species.

Abudefduf troschelii ▼

Azurina eupalama ▼

Photograph: John McCosker

PANAMANIAN SERGEANT
Abudefduf troschelii (Gill, 1862)

Dorsal rays XIII,12-13; anal rays II,10-13; pectoral rays 16-20; lateral-line scales 19-23; gill rakers on first arch 23-33; greatest body depth 1.6-2.0 in standard length; generally whitish or pale silvery green, with *5 black bars on side* and additional, less distinct bar across caudal peduncle; back usually bright yellow. Similar in appearance to the Atlantic *A. saxatilis* and Indo-West Pacific *A. vaigiensis*, but unlike these species the adults (over 70 mm standard length) possess a patch of scales on the inner face of the pectoral fin base. Eastern Pacific Ocean from Baja California to Peru, including the Galapagos and other offshore islands. Occurs along rocky shores in 1-12 m depth. Grows to 18 cm.

GALAPAGOS DAMSEL
Azurina eupalama Heller & Snodgrass, 1903

Dorsal rays XIII or XIV,10-12; anal rays II,12-13; *pectoral rays 17-18*; lateral-line scales 29-31; gill rakers on lower limb of first arch 21-23; greatest body depth 3.0-3.7 in standard length; overall gray, grading to whitish on underside of head and on belly; scale centers usually lighter; a whitish spot on upper back below soft dorsal fin; each caudal fin lobe with a blackish streak, and a black blotch covering pectoral fin base. Similar in overall appearance and habits to *Chromis atrilobata*, but is more slender-bodied and grows to a larger size; also lateral line continues to upper part of caudal peduncle, nearly to base of caudal fin. Galapagos Islands; usually seen in open water over rocky habitats in 5-30 m depth. Attains 16 cm.

SWALLOW DAMSEL

Azurina hirundo Jordan & McGregor, 1898

Dorsal rays XII,10-12; anal rays II,11-12; *pectoral rays 20-21*; lateral-line scales 27-31; gill rakers on lower limb of first arch 25-27; greatest body depth 3.3-4.0 in standard length; steel blue to olive on upper portion, paler below, with orange hue on throat. A poorly-known species known only from Alijos Rocks (about 240 km off Baja California), Guadalupe Island, and the Revillagigedo Islands. Reaches 16 cm.

OVAL CHROMIS

Chromis alta Greenfield & Woods, 1980

Dorsal rays XIII,12-14; anal rays II,12-13; pectoral rays 19-21; lateral-line scales 16-19; gill rakers on first arch 27-33; greatest body depth 1.8-2.2 in standard length; *adults overall light brown* with darker scale margins; *juveniles blue with neon-blue stripe above and below eye, and broad whitish band along base of dorsal fin* that joins diffuse pale bar across caudal peduncle. Eastern Pacific Ocean, including west coast of Baja California (Isla Catalina south to Cabo San Lucas) and the Galapagos Archipelago; inhabits rocky reefs, usually in about 30-150 m depth, but in shallower water in the Galapagos (5-43 m). Maximum size to 15 cm.

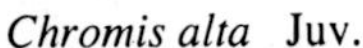

Chromis alta Juv.

SCISSORTAIL CHROMIS

Chromis atrilobata Gill, 1862

Dorsal rays XII,12 or 13; anal rays II,10-12; pectoral rays 18-19; lateral-line scales 20-21; gill rakers on lower limb of first arch 20-23; greatest body depth 2.3-2.7 in standard length; overall *metallic gray with prominent white spot just below base of soft dorsal fin; each caudal fin lobe with intensely blackish streak*; also a blackish mark at base of pectoral fin. Eastern Pacific Ocean from the upper Gulf of California to northern Peru and the Galapagos Islands. A common species that forms large aggregations around rocky reefs in 6-76 m depths. Attains 12 cm.

LIMBAUGH'S CHROMIS
Chromis limbaughi Greenfield & Woods, 1980

Dorsal rays XIII,11 or 12; anal rays II,11; pectoral rays 18-19; lateral-line scales 15-18; gill rakers on first arch 27-31; greatest body depth 1.8-2.2 in standard length; *bluish gray with a largely bright blue head, and with yellow dorsal, caudal, and anal fins*; yellow coloration of dorsal fin usually continuous with pale yellowish or whitish caudal peduncle; *juveniles largely bright blue on anterior two-thirds and bright yellow posteriorly*, the yellow region confluent with yellow dorsal, caudal, and anal fins. Gulf of California; inhabits rocky areas in 5-75 m depths. Maximum size to 12 cm.

Chromis limbaughi Juv.

Microspathodon bairdii Adult ▼ Juv. ▲

BUMPHEAD DAMSEL
Microspathodon bairdii (Gill, 1862)

Dorsal rays XII,15-16; anal rays II,13-14; pectoral rays 21-23; lateral-line scales 20-23; gill rakers on lower limb of first arch 17-21; greatest body depth 1.7-2.0 in standard length; dusky, dark gray-brown; median fins with slight golden brown hue; iris bright blue; *juveniles mainly bright blue on upper half of head and body, and yellow on lower half; adults have a pronounced bump on forehead.* Eastern Pacific Ocean from the Gulf of California to Ecuador, including offshore islands such as the Revillagigedo and Galapagos groups; inhabits rocky shores in 3-10 m depth. Grows to 30 cm.

GIANT DAMSELFISH

Microspathodon dorsalis (Gill, 1862)

Dorsal rays XII,15-16; anal rays II,13-14; pectoral rays 22-23; lateral-line scales 20-22; gill rakers on first arch 24-25; greatest body depth 1.7-2.0 in standard length; overall dark gray-blue, sometimes nearly bluish black; head and anterior part of body often slightly lighter with dusky scale margins; margins of soft dorsal, caudal, and anal fins narrowly white or light blue; nuptial males can quickly assume a pattern in which the anterior half to two-thirds of the body (and head) is very pale, nearly whitish, sometimes with a darker bar below the front part of the dorsal fin; *juveniles normally blue, with lighter blue scale centers and several scattered, neon-blue spots on forehead,* back, and base of last dorsal ray; an unusual juvenile variety from the Gulf of Chiriqui and Perlas Islands in Panama is mainly lime-green with a blue iris and scattered blue spots along the back, and a blackish, saddle-like marking on the dorsal edge of caudal peduncle; *adults are readily recognized by their pale-edged, filamentous dorsal and anal fins.* Eastern Pacific Ocean from the Gulf of California to Ecuador, including offshore islands such as the Galapagos and Mapelo; inhabits rocky reefs in 1-10 m depths. Maximum size to 30 cm.

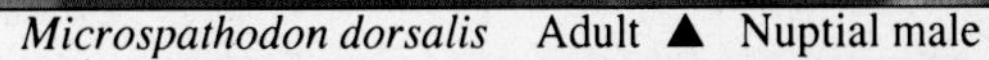

Microspathodon dorsalis Adult ▲ Nuptial male ▼

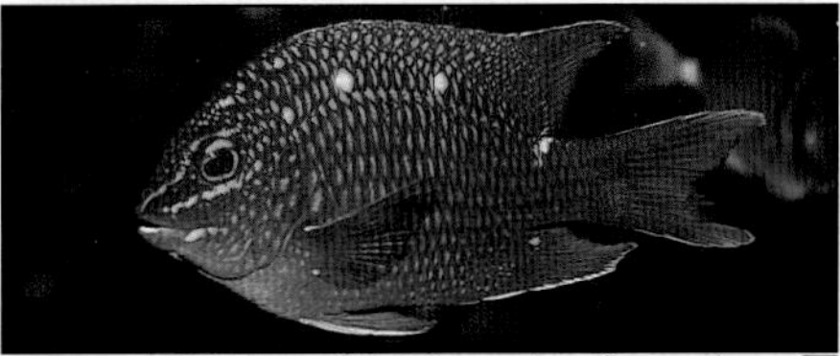

M. dorsalis Juv. ▲ Green juv. ▼

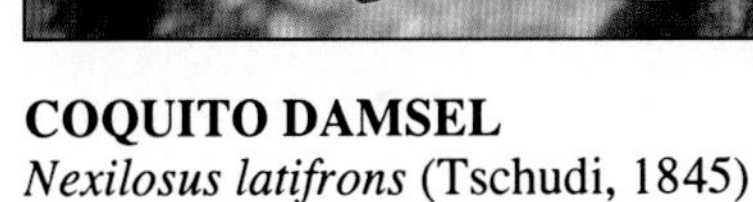

COQUITO DAMSEL

Nexilosus latifrons (Tschudi, 1845)

Dorsal rays XIII,17; anal rays II,13; pectoral rays 20-23; lateral-line scales 19-21; gill rakers on first arch 20-21; greatest body depth 1.8-2.0 in standard length; overall brown with darker scale margins; *an incomplete yellow bar on middle of side; juveniles dark brown with blue streak or spot on most scales, blue spots and lines on head, and narrow blue margins on most fins.* Galapagos Islands, Peru, and northern Chile; inhabits rocky shores in 1-10 m depths. Attains a maximum length of 30 cm.

Nexilosus latifrons Juv.

ACAPULCO GREGORY

Stegastes acapulcoensis (Fowler, 1944)

Dorsal rays XII,15 (rarely 16); anal rays II,13; pectoral rays usually 21 or 22 (rarely 20 or 23); lateral-line scales 20; gill rakers on lower limb of first arch 11-13; greatest body depth 1.8-2.0 in standard length; overall brown, lighter on head and anterior part of body, most of scales with blackish margin; *white uppermost pectoral rays are diagnostic*, also has prominent white band across base of rays on outer face of pectoral fin axil; *juveniles are bright blue with a prominent ocellus at the base of the soft dorsal fin and an ocellated black spot on the dorsal edge of the caudal peduncle.* Pacific coast of Central and South America from Mexico (Baja California and Sinaloa Province) to Lobos de Afuera, Peru; occurs along rocky shores in about 2-15 m. Reaches a length of 17 cm.

Stegastes acapulcoensis ▲

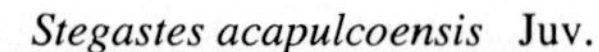

Stegastes acapulcoensis Juv.

GALAPAGOS GREGORY

Stegastes arcifrons (Heller & Snodgrass, 1903)

Stegastes arcifrons ▼

Dorsal rays XII,15 (rarely 16); anal rays II,13 (rarely 12); pectoral rays 20 (rarely 19); lateral-line scales 20 (rarely 19); gill rakers on lower limb of first arch 12 or 13 (rarely 14); greatest body depth 1.8-2.1 in standard length; overall dark brown, head and breast lighter, often with violet color on gill cover; *distinguished by pale yellowish lips and bright yellow caudal fin; juveniles somewhat similar, but yellow color of tail extends forward onto posteriormost part of body and rear of dorsal and anal fins, and there is also a large, ocellated black spot near the base of the soft dorsal junction.* Galapagos Islands, Malpelo Island, and Isla del Coco; abundant along rocky shores to about 5 m depth. To 15 cm.

Stegastes arcifrons Juv.

CLIPPERTON GREGORY

Stegastes baldwini Allen & Woods, 1980

Stegastes baldwini ▼

Dorsal rays XII,15 (rarely 16); anal rays II,13; pectoral rays 21 (rarely 22); lateral-line scales 20 (rarely 21); gill rakers on lower limb of first arch 12 or 13; greatest body depth 1.8-2.1 in standard length; dark brown with darker scale outlines; *distinguished by broad white band on caudal peduncle and pale margin on pectoral, dorsal, and anal fins; lips yellow; yellow head color can be switched on and off.* Known only from Clipperton Island; abundant in shallow (from 0-30 m) water. Maximum length to 12.5 cm.

BEAUBRUMMEL GREGORY
Stegastes flavilatus (Gill, 1863)

Dorsal rays XII,14 (rarely 15); anal rays II,12; pectoral rays 22 or 23 (rarely 24); lateral-line scales 20; gill rakers on lower limb of first arch 11 (rarely 12); greatest body depth 1.8-2.0 in standard length; light brown with blackish scale outlines, head often with violaceous markings; *caudal fin, rear part of dorsal and anal fins, pelvic fins, and pectoral fins yellowish; juveniles bright yellow with broad blue area on back and upper part of head, and also with black ocellated spot at soft dorsal junction.* Widely distributed on the Pacific coast of Central America from Cape San Lucas, Mexico, and the Lower Gulf of California, to Bahia de Santa Elena, Ecuador; inhabits rocky shores to 12-15 m depth. Attains 14 cm.

Stegastes flavilatus Juv.

WHITETAIL GREGORY
Stegastes leucorus (Gilbert, 1891)

Dorsal rays XII,15 or 16 (occasionally 14); anal rays II,13 (rarely 12 or 14); pectoral rays 20 or 21 (rarely 22); lateral-line scales 20 (rarely 19, 21, or 22); gill rakers on lower limb of first arch 10-12; greatest body depth 1.8-2.0 in standard length; overall dark brown with darker scale outlines, often with white band on caudal peduncle; *distinguished by blue iris and white outer edge of pectoral fin*; similar to *S. baldwini* from Clipperton Island, but has slightly different color pattern, and adult males of *S. leucorus* have dense covering of microscopic papillae on outer half of the pectoral, which is lacking in *S. baldwini*. Two subspecies are recognized: *S. leucorus leucorus* from the Revillagigedo Islands, Guadalupe Island, and the lower Gulf of California, and *S. leucorus beebei* from Panama (Perlas Islands), Malpelo Island, and the Galapagos Islands. Allen and Woods (1980) gave detailed information on the differences between these subspecies. Common along rocky shores to about 15 m depth. Grows to 17 cm.

Stegastes leucorus beebei ▲

Stegastes leucorus leucorus ▼

Stegastes leucorus beebei Juv. ▼

Stegastes leucorus leucorus Juv. ▼

CORTEZ GREGORY
Stegastes rectifraenum (Gill, 1862)

Dorsal rays XII,15 (rarely 16); anal rays II,13 (rarely 12 or 14); pectoral rays usually 20 (occasionally 19 or 21); lateral-line scales 20 (rarely 19); gill rakers on lower limb of first arch 12 (rarely 11); greatest body depth 1.9-2.1 in standard length; *generally dark brown, grading to lighter brown on head*; most of body scales with blackish margins; fins mainly dark brownish, except pectorals slightly yellowish; adults, in contrast with juveniles, are drab in appearance, without conspicuous markings; *juveniles bright blue with darker scale margins*, a pair of neon-blue stripes on upper part of head and nape, black ocellated spot at base of soft dorsal junction, and small, blue-edged black spot on upper surface of caudal peduncle. Magdalena Bay on the west coast of Baja California and the entire Gulf of California; not known south of the state of Sonora on the Mexican mainland; inhabits rocky reefs to about 10 m depth. Attains length of 12 cm.

Stegastes rectifraenum Adult ▲ Juv. ▼

Stegastes redemptus Adult ▼ Juv. ▲

CLARION GREGORY
Stegastes redemptus (Heller & Snodgrass, 1903)

Dorsal rays XII,15 (rarely 14); anal rays II,13; pectoral rays 20 (rarely 19 or 21); lateral-line scales 20 (rarely 19 or 21); gill rakers on lower limb of first arch 13-14 (rarely 12 or 15); greatest body depth 1.9-2.0 in standard length; *head and front of body light brown, middle part of body gradually darker brown, and posterior part whitish*; most of body scales with blackish margins; soft dorsal and anal fins mostly whitish with pale yellow on outer portion; caudal fin pale yellow; *juveniles overall bright yellow with dusky back, a prominent dark spot at base of anterior soft dorsal rays and smaller dark spot on upper edge of caudal peduncle.* Known only from the Revillagigedo Islands and Cape San Lucas, Mexico. Grows to 14.5 cm.

HAWKFISHES (HALCONES, MERITOS)
FAMILY CIRRHITIDAE

Hawkfishes are colorful reef dwellers that occur in most tropical seas. They are largely sedentary, remaining motionless on the bottom for long intervals, periodically swimming to a new vantage point on the reef surface. They are characterized by thickened pectoral rays and frilly threads at the tip of each dorsal spine. The stout pectoral rays are adapted for perching on the bottom or wedging among coral branches. This thickening is readily apparent in the Giant Hawkfish, often found in surge areas, and is weakly developed in the Longnose Hawkfish, which lives in deeper water. Hawkfishes feed chiefly on small fishes, but also consume a variety of crabs, shrimps, and other crustaceans. Worldwide the family is represented by 35 species in nine genera. The group was revised by Randall (1963).

LONGNOSE HAWKFISH
Oxycirrhites typus Bleeker, 1857

Dorsal rays X,13; anal rays III,7; lower 5 or 6 pectoral rays unbranched; lateral-line scales 51-53; 4 rows of large scales above lateral line in middle of body; palatine teeth absent; body depth 4.4-4.6 in standard length; *snout extremely long, its length about 2.0 in head*; 2-4 cirri near tip of each dorsal spine; *whitish with horizontal and vertical red bands forming a crosshatch pattern*. Widely distributed in the Indo-Pacific; in the eastern Pacific it ranges from the Gulf of California to Colombia and the Galapagos; generally seen perched on gorgonians or black coral at depths below 20 m. To 13 cm.

GIANT HAWKFISH
Cirrhitus rivulatus Valenciennes, 1855

Dorsal rays X,11-12; anal rays III,6; lower 7 pectoral rays unbranched; lateral-line scales 45-49; 5 rows of large scales above lateral line in middle of body; body depth 2.6-2.8 in standard length; interorbital space of adults without scales, covered with tiny papillae; longest dorsal spine about 3.3 in head length; a tuft of cirri near tip of each dorsal spine; *grayish brown with 5 "bars", each composed of a "maze" of golden brown markings with black margins narrowly surrounded by blue; head with broad, golden brown, spoke-like bands radiating from eye*, these also with black margin narrowly outlined with blue; *a pair of white spots on upper back posteriorly*; juveniles white with dark brown bars. Gulf of California to Colombia and the Galapagos; juveniles found in shallow surge zone, adults on rocky reefs to at least 20 m. Attains 52 cm; it is the largest member of the family.

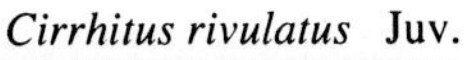

Cirrhitus rivulatus Juv.

CORAL HAWKFISH
Cirrhitichthys oxycephalus (Bleeker, 1855)

Dorsal rays X,12, the first soft ray prolonged; anal rays III,6; lower 6 pectoral rays unbranched; lateral-line scales 41-45; body depth 2.8-3.0 in standard length; bony interorbital space of adults 2.0 in eye; maxilla reaching level of front edge of eye; fifth dorsal spine longest, 1.8-2.0 in head; a tuft of cirri near tip of each dorsal spine; *whitish or pink, with squarish red to brown blotches covering sides*; head with smaller red to brown spots and blotches. Widely distributed in the tropical Indo-Pacific; in the eastern Pacific it ranges from the Gulf of California to Colombia and the Galapagos; usually seen perched on coral heads in 3-15 m. Grows to 8.5 cm.

MULLETS (LISAS, LEBRANCHES)
FAMILY MUGILIDAE

Mullets are silvery gray fishes with a small mouth, moderately elongate body, two widely spaced dorsal fins, and emarginate or weakly forked caudal fin. The mouth is small, and the lips are relatively thin in most species; the teeth are usually either very small or absent. In addition, the lateral line is lacking. Mullets occur in all tropical and temperate seas, usually near shore, frequently in brackish estuaries and fresh water. They commonly form schools containing up to several hundred fish. Their diet consists mainly of detritus and algae, but insects, fish eggs, and plankton are also consumed. Fresh mullet flesh is excellent eating, and they are commercially important in many areas. In southeastern Asia, mullets are frequently cultivated in ponds. The family contains approximately 95 species in about 13 genera. The eastern Pacific species were treated by Ebeling (1957 and 1961).

THICK-LIPPED MULLET
Chaenomugil proboscidens (Günther, 1861)
(Plate XIII-1, p. 217)

Dorsal rays IV+I,8; *anal rays III,10*; scales in midlateral series about 38; *eye without covering of fatty (adipose) tissue; upper lip extremely thick; a band of soft, pavement-like papillae on each lip; all teeth in jaws bifid, shaped like tiny crowbars*; pectoral fin about 80 percent of head length; caudal fin emarginate; greenish or olive on back, silvery on sides, white below; faint brownish stripes corresponding with scale rows on side. Mexico to Panama, including some offshore islands; occurs along rocky shorelines, also in tide pools. Grows to 16 cm.

THOBURN'S MULLET
Xenomugil thoburni (Jordan & Starks, 1896)
(Plate XIII-6, p. 217)

Dorsal rays IV+I,7; anal rays III,9; *scales in midlateral series 45-47*; fatty (adipose) tissue covering most of eye; *upper lip relatively thick, lower lip in large subadults and adults turned under*; teeth of jaws simple (not bifid); mouth small, rear end of upper jaw reaching level of front of eye; bluish on back, silvery on sides, white below; faint darkish stripes on side corresponding with scale rows; a dusky bar across base of pectoral fin. Guatemala to Peru and Galapagos. Reaches at least 20 cm.

STRIPED MULLET
Mugil cephalus Linnaeus, 1758
(Plate XIII-2, p. 217)

Dorsal rays IV+I,8; *anal rays III,7-8 (usually 8)*; scales in midlateral series 39-44; fatty (adipose) tissue covering most of eye; lips thin, lower lip with a prominent median knob; mouth small, rear end of upper jaw reaching to level of anterior rim of eye; *outer row (primary) teeth in jaws mainly simple (i.e. not bifid)*; pectoral fin length about 66-74 percent of head length; olive on back, silvery on sides, white below; 6-7 brownish stripes on side; fins yellow in Galapagos-Peru subspecies, *M. cephalus rammelsbergii*. Circumglobal in tropical and temperate seas; Southern California to Peru in the eastern Pacific; forms schools in sandy bottom areas, but large adults sometimes seen near the surface in clear water near reefs. Grows to 60 cm.

WHITE MULLET
Mugil curema Valenciennes, 1836
(Plate XIII-4, p. 217)

Dorsal rays IV+I,8; *anal rays III,9*; scales in midlateral series 35-40; fatty (adipose) tissue covering most of eye; lips thin; lower lip with a prominent median knob; mouth small, rear end of upper jaw reaching to level between posterior nostril and front edge of eye; *outer (primary) teeth in jaws easily visible with naked eye, distinctly curved; upper jaw with broad, toothless gap between outer (primary) row and inner (secondary) row of teeth; pectoral axillary scale 5-8 percent of standard length*; olive to bluish on back, silvery on sides, white below; a small black blotch at upper base of pectoral fin; anal and pelvic fins yellowish, other fins whitish. Atlantic and Pacific sides of the Americas, mainly in tropical seas; in the eastern Pacific from Baja California to Chile; occurs in mangrove estuaries and brackish lagoons, and also near the surface in clear water. To 45 cm. *M. setosus* is an additional mullet known only from the Revillagigedos (Clarion Island). It is similar to *M. curema*, but has scattered teeth between the outer (primary) and inner (secondary teeth); the anal fin is also taller (15-18 vs 10-15 percent of standard length).

GALAPAGOS MULLET
Mugil galapagensis Ebeling, 1961
(Plate XIII-3, p. 217)

Dorsal rays IV+I,8; *anal rays III,8-9 (usually 8)*; scales in midlateral series 43-45; fatty (adipose) tissue covering most of eye in adults; lips thin, lower lip without median knob; mouth small; rear end of upper jaw reaching to level between posterior nostril and front of eye; *outer row (primary) teeth in jaws bifid*; pectoral fins very short, about 65-70 percent of head length; caudal fin emarginate; dusky brownish on back, silvery on sides; faint brown stripes corresponding with scale rows on sides. Galapagos Islands; forms schools over sandy bottoms. Grows to about 40 cm.

HOSPE MULLET
Mugil hospes Jordan & Culver, 1895
(Plate XIII-5, p. 217)

Dorsal rays IV+I,8; *anal rays III,9*; scales in midlateral series 38-40; fatty (adipose) tissue covering most of eye; lips thin, lower lip with a prominent median knob; mouth small, rear end of upper jaw reaching to level between posterior nostril and front of eye; *outer (primary) teeth in jaws barely visible with naked eye, not distinctly curved*; pectoral fin length about 87-89 percent of head length; *pectoral axillary scale 9-11 percent of standard length*; dark greenish on back and upper sides, silvery on lower sides, white below; a small black spot on upper base of pectoral fin. Atlantic and Pacific sides of the Americas, mainly in tropical seas; in the eastern Pacific from Mexico to Ecuador; occurs over sand and mud bottoms, sometimes entering estuaries. Maximum size 30 cm.

BARRACUDAS (PICUDAS, ESPETONES)
FAMILY SPHYRAENIDAE

Barracudas are very elongate fishes with a cylindrical body anteriorly, pointed snout, and protruding lower jaw. The large mouth is equipped with an awesome array of long, sharp-edged teeth of unequal size. Other features include a well-developed, straight lateral line; two widely separated dorsal fins; small pectoral, pelvic, and anal fins; and a forked caudal fin. The overall coloration is silvery, often with darker bars, saddles, or chevron markings. The family contains a single genus, *Sphyraena,* represented by about 20 species. Several are commonly seen in the vicinity of reefs. They frequently occur in small to large schools, often on the edge of outer reef drop-offs. However, *S. barracuda*, the largest species, is often encountered alone. It is frequently attracted to divers and may approach at close range. Barracudas feed primarily on fishes. They are a favorite target of anglers, and many are caught by trolling artificial lures. The flesh is excellent eating, but large specimens should be avoided because of the risk of ciguatera poisoning.

PACIFIC BARRACUDA
Sphyraena ensis Jordan & Gilbert, 1882
(Plate XIII-10, p. 217)

Dorsal rays V+I,9; anal rays II,8; *lateral-line scales about 108-116; body elongate, the depth about 8.0-9.0 in standard length; eye 6.0-7.0 in head length*; canine teeth of lower jaw, palatines, and inner row of premaxillary very large; generally silvery with series of chevron-shaped bars on upper two-thirds of side. Gulf of California to Peru. Grows to about 70 cm.

MEXICAN BARRACUDA
Sphyraena lucasana Gill, 1863
(Plate XIII-12, p. 217)

Dorsal rays V+I,9; anal rays II,8-9; scales very small, about 200+ in midlateral series and *160-170 in lateral line; body elongate, the depth 7.0-8.0 in standard length; eye about 10.0 in head length*; teeth of moderate length; generally silvery with about 20 dark bars on upper side. Outer coast of Baja California and Gulf of California. Reaches 70 cm.

THREADFINS (BOBOS, BARBETAS)
FAMILY POLYNEMIDAE

Threadfins have a bluntly rounded snout and ventral mouth. There are two separate dorsal fins, and the caudal fin is deeply forked. The common name is derived from their peculiar pectoral rays. The pectoral fin is divided into a "normal" upper section, with the rays attached to the fin membrane, and a detatched lower section with 3-7 free, thread-like rays. The number of free pectoral rays is diagnostic for separating many of the species. Threadfins are usually encountered near the coast, often in river mouths or brackish mangrove estuaries. Their diet consists of shrimps, crabs, polychaete worms, and other benthic invertebrates. They are considered good eating and are often seen in fish markets throughout the region. The family contains about 32 species divisible into seven genera, occuring in all tropical and subtropical seas. Only two species are found in our area.

BLUE BOBO
Polydactylus approximans Lay & Bennett, 1839
(Plate XIII-7, p. 217)

Dorsal rays VIII+I,12-13; anal rays III,14; *pectoral rays 13-14, lower portion consisting of 6 free filaments; lateral-line scales 55-59*; gill rakers on first arch 11-12 + 15-16; snout projecting about three-fourths its length beyond tip of lower jaw; origin of anal fin below origin of second dorsal fin; bluish on back, silvery whitish on sides; *main part of pectoral fin black.* Central California to Peru and Galapagos; inhabits sand-mud bottoms, often near river mouths or along sandy beaches where surf is breaking. Reaches 35 cm.

YELLOW BOBO
Polydactylus opercularis (Gill, 1863)
(Plate XIII-8, p. 217)

Dorsal rays VIII+I,12; anal rays III,14; *pectoral rays 15, lower portion consisting of 9 free filaments; lateral-line scales 67-75*; gill rakers on first arch 16 + 21; snout projecting about two-thirds its length beyond tip of lower jaw; origin of anal fin below middle of second dorsal fin; bronzy or brownish on back, silvery yellow on sides; *fins yellowish, particularly pectoral fins, which are bright yellow.* Southern California to Peru; inhabits sand-mud bottoms of bays, estuaries, and river mouths, also occurs along sandy beaches. Grows to 45 cm.

Mexican Barracuda (*Sphyraena lucasana*), Gulf of California, Mexico

WRASSES (PECES PERRO, VIEJAS, DONCELLAS, COLLAREJOS, SEÑORITAS)

FAMILY LABRIDAE

Wrasses are a diverse family occurring in a variety of environmental conditions including tide pools, rocky or coral reefs, weed beds, and open sand bottoms. Most inhabit tropical and subtropical latitudes, but a significant number of species also dwell in cooler temperate seas. This large family contains approximately 60 genera and about 500 species, well over half of which are distributed in the Indo-Pacific region. Australia has the world's largest labrid fauna, with about 165 species in 42 genera. Diagnostic features include a single continuous dorsal fin, not obviously notched between the spinous and soft portions (except in some *Xyrichtys* there are a few isolated spines on top of the head), cycloid scales, and a continuous or interrupted lateral line. Dorsal spine counts of tropical species generally range from VIII-XIV, and vary from flexible to stout and sharp-tipped. There are usually III anal spines (rarely II). The mouth is terminal in position and varies from small to moderate in size; the maxilla is not visible on the cheek; there are usually well-developed canine teeth, which may project forward, at the front of the jaws; there are rarely teeth on the roof of the mouth, but the pharyngeal dentition is well developed. Wrasses feed on a wide variety of items, including zooplankton, fishes, and invertebrates such as polychaete worms, brittle stars, crabs, and shrimps. Many also consume hard-shelled items such as molluscs and sea urchins, which are crushed with their pharyngeal teeth. Many of the larger wrasses are considered good eating. Labrids have an interesting reproductive biology that involves sex reversal. Initial stage adults are generally female, but are capable of changing into males. Many species have very different color patterns depending on the stage of the life cycle. Juveniles either have cryptic camouflage patterns or exhibit bright "poster" colors. They frequently possess one or more false eye spots (ocelli) on the median fins. Initial phase females are often somber shades of brown or green, or if a distinct pattern is present it is usually less attractive than that of the brightly colored, terminal-phase males. Spawning occurs in groups or in pairs. Males may instantaneously "flash" special nuptial colors and engage in aggressive chasing behavior in order to attract prospective mates. The resultant spawning frenzy culminates when eggs and sperm are released at the apex of a rapid ascent towards the surface.

Wounded Wrasse (*Halichoeres chierchiae*), terminal male, Isla Uva, Panama

MEXICAN HOGFISH
Bodianus diplotaenia (Gill, 1862)

Dorsal rays XII,10; anal rays III,12; pectoral rays 17; lateral line with 31 pored scales; large *adults with pronounced hump between eyes*; jaw teeth distinctly caniniform, 2 pairs at front of each jaw enlarged and somewhat curved; *adult males with elongate filaments on caudal fin lobes and prolonged rays posteriorly on dorsal and anal fins*; initial phase reddish, grading to yellow on posterior part of body and caudal fin; *a pair of blackish stripes (may be broken) on upper half of side,* and individual scale margins brown to reddish; terminal phase bluish green with brown head (except lower jaw white) and narrow *yellowish bar on middle of side*; juveniles similar to initial phase, but with yellow base color. Baja California to northern Chile, including offshore islands; commonly seen on rocky reefs in 4-20 m depth. Maximum size to 76 cm; common to 35 cm.

Bodianus diplotaenia Terminal phase ▲ Initial phase ▼

VIEJA OR GALAPAGOS HOGFISH
Bodianus eclancheri (Valenciennes, 1855)

Dorsal rays XII,10; anal rays III,11 or 12; pectoral rays 17; lateral line with 31-33 pored scales; large adults with pronounced hump on forehead and elongate upper and lower lobes on caudal fin; 2 pairs of enlarged, somewhat incisiform teeth at front of each jaw; *adult coloration extremely variable from uniformly dark gray or brownish to bright orange or white, with highly irregular and variable pattern of black blotching*; juveniles pale yellow to white, with 3 black stripes on head and body, and margins of median fins black. Ecuador to central Chile, and also Galapagos Islands; in the Galapagos it is usually seen on rocky reefs in the cooler waters of the western islands (Isabela and Fernandina) in 5-30 m. Grows to about 60 cm.

Bodianus eclancheri Mottled variety ▲ Drab variety ▼

Bodianus eclancheri Juv.

BLACKSPOT WRASSE
Decodon melasma Gomon, 1974

Dorsal rays XI,10; anal rays III,10; pectoral rays 17 (rarely 18); lateral line with 28 pored scales; jaw teeth distinctly caniniform, 2 pairs at front of each jaw enlarged and curved; pinkish red on upper portion of head and sides, white below, with *3 bright yellow, curved stripes on head and an oblong black blotch on each side above tip of pectoral fin*; fins mostly pink except anal white with bright yellow stripe; juveniles with up to 6 dusky bars on side. Gulf of California to Peru; found on rubble and isolated rocky reefs in sandy areas at depths of 40-160 m.

Photograph: Alex Kerstitch

BLACK WRASSE
Halichoeres adustus (Gilbert, 1890)

Dorsal rays IX,11; anal rays III,12; pectoral rays 12; lateral line with 27 pored scales; jaw teeth caniniform; 2 pairs of enlarged canines at front of lower jaw, and a pair of similar teeth at front of upper jaw, all directed obliquely forward; rear of upper jaw without enlarged canines; *initial phase dark brown, nearly blackish, except lighter on head and breast; head and body with numerous small blue or white spots arranged in longitudinal rows; caudal fin with broad white posterior marking; terminal phase uniformly dark brown or blackish* with narrow bluish margin on median fins; juveniles similar to initial phase, except a white bar at level of posterior part of pectoral fin, a white-edged black spot on middle of dorsal fin, and several large white blotches scattered on side. Revillagigedos and Isla del Coco; common on rocky reefs. Reaches at least 14 cm.

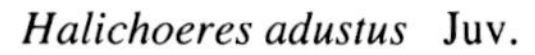

Halichoeres adustus Juv.

Halichoeres adustus Female ▲

Halichoeres aestuaricola Terminal phase ▲ Initial phase ▼

ESTUARY WRASSE
Halichoeres aestuaricola Bussing, 1972

Dorsal rays IX (rarely VIII),11; anal rays III,11 or 12; pectoral rays 12-14; lateral line with 27-28 pored scales; teeth caniniform, 2 pairs of enlarged canine teeth anteriorly in upper jaw, and forward-directed canine near rear corner of each side of upper jaw; 2 pairs of canines in lower jaw; initial phase mauve grayish with yellowish suffusion on middle of side, dorsal and anal fins yellowish; *a black spot on middle of dorsal fin and upper base of caudal fin*; terminal phase mauve to gray with narrow yellow scale margins; *head blue-gray; most of fins yellowish, base of dorsal fin and lower fourth of caudal fin dusky mauve*. Gulf of California to Colombia; inhabits shallow mangrove areas and tidal creeks. Grows to at least 20 cm.

WOUNDED WRASSE
Halichoeres chierchiae Caporiacco, 1947

Dorsal rays IX,11; anal rays III,12; pectoral rays 13; lateral line with 26-27 pored scales; jaw teeth caniniform, 2 pairs of enlarged curved canines at front of each jaw; rear of upper jaw with prominent, forward-directed canine on each side; *initial phase with greenish brown midlateral stripe intersecting with dark bars and blotches on upper half of body; lower portion of head and body pale yellow-green; terminal phase mainly bluish on upper half with green blotches, yellow on lower half; a black blotch above tip of pectoral fins, this mark adjacent to similar-sized patch of bright red*; fins reddish; juvenile color pattern is similar to that of *H. nicholsi*, except it is duskier overall, and the dorsal fin ocellus is smaller and less defined. Gulf of California to Panama; common on shallow rocky reefs with occasional sand-rubble patches; ranges to 70 m depth. Attains at least 20 cm.

Halichoeres chierchiae Terminal male (top) and initial phase ▲

COCOS WRASSE
Halichoeres discolor Bussing, 1983

Dorsal rays IX,11; anal rays III,12; pectoral rays 13; lateral line with 27 pored scales; teeth caniniform; 2 pairs of curved canines anteriorly in upper jaw and 1-2 canine teeth at rear corner of upper jaw; a pair of curved canines anteriorly in lower jaw; *initial phase white (except head yellowish) with pair of blackish stripes on head and anteriormost part of body; 3-4 large, irregular, blackish blotches on middle to lower half of side, and smaller blotches at base or on dorsal fin; terminal phase bright yellow on lower half of head and body, mainly charcoal gray on upper half, caudal fin with yellow margin and pair of yellow spots on basal part*; juveniles similar to initial phase, but with more white coloration and black, ocellated spot at middle of dorsal fin. Known only from Isla del Coco; usually seen in areas of live coral in 4-30 m depth. Grows to 15 cm; common to 10 cm.

Halichoeres discolor Terminal male ▲ Initial phase ▼

Halichoeres dispilus Terminal phase (top) and initial phase, Gulf of California ▲

Halichoeres dispilus Terminal male, Galapagos ▲ Initial phase, Galapagos ▼

Halichoeres insularis Terminal male ▲ Initial phase ▼

CHAMELEON WRASSE
Halichoeres dispilus (Günther, 1864)

Dorsal rays IX,11; anal rays III,12; pectoral rays 13; lateral line with 27 pored scales; teeth caniniform; 2 pairs of enlarged and curved canines at front of each jaw; rear of upper jaw with prominent forward-directed canine on each side; ground color has 2 distinct phases, reddish or greenish, which can be interchanged quickly; *initial phase primarily reddish or greenish above, white below; terminal phase with blue stripes on head, blue stripes or horizontal rows of blue spots on side, and usually a prominent bluish-black spot on lateral line below anterior part of dorsal fin*; juveniles greenish on upper half, with white midlateral stripe, whitish below, with small black spot at midbase of caudal fin. Gulf of California to Peru, including Galapagos Islands; common on rocky reefs in 1-70 m depth. Reaches 25 cm.

SOCORRO WRASSE
Halichoeres insularis Allen & Robertson, 1992

Dorsal rays IX,12; anal rays III,12; pectoral rays 12; lateral line with 26-27 pored scales; jaw teeth caniniform; a pair of enlarged, curved canines at front of upper jaw and 2 pairs at front of lower jaw; rear of upper jaw without enlarged canines; *initial phase greenish brown on back, whitish on lower two-thirds of side; a brown to greenish zig-zag stripe from pectoral base to lower caudal base* and similar, but darker stripe on upper edge of whitish area from rear margin of eye to upper caudal base; also a thin dark stripe along base of dorsal fin; snout yellowish; fins mainly translucent; *terminal phase is much more colorful, with a yellow stripe passing below the eye and continued midlaterally to the caudal base*; other markings are similar to the initial phase, except the stripe extending back from the eye is bright blue and the ventral portion of the body is reddish; median fins pinkish to translucent, caudal fin yellowish. Known only from the Revillagigedos; inhabits shallow reefs. Attains at least 8.2 cm.

MALPELO WRASSE

Halichoeres malpelo Allen & Robertson, 1992

Dorsal rays IX,12; anal rays III,12; pectoral rays 13; lateral line with 27 pored scales; teeth caniniform; pair of enlarged and curved canines at front of upper jaw, with a smaller canine just behind on each side; tip of of lower jaw with single large projecting canine that fits in slot between large pair of canines at front of upper jaw; no canines present at rear corner of jaws; *terminal adult males mainly pale greenish blue in life; initial phase adults largely pinkish with mainly yellow head.* Known only from Isla Malpelo, Colombia; inhabits rocky reefs. Attains at least 18 cm.

Halichoeres malpelo Terminal male ▲ Initial phase ▼

GOLDEN WRASSE

Halichoeres melanotis (Gilbert, 1890)

Dorsal rays IX,12; anal rays III,12; pectoral rays 12-13; lateral line with 27 pored scales; jaw teeth caniniform; 2 pairs of enlarged curved canines at front of each jaw; rear of upper jaw without enlarged canines; adults greenish to reddish brown on upper half, whitish to pink on lower half; sometimes with 5-6 broad bars on upper half; *distinguished by prominent black "ear" spot on upper edge of gill cover and blackish blotch or spot at base of caudal fin*; juveniles bright yellow with broad, black, midlateral stripe, and narrower black stripe along base of dorsal fin. Gulf of California to Panama; often seen in sand-rubble patches near rocky inshore reefs. Maximum size to 15 cm.

Halichoeres melanotis Juv.

Halichoeres melanotis Terminal males ▼

SPINSTER WRASSE

Halichoeres nicholsi (Jordan & Gilbert, 1881)

Dorsal rays IX,11; anal rays III,12; pectoral rays 13; lateral line with 28 pored scales; jaw teeth caniniform, 2 pairs of enlarged, curved canines at front of each jaw; rear of upper jaw with prominent forward-directed canine on each side; initial phase light green on back, whitish below; *a broad diffuse blackish stripe along middle of side, joined by short blackish bar below base of anterior part of dorsal fin; terminal phase mainly bluish or green, with broad blackish bar behind head and bright yellow spot just above pectoral fin*; juveniles whitish to yellowish, with irregular blackish stripes and blotches on side, and ocellated black spot on middle of dorsal fin. Gulf of California to Panama, and also offshore islands including the Galapagos. To 38 cm.

Halichoeres nicholsi Terminal phase ▲ Initial phase ▼

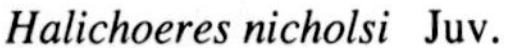

Halichoeres nicholsi Juv.

BANDED WRASSE

Halichoeres notospilus Günther, 1864

Dorsal rays IX,11; anal rays III,11; pectoral rays 13; lateral line with 27 pored scales; jaw teeth caniniform; a pair of enlarged, curved canines at front of upper jaw and 2 pairs at front of lower jaw; rear of upper jaw without enlarged canines; initial phase dark green on upper half, light green below, about *5-6 alternating yellow and black patches along upper back; terminal phase with 7-8 blackish bars on upper half of body, with narrow yellow bars between, a blackish patch behind pectoral fin*, and red caudal fin; juvenile color pattern similar to that of *H. chierchiae*, except very small fish are almost entirely black with a clear caudal fin and narrow white bar behind the pectoral fin. Gulf of California to Peru, including offshore islands; common in shallows (0.5-10 m) exposed to surge. Maximum size to 25 cm.

Halichoeres notospilus Terminal male ▲ Initial phase ▼

Halichoeres semicincta Terminal male ▲ Initial phase ▼

Novaculichthys taeniourus Terminal male ▼

ROCK WRASSE

Halichoeres semicincta (Ayres, 1859)

Dorsal rays IX,12; anal rays III,12; pectoral rays 13; lateral line with 27 pored scales; jaw teeth caniniform, 2 pairs of enlarged, curved canines at front of each jaw; rear of upper jaw with prominent forward-directed canine on each side; pale green to pale yellow-orange; *initial phase with scattered black dots on upper back; terminal phase with prominent, pale-edged black bar on lower two-thirds of side,* a short distance behind base of pectoral fin; juveniles with white midlateral stripe and black ocellated spot at middle of dorsal fin. Baja California and Gulf of California; occurs over clean white sand adjacent to rocky reefs, usually in 15-40 m depths. Reaches a maximum length of 38 cm.

Halichoeres semicincta Juv.

ROCKMOVER WRASSE

Novaculichthys taeniourus Lacepède, 1801

Dorsal rays IX,12; anal rays III,13; pectoral rays 13; lateral line interrupted, 19-20 pored scales on anterior upper portion and 5-6 on base of caudal fin; head mostly scaleless; a pair of large curved canine teeth at front of each jaw; *first 2 dorsal spines greatly prolonged in juveniles; adults dark greenish brown with white spot on each scale; head gray with several brown bands radiating from eye*; juveniles mottled and banded, either green, reddish or brown, and spotted with white. Widespread in the Indo-Pacific from East Africa to the Americas; in the eastern Pacific from the Gulf of California to Panama and the Galapagos; can overturn large rocks with its powerful jaws when searching for invertebrate prey; the young mimic drifting masses of algae. Grows to 30 cm.

Novaculichthys taeniourus Juv.

SMALLTAIL WRASSE

Pseudojuloides cerasinus (Snyder, 1904)

Dorsal rays IX,11; anal rays III,12; pectoral rays 13; lateral line with 27 pored scales; head scaleless; a single pair of outwardly curved canine teeth at front of each jaw, remaining teeth in a single row, somewhat incisiform; no canines at rear of upper jaw; *initial phase mainly light red; terminal phase green above, bluish on belly, a blue-edged yellow midlateral stripe and blue-edged black spot posteriorly on caudal fin.* Widely distributed in the Indo-Pacific from East Africa to the Americas, but rare in the eastern Pacific; usually seen over rubble bottoms in 3-60 m. Largest specimen 12 cm.

Photograph: John Randall *Pseudojuloides cerasinus* Terminal male ▼ Initial phase ▲

Photograph: John Randall

GOLDSPOT SHEEPSHEAD

Semicossyphus darwini (Jenyns, 1842)

Dorsal rays XII,10; anal rays III,11 or 12; pectoral rays 18; scales relatively small, *53-56 pored scales in lateral line*; 2 pairs of enlarged, curved canines at front of each jaw; rear of upper jaw with moderately enlarged canine on each side; *initial phase red with large yellow patch just behind head and black spot at front of dorsal fin; terminal phase bluish gray with large yellow patch behind head; both phases with white chin.* Ecuador to central Peru, including Galapagos Islands; found along rocky shores of the cooler southern and western regions of the Galapagos. Maximum size to about 75 cm.

Semicossyphus darwini Terminal male ▲ Initial phase ▼

CALIFORNIA SHEEPSHEAD
Semicossyphus pulcher (Ayres, 1854)

Dorsal rays XII,10; anal rays III,12; pectoral rays 17-19; scales relatively small, *52-57 pored scales in lateral line*; 2 pairs of enlarged, curved canines at front of each jaw; rear of upper jaw with prominent canine on each side; *large adults develop hump on forehead; initial phase brownish red to rose with white chin; terminal phase with black head and white chin, pink to dusky red midsection, and black rear portion of body*; juveniles reddish orange with whitish midlateral stripe and large black spots on dorsal, anal and pelvic fins, and also on caudal fin base. Monterey, California, to the Gulf of California; inhabits kelp beds and rocky shores, usually in 3-30 m. Attains 90 cm.

Photograph: Alex Kerstitch *S. pulcher* Juv. ▲

Photograph: Alex Kerstitch *Semicossyphus pulcher* Terminal male ▲

Photograph: John Randall *Stethojulis bandanensis* Terminal male ▼ Initial phase ▲

Photograph: John Randall

BLUELINED WRASSE
Stethojulis bandanensis (Bleeker, 1851)

Dorsal rays IX,11; anal rays III,11; pectoral rays 14 or 15; lateral line with 25 pored scales; head scaleless; small incisiform teeth in jaws, none noteably enlarged, and a canine present at corner of mouth; *initial phase with numerous small white spots on upper half of side and pair of black spots on caudal fin base; terminal phase green on upper half with bright red patch above pectoral fin base and several thin blue stripes on head and upper half of body*. Widespread in the tropical Pacific Ocean from Australia to southern Japan, and eastward to islands of the eastern Pacific, including Isla del Coco, Clipperton Island, and the Galapagos; usually associated with coral reefs. Grows to 12.5 cm.

ISLAND WRASSE

Thalassoma grammaticum Gilbert,1890

Dorsal rays VIII, 12-13; anal rays III,11; pectoral rays 15-16; lateral-line scales 26-27; total gill rakers on first arch 19-22; head scaleless; a pair of enlarged canines at front of upper and lower jaws; caudal fin lunate with pointed lobes; *green or blue-green, with narrow red streak on each scale; head and breast rose pink, also pink strip on belly; head with about 5 curved pink bands; inner half of dorsal and anal fins, and upper and lower edges of caudal fin, rose pink; terminal phase similar, but may have bright blue ground color.* Gulf of California to Panama, also Galapagos and other offshore islands; inhabits rock and coral reefs. Grows to about 24 cm. Closely related to *T. lutescens* of the Indo-West Pacific.

CORTEZ RAINBOW WRASSE

Thalassoma lucasanum (Gill, 1863)

Dorsal rays VIII, 13; anal rays III,11; pectoral rays 15; lateral line with 26-27 pored scales; head scaleless; each jaw with 1 pair of canine teeth anteriorly followed by progressively shorter conical teeth; no canine at rear of jaw; *initial phase with pair of bright yellow stripes bordering broad, dark brown to greenish area from snout, and encompassing most of upper side; terminal phase mainly blue except for broad yellow saddle-like marking just behind head.* Gulf of California to Galapagos Islands; forms aggregations that feed on crustaceans, soft corals, and algae; the young sometimes "clean" parasites from other fishes. Maximum size to 15 cm.

Thalassoma lucasanum Terminal male ▲ Initial phase ▼

CLIPPERTON WRASSE
Thalassoma sp.

Dorsal rays VIII,13; anal rays III,11; pectoral rays 13-15; lateral line 25-27; gill rakers on first arch 16-19; head scaleless; each jaw with 1 pair of canine teeth anteriorly followed by gradually shorter, conical teeth; no canine at rear of jaw; caudal fin truncate to slightly emarginate; *initial phase white on lower half and mainly dark brown above; terminal phase mainly rose pink with broad yellow bar just behind head; a broad green stripe on lower part of head; caudal fin bluish*. This recently discovered species is known only from Clipperton Island; inhabits coral reefs. Maximum size to about 12 cm.

Thalassoma sp. Initial phase ▲ Terminal male ▼

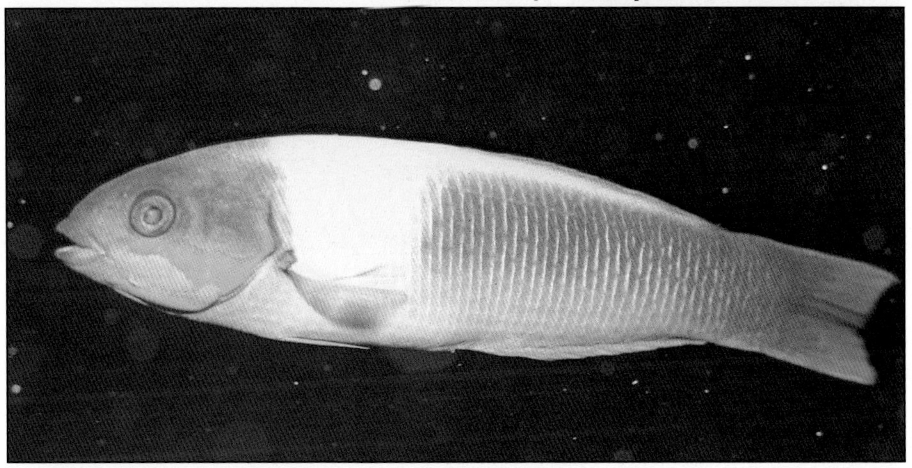

LONG-FINNED WRASSE
Thalassoma virens Gilbert, 1890

Dorsal rays VIII, 12-13; anal rays III,10-11; pectoral rays 15-16; lateral-line scales 26-27; total gill rakers on first arch 21; head scaleless; a pair of enlarged canines at front of upper and lower jaws; caudal fin with prolonged pointed lobes; *initial phase dull greenish with reddish streak in each scale; body with about 6 irregular, blotchy bars; head with numerous red spots or spoke-like lines radiating from eye; terminal phase entirely green, occasionally blue with reddish belly*. Revillagigedo Islands and Clipperton Island. Grows to at least 29 cm. Similar to *T. purpureum* of the Indo-West Pacific.

Thalassoma virens Initial phase ▲ Terminal male ▼

PEACOCK WRASSE
Xyrichtys pavo (Valenciennes, 1840)

Dorsal rays II-VII,12; anal rays III,12; pectoral rays 12; *first 2 dorsal spines separated from rest of fin, greatly prolonged in juveniles*; lateral line interrupted, the pored scales 20-22 + 4-5; snout steep and laterally compressed; suborbital region very broad; a pair of long, curved canine teeth anteriorly in each jaw; caudal fin small, slightly rounded; *overall white to yellowish white, with 4 indistinct, broad, dark bars on body, and small black spot on upper side above eighth scale of lateral line*; juveniles whitish with 4 diffuse brown bars, thin brown lines radiating from eye, and pelvic fins entirely brown. Widespread in the Indo-Pacific from East Africa to the Americas (Gulf of California to Panama and the Galapagos); found on open sand near reefs; dives into the sand when danger approaches. To 35 cm.

Xyrichtys pavo Juv.

GALAPAGOS RAZORFISH
Xyrichtys victori Wellington, 1992

Dorsal rays IX,12; anal rays III,12; pectoral rays 12; lateral line interrupted, the pored scales 19-20 + 5; snout steep and laterally compressed; suborbital region very broad; a pair of long, curved canine teeth anteriorly in both jaws; caudal fin relatively small, slightly rounded; *males iridescent dark blue-green with scattered, scale-sized black blotches on sides; females mainly pinkish orange, without dark markings; juveniles either uniform sandy white or yellowish brown with dark stripe on upper part of side*. Known only from the Galapagos; inhabits open sand slopes, usually below 10 m depth; adults often share this habitat with the garden eel *Heteroconger klausewitzi*. Reaches about 15 cm.

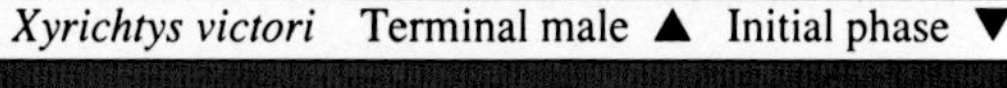

Xyrichtys victori Terminal male ▲ Initial phase ▼

PANAMANIAN RAZORFISH
Xyrichtys sp.

Dorsal rays IX,12; anal rays III,12; pectoral rays 12; lateral line interrupted, the pored scales 20 + 5; snout steep and laterally compressed; suborbital region very broad; a pair of long, curved canine teeth anteriorly in both jaws; caudal fin relatively small, slightly rounded; *generally pale pinkish, except operculum and belly region white; a diffuse, pale yellowish midlateral stripe on side of body; a reddish bar just behind rear margin of preopercle; fins uniformly pale.* Known only from a single collection site at Isla San Jose in the Perlas Islands, Panama; inhabits clean sand bottoms in 12-15 m depth. To at least 6.5 cm.

PARROTFISHES (PECES LORO, PERICOS, POCOCHOS DE MAR)
FAMILY SCARIDAE

Parrotfishes are common inhabitants of tropical reefs. Most are easily recognized by the fused teeth which form a parrot-like beak. The family is distributed worldwide and contains 11 genera with about 70 species. Most parrotfishes have a very similar morphology. Typical features include an unnotched dorsal fin with IX,10 rays, and an anal fin of III,9 rays. Scales are cycloid and large, with 22-24 in the lateral line. The characteristic dental plates have a median suture, and the adults of some species possess one or two short canine teeth posteriorly. Parrotfishes feed on corals, and are also one of the main algal grazers that inhabit tropical reefs. Their characteristic "chisel" marks are seen everywhere, and foraging schools emit biting sounds that are clearly audible. The unusual pharyngeal dentition consists of interlocking molar-type teeth on the upper and lower surface of the throat. These are used for grinding bits of coral and rock, from which their algal food is extracted. This process results in a fine powder and is voided as a white cloud with the feces. This constitutes one of the most important sources of fine bottom sediments. Spawning often occurs at dusk. Eggs and sperm are released at the apex of a rapid rush towards the surface. Both pair spawning and group spawning are common. Sex reversal, similar to that of wrasses, is typical; juveniles, females, and males of the same species may display strikingly different color patterns. The initial adult phase, which is generally female (but also male in some species), is usually characterized by a relatively drab pattern. Females of most parrotfishes in this phase are able to transform to functioning males and undergo a change in color; the terminal male is usually bright colored. At night, parrotfishes seek out caves and ledges, where they rest on the bottom and secrete a strange mucous cocoon that completely envelops them.

STAREYE PARROTFISH
Calotomus carolinus (Valenciennes, 1840)

Dorsal rays IX,10; anal rays III,9; pectoral rays 13; median predorsal scales 3-4 (usually 4); a single row of 4 or 5 scales on cheek below eye; *teeth not fully fused to form plates, the individual flattened teeth readily apparent on outer surface of jaws, imbricate, the tips of outer row form a jagged cutting edge*; dorsal spines flexible; caudal fin slightly rounded in juvenile, and truncate with lobes prolonged in adults; initial phase mottled dark orangish brown, shading ventrally to pale orangish; base of pectoral fins dark; posterior margin of caudal fin narrowly white; body of terminal males a mixture of brownish red and blue-green; head blue-green with orange-pink bands radiating from eye. Indo-Pacific and tropical eastern Pacific. Reaches 50 cm.

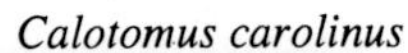

Calotomus carolinus

Photograph: John Randall

LOOSETOOTH PARROTFISH

Nicholsina denticulata (Evermann & Radcliffe, 1917)

Dorsal rays IX,10; anal rays III,9; pectoral rays 13; median predorsal scales 4-5; a single row of 4-5 scales on cheek below eye; *teeth not fully fused to form plates, the individual flattened teeth readily apparent on outer surface of jaws, their tips forming a jagged cutting edge*; drab brown or gray to reddish, but always strongly mottled, effectively blending in with surroundings; terminal males with red throat and caudal fin. Baja California to Peru, including the Galapagos and other offshore islands; often common on algae-covered reefs. To about 35 cm.

Nicholsina denticulata Terminal male ▲ Green initial phase ▼

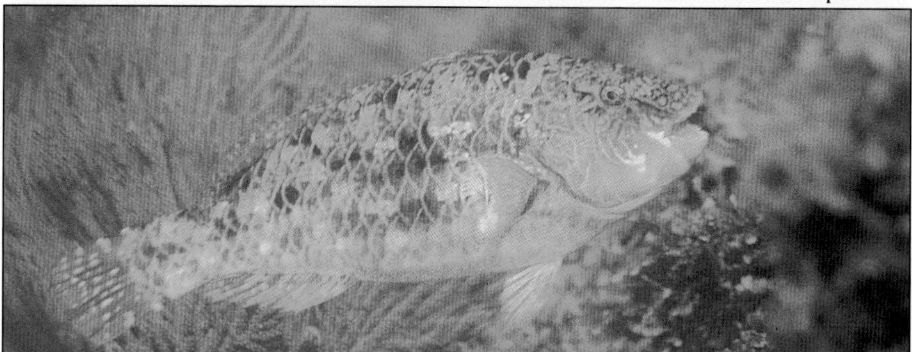

Nicholsina denticulata Red initial phase ▼

AZURE PARROTFISH

Scarus compressus (Osburn & Nichols, 1916)

Dorsal rays IX,10; anal rays III,9; median predorsal scales 6; 3 rows of scales on cheek; body relatively deep and compressed; caudal fin emarginate; *initial phase brownish green, usually lighter on head with bluish or green spoke-like lines radiating from eye and about 5 diffuse, whitish bars on side; terminal male mainly green with orange-brown scale outlines, snout green, and green, irregular, spoke-like bands radiating from eye*. Gulf of California to Ecuador, including Galapagos; generally the least abundant and most wary of the eastern Pacific parrotfishes; occurs on rock or coral reefs to at least 25 m. Grows to at least 50 cm.

Scarus compressus Terminal male ▲ Initial phase ▼

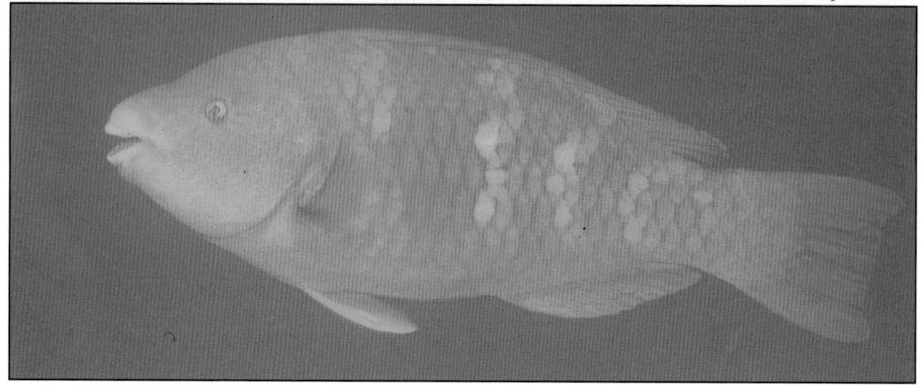

BLUECHIN PARROTFISH

Scarus ghobban Forsskål, 1775

Dorsal rays IX,10; anal rays III,9; pectoral rays 15-16; median predorsal scales usually 6; 3 rows of scales on cheek; caudal fin slightly emarginate in small initial phase fish to lunate in terminal males; *initial phase overall yellowish to yellowish green, with 5 irregular blue to whitish bars on side; terminal males blue greenish on back with salmon-pink scale edges, lower side largely salmon pink* with green scale margins; a pair of blue bands on chin. Widespread in the tropical Indo-Pacific from East Africa to the Americas; in the eastern Pacific it ranges from the Gulf of California to Ecuador, including the Galapagos; usually seen on shallow reefs and adjacent sand-rubble areas. Grows to 75 cm.

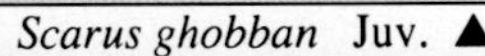

Scarus ghobban Juv. ▲

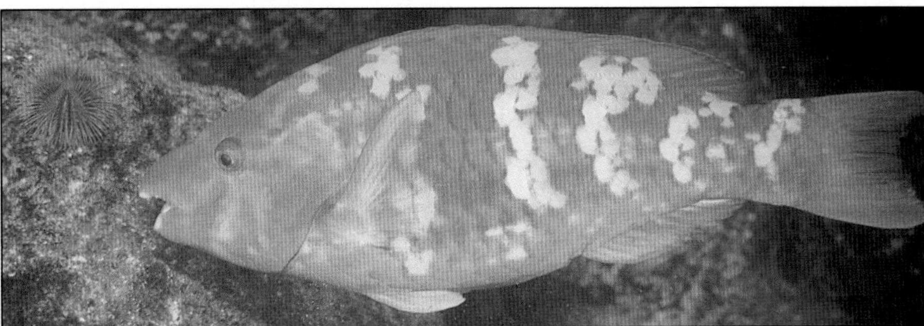

Scarus ghobban Initial phase ▲ Terminal phase ▼

BUMPHEAD PARROTFISH

Scarus perrico Jordan & Gilbert, 1882

Dorsal rays IX,10; anal rays III,9; pectoral rays 14-16; median predorsal scales 5; 3 rows of scales on cheek; *caudal fin slightly rounded; large individuals develop prominent hump on forehead; overall green to bluish green, with dark blue fins*; dark, spoke-like lines radiating from eye; dental plates blue. Gulf of California to Peru; usually seen grazing on rocky reefs in small aggregations in 3-20 m. Attains a length of 80 cm.

Scarus perrico Terminal male

BICOLOR PARROTFISH
Scarus rubroviolaceus Bleeker, 1847

Dorsal rays IX,10; anal rays III,9; pectoral rays 14-16; median predorsal scales 6; 3 rows of scales on cheek; body moderately elongate, its depth 2.7-3.1 in standard length; *snout profile steep and blunt*; caudal fin of initial phase adults slightly emarginate, that of terminal males lunate; *initial phase gray on anterior half and whitish posteriorly, with small black spots and dark lines on sides, and red fins; terminal males mainly green to bluish green, sometimes purplish on anterior half, giving strong bicolor effect*; a pair of blue bands on chin, and dental plates blue-green. Widespread in the tropical Indo-Pacific from East Africa to the Americas; in the eastern Pacific it ranges from the Gulf of California to the Galapagos; occurs on shallow reefs. To 70 cm.

rubroviolaceus Terminal male ▲ Initial phase ▼

JAWFISHES (BOCAS GRANDES)
FAMILY OPISTOGNATHIDAE

These are small to medium-sized fishes with narrow tapering bodies and a noticeably enlarged head and mouth. The group occurs in all warm seas, and contains an estimated 70 species in three genera. They construct elaborate burrows by scooping sand or small stones with the mouth, and can shift larger rocks by using their powerful jaws. The burrows are frequently lined and reinforced with pebbles and shell fragments. These fishes exhibit the unusual habit of oral egg incubation. They feed chiefly on benthic and planktonic invertebrates.

Giant Jawfish (*Opistognathus rhomaleus*), Gulf of California, Mexico

LONGTAILED JAWFISH
Lonchopisthus sinuscalifornicus Castro-Aguirre & Villavicencio-Garayzar, 1989
(Plate XIII-11, p. 217)

Dorsal rays XI,18 (occasionally 17 or 19); anal rays III,17 (rarely 18); pectoral rays 16-20; scales in longitudinal series 67-74; *caudal fin lanceolate, much longer than head length*; posterior end of bony maxilla bluntly notched; head and body tan or cream-colored, with scattered pale blue spots; a larger, intense, bluish violet spot on opercle; median fins violet. Closely related to *L. higmani* of the western Atlantic. Gulf of California; most specimens have been caught by trawls on flat sandy bottoms in depths to 15 m. Grows to about 30 cm.

Opistognathus galapagensis ▲

GALAPAGOS JAWFISH
Opistognathus galapagensis Allen & Robertson, 1991

Dorsal rays XI,16-17; anal rays II,16; pectoral rays 21; *gill rakers on first arch 11 + 22 = 33; body with about 106-109 oblique scale rows*; end of maxilla truncate; upper jaw 71.1-80.0 percent of head length; whitish or pale tan, with 6 irregular brown bars, the middle portion of bars somewhat expanded and may coalesce to form a longitudinal stripe; 6 large brown blotches, corresponding with each bar, on back and extending onto dorsal fin; *a prominent black spot on distal part of first 3 dorsal spines, and a nearly equal-sized white patch immediately below*. Galapagos Islands; inhabits sand-rubble areas. Reaches at least 16 cm.

Opistognathus mexicanus ▲

MEXICAN JAWFISH
Opistognathus mexicanus Allen & Robertson, 1991

Dorsal rays XI,17; anal rays II,17; pectoral rays 21; *gill rakers on first arch 15 + 32 = 47*; body with *about 150 oblique scale rows*; maxilla rounded on lower rear edge, with slight indentation on upper rear margin; upper jaw about 66 percent of head length; *light tan with brown spots, those of head generally smaller and more numerous*. Gulf of California; inhabits sand bottoms to at least 14 m depth. Grows to at least 12 cm.

O. rosenblatti Female ▲ Nuptial male ▼

Opistognathus rosenblatti Juv. ▼

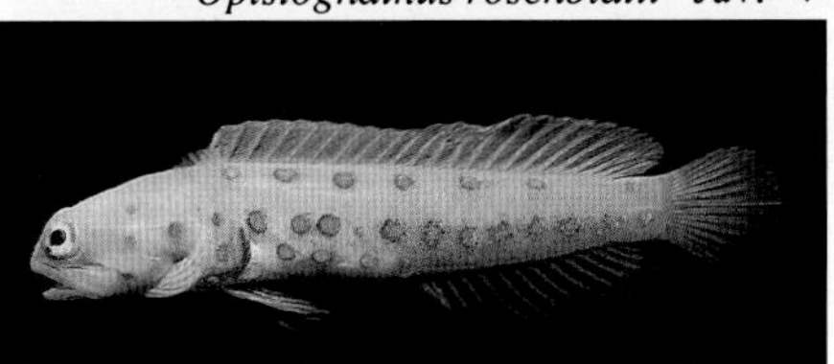

BLUE-SPOTTED JAWFISH
Opistognathus rosenblatti Allen & Robertson, 1991

Dorsal rays XI,13; anal rays II,13; *pectoral rays 18* (rarely 17); gill rakers on first arch 12-13 + 22 = 34-35; body with about 65-70 oblique scale rows; end of maxilla truncate; upper jaw 51-53 percent of head length; *courting males white anteriorly and blackish on posterior half with large brown spot on anterior part of dorsal fin; females with orange-brown head and dark brown to blackish body, covered with relatively large blue spots*; small juveniles overall yellow with blue spots. Gulf of California; found in colonies of up to several hundred fish, with a minimal spacing of about 1 m between individuals; each fish constructs a lid over their burrow entrance at dusk and rebuilds the opening each morning. Maximum size about 10 cm.

PANAMANIAN JAWFISH
Opistognathus panamaensis Allen & Robertson, 1991

Dorsal rays XI,13; anal rays II,13; *pectoral rays 20*; gill rakers on first arch 13 + 24 = 37; end of maxilla truncate; upper jaw 51-53 percent of head length; *very similar in coloration and habits to* O. rosenblatti; dorsal and caudal fins are generally bright yellow, and the blue spots on the body sometimes coalesce to form stripes. Isla del Coco, Panama, and Malpelo Islands; inhabits burrows in sand-rubble areas. Reaches about 10 cm.

FINESPOTTED JAWFISH
Opistognathus punctatus Peters, 1869
(Plate XIII-9, p. 217)

Dorsal rays XI,17; anal rays III,17; pectoral rays 20-21; total gill rakers on first arch 41-49; body with about 125 oblique scale rows; *maxillary elongate, as long as head in large adults; light gray or brownish, with numerous fine dark spots on head and body, and larger, blackish spots and blotches on side and fins*; juveniles become increasingly spotted with growth; juveniles under about 15 cm with elongate black spot on middle of dorsal fin between fourth and eighth spines, this spot an ocellus in very small individuals. Baja California to Panama; inhabits sand-rubble bottoms. Maximum size to 40 cm.

GIANT JAWFISH
Opistognathus rhomaleus Jordan & Gilbert, 1881

Dorsal rays XI,12-14; anal rays II,13; pectoral rays 20-21; total gill rakers on first arch 28-33; body with about 100-105 oblique scale rows; end of maxilla truncate; *maxillary extending well beyond eye, about 1.7 in head length, but not reaching margin of preopercle; light gray with numerous small black spots on head and upper back; sides of adult devoid of markings, but juvenile with 5-6 large, dark brown blotches* and corresponding spots on base of dorsal fin. Outer Baja California, Gulf of California, and Revillagigedo Islands; inhabits sand-rubble bottoms; often caught by anglers and trawlers. Reaches 51 cm.

Opistognathus rhomaleus Adult ▼ Juv. ▲

BULLSEYE JAWFISH
Opistognathus scops (Jenkins & Evermann, 1889)

Dorsal rays X,16-17; anal rays II,15-17; pectoral rays 20; gill rakers on first arch 11-12 + 23-24 = 34-36; body with about 120-125 oblique scale rows; maxillary extending beyond eye; *generally brownish with 3-4 longitudinal dark stripes on side, and scattered, elongate white blotches and spots giving appearance of broken, wavy stripes; a prominent ocellus at front of dorsal fin.* Gulf of California to Panama; inhabits sand-rubble bottoms. Reaches at least 12 cm.

STARGAZERS (MIRACIELOS)
FAMILY URANOSCOPIDAE

Stargazers are similar in general appearance to the toadfishes (Batrachoididae). They have a large, robust, somewhat dorsally flattened head, and a body that tapers in width posteriorly. The mouth is large with fringed lips and the eyes are directed upwards. There are a pair of large, double-grooved poison spines, with a venom gland at the base, immediately above the pectoral fin and behind the opercle. Venom from these fishes has caused death in humans, therefore they should not be handled. In addition, members of the genus *Astroscopus* have electric organs formed from modified eye muscles. They are capable of delivering a shock of up to 50 volts. The family is distributed worldwide in tropical and warm temperate seas. They occur in shallow coastal waters as well as deeper parts of continental shelves. There are approximately nine genera and 25 species. Most grow to about 30 cm, but some species may reach 46 cm.

ZEPHYR STARGAZER
Astroscopus zephyreus Gilbert & Starks, 1896

Dorsal rays IV+I,12; anal rays 13; scales in midlateral series about 80; head large, broad, partly covered with bony plates, but without spines; *a pair of large venomous spines above pectoral fin base and behind opercle; grayish brown on dorsal half of head and body, with numerous white spots*; whitish on operculum and lower half of body; fins dusky to blackish, *caudal with longitudinal white stripes.* Baja California to Peru; inhabits sandy bottoms. Grows to about 30 cm.

Astroscopus zephyreus

Photograph: Alex Kerstitch

PLATE XIII

1 **THICK-LIPPED MULLET** (*Chaenomugil proboscidens*)

2 **STRIPED MULLET** (*Mugil cephalus rammelsbergii*)

3 **GALAPAGOS MULLET** (*Mugil galapagensis*)

4 **WHITE MULLET** (*Mugil curema*)

5 **HOSPE MULLET** (*Mugil hospes*)

6 **THOBURN'S MULLET** (*Xenomugil thoburni*)

7 **BLUE BOBO** (*Polydactylus approximans*)

8 **YELLOW BOBO** (*Polydactylus opercularis*)

9 **FINESPOTTED JAWFISH** (*Opistognathus punctatus*)

10 **PACIFIC BARRACUDA** (*Sphyraena ensis*)

11 **LONGTAILED JAWFISH** (*Lonchopisthus sinuscalifornicus*)

12 **MEXICAN BARRACUDA** (*Sphyraena lucasana*)

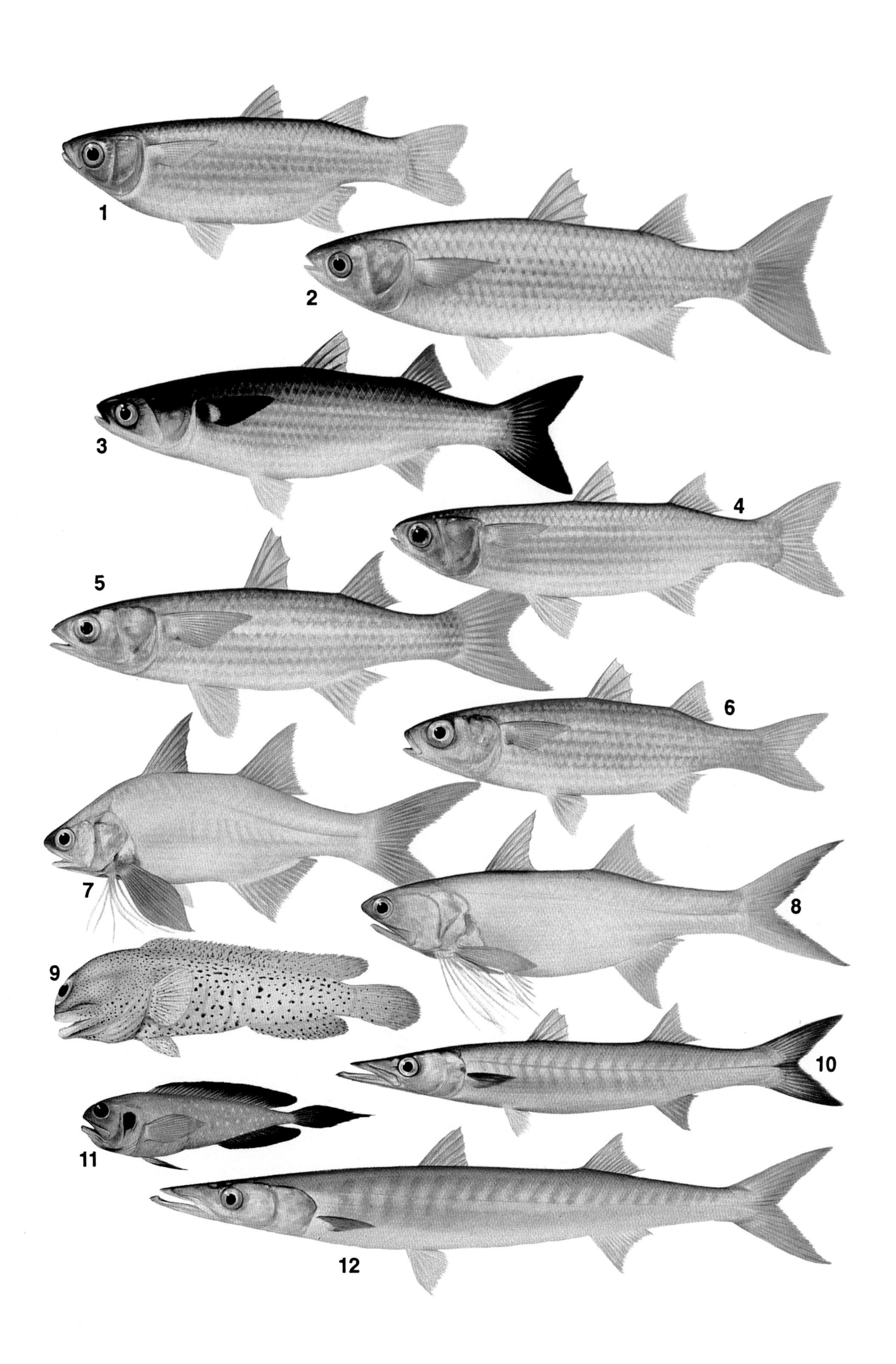
1
2
3
4
5
6
7
8
9
10
11
12

TRIPLEFINS (TRES ALETAS)
FAMILY TRIPTERYGIIDAE

The triplefins are small (usually 3-5 cm), relatively inconspicuous inhabitants of tropical and temperate reefs. The family occurs worldwide, but most of the estimated 150 species occur in the Indo-Pacific. The taxonomy of this group has long been neglected, consequently the identification of many species is problematical and there are a number of undescribed forms. The common name is derived from the division of the dorsal fin into three parts: the first consists of III or IV spines, the second of VIII-XVI spines, and the third of 8-12 soft, segmented rays. Many of the species are cryptically colored or semitransparent; this, in combination with their small size, often makes them difficult to detect. They dwell on the reef's surface, often in weedy areas, on algae-covered rocks, or on rubble. Their food consists primarily of tiny invertebrates and algae.

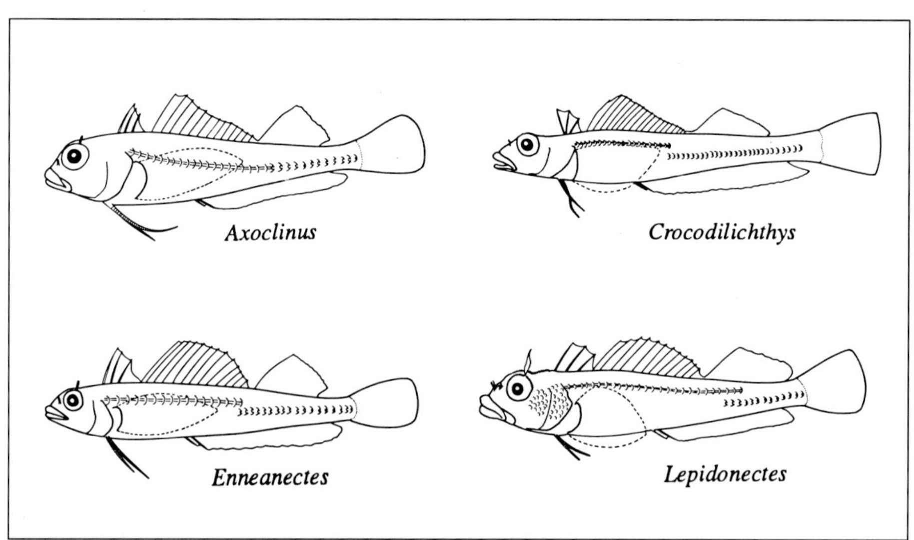

Eastern Pacific genera of triplefins (Trypterygiidae) ▲

Coral Triplefin (*Lepidonectes corallica*), North Seymour Island, Galapagos

CARMINE TRIPLEFIN
Axoclinus carminalis (Jordan & Gilbert, 1882)

Dorsal rays III+XII+9-10, segmented dorsal rays unbranched, except last ray may be branched at base; anal rays II,16-17; pectoral rays 15-16; pelvic rays I,2; *lateral line gradually descending from upper edge of gill cover to midlateral axis; lateral line with 22-25 tubed scales followed by 0-2 weakly notched scales; total scales in longitudinal series 33-36; a small, lanceolate cirrus above eye*; scales absent on head, breast, belly, and base of pectoral fins; upper half brown or greenish, lower half reddish (particularly intense in breeding males); sides with 3 reddish brown bars, and additional black band around tail base; caudal fin black in males, mostly pale except blackish posteriorly in females. Gulf of California and outer coast of Baja California; inhabits weedy shallows in rocky areas. Grows to 3 cm.

COCOS TRIPLEFIN
Axoclinus cocosensis Bussing, 1991

Dorsal rays III+XIII+9, segmented rays unbranched, except last ray may be branched at base; anal rays II,18; pectoral rays 16-17; pelvic rays I,2; *lateral line gradually descending from upper edge of gill cover to midlateral axis; lateral line with 21-26 tubed scales followed by 8-12 notched scales; total scales in longitudinal series 33-35; a small, lanceolate cirrus on upper edge of eye*; scales absent on head, breast, belly, and base of pectoral fins; head pink to reddish; side with 3 broad, dark brown bars with narrow, whitish spaces between them; a black band around base of caudal fin; caudal fin frequently black. Isla del Coco; inhabits shallow rocky areas. Reaches about 3.5 cm. A very similar but possibly undescribed species occurs at the Revillagigedos Islands.

PANAMA TRIPLEFIN
Axoclinus lucillae Fowler, 1941

Dorsal rays III+XII+9, segmented rays unbranched, except last 1-2 rays may be branched; anal rays II,17; pectoral rays 15-16; pelvic rays I,2; *lateral line gradually descending from upper edge of gill cover to midlateral axis; lateral line with 19-22 tubed scales followed by about 12 weakly notched scales; total scales in longitudinal series 33-34; a short, lanceolate cirrus on upper part of eye*; greenish on back, red on ventral surface; narrow, dark brown bar below eye, and 3 broad brown bars on side of body (these may be split, forming double bars); caudal peduncle with pearl-white bar followed by intense black bar; caudal fin mostly blackish with narrow white bar at base; pectoral fins with several dark bars. Mexico to Colombia; common in rocky shallows. Reaches 3 cm.

MULTIBARRED TRIPLEFIN

Axoclinus multicinctus Allen & Robertson, 1992

Dorsal rays III+XII+9-10, segmented rays unbranched except last ray branched at base; anal rays II,18; pectoral rays 15; pelvic rays I,2; *lateral line gradually descending from upper edge of gill cover to midlateral axis; lateral line with 23-24 tubed scales followed by 11-12 notched scales; total scales in longitudinal series 34-35; a small lanceolate cirrus on upper edge of eye*; scales absent on head, breast, belly, and base of pectoral fins; body with 3 sets of "double", reddish brown bars and intense black band on caudal peduncle; pale interspaces between bars with narrow, faint brown bar through its center; a small, dark brown spot just posterior to upper pectoral base; outer half of first dorsal fin and caudal fin of males dusky blackish. Known only from the Revillagigedos Islands. Reaches about 3 cm.

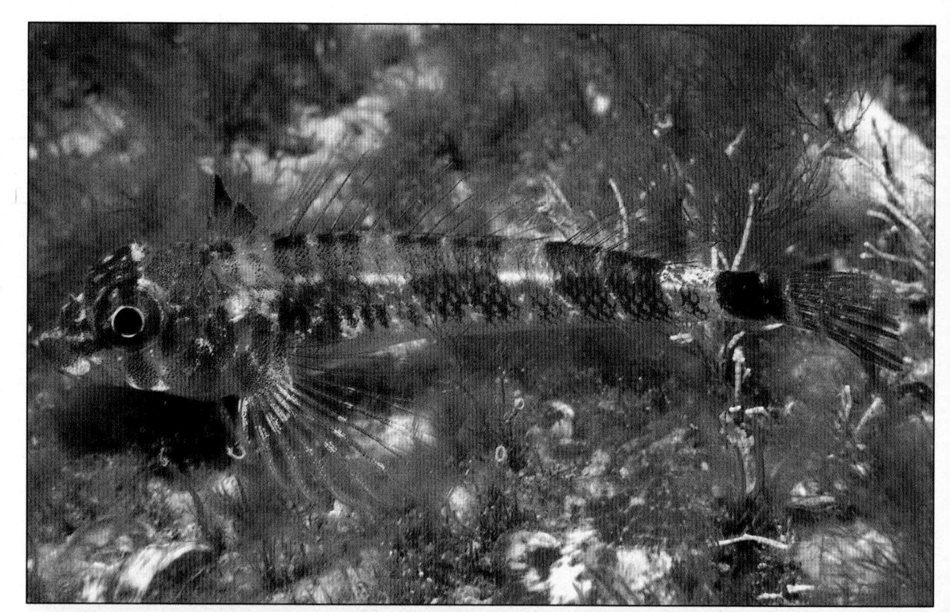

CORTEZ TRIPLEFIN

Axoclinus nigricaudus Allen & Robertson, 1991

Dorsal rays III+XII or XIII+10 or 11, segmented rays unbranched; anal rays II,18; pectoral rays 15; pelvic rays I,2; *lateral line gradually descending from upper edge of gill cover to midlateral axis; lateral line with 15 or 16 tubed scales followed on row below by 18 or 19 notched scales; total scales in longitudinal series 36; orbital cirrus absent*; scales absent on head, breast, belly, and base of pectoral fins; green to brownish on back, whitish below; 5 oblique, brown to blackish bars on side, including black band on caudal peduncle; caudal fin blackish. Gulf of California; common in intertidal zone. To 4.5 cm.

Axoclinus nigricaudus ▲

Axoclinus rubinoffi ▼

RUBINOFF'S TRIPLEFIN

Axoclinus rubinoffi Allen & Robertson, 1992

Dorsal rays III+XII+10, segmented rays unbranched, except last ray branched at base; anal rays II, 17-18; pectoral rays 15; pelvic rays I,2; *lateral line gradually descending from upper edge of gill cover to midlateral axis; lateral line with 22-23 tubed scales followed by 12-13 notched scales; total scales in longitudinal series 35-36; a small, lanceolate cirrus on upper edge of eye*; scales absent on head, breast, belly, and base of pectoral fins; body with 3 broad, brown to blackish bars, and blackish band on caudal peduncle, bars narrowly margined with pearl white, particularly posterior ones; narrower, pale brownish bars between dark bars; male with blackish caudal fin that is confluent with dark peduncular band. Known only from Isla Malpelo, Colombia; inhabits rocky areas. Grows to about 3.5 cm.

LIZARD TRIPLEFIN

Crocodilichthys gracilis Allen & Robertson, 1991

Dorsal rays III+XVII+12 or 13, segmented rays branched except for first and last 3-5 rays; anal rays I or II,25 or 26; pectoral rays 15; *lateral line more or less horizontal; lateral line with 19-20 tubed scales followed 2 rows lower by 17-23 notched scales; total scales in longitudinal series 44-47; orbital cirrus absent, but microscopic papillae present on upper rim of eye*; scales absent on head, breast, belly, and pectoral fin base; pale tan or pinkish, with 4 "double" brown bars on side, and longitudinal row of dark brown spots or dashes along middle of side; a prominent black band, with white anterior border, around base of caudal fin. Gulf of California; inhabits rocky shores and steep slopes to at least 38 m. Attains 6.4 cm.

NETWORK TRIPLEFIN

Enneanectes reticulatus Allen & Robertson, 1991

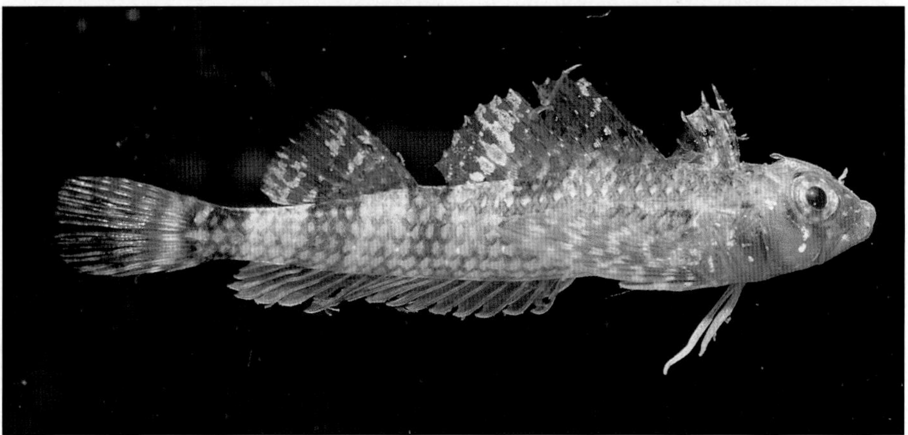

Enneanectes reticulatus Adult, 5 cm ▲ Underwater photograph of adult ▼

Enneanectes sexmaculatus ▼

Dorsal rays III+XII+9, segmented rays branched except for first 1-2 and last ray; anal rays II,18; pectoral rays 15; pelvic rays I,2; *lateral line more or less horizontal; lateral line with 15 tubed scales followed 2 rows lower by 18 or 19 notched scales; total scales in longitudinal series 33-34; a short, broad cirrus above eye*; scales ctenoid, except cycloid on side of belly; head scaleless except for *small patch of scales on upper opercle*; scales absent on middle of belly and on base of pectoral fins; whitish or tan, with 4 slightly diagonal brown bars on side, and black band around base of caudal fin; scale margins dusky, imparting network pattern; head and breast sometimes pink to reddish. Gulf of California; occurs in 0-5 m along rocky shores. Grows to about 4.5 cm.

DELICATE TRIPLEFIN

Enneanectes sexmaculatus (Fowler, 1944)

Dorsal rays III+XI-XII+7-8, last 4-5 segmented rays branched; anal rays II,14-15; pectoral rays 15; pelvic rays I,2; *lateral line more or less horizontal; lateral line with 14-15 tubed scales followed 2 rows lower by about 16 strongly notched scales; total scales in longitudinal series 30-31; a short, broad cirrus above eye*; scales ctenoid, except cycloid just in front of anus; head scaleless except for *several scales on upper opercle*; scales absent on pectoral fin base; yellowish tan with 5-6 relatively narrow dark bars on side, including intensely black bar on caudal peduncle; brown bar below eye, and prominent dark brown spot or blotch at base of lower pectoral rays; pectoral fin with narrow dark bars. Baja California to Panama; inhabits shallow reefs. Reaches 2.5 cm.

TWINSPOT TRIPLEFIN
Lepidonectes bimaculata Allen & Robertson, 1992

Dorsal rays III+XIII+10-11, segmented rays branched except first 2 and penultimate ray sometimes unbranched; anal rays II,19; pectoral rays 17; pelvic rays I,2; *lateral line more or less horizontal; lateral line with 24-26 tubed scales followed 3 rows below by 8-11 notched scales; total scales in longitudinal series 35; orbital cirrus lanceolate*; scales ctenoid, except cycloid on belly; *upper half of preopercle covered with small scales*; scales present on pectoral fin base; small spinules on edge of orbits, margin of preopercle, nape, and dorsal surface of snout; brown to bright red bars or saddles on dorsal half of body, the posterior ones extending onto lower side; the bars frequently join to form wide midlateral stripe; a dark brown to red band at base of caudal fin; *a pair of brilliant, pearl white saddles on back on posterior half of body*; fins mainly pale, except male first dorsal fin black with 2 pale spots basally. Known only from Isla Malpelo, Colombia; inhabits rocky reefs. Reaches 6.8 cm.

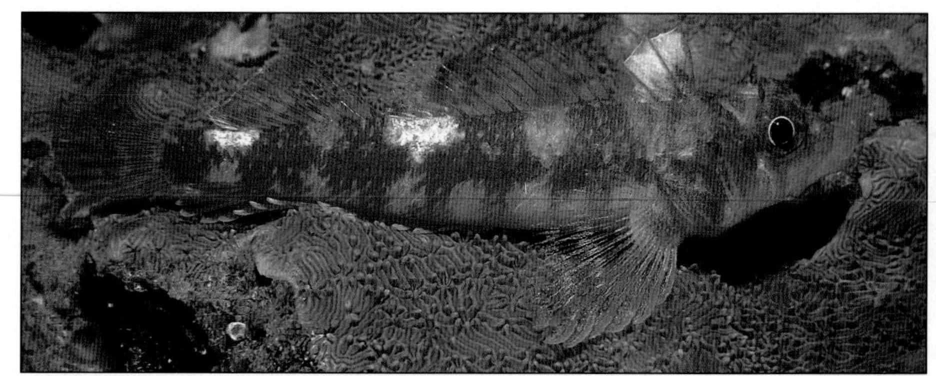

Lepidonectes bimaculata Female ▲

Lepidonectes bimaculata Juv.

Lepidonectes corallicola ▲

CORAL TRIPLEFIN
Lepidonectes corallicola (Kendall and Radcliffe,1912)

Dorsal rays III+XIII+12-13, segmented rays branched except first; anal rays II,21-22; pectoral rays 16-17; pelvic rays I,2; *lateral line more or less horizontal; lateral line with 27-29 tubed scales followed 2 rows below by 8-10 notched scales; total scales in longitudinal series 33-38; orbital cirrus lanceolate*; scales ctenoid, except cycloid on belly; *preopercle mostly covered with small scales*; scales present on pectoral fin base; *small spinules on edge of orbits, margin of preopercle, nape, and dorsal surface of snout*; light brown to reddish pink with 4 H-shaped, red-brown bars on side, and black band at base of caudal fin; a large whitish blotch between each set of bars on back, that in front of peduncular bar silvery white and particularly prominent; first dorsal fin with red and black markings. Galapagos Islands; inhabits rocky reefs to at least 10 m depth. Grows to 7 cm.

Lepidonectes clarkhubbsi Male ▲ Female ▼

SIGNAL TRIPLEFIN
Lepidonectes clarkhubbsi Bussing, 1991

Dorsal rays III+XIII+10 to 13, segmented rays branched except first; anal rays II,18-21; pectoral rays 17; pelvic rays I,2; *lateral line more or less horizontal; lateral line with 24-26 tubed scales followed 2 rows below by 8-12 notched scales; total scales in longitudinal series 34; orbital cirrus lanceolate*; scales ctenoid, except cycloid on belly; *preopercle and opercle mostly covered with small scales*; scales present on pectoral fin base; *small spinules on edge of orbits, margin of opercle, nape, and dorsal surface of snout*; whitish with 4 "double" red-brown bars on side, and large black spot at base of caudal fin; first dorsal of female pale yellowish, that of male black with bright orange area on basal portion. Known only from Costa Rica and Panama; occurs along rocky shores in 2-12 m. Reaches 6.5 cm. *Taboguilla signata* Allen and Robertson is a synonym.

SAND STARGAZERS (PECES QUIJADA, MIRAESTRELLAS)
FAMILY DACTYLOSCOPIDAE

Sand Stargazers are small, sand-dwelling fishes found both on the Pacific and Atlantic coasts of the Americas. The common name is derived from the position of the eyes which are on top of the head, sometimes protruding or on stalks. Other features include an upturned, fringed mouth; a protruding lower jaw; tubular nostrils; dorsal fin continuous or divided with VII-XXIII spines and 12-36 soft rays; upper part of operculum with finger-like fimbriae; and each pelvic fin with three thickened rays that are free at the tips. In some species the males are known to incubate the eggs in two balls, one under each pectoral fin. Sand stargazers are "lie and wait" predators of small fishes and invertebrates. They frequently burrow into the sandy substratum, leaving only the eyes, snout, and mouth protruding. The family contains nine genera and about 46 species; 24 are known from the eastern Pacific. We include only nine of the more common representatives. These fishes are seldom seen except by scientific collectors using chemicals. The various genera were revised by Dawson (1974, 1975, 1976, 1977, and 1982).

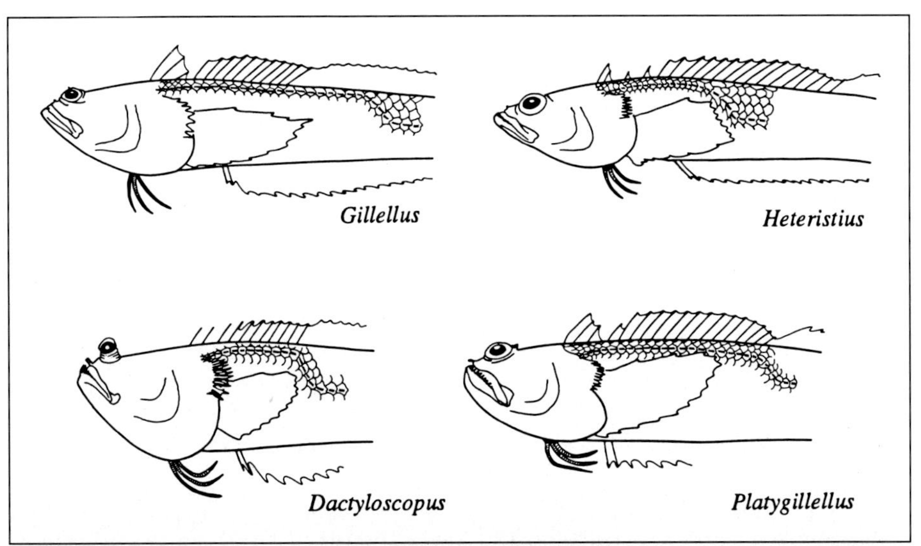

Eastern Pacific genera of sand stargazers (Dactyloscopidae) ▲

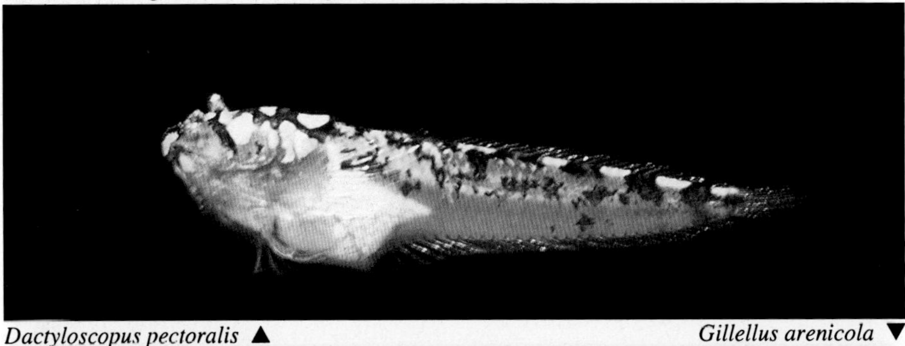

Dactyloscopus pectoralis ▲

Gillellus arenicola ▼

WHITESADDLE STARGAZER
Dactyloscopus pectoralis Gill, 1861

Dorsal rays III or IV+VII-X,23-27; total dorsal fin elements 34-39; segmented anal rays 26-32 (usually 28-30); pectoral rays usually 13; total lateral-line scales 39-44; scales in straight (posterior) part of lateral line 23-28; *eyes on protrusile stalks; caudal peduncle with a notch-like depression in ventral margin; segmented caudal rays unbranched; both lips fringed with branched or unbranched fimbriae*; top of head and back with dark mottling and blackish blotches; *a broad white saddle across top of rear portion of head, with forward extension nearly reaching interorbital space; about 7-9 prominent white spots or saddles along dorsal midline of back.* Baja California to Ecuador; inhabits sand bottoms from tide pool depths to 45 m. Attains 5.2 cm. Dawson (1975) recognized 3 subspecies: *D. pectoralis pectoralis* from the Gulf of California and outer coast of Baja California, *D. pectoralis fallax* Dawson (illustrated here) from Sinaloa, Mexico, to Ecuador, including Isla del Coco, and *D. pectoralis insulatus* Dawson from Islas Revillagigedos.

SANDY STARGAZER
Gillellus arenicola Gilbert, 1890

Dorsal rays II+XI-XIII,27-31; segmented anal rays 34-38 (usually 35-36); pectoral rays 12-13; total lateral-line scales 51-55; scales in straight (posterior) part of lateral line 27-29; *lateral line deflected ventrally below dorsal fin elements 14-17; principal segmented caudal rays 10; tip of lower jaw fleshy, pointed, and strongly protruding; isolated dorsal finlet on nape composed of 2 spines*; overall whitish, with series of brown saddles and bars across back and on upper side; saddle-like marks on back alternating with hourglass-shaped bars that extend to middle of side. Southern tip of Baja California to Panama; depth range 8-137 m. Grows to 5.5 cm.

COCOS STARGAZER
Gillellus chathamensis Dawson, 1977

Dorsal rays III+VII-XI,28-32 (usually 29-30); *segmented anal rays 28-32 (usually 29-30); pectoral rays 12-14 (usually 13); total lateral-line scales 50-56, scales in straight (posterior) part of lateral line 23-25*; lateral line deflected ventrally below dorsal fin elements 17-21; principal segmented caudal rays 10; *isolated dorsal finlet on nape composed of 3 spines*; overall white, with 7 dark-edged, light gray to brownish saddles on upper sides. Known only from Isla del Coco; depth range 5-12 m. Grows to 3.8 cm.

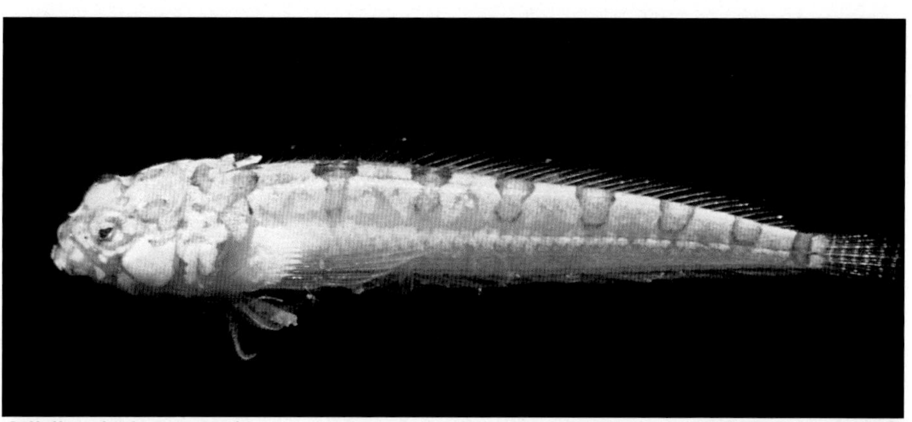

Gillellus chathamensis ▲

Gillellus searcheri ▼

Gillellus semicinctus ▼

Heteristius cinctus ▼

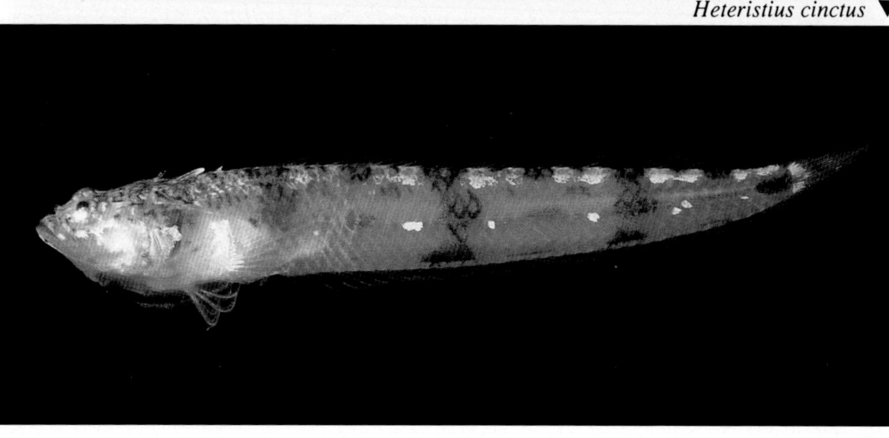

SEARCHER STARGAZER
Gillellus searcheri Dawson, 1977

Dorsal rays III+XVII-XXI,16-19; segmented anal rays 30-32 (usually 31); pectoral rays 12-14 (usually 13); *total lateral-line scales 64-73; scales in straight (posterior) part of lateral line 9-13; lateral line deflected ventrally below dorsal fin elements 27-32*; principal segmented caudal rays 10; *isolated dorsal finlet on nape composed of 3 spines*; overall whitish, with blackish to brown mottled saddles and bars across back and on upper half of sides. Mexico (Islas Tres Marias and Nayarit) to Panama; depth range 1-15 m. Reaches 3.5 cm. *G. ornatus* from the Gulf of California is similar in general appearance and coloration, but has only 2 spines in the first dorsal fin.

HALF-BANDED STARGAZER
Gillellus semicinctus Gilbert, 1890

Dorsal rays III+VIII-XI,25-30; *segmented anal rays 30-35 (usually 32-33); pectoral rays 11-14 (usually 12); total lateral-line scales 44-53; scales in straight (posterior) part of lateral line 18-23 (usually 19-21)*; lateral line deflected ventrally below dorsal fin elements 17-24; principal segmented caudal rays 10; *isolated dorsal finlet on nape composed of 3 spines*; overall whitish, with dark grayish bars on upper half of side, and smaller dark saddles on back between each bar. Gulf of California to Colombia; also Galapagos and other offshore islands; depth range 5-137 m. Maximum length to 5.2 cm.

BANDED STARGAZER
Heteristius cinctus (Osburn & Nichols, 1916)

Dorsal rays II+XVII-XX,18-22; segmented anal rays 33-36 (usually 34-35); pectoral rays 13-15 (usually 14); total lateral-line scales 52-56; scales in straight (posterior) part of lateral line 31-35 (usually 32-33); *lateral line deflected ventrally below dorsal fin elements 8-12; principle segmented caudal rays 12; isolated dorsal finlet on nape composed of 2 spines*; somewhat translucent with 3 blackish bars on side and about 12 blackish bars across back; white, saddle-like markings also present on back, several white spots on side. Baja California to Ecuador; depth range 1-27 m. To 4.5 cm.

SAILFIN STARGAZER
Platygillellus altivelis (Kendall & Radcliffe, 1912)

Dorsal rays III+XII-XIV,15-17 (*total elements 30-33*); segmented anal rays 24-26 (usually 24 or 25); pectoral rays 13-15 (usually 14); total lateral-line scales 39-42; *scales in straight (posterior) part of lateral line 11-14* (usually 12 or 13); lateral line deflected ventrally below third to fifth segmented dorsal rays; first 3 dorsal spines forming isolated or semi-isolated finlet; *dorsal finlet relatively tall, its height about 87-107 percent of predorsal length*; overall whitish with large pair of Y-shaped blackish bars on middle of side; narrower blackish bars also present just behind head and across base of caudal fin. Costa Rica and Panama; depth range 3-37 m. Reaches 4.4 cm.

Platygillellus bussingi ▲

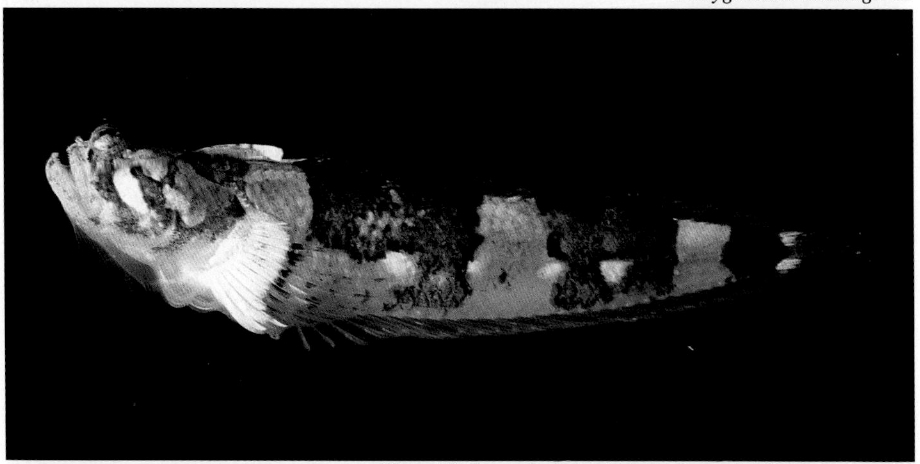

Platygillellus rubellus Pale variety ▲ Dark variety ▼

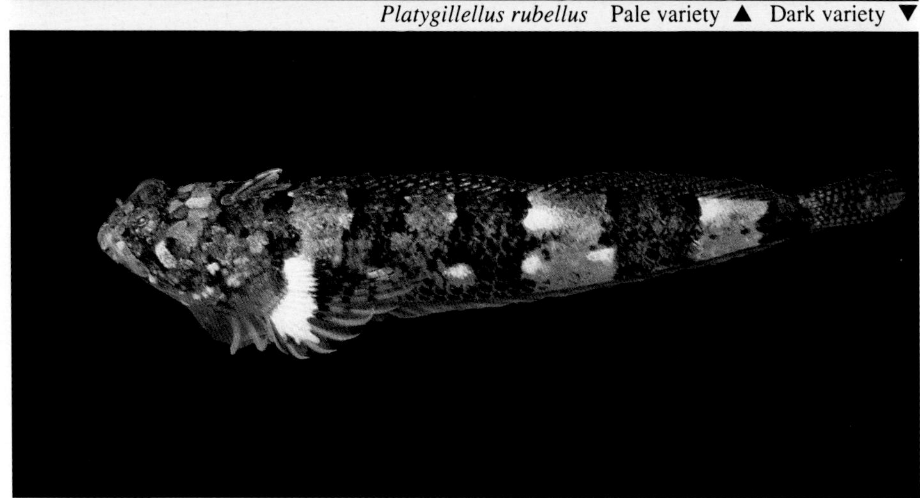

BUSSING'S STARGAZER
Platygillellus bussingi Dawson, 1974

Dorsal rays III+XIV-XVI,15-17 (*total elements 33-34*); segmented anal rays 26-28; pectoral rays 13-15 (usually 14); total lateral-line scales 45-48; *scales in straight (posterior) part of lateral line 16-17*; lateral line deflected ventrally below junction between last dorsal spine and first segmented dorsal ray; first 3 dorsal spines forming isolated or semi-isolated finlet; *dorsal finlet low, its height about 40-63 percent of predorsal length; lower lip fimbriae 21-33; no scales on preopercle; no papillae on eye*; white with pair of broad blackish bars on middle of side, and smaller bars or saddles below dorsal finlet and across base of caudal fin; also about 5 white spots interrupting bars along lower side. Costa Rica and Panama; depth range 1-15 m. To 5 cm.

GALAPAGOS STARGAZER
Platygillellus rubellus (Kendall & Radcliffe, 1912)

Dorsal rays III+XIV-XVI,16-18 (*total elements 33-35*); segmented anal rays 26-29 (usually 27-28); pectoral rays 13-15 (usually 14); total lateral-line scales 41-44; *scales in straight (posterior) part of lateral line 16-19*; lateral line deflected ventrally below junction between last dorsal spine and first segmented dorsal ray; first 3 dorsal spines forming isolated or semi-isolated finlet; *dorsal finlet low, its height about 40-63 percent of predorsal length; lower lip fimbriae 5-17; patch of several scales on upper part of preopercle; eye papillae present*; white with 4-5 blackish bars on side and longitudinal row of silvery white spots on lower side; some specimens have broadly expanded dark bars with relatively narrow white spaces between them. Galapagos Islands; depth range 0-15 m. Attains 6.5 cm.

WEED BLENNIES (TRAMBOLLOS, CHALACOS)
FAMILY LABRISOMIDAE

This family contains small, elongate fishes that live on the reef's surface or among weeds. Their diet consists of various invertebrates including crabs, gastropod molluscs, chitons, brittle stars, urchins, and polychaetes. They are considerably diverse in shape, and most are cryptically colored to blend with the background. It is difficult to characterize these fishes on the basis of a particular feature, but they possess cycloid scales with radii only on the anterior part and most species have more spinous elements than soft rays in the dorsal fin (a few species lack soft rays entirely). There are often cirri (filamentous appendages) on the nape, nostrils, and above the eye. The family contains 16 genera and approximately 100 species, most of which occur in the North American tropics. The genera *Starksia* (Pacific species only) and *Xenomedea* are unusual in that they are viviparous. The group has attracted the attention of several researchers, and there are published reviews of eastern Pacific *Labrisomus* (Hubbs, 1953), *Malacoctenus* (Springer, 1958), *Paraclinus* (Rosenblatt and Parr, 1969), and *Starksia* and *Xenomedea* (Rosenblatt and Taylor, 1971). In addition, Hubbs (1952) provided a general review of this family (formerly recognized as part of the Clinidae) and a partial revision of the eastern Pacific species.

Eastern Pacific genera of weed blennies (Labrisomidae) ▼

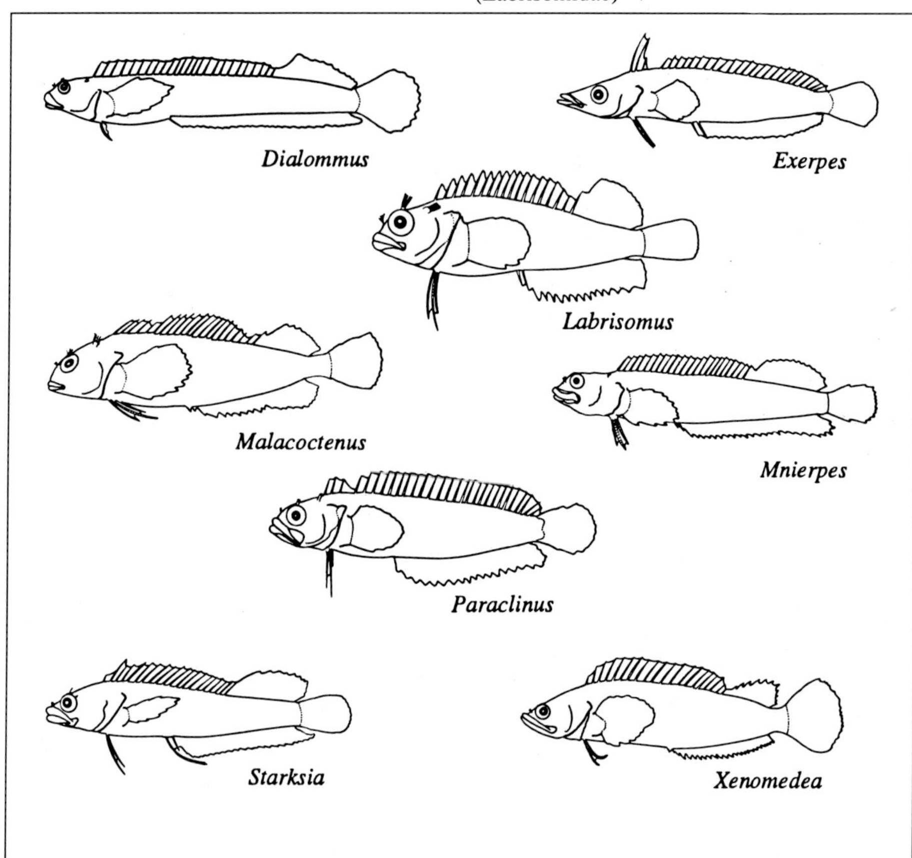

Dendritic Blenny (*Labrisomus dendriticus*), Isla Malpelo, Colombia ▼

FOUR-EYED BLENNY
Dialommus fuscus Gilbert, 1890

Dorsal rays XXV,13 to 15; anal rays I,28; scales small, about 60 in lateral series; greatest body depth 6.0-7.0 in standard length; *cornea divided by an oblique, pigmented band into an anterior lower half and posterior upper half; upper edge of eye with 1 or more slender cirri; each side of nape with a single short cirrus*; teeth in jaws forming narrow bands; vomer with a single series of teeth, none on palatines; generally light greenish to nearly black, with series of 7-9 white blotches along base of dorsal fin, and diffuse white blotches along lower side; lips white; scattered white markings on head and pectoral fin, including a pair of large whitish areas on base of pectoral fin. Galapagos Islands; found in rocky areas of the intertidal zone, often seen above the waterline, resting on rocks. Grows to at least 7 cm. Similar in general appearance to *Mnierpes macrocephalus* from the eastern Pacific mainland (see page 234).

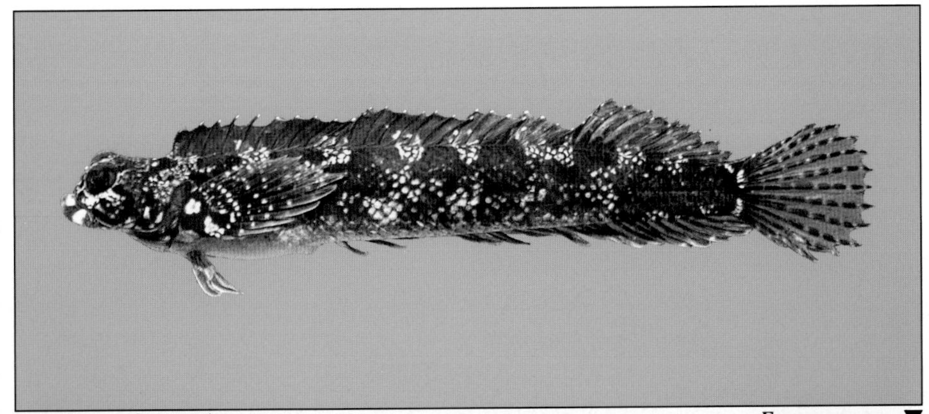

Dialommus fuscus ▲ *Exerpes asper* ▼

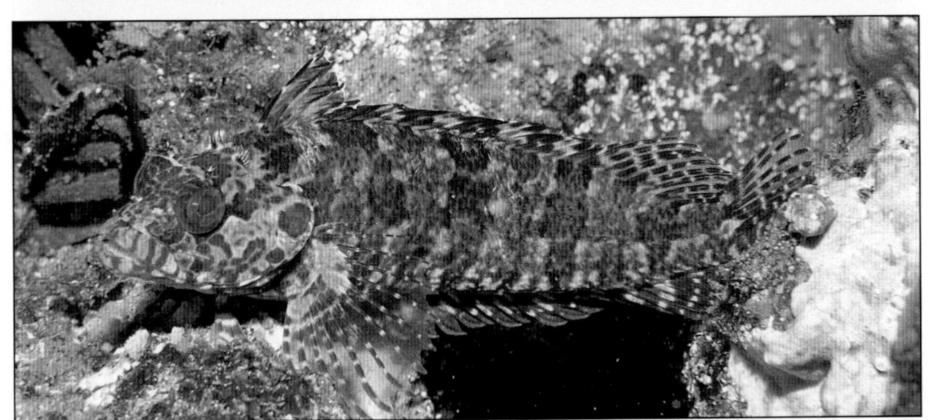

SARGASSUM BLENNY
Exerpes asper (Jenkins & Evermann, 1889)

Dorsal rays III or IV+XXIV or XXV; anal rays II,17-20; lateral-line scales about 40; teeth in jaws biserial with larger outer series; vomerine teeth present; palatine teeth absent; no cirri on head; *easily recognized by the tall, isolated first dorsal fin, long pointed snout, lack of soft rays in the second dorsal fin (except last ray sometimes segmented), and habitat preference*; overall green with midlateral white stripe from snout to caudal fin base (may be broken into series of spots); a pair of ocellated dark spots on rear half of dorsal fin. Upper and central Gulf of California; common among sargassum weed or eelgrass. Maximum length 6.5 cm.

Labrisomus dendriticus Adult ▲ Nuptial male ▼

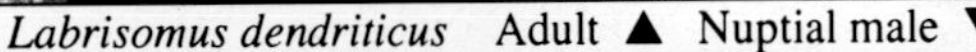

DENDRITIC BLENNY
Labrisomus dendriticus (Reid, 1935)

Dorsal rays XX,10 to 12; anal rays II,19-20; pectoral rays 13; lateral-line scales 59-65; head scaleless; palatine teeth larger than vomerine teeth; an enlarged canine behind patch of small teeth on each side of upper and lower jaw; cirri present on nostrils, above eye, and on nape; nuchal cirri about half the width of gill cover; first dorsal very tall, longer than or equal to the length of the longest soft rays; *dorsal fin indented at fourth to fifth spines, and again at junction between spinous and soft rays; the combination of 20 dorsal spines, presence of enlarged canines, and tall first dorsal spine are distinctive for the genus*; overall greenish brown with strongly mottled pattern, usually a series of diffuse dark bars on side and *relatively large blackish spot on upper portion of gill cover*; median fins with oblique dark bands or small spots; nuptial males assume a drastically different pattern, and are overall white with a series of black double bars on side, and bright orange markings on the gill cover. Galapagos and Malpelo Island; the most abundant member of the genus at the Galapagos, it occurs on shallow, weed-covered rocky reefs. Reaches at least 13 cm.

JENKINS' BLENNY

Labrisomus jenkinsi (Heller & Snodgross, 1903)

Dorsal rays XVIII or XIX,10-11; anal rays II,17; pectoral rays 14 or 15; *lateral-line scales 58-62*; a few scales present along upper margin of gill cover; no enlarged canines behind smaller teeth in jaws; cirri present on nostrils, above eye, and on nape; no pronounced indentation at front part of spinous dorsal fin; *dorsal fin indented at junction between spinous and soft rays*; overall mottled greenish brown with 5-6 diffuse blackish bars on side; 2 rows of white spots may also be present on side, and white streak at midbase of caudal fin; ocellus often present at front of dorsal fin; head with numerous dark spots, bands, or broken lines; nuptial males lack strong mottling on side and have orange head. Galapagos Islands; infrequently encountered in rocky shallows. To at least 13 cm.

Labrisomus jenkinsi Female ▲ Male ▼

Labrisomus jenkinsi Male ▼

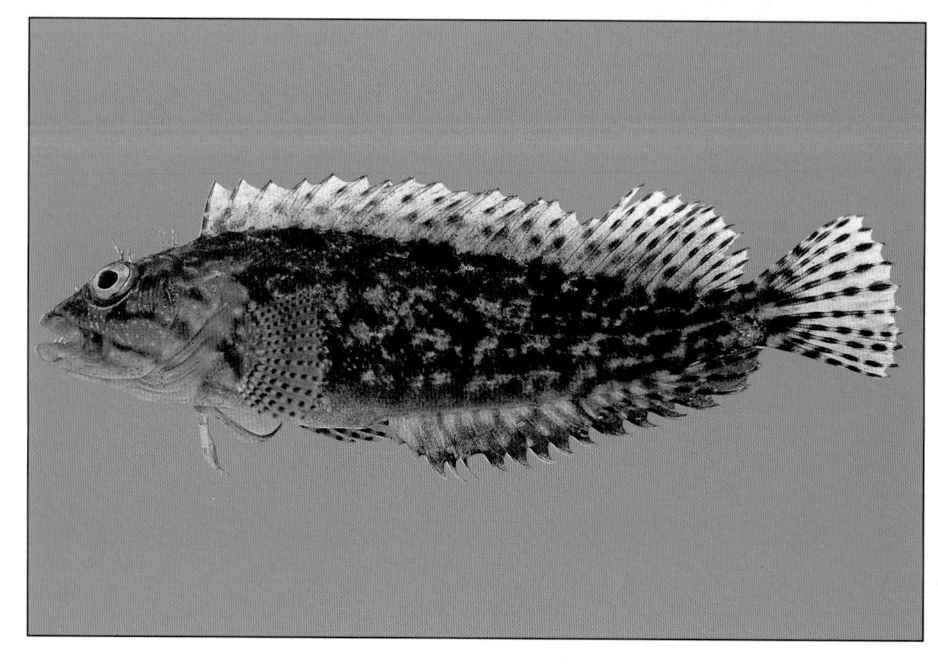

POREHEAD BLENNY

Labrisomus multiporosus Hubbs, 1953

Dorsal rays XVII to XIX,11-13; anal rays II,17 or 18; pectoral rays 13-15; *lateral-line scales 65-69; a few scales present along upper margin of gill cover; cheek scaleless*; no enlarged canines behind smaller teeth in jaws; cirri present on nostrils, above eye, and on nape; *a dense covering of sensory pores on head, especially behind eye; no pronounced indentation at front part of spinous dorsal fin; dorsal fin indented at junction between spinous and soft rays;* overall brownish with heavy, dark brown mottling on sides; a pair of oblique brown bands across cheek; prominent dark spotting on dorsal, caudal, and pectoral fins. Baja California to Peru, including Galapagos Islands; inhabits weed-covered rocky reefs. Attains at least 13 cm.

Labrisomus striatus Male ▲ Female ▼

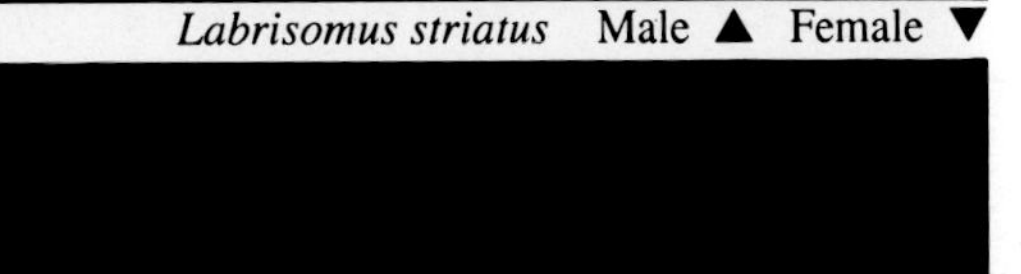

GREEN BLENNY
Labrisomus striatus Hubbs, 1953

Dorsal rays XVIII,10; anal rays II,17; pectoral rays 13; *lateral-line scales 42-44*; a few scales present on upper margin of gill cover and on upper cheek; no enlarged canines behind smaller teeth in jaws; cirri present on nostrils, above eye, and on nape; membranes between anterior dorsal spines deeply incised; first dorsal spine short, less than width of eye diameter; dorsal fin indented at junction between spinous and soft rays; body light brownish or greenish, with *series of thin, dark, longitudinal lines between scale rows; large red-brown spots scattered on sides*; head yellowish with pair of brown bands across cheek. Baja California and Gulf of California to southern Mexico; relatively common in weed-covered rocky shallows in the vicinity of Cabo San Lucas, Baja California. Grows to at least 6 cm.

LARGEMOUTH BLENNY
Labrisomus xanti (Gill, 1860)

Dorsal rays XVII to XIX,10-12; anal rays II,17-19; pectoral rays 13-15; lateral-line scales 64-69; *10 or more scales (embedded) on upper margin of gill cover; cheek scaleless*; no enlarged canines behind smaller teeth in jaws; *palatine teeth absent*; cirri present on nostrils, above eye, and on nape; no pronounced indentation at front part of spinous dorsal fin; dorsal fin indented at junction between spinous and soft rays; overall greenish brown with about 8 darker bars along side, including diagonal band from snout to first few dorsal spines; *numerous small white spots on lower part of head and body*. Baja California and Gulf of California; the most common member of the genus at most localities in the Gulf of California. Maximum size 18 cm.

FISHGOD BLENNY

Malacoctenus ebisui Springer, 1958

Dorsal rays XIX to XXI,10-12 (usually XX,11); anal rays II,18-21 (usually 20); pectoral rays 14; lateral-line scales 52-58; scales present in front of dorsal fin, also on breast of males; scales absent from prepectoral region; jaws with single row of large conical teeth; vomerine teeth present; palatine teeth absent; cirri present on nostrils, above eye, and on nape; *similar to* M. costaricanus *Springer (not illustrated), but has shorter first dorsal spine (less than 10.5 percent of standard length vs more than 11 percent), more dorsal spines and anal rays (usually 20 vs 19), and has only 1 or 2 symphysial pores (vs 4 or more)*; males reddish brown on back with 2-3 large black patches posteriorly, belly yellowish green with tiny white spots; females greenish brown on upper half, pale green or whitish below, with several diffuse brownish bars across side. Gulf of California to Panama. Attains 6.5 cm.

Malacoctenus ebisui Male ▲ Female ▼

SONORA BLENNY

Malacoctenus gigas Springer, 1958

Dorsal rays XIX to XXIII,12-14 (usually XXI,12); anal rays II,20-24 (usually 22-23); pectoral rays 13-15 (usually 14); lateral-line scales 53-65; scales present or absent on midline in front of dorsal fin; scales present on prepectoral and breast areas of specimens in excess of 5 cm (caudal fin excluded); jaws with single row of large conical teeth; vomerine teeth present; palatine teeth absent; cirri present on nostrils, above eye, and on nape; *olive greenish on upper half, whitish on breast and belly; 5 irregular-shaped, dark brown bars on side; dorsal, anal and caudal fin of males uniformly dusky, but spotted in females; females with reticular pattern of dark lines on lower part of sides.* Northern part of Gulf of California, south to about Guaymas and Isla Espirito Santo; very common in the northern Gulf, on shallow reefs among seaweed. To 13 cm.

Malacoctenus gigas Male (top) and female ▼

Photograph: Alex Kerstitch

REDSIDE BLENNY
Malacoctenus hubbsi Springer, 1959

Dorsal rays XIX to XXI,9-13 (usually XX,11); anal rays II,18-23 (usually 19-21); pectoral rays 13-15 (usually 14); lateral-line scales 49-61; scales absent on midline in front of dorsal fin; scales absent or present on prepectoral and breast areas of adults; jaws with single row of large conical teeth; vomerine teeth present; palatine teeth absent; cirri present on nostrils, above eye, and on nape; *males light greenish on upper half, reddish or pink on lower half, 5-6 irregular-shaped, dark brown bars on side; a small dark spot in front of each pelvic fin base; females similar, except lacking reddish or pink hue on belly*. Baja California, Gulf of California, and southward to Acapulco; the most common blenny in rocky shallows of the Gulf of California; usual depth range about 1-4 m. Springer (1958) recognized 2 subspecies: *M. hubbsi hubbsi* from Baja California and the Gulf, and *M. hubbsi polyporosus* from the Mexican coast south of the Gulf. Grows to 9 cm.

Male ▲ Female ▼

Female ▼

MARGARITA BLENNY
Malacoctenus margaritae Fowler, 1944

Dorsal rays XX to XXII,10-12 (usually XXI,11); anal rays II,20-22 (usually 21); pectoral rays 14-16 (usually 15 or 16); lateral-line scales 51-56; scales present on midline in front of dorsal fin, on breast, and on prepectoral area; jaws with single row of large conical teeth; vomerine teeth present; palatine teeth absent; cirri present on nostrils, above eye, and on nape; females whitish with numerous irregular-shaped yellow-brown blotches on sides, belly white; a large, pale-edged brown spot covering most of operculum; nuptial males overall dark greenish to gray, with faint hint of several dark bars on side, and 3 black spots posteriorly; *the combination of 15 or more pectoral rays and color pattern are distinctive*. Baja California, Gulf of California, and southward to Panama. Springer (1958) recognized a separate subspecies from Baja California and the Gulf, *M. margaritae mexicanus*. Reaches 6.5 cm.

Female ▲ Male ▼

Courting pair ▼

THROAT-SPOTTED BLENNY

Malacoctenus tetranemus (Cope, 1877)

Dorsal rays XVIII to XX,9-12 (usually XIX,11); anal rays II,18-20 (usually 19); pectoral rays 13-15 (usually 14); lateral-line scales 53-61; no scales on midline in front of dorsal fin, present on breast of males only, and sometimes present on the prepectoral area; jaws with single row of large conical teeth; vomerine teeth present; palatine teeth absent; cirri present on nostrils, above eye, and on nape; a double row of dark, greenish brown blotches on side; *head and lower half of body with numerous dark spots*. Baja California and Gulf of California to Peru, including Galapagos Islands. *M. afuerae* (Hildbrand) is a synonym. Springer (1958) recognized the population from Baja California and the Gulf as a separate subspecies, *M. afuerae multipunctatus*. Maximum length to about 7.5 cm.

Malacoctenus tetranemus Male ▲ Female ▼

ZACA BLENNY

Malacoctenus zacae Springer, 1958

Dorsal rays XX or XXI,9-11 (usually XX,10); anal rays II,18-20 (usually 19); pectoral rays 14; lateral-line scales 50-55; no scales on midline in front of dorsal fin or on prepectoral area, sometimes a few scales on breast; jaws with single row of large conical teeth; vomerine teeth present; palatine teeth absent; cirri present on nostrils, above eye, and on nape; *closely related to* M. hubbsi, *but usually lacks dark spot at pelvic-fin base and is much smaller*; generally pale olive, either with dark greenish stripe from eye to caudal fin base, or with about 6 greenish brown bars on side; belly and pectoral fins often yellowish. Southern Baja California to Acapulco. Attains a length of 6.5 cm.

Malacoctenus zacae Male (top) and female ▲ Female ▼

GLOSSY BLENNY
Malacoctenus zonifer (Jordan & Gilbert, 1882)

Dorsal rays XIX to XXI,10-12 (usually XX,11); anal rays II,17-21 (usually 19 or 20); pectoral rays 13 or 14 (usually 14); lateral-line scales 52-60; scales present on midline in front of dorsal fin, absent on prepectoral region, sometimes present on breast; jaws with single row of large conical teeth; vomerine teeth present; palatine teeth absent; cirri present on nostrils, above eye, and on nape; *olive on back, with series of blackish saddles dorsally; an irregular, dark brown to blackish midlateral stripe; lower side whitish with brown to blackish blotches and spots; a prominent dark spot on head, just behind eye; spotting on median fins.* Mexico (Mazatlan southwards) to Ecuador. Springer (1958) recognized a separate subspecies for the population occurring south of Mexico, *M. zonifer sudensis*. Attains 8 cm.

BANDED BLENNY
Malacoctenus zonogaster (Heller & Snodgrass, 1903)

Dorsal rays XX to XXII,10-12 (usually XXI,11); anal rays II,20-22 (usually 21); pectoral rays 14; lateral-line scales 57-64; scales present on midline in front of dorsal fin and on breast, absent from prepectoral region; jaws with single row of large conical teeth; vomerine teeth present; palatine teeth absent; cirri present on nostrils, above eye, and on nape; greenish on upper half and white below, 5-6 irregular-shaped dark bars on side; *a large (eye-sized) dark spot on upper half of gill cover, and greenish brown bands on lower part of head and on breast region.* Galapagos Islands; the second most abundant blenny in the Galapagos after *Labrisomus dendriticus*; common in rocky shallows, including tide pools. To 8.5 cm.

FOUREYE ROCKSKIPPER
Mnierpes macrocephalus (Günther, 1861)

Dorsal rays XXII,12 or 13; anal rays II,24; *cornea divided into halves, similar to* Dialommus*; scales very small and cycloid, about 75 in lateral series*; greatest body depth 5.6-7.3 in standard length; *no cirri over eye or on nape*; upper lip thickened; teeth in jaws forming villiform bands, those in outer series enlarged; a band of teeth on vomer, none on palatines; overall dark grayish with light and dark mottlings; head blackish; small white spots present on head, pectoral fin base, and upper back. Baja California to Colombia; occurs in very shallow water along rocky shores. Maximum recorded size 11 cm.

MEXICAN BLENNY
Paraclinus mexicanus (Gilbert, 1904)

Dorsal rays XXVII to XXXI,0-2; anal rays 17-20; pectoral rays 12-14; lateral-line scales 31-36; a distinct spine on upper opercle; cirri present on nostril, above eye, and on nape; *gap between bases of dorsal spines 3 and 4 less than distance between dorsal spines 1 and 3; first dorsal spine less than 10 percent of standard length; membrane between third and fourth dorsal spines usually attached above midpoint of fourth spine*; strongly mottled greenish brown, sometimes with 6-7 diffuse bars on side; a prominent dark band below eye; a black ocellated spot on rear half of dorsal fin. Baja California and Gulf of California to Ecuador; inhabits shallow, weed-covered rocky reefs. Reaches 6 cm.

ONE-EYED BLENNY
Paraclinus monophthalmus (Günther, 1861)

Dorsal rays XXVIII to XXX,1; anal rays II,19-21; pectoral rays 11-14; lateral-line scales 36-40; a distinct spine on upper opercle; cirri present on nostril, above eye, and on nape; *nape cirrus a narrow triangular flap; gap between bases of dorsal spines 3 and 4 equal to or greater than distance between dorsal spines 1 and 3; first dorsal spine moderately elevated; membrane between third and fourth dorsal spines usually attached well below midpoint of fourth spine*; grayish with faint dark bars on side; *a black spot on upper part of gill cover; dorsal fin with broad white margin* and ocellus on rear half. Costa Rica and Panama; relatively rare; found in shallow weedy areas. Attains 8.5 cm.

FLAPSCALE BLENNY
Paraclinus sini Hubbs, 1952

Dorsal rays XXVIII to XXX (no segmented rays); anal rays II,16-20; pectoral rays 12-14; lateral-line scales 32-37; a distinct spine on upper opercle; cirri present on nostril, above eye, and on nape; *differs from all other Paraclinus in having membranous, flap-like projections on the posterior margins of the scales, except those along the lateral line*; usually mottled reddish brown or greenish brown, with white blotches and flecks, sometimes with diffuse dark bars on side; color may also be nearly all white, pinkish, orange or dark brown; a small ocellated spot on posterior half of dorsal fin. Baja California and Gulf of California. Other Gulf *Paraclinus* (not illustrated) include *P. altivelis* (first dorsal spine very long, over 16 percent of standard length), and *P. beebei* (broad scaleless area along base of dorsal fin). To 6 cm.

Paraclinus mexicanus Female (top) and male ▲ *Paraclinus monophthalmus* ▼

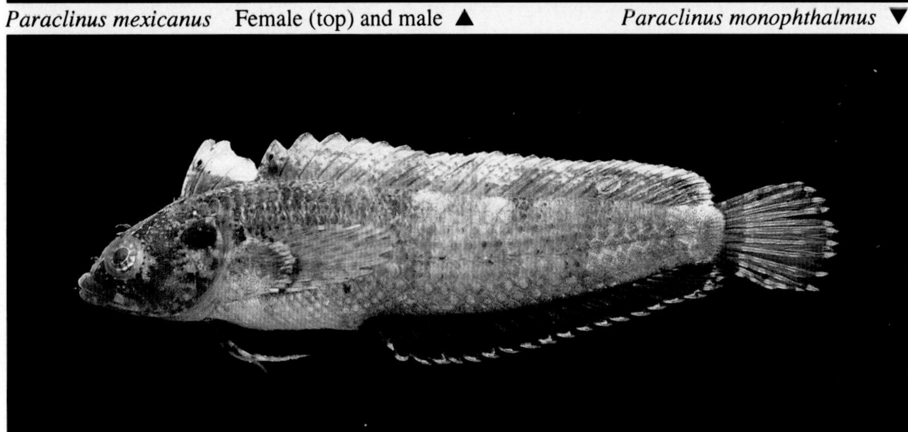

Paraclinus sini Red variety ▲ Brown variety ▼

LONGJAW BLENNY
Paraclinus tanygnathus Rosenblatt & Parr, 1969

Dorsal rays XXVI to XXVIII (no segmented rays); anal rays II,16-18; pectoral rays 12-13; lateral-line scales 31-33; a distinct spine on upper opercle; cirri present on nostril, above eye, and on nape; *gap between bases of dorsal spines 3 and 4 equal to or greater than distance between dorsal spines 1 and 3; first dorsal spine moderately elevated, more than 10 percent of standard length; membrane between third and fourth dorsal spines attached to base of fourth spine; last dorsal element a spine, rather than a soft ray*; reddish brown with whitish blotches and flecks; a broad, pale band behind eye and prominent white blotches on pectoral fin base; sometimes a black spot on rear half of dorsal fin. Gulf of California to southern Mexico. Reaches length of 3.5 cm. Other *Paraclinus* (not illustrated) in the central-southern Mexico region include *P. ditrichus* (2 pelvic soft rays, 3 rays in all other *Paraclinus*), and *P. stephensi* (similar to *P. monophthalmus*, but nape cirrus a broad, leaf-like flap instead of a narrow, triangular flap).

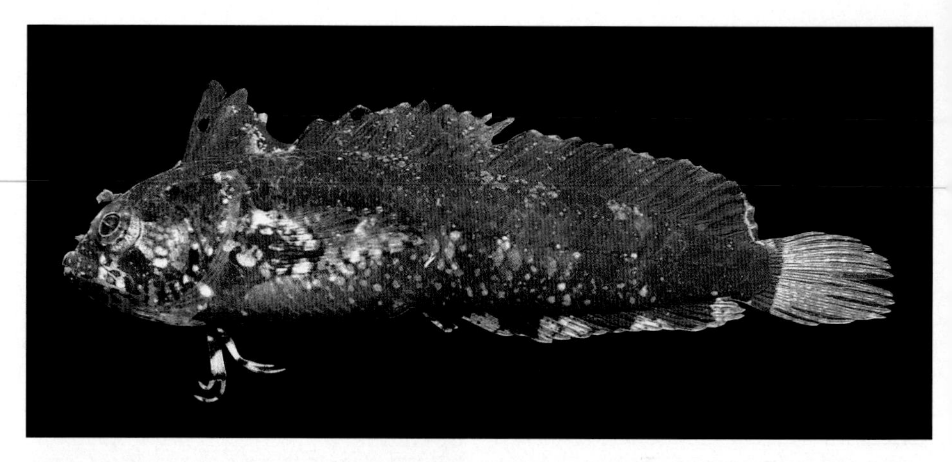

Starksia fulva Male ▲ Female ▼

Starksia galapagensis ▼

YELLOW BLENNY
Starksia fulva Rosenblatt & Taylor, 1971

Dorsal rays XIX to XXI,8-10; anal rays II,17-18; total (right and left) pectoral rays 26-30 (usually 28); lateral-line scales 36-40; a short, unbranched cirrus on each nostril, above eye, and on each side of nape; teeth of jaws conical, curved, and pointed; vomerine teeth present; palatine teeth absent; first anal spine of males separate from anal fin and modified as an intromittent organ; tip of male genital papilla extending slightly beyond fleshy tip of first anal fin; mottled dark brown to yellow-brown; *prominent black spots along bases of dorsal and anal fins; 2 round black spots on pectoral fin base; median fins with small black spots.* Costa Rica to Ecuador; occurs in sandy or weedy areas in 1-3 m depth. Maximum size, 4.5 cm.

GALAPAGOS BLENNY
Starksia galapagensis Rosenblatt & Taylor, 1971

Dorsal rays XX to XXII,8-9; anal rays II,16-19; total (right and left) pectoral rays 25-28 (usually 26); lateral-line scales 38-42; a short, unbranched cirrus on each nostril, above eye, and on each side of nape; teeth of jaws conical, curved, and pointed; vomerine teeth present; palatine teeth absent; first anal spine of males separate from anal fin and modified as an intromittent organ; tip of male genital papilla not extending beyond fleshy tip of first anal fin; yellowish to reddish brown, with scattered white flecks on side, *usually diffuse brown bars on side, and double row of 4-5 small dark spots on posterior half of body.* Galapagos Islands; inhabits rocky, weed-covered reefs in 1-4 m depth. Attains 4.5 cm.

BROWN-SPOTTED BLENNY
Starksia posthon Rosenblatt & Taylor, 1971

Dorsal rays XXI,9; anal rays II,19; total (right and left) pectoral rays 26; lateral-line scales 39; a short, unbranched cirrus on each nostril, above eye, and on each side of nape; teeth of jaws conical, curved, and pointed; vomerine teeth present; palatine teeth absent; first anal spine of males separate from anal fin and modified as intromittent organ; tip of male genital papilla extending beyond fleshy tip of first anal fin; *yellowish or pale tan, with 2-3 longitudinal rows of large brown spots or blotches; head yellowish; a dark spot between first and second dorsal spines, otherwise fins plain.* Costa Rica and Panama; occurs in shallow sand-weed areas close to shore. Attains 4 cm.

Starksia posthon ▲ *Starksia spinipenis* ▼

PHALLIC BLENNY
Starksia spinipenis Al-Uthman, 1960

Dorsal rays XIX to XXI,8-10; anal rays II,16-19; total (right and left) pectoral rays 27-30 (usually 28); lateral-line scales 38-42; a short, unbranched cirrus on each nostril, above eye, and on each side of nape; teeth of jaws conical, curved, and pointed; vomerine teeth present; palatine teeth absent; first anal spine of males separate from anal fin and modified as intromittent organ; tip of male genital papilla extending beyond fleshy tip of first anal fin; *head reddish, body whitish with numerous clusters of brown spots.* Gulf of California to Acapulco, Mexico; the most common member of the genus in the Gulf where it occurs in shallow sand-weed areas. Other Gulf *Starksia* (not illustrated) include *S. hoesei* (9 dark bars on side and long curved first anal spine in males), *S. cremnobates* (no body markings, scales in straight part of lateral line with simple pores instead of tubes, found below 30 m depth), and *S. grammilaga* (6-8 pale, horizontal lines with dark borders on rear half of body). Grows to 5 cm.

Xenomedea rhodopyga Plain variety ▲ Barred variety ▼

REDRUMP BLENNY
Xenomedea rhodopyga Rosenblatt & Taylor, 1971

Dorsal rays XX to XXIII,8-11; anal rays II,18-22; total (right and left) pectoral rays 24-28 (usually 26); lateral-line scales 39-45; a short, unbranched cirrus on each nostril, above eye, and on each side of nape; teeth conical, curved, and pointed; vomerine teeth present; palatine teeth absent; *first anal spine of males very short and separated from anal fin* and modified as intromittent organ, surrounded by fleshy folds that extend posteriorly along anal fin base during spawning periods; color of body variable, from plain pinkish to strongly marked with brown bars or squarish brown to red blotches and white flecks; *an oblique brown band with white band immediately above just below rear corner of eye; a small black spot at front of dorsal fin.* Gulf of California; relatively common on rocky, weed-covered reefs from tide pools to 8 m depth. Maximum size 6.5 cm.

TUBE BLENNIES (TRAMBOLLITOS, TRAMBOLLOS ALARGADOS)

FAMILY CHAENOPSIDAE

The tube blennies are small, scaleless fishes that are common in shallow seas throughout the region and in the western Atlantic. They are one of the few families that are confined to the tropical and subtropical Americas. Their common name is derived from the interesting habit of seeking permanent shelter in the unoccupied shells of barnacles or in tubes of certain worms and molluscs. Most species are inhabitants of rocky reefs or coral areas, although members of *Chaenopsis* and *Emblemaria* are associated with rubble or sandy bottoms. They feed mainly on zooplankton including copepods, isopods, and amphipods. They also consume barnacle cirri. Marked sexual dimorphism is present in most species. Males are usually brighter colored and often have better-developed supraorbital cirri, longer jaws, and higher dorsal fins. Other features include the absence of a lateral line, usually small conical teeth, at least half as many soft rays as spines in the dorsal fin, and all fin rays are unbranched. They seem to be most closely related to and possibly derived from the Weed Blennies (family Labrisomidae). Although very tiny, these fishes are a favorite subject of underwater photographers. Prior to spawning, the males periodically emerge from their refuge and display to nearby females. In species of *Chaenopsis* and *Emblemaria* this display is very spectacular and consists of thrusting the fully erect, sail-like dorsal fin far forward. The male nuptial display is further enhanced by dramatic changes in fin and body coloration. Once attracted to the male's lair, the female deposits her eggs. The nest is subsequently guarded by the male for several days until hatching. The family contains 12 genera and approximately 60 species; about 30 are known from the tropical eastern Pacific. We include *Stathmonotus*, formerly classified as a labrisomid, following the advice of P. Hastings and V. Springer.

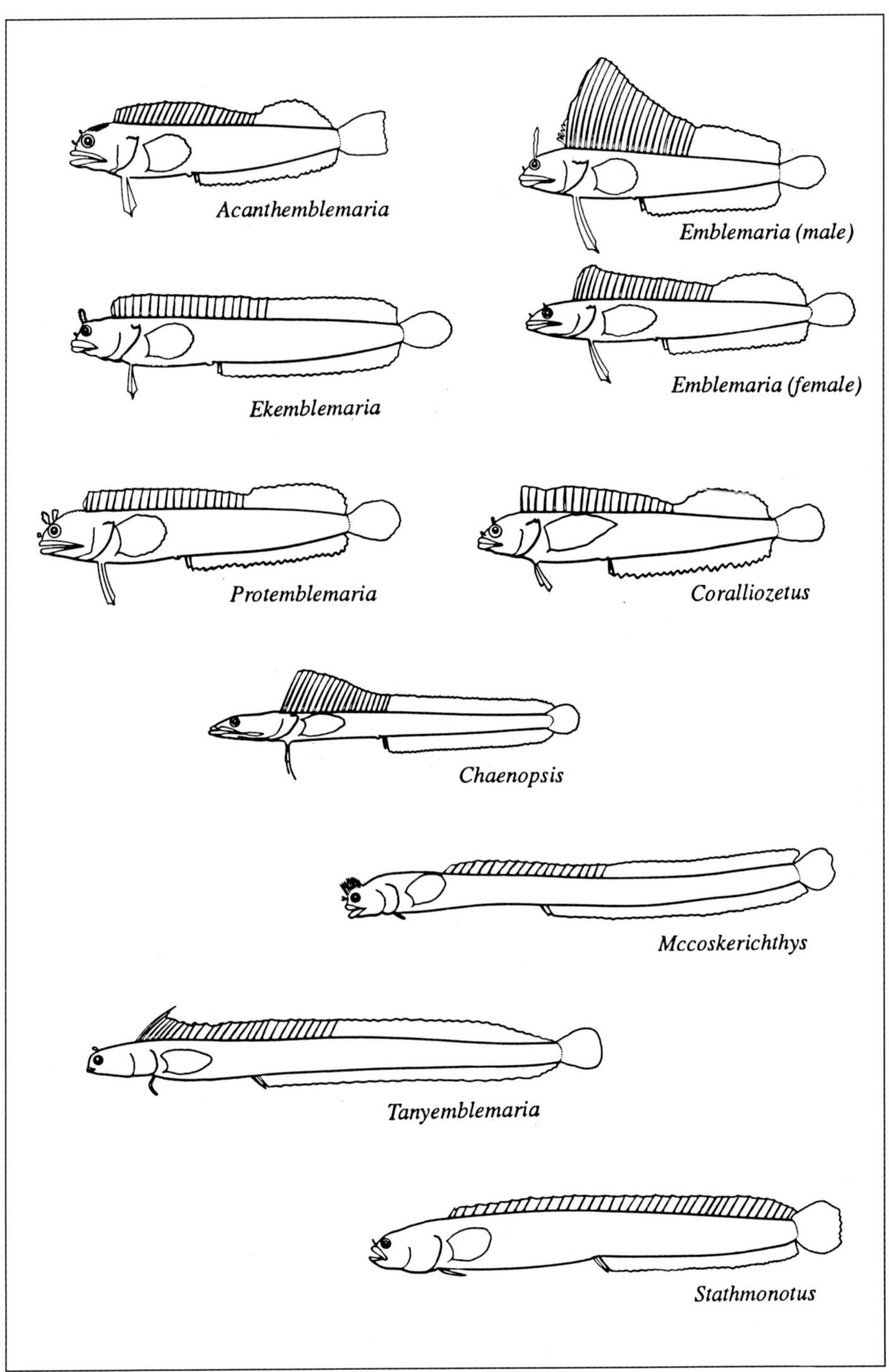

Eastern Pacific genera of tube blennies (Chaenopsidae)

CLUBHEAD BLENNY

Acanthemblemaria balanorum Brock, 1940

Dorsal rays XXIV,13, total dorsal elements 37; anal rays II,24-25; pectoral rays 13; *bony spines or tubercles on frontal bones in 2 triangular patches separated on middle of head by narrow spineless groove, spines large, blunt, expanded at tips, almost bridging spineless gap on midinterorbital; a single, unbranched cirrus above eye*; overall brownish, head often darker; *a single row of 6-8 large, dark brown to blackish spots along middle of sides*. Gulf of California; lives exclusively in empty barnacle shells on rocky reefs, in 1-5 m depth. Attains 4.5 cm.

CASTRO'S BLENNY

Acanthemblemaria castroi Stephens & Hobson, 1966

Dorsal rays XXIII-XXV,12-14, total dorsal elements 36-38; anal rays II,24-26; pectoral rays 13 or 14 (usually 13); *head spines well developed only on snout and infraorbital bones; top of head behind eyes with rugose ridges*, but weak spines sometimes present; a single unbranched or branched cirrus above eye; *primarily reddish with about 10 white spots (decreasing in size posteriorly) along lower side; a second row of somewhat diamond-shaped, pale marks along base of dorsal fin*; lower portion of head and chin whitish. Galapagos Islands; dwells in empty barnacle shells on rocky reefs in 1-5 m depth; seldom seen in the open. Grows to 6 cm.

Acanthemblemaria castroi ▲

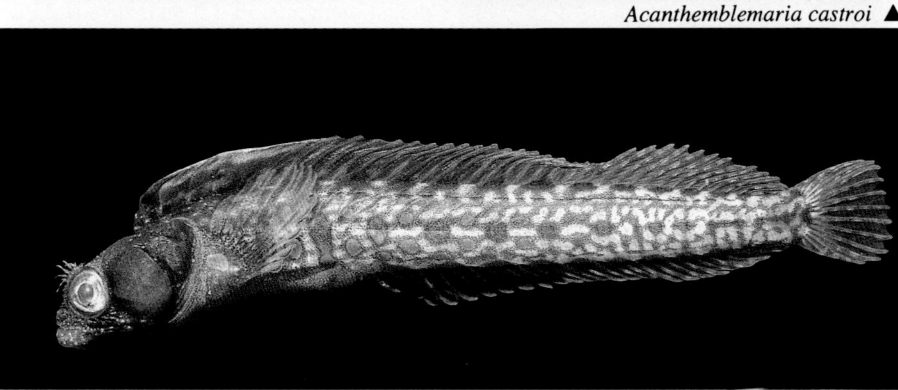

Acanthemblemaria crockeri Pale variety ▲ Dark variety ▼

BROWNCHEEK BLENNY

Acanthemblemaria crockeri Beebe & Tee-Van, 1938

Dorsal rays XXIII-XXVII,12-15, total dorsal elements 38-41; anal rays II,24-28; pectoral rays 13-14; *head spines strongly developed in 2 triangular patches on frontals separated by deep, spineless depression*; spines also present in 3 longitudinal rows on each side of interorbital groove; *a single pair of multifid, bush-like cirri above eye*; 2 color varieties occur that are primarily related to sex: a largely dark brownish male variety with numerous small white spots and lines; and a pale (whitish) variety with red to orange blotches and an orange throat region (this variety includes males and females, however pale males are apparently absent from the central and northern Gulf of California, and dark males do not occur at the southern tip of Baja California); both varieties have *a large, dark brown mark covering most of gill cover*. Gulf of California; inhabits empty worm and mollusc tubes on rocky reefs, in 1-60 m depth. Attains 6 cm.

Acanthemblemaria crockeri Subadult

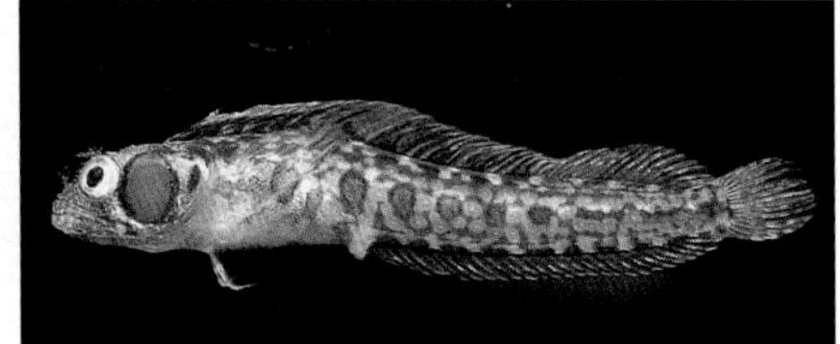

BLUNTSPINE BLENNY
Acanthemblemaria exilispinus Stephens, 1963

Dorsal rays XXIII-XXIV,14-16, total elements 37-39; anal rays II,24-25; pectoral rays 13; *head spines weakly developed, in 2 triangular patches on frontals; top of head behind eyes smooth or serrate, but never with spines*; spines also present on midline of snout; *a single, pinnately multifid cirrus above eye; head brownish with numerous small white spots on lower half*; body brown on upper half, whitish below, a row of dark brown blotches along middle of side, and small whitish blotches on upper half of body. Costa Rica to Ecuador; inhabits empty worm and mollusc tubes on rocky reefs in 1-5 m depth. Reaches 5.5 cm.

Acanthemblemaria exilispinus 4 cm specimen ▲ Underwater photograph ▼

HANCOCK'S BLENNY
Acanthemblemaria hancocki Myers & Reid, 1936

Dorsal rays XXIII-XXIV,12-14, total dorsal elements 36-37; anal rays II,23-25; pectoral rays 12-13; *head spines short, in a single triangular patch on frontals*; a single pair of supraorbital cirri, unbranched or bifid at tip; generally pale tan, nearly translucent, with midlateral row of horizontally elongate, dark brown blotches and horizontal row of small dark brown spots above and below midlateral row; *iris red; a large, dark brown spot on upper part of gill cover; outer edge of anterior dorsal rays red*. Costa Rica and Panama; inhabits empty worm and mollusc tubes on rocky reefs in 1-5 m depth. To 4.5 cm.

Acanthemblemaria hancocki 5 cm specimen ▲ Underwater photograph ▼

BARNACLE BLENNY

Acanthemblemaria macrospilus Brock, 1940

Dorsal rays XXIII-XXV,12-14, total dorsal elements 36-38; anal rays II,24-26; pectoral rays 12-13; *head spines short, in a single triangular patch on frontals; a single, unbranched cirrus above eye*; head dark brown to grayish; *body brown or tan, with midlateral row of large brown spots or blotches*, and dark brown, saddle-like markings along base of dorsal fin; *a bright red-orange band covering most of first 5-6 dorsal spines*, with narrow black basal band. Southern portion of Baja California between Santa Rosalia and Cabo San Lucas, Revillagigedo Islands, and Mexican coast from Mazatlan to Acapulco; inhabits empty barnacles and worm or mollusc tubes on rocky reefs in 2-15 m depth. Maximum size 6 cm.

The population from the Revillagigedo Islands exhibits a distinctive coloration, and is extremely variable regarding the branching of the supraorbital cirri (these may be brush-like with numerous branches).

Acanthemblemaria stephensi Male ▲ Female ▼

MALPELO BARNACLE BLENNY

Acanthemblemaria stephensi Rosenblatt & McCosker, 1988

Dorsal rays XXII-XXV,11-13 (usually XXIV,12), total dorsal elements 34-37; anal rays II,24-25; pectoral rays 13; *bony spines on head well developed, forming diamond-shaped patch above eye; spines and bony knobs also present on interorbital and snout regions; a single, unbranched cirrus above each eye; translucent blue-gray with bright red bars on lower half of body*, and corresponding diffuse red spots or blotches on upper half; a midlateral row of horizontally elongate white marks; head and body covered with numerous small, light blue spots; *males with dark bluish black head, a bright red anterior dorsal fin with a black spot, red lips, and red nasal-orbital cirri*. Known only from Isla Malpelo; inhabits empty barnacles and worm or mollusc tubes on rocky reefs in 9-10 m depth. Reaches 5 cm.

COCOS BARNACLE BLENNY
Acanthemblemaria sp.

Dorsal rays XXIII-XXV,12-14, total dorsal elements 36-37; anal rays II,24-25; pectoral rays 13 (rarely 12); *head spines short, in a single, triangular patch on frontals; a single, unbranched cirrus above eye; color variable, from nearly solid dark brown or blackish to light gray with series of hourglass-shaped blackish bars on side; also a network of tiny pale spots on head and body; snout, middle portion of lips, and tips of orbital cirri whitish, except red in nuptial males; a large dark blotch just behind eye*; juveniles somewhat translucent with about 5 brown bars on side. Known only from Isla del Coco; inhabits empty barnacles and worm or mollusc tubes on rocky reefs in 1-10 m depth. Maximum size to 4.5 cm.

Acanthemblemaria sp. Female ▲ ▼

Acanthemblemaria sp. Male ▼

Chaenopsis alepidota ▼

ORANGETHROAT PIKEBLENNY
Chaenopsis alepidota (Gilbert, 1890)

Dorsal rays XVIII-XX,*34-36*; anal rays II,34-37; pectoral rays 12-13; body and snout very elongate; lower lip extending beyond snout tip; caudal fin rounded, continuous with dorsal and anal fins in adults; no cirri on nostrils or above eyes; *at least a few teeth present on vomer; tip of tongue extending forward at least as far as vomerine tooth patches; anterior half of dorsal fin only slightly elevated in adult males; first 3 mandibular pores (counting from front of jaw) equally spaced*; overall green, with longitudinal row of white spots just below middle of side, and second row along base of dorsal fin; 8-10 dark brown to green bars usually evident on side. Southern California to the Gulf of California; inhabits worm tubes on sand-rubble bottoms in 2-23 m depths. Feeds mainly on zooplankton. To 15 cm. *C. coheni* (not illustrated) is another Gulf species; it has only 28-30 soft dorsal rays and lacks vomerine teeth. Böhlke (1952) recognized a separate subspecies, *C. alepidota californiensis*, from California.

Chaenopsis alepidota Subadult

DELTA PIKEBLENNY
Chaenopsis deltarrhis Böhlke, 1957

Dorsal rays XVIII-XX,*28-30*; anal rays II,32-36; pectoral rays 14; body and snout very elongate; lower lips extend beyond snout tip; caudal fin rounded, continuous with dorsal and anal fins in adults; no cirri on nostrils or above eyes; *no teeth on vomer; anterior half of dorsal fin of adult males very tall and sail-like*; light green or olive, white below, often with greenish midlateral stripe and/or with series of 8-10 dark bars; females with black ocellated spot at front of dorsal fin. Costa Rica to Colombia; inhabits worm tubes on sand-rubble bottoms in 5-30 m depths. Attains 7.5 cm.

Chaenopsis deltarrhis Male ▲ Female ▼

SOUTHERN PIKEBLENNY
Chaenopsis schmitti Böhlke, 1957

Dorsal rays XVIII-XX,*33-38*; anal rays II,34-36; pectoral rays 12-13; body and snout very elongate; lower lip extending beyond snout tip; caudal fin rounded, continuous with dorsal and anal fins in adults; no cirri on nostrils or above eyes; *at least a few teeth present on vomer; tip of tongue not extending forward as far as palatine tooth patches; first 3 mandibular pores (counting from front of jaw) equally spaced; first half of dorsal fin moderately tall and sail-like in adult males*; generally light olive to whitish, with about 10 narrow, greenish bars on side; branchiostegal membrane black; head and anterior part of dorsal fin of males often brownish to black; females with black ocellated spot at front of dorsal fin. Costa Rica to Colombia and Galapagos Islands; inhabits worm tubes on sand-rubble bottoms in 5-30 m depths. Reaches 8 cm.

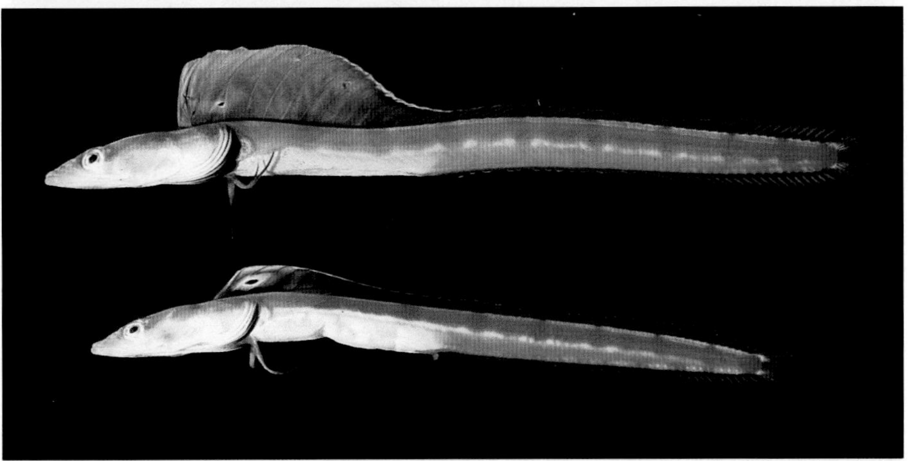

Chaenopsis schmitti Male (top) and female ▲ Nuptial male ▼

ANGEL BLENNY
Coralliozetus angelica (Böhlke & Mead, 1957)

Dorsal rays XVIII-XX,10-12, *total dorsal elements 29-31; anal rays II,19-20*; pectoral rays 13; body robust; *cirrus above eye long and fleshy (as long or longer than eye width) in males, short and slender in females; dorsal fin relatively low in both sexes*; body whitish to yellowish, with diffuse brown blotches or spots; head and spinous dorsal fin of males dark brown; head of female whitish with brown markings, also with prominent brown marks at base of pectoral fin. Gulf of California to southern Mexico (Acapulco); inhabits empty barnacle shells on rocky reefs in 1-2 m depth. Maximum length to 3.5 cm.

Coralliozetus angelica Male ▲ Female ▼

SCARLETFIN BLENNY
Coralliozetus micropes (Beebe & Tee-Van, 1938)

Dorsal rays XIX-XXI,11-13, *total dorsal elements 31-33; anal rays II,21-24*; pectoral rays 12-14; body elongate and slender; *cirrus above eye longer than eye diameter in males, about half eye diameter in females; dorsal fin elevated anteriorly in males*, uniformly low in females; male overall dark brown with yellowish caudal fin, several narrow diagonal white lines across cheek and gill cover; females yellowish white, nearly translucent, with white and brown flecks on side, and several brown bars on lower half of head. Baja California and Gulf of California; inhabits mollusc tubes and barnacles on rocky reefs in 1-3 m depth. Grows to 4 cm.

C. micropes Female (top) and male ▼

SPIKEFIN BLENNY
Coralliozetus rosenblatti Stephens, 1963

Dorsal rays XVIII-XIX,12-13, total dorsal elements 31-33; anal rays II,21-22; pectoral rays 13; body elongate and slender; *cirrus above eye short and flat in both sexes; third dorsal spine of males elongate, twice as long as following spines; dorsal fin of females low, but first 3 spines slightly elongate*; overall semitranslucent with numerous white flecks and brown dashes on side; males with brownish head and ocellus at front of dorsal fin; females with several distinct lines of small brown dots below mouth. Baja California and Gulf of California; inhabits empty mollusc tubes on rocky reefs in 1-4 m depth. Attains 3.5 cm. *C. boehlkei* (not illustrated), which ranges from the Gulf of California to Costa Rica, is very similar in appearance to *C. rosenblatti*, but has a low dorsal fin in both sexes.

Coralliozetus rosenblatti ▼

SPRINGER'S BLENNY
Coralliozetus springeri Stephens & Johnson, 1966

Dorsal rays XVII-XVIII,11-12, *total dorsal elements 28-29; anal rays II,18-19*; pectoral rays 13; body relatively robust; *usually a pair of unbranched cirri above each eye, the posterior 1 very tiny*; overall semitranslucent to whitish; males with reddish brown head and small brown spots on anterior dorsal fin and anterodorsal part of body, and also a series of brown blotches above base of anal fin. Costa Rica to Ecuador, including Isla del Coco; inhabits empty worm and mollusc tubes on rocky reefs in 5-15 m depths. To 2.5 cm. Specimens from Isla del Coco differ in having 1 pair instead of 2 pairs of cirri above the eye. They may prove to be a separate species. This problem is being studied by chaenopsid expert Philip A. Hastings.

Coralliozetus springeri Male ▲ Female ▼

REEFSAND BLENNY
Ekemblemaria myersi Stephens, 1963

Dorsal rays XIX-XXI,19-21, total dorsal elements 38-42; anal rays II,24-27; pectoral rays 13-14; *spines absent on top of head, but frontals strongly rugose; no longitudinal bony ridges on snout; a single, pinnately multifid cirrus above each eye*; color variable, either entirely dark brown, or brownish with 10-12 darker brown bars on side; usually with brown bars on lips, and *large brown ocellated spot on gill cover; distal portions of soft dorsal, caudal, and posterior part of anal fin clear*. Gulf of California to Colombia; inhabits mollusc tubes on rocky outcrops in sand-weed areas in 1-4 m depths. Reaches 7 cm.

GULF SIGNAL BLENNY

Emblemaria hypacanthus (Jenkins & Evermann, 1889)

Dorsal rays XX-XXII,13-16, total dorsal elements 34-38 (usually 35-36); anal rays II,22-25; pectoral rays 13; spines absent on top of head; 2 longitudinal bony ridges on snout; pelvic fins longer than pectoral fins; spinous dorsal fin strongly elevated and sail-like in males, low or with first 1 or 2 spines very slightly elevated in females; *males with small, weakly developed, flag-like flap on anterior edge of first dorsal spine, near base; bony flange below eye not expanded and not overlapping upper lip in males*; males usually dark brown to tan, with dark brown bars or blotches along middle of side, cirri above eyes red with black tip; *females with highly variegated pattern, usually a row of small brown spots along middle of side, and brown, saddle-like marks below base of dorsal fin.* Gulf of California; inhabits empty mollusc tubes on sand-rubble bottoms in 1-10 m depths. To 5 cm.

Nuptial male ▲ Young male ▼

Female ▼

WHITEBACK SIGNAL BLENNY
Emblemaria nivipes Jordan & Gilbert, 1883

Dorsal rays XXII,13-14, total dorsal elements 35-36; anal rays II,23-24; pectoral rays 14-15; spines absent on top of head; *2 longitudinal bony ridges on snout, smooth and not broken into tubercles; spinous dorsal fin strongly elevated and sail-like in males, only first spine elevated (about twice as long as second ray) in females; cirrus above eye often with filamentous branches*, but may also be simple (or with tiny terminal branches); males dark brown with pronounced white back; back with numerous small brown spots; also midlateral row of white spots or streaks; dorsal fin with blue oblique lines and blue or white spots; females overall whitish with midlateral row of rectangular orange-brown markings; back and upper half of head with numerous small brown spots; a prominent ocellus at front of dorsal fin. Isla del Coco and Costa Rica to Panama and Gorgona Island, Colombia; inhabits empty mollusc tubes on sand-rubble bottoms in 5-30 m. Grows to 4.2 cm.

Emblemaria nivipes Male ▲ Female ▼

Emblemaria piratica Male ▲ Nuptial male ▼

SAILFIN SIGNAL BLENNY
Emblemaria piratica Ginsburg, 1942

Dorsal rays XX-XXI,13-14, total dorsal elements 34-35; anal rays II,21-23; pectoral rays 13; spines absent on top of head; *2 longitudinal bony ridges on snout broken into distinct tubercles*; pelvic fins longer than pectoral fins; *spinous dorsal fin strongly elevated and sail-like in males, only first 3 spines elongated in females; 3-5 small, flag-like flaps on anterior edge of basal part of first dorsal spine of adult males*; membranes between first 4 dorsal spines of males slightly to deeply incised; males with dark brown head and lighter brown body, sides with faint brown bars and white flecks; *dorsal fin of males with broad yellow or red zones between diagonal black bands*; females generally pale, somewhat translucent, with midlateral row of brown spots and numerous white flecks. Southern Mexico to Panama; inhabits empty mollusc tubes on sand-rubble bottoms in 5-30 m depths. Attains 4 cm.

ELUSIVE SIGNAL BLENNY
Emblemaria walkeri Stephens, 1963

Dorsal rays XXI-XXII,15-18, total dorsal elements 37-39; anal rays II,25-26; pectoral rays 13; spines absent on top of head; 2 longitudinal bony ridges on snout; pelvic fins longer than pectoral fins; spinous dorsal fin strongly elevated and sail-like in males, low and even in females; *males with single, relatively large, flag-like flap on anterior edge of first dorsal spine, near base; males with bony flange below eye expanded, overlapping upper lip*; males brown on head, dark reddish brown on body, with faint, large, dark blotches along middle of side, and whitish saddles below base of spinous dorsal fin; *females reddish with whitish saddles below base of dorsal fin*. Gulf of California; inhabits empty mollusc tubes on sand-rubble bottoms in 5-20 m depths. Grows to 6.5 cm.

Emblemaria walkeri Male ▲ Nuptial male ▼

Emblemaria walkeri Female ▼

TUFTED BLENNY
Mccoskerichthys sandae Rosenblatt & Stephens, 1978

Dorsal rays XVII-XX,31-34 (usually XIX,32 or 33), total dorsal elements 50-53; anal rays II,32-36 (usually 34 or 35); pectoral rays 12-14 (usually 13); jaws with outer row of spatulate teeth; *supraorbital cirri forming a dense, bushy mass between eyes, composed of 5 pairs of cirri*; body long and slender, depth at dorsal origin 10-12 in standard length; head tan, remainder of body reddish brown. Costa Rica and Panama; inhabits mollusc and worm tubes on near-vertical rock faces in 1-30 m depths. Feeds on zooplankton, including copepods, amphipods, and ostracods. To 8 cm.

Mccoskerichthys sandae 5.5 cm specimen ▲ Underwater photograph ▼

WARTHEAD BLENNY

Protemblemaria bicirris (Hildebrand, 1946)

Dorsal rays XX-XXII,14-16, total dorsal elements 36-37; anal rays II,23-25; pectoral rays 14; *top of head covered with wart-like, fleshy protuberances; a pair of fleshy ridges between eyes; a pair of relatively large, branched cirri above each eye*; body color variable, from nearly whitish translucent to pale brown or yellowish orange; usually 8-10 brownish saddles below base of dorsal fin, and sometimes a corresponding mid-lateral row of brownish blotches; *a small dark spot at front of dorsal fin*. Gulf of California to Peru; inhabits empty barnacles and mollusc tubes on rocky reefs in 1-15 m depths. To 4.5 cm.

Protemblemaria bicirris Juv.

Protemblemaria bicirris Barred variety ▲ Orange variety ▼

PLUME BLENNY

Protemblemaria lucasana Stephens, 1963

Dorsal rays XXI-XXII,15-16, total dorsal elements 36-38; anal rays II,25-27; pectoral rays 14; *no wart-like projections on top of head, but males with long, cirrus-like flaps; a pair of branched cirri above each eye, anterior cirrus of male at least twice as long as eye width, female with shorter cirri*; translucent whitish to yellowish, with brown saddle-like markings below base of dorsal fin; a pair of thin, oblique brown bands across cheek (at least on males). Gulf of California; inhabits empty mollusc tubes on rocky reefs in 10-30 m depths. Attains 4 cm.

Protemblemaria lucasana Pale variety ▲ Yellow variety ▼

CALIFORNIA WORM BLENNY
Stathmonotus sinuscalifornici
(Chabanaud, 1942)

Dorsal rays XXX to XXXV; anal rays II,22-26; pectoral rays 7-9; greatest body depth 7.1-9.9 (mean 8.2) in standard length; head length 4.9-5.8 (mean 5.4) in standard length; *cirrus present above eye; no cirrus on nostril*; color highly variable, white, green, yellow, brown, or black, usually with series of white bars on dorsal fin; green or yellow variety often with 1 or more longitudinal rows of black spots on side. Gulf of California; occurs in weedy areas along rocky shores. A similar species from southern Mexico, *S. lugubris* (not illustrated), differs by having a cirrus or skin flap on the anterior nostril and 37-38 dorsal spines). Maximum size 6.5 cm.

Green variety ▲ Black-spotted variety ▼

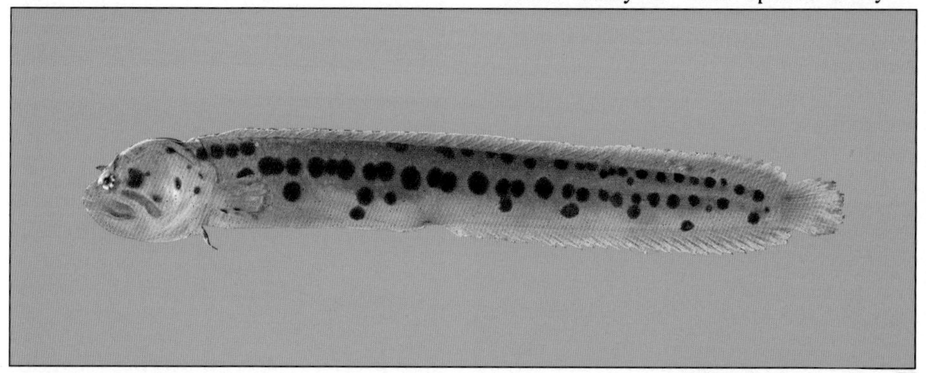

White-flecked variety ▼

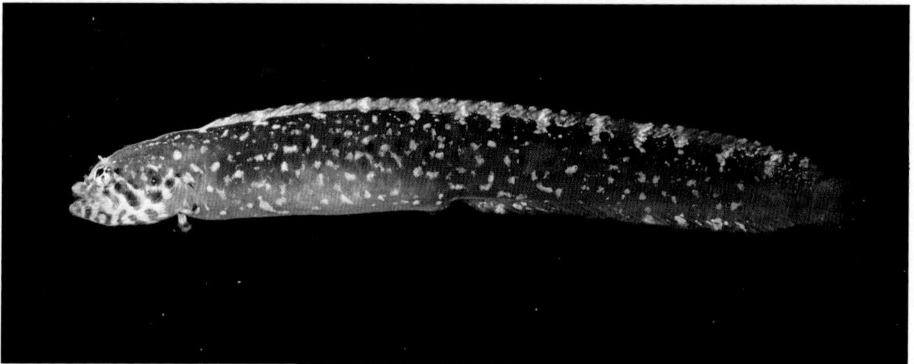

Yellow variety ▼

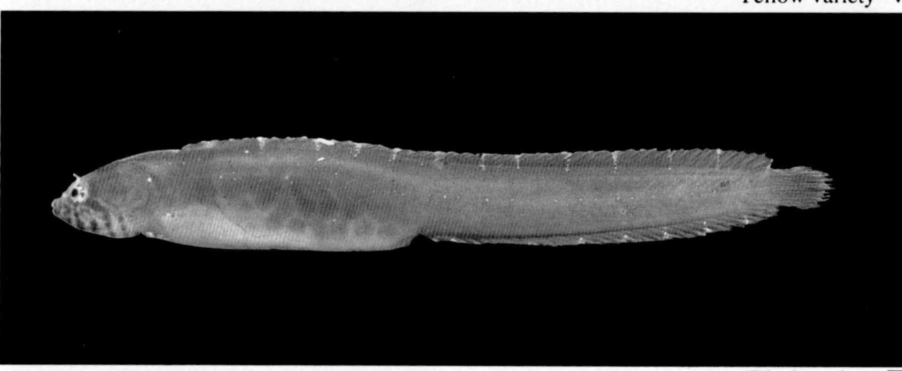

Black variety ▼

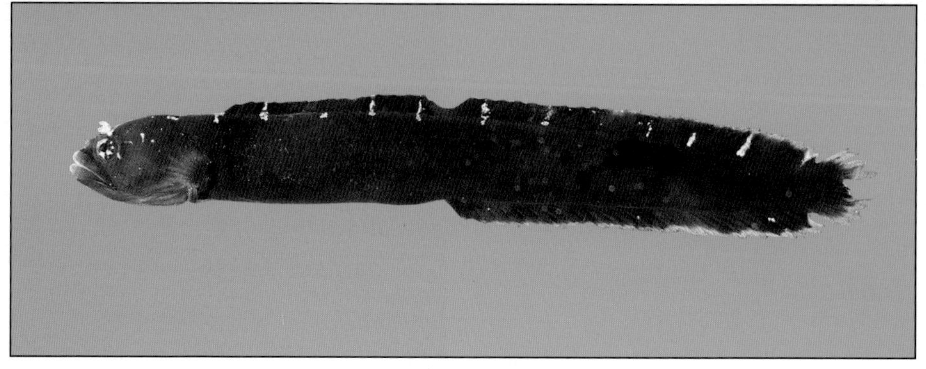

PANAMANIAN WORM BLENNY
Stathmonotus culebrai Seale, 1940

Dorsal rays XXXIX to XXXXII; anal rays II,20-23; pectoral rays 7-10; greatest body depth 7.2-8.1 in standard length; head length 4.9-5.7 in standard length; *no cirrus above eye; cirrus present on anterior nostril*; color highly variable, entirely green, yellowish with leopard-like spotting, or black with dorsal fin and top of head white. Costa Rica and Panama; a cryptic species found among weed on rocky reefs, usually not seen unless flushed out with chemical ichthyocides. Attains 5 cm.

Stathmonotus culebrai Yellow variety ▲ White-back variety ▼

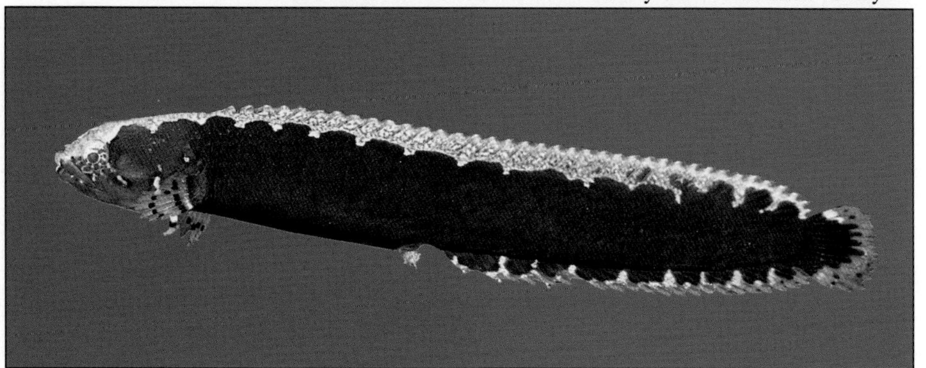

Stathmonotus culebrai Green variety ▼

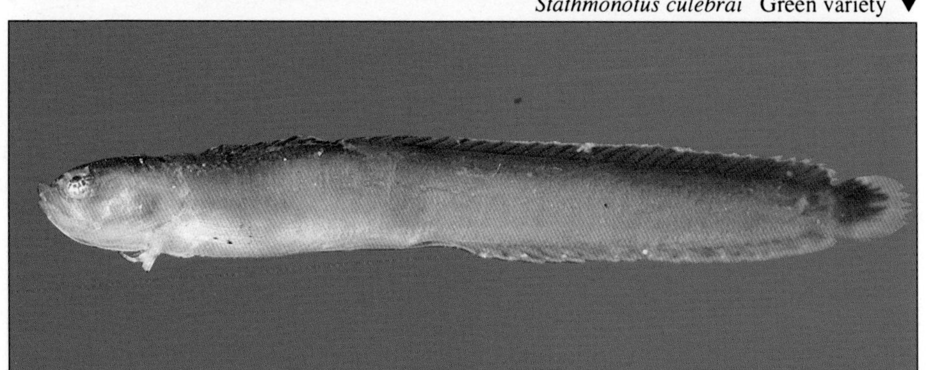

Stathmonotus culebrai Spotted variety ▼

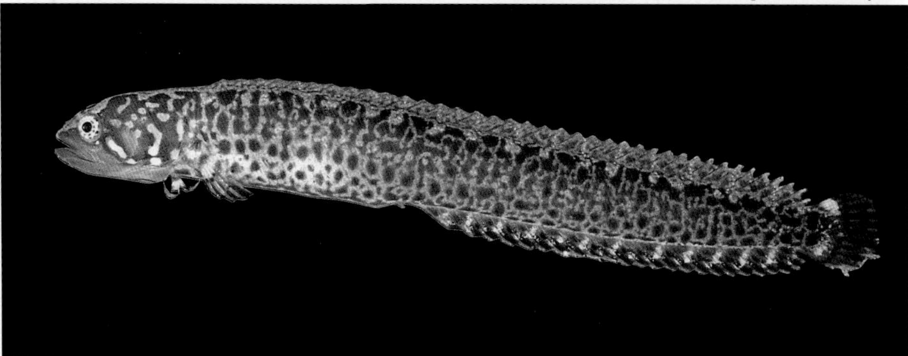

Tanyemblemaria alleni ▼

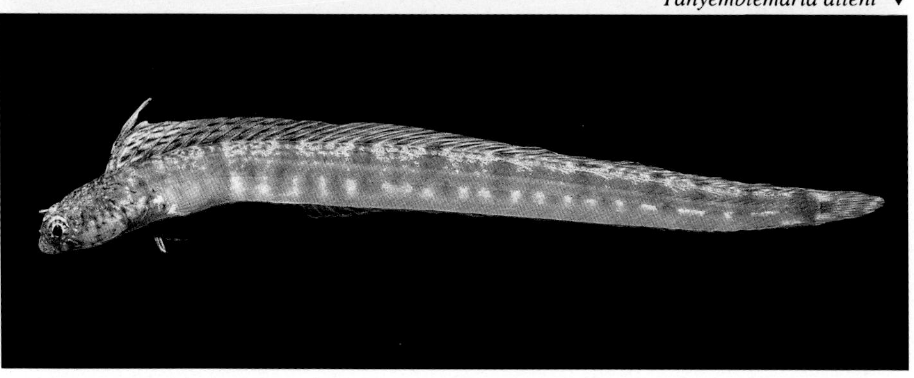

SLENDER BLENNY
Tanyemblemaria alleni Hastings, 1992

Dorsal rays XXVII,24, the first dorsal spine elongate; anal rays II,38; pectoral rays 13; *body elongate, the depth 12.3 in standard length; posterior margin of preopercle with serrations and short spines*; first infraorbital with strong spines on ventral margin; a raised ridge on frontals; widely separated nasal bones with prominent raised ridges and a single row of projecting bony knobs; no cirrus on anterior nostril; 1 pair of flattened, unbranched supraorbital cirri; a small orange blotch on cheek behind upper jaw; several white spots or blotches on head, body, and dorsal fin; opercle, pectoral fin base, and basal portion of pectoral fin rays each with a single large white blotch; entire length of body with alternating white and dark saddles. Known only from a single specimen collected at Isla Rey, Islas Perlas, Panama, in 17 m on a sandy bottom. To at least 5.4 cm.

BLENNIES (BORRACHOS, BLENIOS, CACHUDITOS)
FAMILY BLENNIIDAE

Blennies are well represented in mainly tropical and subtropical seas with 56 genera and approximately 350 species. These small, bottom-dwelling fishes are a conspicuous element of the reef community. Some species are able to traverse short distances between rock pools and are often referred to as "rock-hoppers". Blennies are characterized by a combination of features including an elongate body, lack of scales, and a continuous, long-based dorsal fin (with VII-XIII flexible spines and 13-35 segmented rays in the tropical eastern Pacific species). Unlike the tube blennies (Chaenopsidae) they usually have fewer spines than soft rays in the dorsal fin. Most species are relatively blunt-headed, and many have tentacles, cirri, or a fleshy crest on the upper part of the head. The mouth is low on the head and not protractile; jaw teeth are numerous, slender, and close-set, either fixed or movable. Some species, known as sabertooth blennies, have a pair of enormous canines in the lower jaw, used for defence or feeding. This includes members of the genus *Plagiotremus* (*Runula* is a synonym), which are a nuisance to other fishes due to their habit of feeding on scales, mucus, and dermal tissue. The diet of most non-sabertooth blennies consists mainly of algae. Although most species live on the reef's surface, the sabertooth blennies are free-swimmers, usually seen a short distance above the bottom. However, they frequently retreat to burrows (often a hollow worm tube) on the bottom. Blennies lay demersal eggs, and for the species which have been studied, it appears these are guarded by the male until hatching. Eastern Pacific *Hypsoblennius* was treated by Krejsa (1960, as *Blenniolus*), and Bath (1977); *Entomacrodus* and *Ophioblennius* were treated by Springer (1967; 1962); and Smith-Vaniz (1976) covered *Plagiotremus*. The genus *Scartichthys* Jordan & Evermann was reviewed by Williams (1990); it contains four species that occur mainly in temperate waters of western South America, except for *S. gigas*, which ranges northward to Panama (see text for *O. steindachneri*, page 254).

Panamic Fanged Blenny (*Ophioblennius steindachneri*), Isla Marchena, Galapagos

ROCK BLENNY

Entomacrodus chiostictus (Jordan & Gilbert, 1882)

Dorsal rays XIII (rarely 12 or 14),14-16 (usually 14-15); anal rays II,15-17 (usually 16); pectoral rays 14; total gill rakers on first arch 14-20 (usually 15-18); cirri present on posterior margin of anterior nostril; supraorbital cirri 1-10, number increasing with growth; nape with a single cirrus on each side; each jaw with more than 100 freely movable teeth; teeth present on vomer; dorsal fin deeply notched between spinous and soft portions; olive on back, white on lower sides; *longitudinal row of square to rectangular black blotches along middle of side*; small dark spots anterodorsally on body and sides with white flecks; *thin, brown, vertical bars on lips*. Gulf of California to Panama, also offshore islands including Tres Marias, Revillagigedos, Clipperton, and Cocos; inhabits shallow, weed-covered rocky reef including tide pools. Reaches 7.5 cm.

Entomacrodus chiostictus Female (top) and male ▲ *Hypsoblennius brevipinnis* ▼

BARNACLEBILL BLENNY

Hypsoblennius brevipinnis (Günther, 1861)

Dorsal rays X-XII,15-16; anal rays II,13-14; posterior nostril with a broad cirrus; a relatively long, slender cirrus with 1-2 short basal branches over each eye; no cirrus on nape; teeth in jaws incisor-like, in a single series and not movable; dorsal fin notched between spinous and soft portions; upper sides dark brown, with horizontal row of circular tan patches (often with dark brown centers) just below base of dorsal fin, and second row of smaller patches or spots just below; lower sides whitish to yellow; *head with small brown or red spots and prominent, oblique, tan band behind eye, with dark-edged whitish stripe on its upper and lower margin*. Gulf of California to Peru, including Galapagos Islands and other offshore islands; lives in empty barnacle shells, usually in 1-3 m. Grows to 6.5 cm.

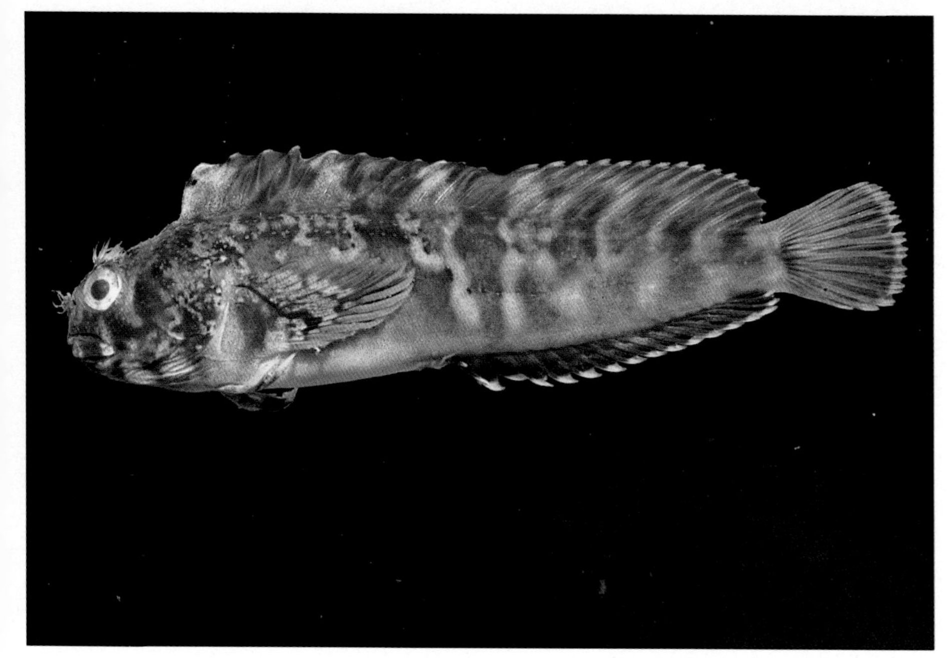

TIDEPOOL BLENNY

Hypsoblennius caulopus (Gilbert, 1898)

Dorsal rays XII,15-16; anal rays II,17; anterior nostril and upper margin of eye each with a tuft of small cirri; no cirrus on nape; teeth in jaws incisor-like, in a single series and not movable; dorsal fin with very slight notch between spinous and soft portions; *wavy, irregular dark brown and light tan bars on side, extending onto dorsal fin; head largely dark brown, with light and dark bars on lips and chin*. Nicaragua to Peru; inhabits shallow, weed-covered rocky reefs, including tide pools. Attains 9 cm. *H. lignus* Meek & Hildebrand is a synonym.

BAY BLENNY

Hypsoblennius gentilis (Girard, 1854)

Dorsal rays XI-XIII,16-18; anal rays II,16-19; pectoral rays 11-12; a small cirrus above posterior nostril; an unbranched cirrus present above eye, very elongate in adult males; no cirri on nape; teeth in jaws incisor-like, in a single series and not movable; dorsal fin very slightly notched between spinous and soft portions; *overall brown to greenish, white on throat and belly region; upper back with series of dark brown, saddle-like marks, and horizontal row of white blotches along side at level of pectoral fin base*; adult males with red bar across throat, most evident during the breeding season. Monterey, California to Magdelena Bay, Baja California, and also the central and northern Gulf of California; inhabits intertidal shallows. To 9 cm.

Hypsoblennius gentilis Female (top) and male ▲

Hypsoblennius striatus ▼

Ophioblennius steindachneri ▼

STRIATED BLENNY

Hypsoblennius striatus (Steindachner, 1876)

Dorsal rays XII,16; anal rays II,17-18; a small cirrus above posterior nostril; a branched cirrus above eye, relatively long and slender, at least in some specimens; no cirri on nape; teeth in jaws incisor-like, in a single series and not movable; dorsal fin not notched between spinous and soft portions; *mottled olive on back with 5-6 short, broad, dark bars below base of dorsal fin, lower half white with many narrow vertical black lines; a large, dark brown to blackish blotch behind eye*. Costa Rica and Panama; inhabits shallow reefs and tide pools. Attains 8 cm. Two other species (not illustrated) formerly placed in *Hypsoblennius* are now in the genus *Parahypsos* Bath (1987). These include *P. piersoni* (dorsal rays IX-X,25; anal rays II,24), and *P. paytensis* (grayish yellow with many small dark saddles along dorsal fin base); both species range from Costa Rica to Peru.

PANAMIC FANGED BLENNY

Ophioblennius steindachneri Jordan & Evermann, 1898

Dorsal rays XI-XIII,21-23 (usually 22); anal rays II,22-24 (usually 23); pectoral rays usually 15; bushy tuft of cirri above posterior nostrils; cirrus above eye slender, unbranched; about 5 short cirri on each side of nape; teeth in jaws numerous and movable, *a pair of large, curved canines posteriorly on lower jaw; lateral line in 2 disconnected, overlapping portions, with 10 or fewer tubes in the posterior section*; dorsal fin not notched between spinous and soft portions; *mainly dark brown, often with yellowish bars on head and anterior part of body*; a dark brown or blackish spot behind eye, sometimes surrounded by pale area. Gulf of California to Peru; inhabits shallow rocky reefs exposed to surge at depths to about 10 m. Maximum size 18 cm. A similar-shaped blenny, *Scartichthys gigas* (Steindachner), ranges along rocky shores of Chile, Peru, Ecuador, and north to Panama. However, it lacks enlarged canines at the back of the lower jaw, and has a continuous lateral line with 18-22 tubes in the posterior section. It reaches 25 cm.

SABERTOOTH BLENNY
Plagiotremus azaleus (Jordan & Bollman, 1890)

Dorsal rays VII-IX,31-35; anal rays II,27-30; pectoral rays 12-13; body very long and slender; no cirri on head; mouth subterminal, below protruding snout; teeth in jaws close-set and slender, except for *a very large recurved canine on each side of lower jaw*; dorsal fin very low with elongate base, no notch between spinous and soft portions; *generally a broad, brown, midlateral band on side, with narrow yellow to white stripe immediately above*; lower part of head and body white. Gulf of California to Peru, including the Galapagos and other offshore islands; inhabits rocky reefs in 2-20 m. Feeds by rapid attacks on other fishes, removing dermal tissue, mucus, and scales. Reaches 10 cm.

Adult ▲ Sheltering in empty worm tube ▼

SAND LANCES (LANZONES)
FAMILY AMMODYTIDAE

The sand lances are small (to about 25-30 cm), elongate fishes that are found along sandy shores. They are adapted for burrowing in the soft substratum. Other features include an absence of pelvic fins (except in *Embolichthys*), absence of teeth, lateral line placed very high on the sides (just below dorsal fin base), and a single, very long-based, but low dorsal fin. The family contains three genera with approximately 12 species. They inhabit both cold and warm circumglobal seas. Only a single species is recorded from the tropical eastern Pacific.

GILL'S SAND LANCE
Bleekeria gilli Bean,1895

Dorsal rays 46-47; anal rays 22-24; pectoral rays 14-15; lateral-line pores about 90; scales cycloid; *body very elongate, spindle-shaped, tapering toward head and caudal fin*; head relatively sharp and pointed; pectoral fins small and pointed, situated well below median axis of body; *pelvic fins absent*; caudal fin deeply forked; mainly translucent whitish in color. Cape San Lucas, Mexico to Panama; inhabits sandy bottoms, sometimes near shallow rocky reefs. Grows to about 11.5 cm.

GUDGEONS (DORMIENTES, CHAMES, PORROCOS)
FAMILY ELEOTRIDIDAE

Gudgeons are primarily fresh- and brackish-water fishes that occur mainly in the tropics. Worldwide there are about 300 species, of which 250 are found in the Indo-Pacific region. Only a few species inhabit purely marine environments. Gudgeons are close relatives of gobies (Gobiidae), and the two groups share a number of morphological features including a generally elongate body, similar head shape, two separate dorsal fins, and absence of a lateral line. Most gobies, however, have the pelvic fins completely or partially fused to form a disc-shaped structure, whereas the pelvics are completely separated in gudgeons. They are mainly bottom-living fishes, although a few species are free-swimming. The diet consists primarily of benthic invertebrates, particularly crustaceans. A single species is common in brackish habitats of the tropical eastern Pacific.

SPOTTED SLEEPER
Dormitator latifrons (Richardson, 1837)

Dorsal rays VII+I,8-9; anal rays I,9-10; lateral-line scales 30-34; body depth about 3.0 in standard length; mouth slightly oblique, maxillary reaching to level of front of eye; lower jaw projecting slightly; head broad and flat dorsally; overall brown to purplish, about 7-8 narrow (1 scale wide) oblique bars on upper sides; *side of head with several dark brown stripes; a prominent blue "ear" spot behind upper edge of gill cover.* Mexico to Ecuador; widespread in fresh and brackish waters. Grows to at least 35 cm.

Bleekeria gilli ▲

Dormitator latifrons ▼

GOBIES (GOBIOS, BOCONES)

FAMILY GOBIIDAE

The gobies are the largest family in the tropical eastern Pacific, with appoximatey 80 species. Indeed, it is the largest family of marine fishes in the world, with about 220 genera and 1 600 species, of which about 160 genera and 1 200 species inhabit the vast Indo-Pacific region. *However, our coverage in this section is limited to relatively common genera and species that divers and reef walkers are likely to encounter.* Gobies exhibit considerable variation in size, coloration, and body shape. Generally, they are elongate fishes with two dorsal fins (although *Gobioides* has a single dorsal fin). Some of the most important general characters for identifying various species in this region include number of spines in the first dorsal fin (6 vs 7), pelvic fin structure (fins separate or partially to completely fused to form a disc or plate), and scalation (completely absent, or completely to partly scaled). In the following species accounts we indicate both spinous (non-segmented) and soft (segmented) elements in the second dorsal fin ray count. This is contrary to the practice of many previous authors who give only the total number of elements. Gobies are usually under 10 cm in length, and many are considerably smaller. In fact some of the world's smallest known vertebrates are included in this family. Gobies dwell in marine or estuarine habitats, but there are a number of purely freshwater species as well. For example, the genus *Sicydium* inhabits streams and rivers of Central America. Gobies are either bottom-dwellers or hover in the water column, a short distance above the bottom. They are associated with a variety of substrata, but many species exhibit an affinity for sand, silt, or mud bottoms. The depth range includes tidepools or shallow waters next to shore and offshore areas down to at least 150 m. Some members of the genus *Lythrypnus* are parasite "cleaners" and can be seen skittering about the surface of much larger fishes such as moray eels. This group includes some of the most brilliantly colored gobies. Because of their small size, gobies are usually not used for human food, but due to their huge numbers they are an important part of the reef's food chain. Gobies are sometimes kept as aquarium pets, and a number of species have been successfully bred in captivity. Gobies exhibit a wide range of feeding habits, but most species are carnivorous. Much of the diet is composed of crabs, shrimps, smaller crustaceans (such as copepods, amphipods, and ostracods), molluscs, annelids, polychaetes, formaninferans, sponges, and eggs of various invertebrates and fishes.

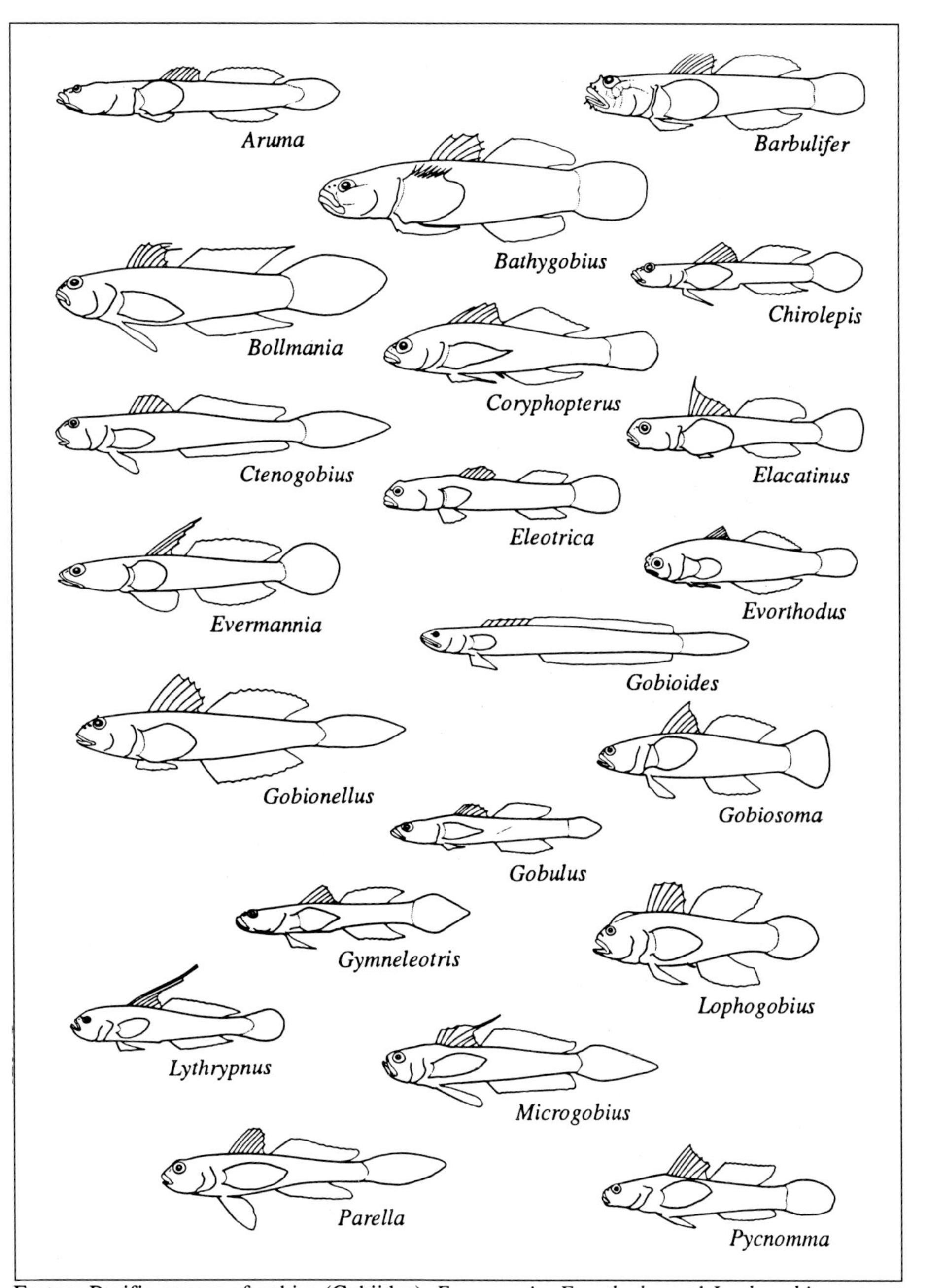

Eastern Pacific genera of gobies (Gobiidae). *Evermannia, Evorthodus* and *Lophogobius* are not treated in the text.

SLOW GOBY

Aruma histrio (Jordan, 1884)

Dorsal rays VII+I,11-12; anal rays I,9-12; pectoral rays 17-20 (rarely 16); pelvic disc present; *scales absent*; tongue bilobed; head without barbels; *brown with 6-7 white bars, these sometimes incomplete, or posterior 1-4 bars weak or missing*; a narrow white bar at pectoral fin base followed by a dark brown bar; fins dusky brown to whitish. Gulf of California; a secretive species that inhabits rocky crevices from tide pools to at least 14 m. Attains 6.5 cm.

PANTHER GOBY

Barbulifer pantherinus (Pellegrin, 1901)

Dorsal rays VII+I,10 (rarely 11); anal rays I,9 (rarely 8 or 10); pectoral rays 17-20 (usually 19); pelvic disc present; *scales absent*; tongue bilobed; *ventral surface of head with 6-12 short (usually less than half of eye diameter) barbels; no median barbel at tip of snout*; light tan or cream-colored, with numerous brown spots on side; *a prominent black bar or blotch on base of pectoral fin*; fins often with faint brown spotting. Gulf of California; inhabits rocky shores from the intertidal zone down to about 20 m. Reaches 5.2 cm. *Barbulifer* is characterized by the presence of whisker-like barbels on the lower jaw. Other species in the region include *B. ceuthoecus* Jordan and Gilbert from Panama (barbels short, equal to half of eye diameter or shorter; a median barbel at tip of snout), and *B. mexicanus* Hoese and Larson from the Gulf of California (barbels equal to or greater than eye diameter; 4-5 dark bars on side).

Barbulifer pantherinus ▲

Bathygobius andrei ▼

ESTUARY FRILLFIN

Bathygobius andrei (Sauvage, 1880)

Dorsal rays VI+I,9-10; anal rays I,8-9; pectoral rays 18-21, *the uppermost rays forming free filaments* (i.e. not connected with adjoining rays by a membrane); mid-lateral scales 36-41; predorsal scales not reaching orbits, remainder of head scaleless; pelvic disc present; overall *gray-brown with light streak on each scale; faint dark bars and blotches evident on side; fins mainly dusky*; juveniles are much lighter, with well-defined, blackish bars and blotches and a dark-spotted caudal fin. Costa Rica to Ecuador; inhabits tide pools and rocky shores of bays and estuaries. Grows to at least 13 cm.

Bathygobius andrei Juv.

GALAPAGOS FRILLFIN

Bathygobius lineatus (Jenyns, 1842)

Dorsal rays VI+I,9; anal rays I,8; pectoral rays 19-21, the *uppermost 2-4 rays forming free filaments* (i.e. not connected with adjoining rays by a membrane); midlateral scales 37-38; predorsal scales not reaching orbits, remainder of head scaleless; pelvic disc present; generally light brown with several diffuse brown saddles across back; *a clearly defined, midlateral, longitudinal row of alternating dark brown and white blotches*; lower portion of head dark brown with scattered white spots. Galapagos Islands and Peru; inhabits rocky shores; common in tide pools. To about 10 cm.

Bathygobius lineatus 5 cm specimen ▲ Underwater photograph ▼

Bathygobius ramosus ▼

PANAMIC FRILLFIN

Bathygobius ramosus Ginsburg, 1947

Dorsal rays VI+I,9; anal rays I,8-9; pectoral rays 17-21, the *uppermost rays forming free filaments* (i.e. not connected with adjoining rays by a membrane); midlateral scales 35-38; predorsal scales not reaching orbits, remainder of head scaleless; pelvic disc present; generally olive-green to brown, with lighter scale centers; often a row of dark spots along midside, and diffuse bars or saddle may be present on back; *caudal and second dorsal fin with wavy brown bands or spots*. Baja California to northern Peru; inhabits the rocky intertidal zone. Grows to 12 cm. *B. arundelii* (Garman) from Clipperton Island and *B. longipinnis* Ginsberg from Socorro are synonyms of *B. ramosus*.

Bollmannia chlamydes ▼

ORANGESPOT GOBY

Bollmannia chlamydes Jordan, 1889

Dorsal rays VII+I,13-14; anal rays I,12-13; pectoral rays 21-23; *scales large, 25-28 midlateral scales and 8 predorsal scales (extending forward to orbits); large scales also present on cheek and upper half of opercle*; pelvic disc present; generally pale tan or whitish, sometimes with faint yellowish-brown bars on side; lower edge of opercular membrane and isthmus dark brown; a diffuse brown spot at middle of caudal fin base; ovate orange spots on dorsal fin and uppermost part of caudal fin; pelvic fins blackish. Mexico to Peru; inhabits sand or mud bottoms; commonly caught by trawlers in Panama Bay at depths between 10-60 m. Maximum size 11-12 cm. Other species of *Bollmannia* in the region include *B. gomezi* Acero (Colombia), *B. longipinnis* Ginsburg (Gulf of California), *B. macropoma* Gilbert (Gulf of California), *B. marginalis* Ginsburg (Ecuador), *B. ocellata* Gilbert (Gulf of California), *B. pawneea* Ginsburg (Gulf of California to Panama), *B. stigmatura* Gilbert (Gulf of California to Costa Rica), and *B. umbrosa* Ginsburg (Panama).

COLOMBIAN GOBY
Bollmannia gomezi Acero, 1981

Dorsal rays VII+I,12; anal rays I,13; pectoral rays 24-26; scales large and deciduous, about 28 midlateral scales and 8-9 predorsal scales; pelvic disc present; dorsal spines somewhat elongate, the longest reaching to middle of second dorsal fin when depressed; generally light tan with faint orange-brown midlateral stripe; *second dorsal with 2 longitudinal rows of large orange spots; caudal fin with similar but slightly smaller spots; a black spot on base of caudal fin; upper lip with dark marginal band.* Panama and Colombia; inhabits sand and mud bottoms to at least 58 m. To 11 cm. This species is provisionally identified as *B. gomezi*. It differs from the original description of *gomezi*, however, in having the least caudal peduncle depth slightly greater than the eye diameter instead of less than the eye diameter.

Bollmannia gomezi ▲

Chirolepis minutillus Male ▲ Female ▼

Chirolepis zebra ▼

RUBBLE GOBY
Chirolepis minutillus Gilbert, 1892

Dorsal rays VII+I,11; anal rays I,10; pectoral rays 17-19; body robust; snout profile blunt, somewhat steep; *pelvic fins completely separate; head pores absent; scales present only on posterior half of body, 2-13 weakly ctenoid, embedded scales on side of caudal peduncle and 4 enlarged scales on base of caudal fin*; spawning and courting males with greatly elongate first dorsal spine; brown to orange-brown, with blackish anal fin; base of pectoral fin dark brown; male with light spots on head. Gulf of California; inhabits sand-rubble bottoms in 12-46 m. Attains 3.2 cm.

GECKO GOBY
Chirolepis zebra Ginsburg, 1938

Dorsal rays VII+I,9-10; anal rays I,8-9; pectoral rays 17-19; *scales present only on posterior half of body*, anteriormost scales small, cycloid, and embedded, becoming larger and ctenoid posteriorly; 4 ctenoid scales on caudal fin base, the upper and lower with excessively long ctenii; head pores absent; *white with 6 brown bars on body, and brown bands also present on head; caudal fin round with 4 brown crossbars*, including 1 at base. Gulf of California; inhabits sand and rubble among rocks in about 2-30 m. Attains 4.4 cm. *C. cuneata* Bussing (Gulf of California to Costa Rica) is similar but is easily separated by its unmarked caudal fin. Other species of *Chirolepis* in the region include *C. dialepta* Bussing (Isla del Coco), *C. lepidota* Findley (Malpelo Island), *C. tagus* Ginsburg (Galapagos), and 2 undescribed species from the Gulf of California.

REDLIGHT GOBY
Coryphopterus urospilus Ginsburg, 1938

Dorsal rays VI+I,8-9; anal rays I,8-9; pectoral rays 19-21 (usually 20); pelvic disc present; *midlateral scales 25-26*; head scaleless; *semitransparent whitish with about 5 horizontal rows of reddish orange to brown spots on side*; a dark brown spot just below base of middle caudal rays; a pair of red-orange to brown spots on base of pectoral fin. Baja California to Galapagos; inhabits sand-rubble fringe of rocky reefs or coral patches. Maximum size, 6.5 cm.

LANCETAIL GOBY
Ctenogobius sagittula (Günther, 1861)

Dorsal rays VI+I,12-13; anal rays I,13; pectoral rays 17; pelvic disc present; *midlateral scales 55-65; caudal fin lanceolate, much longer than head*; predorsal scales extending forward to level of middle of opercle, but *scales absent on mid-dorsal line, anterior to dorsal fin origin*; remainder of head scaleless; head pores terminating above preoperculum; an oblique row of papillae posteriorly on operculum; light tan with 4 dark brown blotches along middle of side, and dark brown spot at middle of caudal fin base; a brown stripe on cheek, and brown blotch just behind and above upper margin of gill cover; second dorsal and caudal fin with brown spots. California to Peru; inhabits sand or mud bottoms of shallow bays and estuaries. To at least 13.5 cm. *C. manglicola* (Jordan & Starks) is a similar species occurring from Mexico to Peru. It has fewer midlateral scales (31-35).

BANDED CLEANER GOBY
Elacatinus digueti Pellegrin, 1901

Dorsal rays VII+11; anal rays I,8-9; pectoral rays 18-20 (usually 19); pelvic disc present; *head and body scaleless*; about 10 pores on head; *head reddish pink to orange-red with 3-4 white bars; body translucent yellow with alternating brown (may be obscure) and white bars*. Gulf of California to Colombia; inhabits coral and rocky reefs; cleans parasites from the surface of other fishes. Grows to 3.2 cm.

INORNATE GOBY

Elacatinus inornatus Bussing, 1990

Dorsal rays VII+I,10 (rarely 9); anal rays I,9 (rarely 8); pectoral rays 18-20 (rarely 21); pelvic disc present; *head and body scaleless*; about 10 pores on head; *head orange-red or pinkish, with 3 white bars; an additional white bar at level of pectoral fin base; body translucent yellow with 5 broad, internal brown bars*. Costa Rica to Colombia; inhabits coral and rocky reefs in 6-28 m. To 3.3 cm.

SPOTBACK GOBY

Elacatinus janssi Bussing, 1982

Dorsal rays VII+I,9; anal rays I,8-9; pectoral rays 18; pelvic disc present; *head and body scaleless*; about 10 pores on head; *head yellowish with brown bar below eye, and brown bands and spots on cheek and opercle; body translucent pale yellow to whitish, with small brown spots on upper half*; body with 6-7 internal, wavy, brown bars. Costa Rica and Panama; inhabits reefs and rock outcrops surrounded by sand to at least 20 m depth. Size to 3 cm.

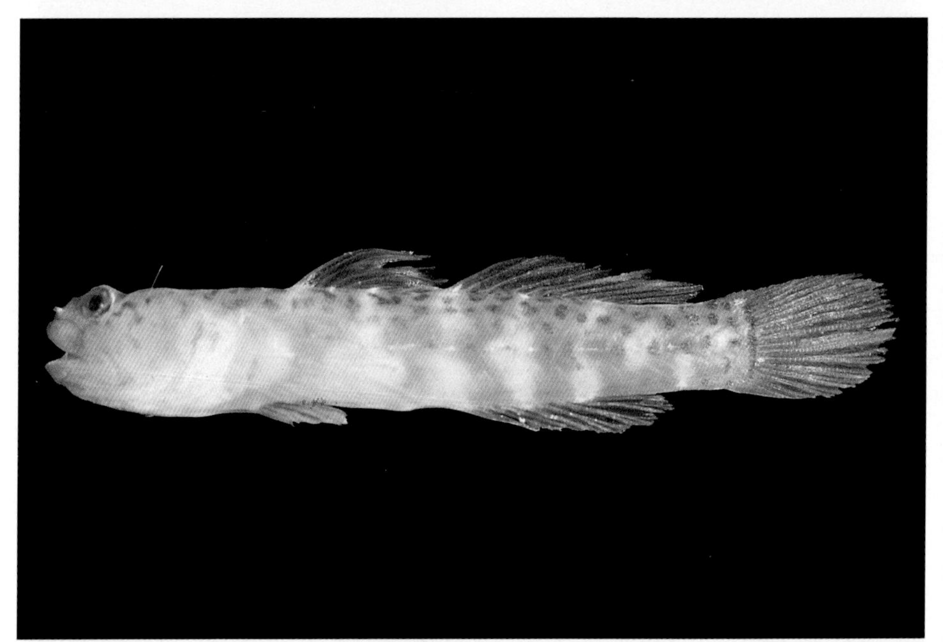

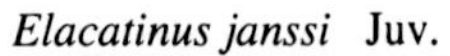

Elacatinus janssi Juv.

REDHEAD GOBY

Elacatinus punticulatus (Ginsburg, 1938)

Dorsal rays VII+I,11 (rarely 10 or 12); anal rays I,10 (rarely 9); pectoral rays 20-23; pelvic disc present; *head and body scaleless*; about 10 pores on head; *head pinkish red with bright red stripe behind each eye and on mid-interorbital*; body translucent pale yellowish with about 8 internal black blotches or spots along middle of side, and broad blackish internal stripe just above abdominal cavity. Gulf of California to Ecuador; inhabits rocky reefs in 1.5-6 m, frequently found with Club Urchins (*Eucidaris thouarsi*). Attains length of 4.4 cm.

WIDEBANDED CLEANER GOBY
Elacatinus sp.

Dorsal rays VII+I,9; anal rays I,9; pectoral rays 20-21; pelvic disc present; *head and body scaleless*; about 10 pores on head; males may develop filamentous prolongation of first dorsal spine; *head and anterior part of body red-orange, grading to dark brown posteriorly; head with 3 pink to whitish bars, and additional 9-10 white bars on body*. Gulf of California; inhabits rocky reefs, cleans parasites from other fishes (frequently moray eels). Reaches 3.2 cm. An additional member of the genus, *E. nesiotes* Bussing (not illustrated) is known from Isla del Coco and the Galapagos. It has about 10 brown, saddle-like bars on the side and the ventral portion of some of these bars is bifurcate.

CABLE'S GOBY
Eleotrica cableae Ginsburg, 1933

Dorsal rays VII+I,9-11; anal rays I,10; pectoral rays 18-19; *pelvic fins separate; scales absent, except for embedded basicaudal scales; head dorsally flattened; fringe of small cirri above rear portion of upper lip; 5 pores on occipital region, no preopercle pores*; brown to nearly white; head with small white spots; body with larger white spots and sometimes blotches or wavy vertical lines; males with midlateral row of double brown spots, and oblique dark stripes on dorsal fins; a distinctive, short, dark brown bar on lower part of gill cover, just behind cheek. Galapagos Islands; inhabits sand-rubble adjacent to rocky reefs. To 5.5 cm.

PERUVIAN EELGOBY
Gobioides peruanus (Steindachner, 1880)

Dorsal rays VIII,14-15; anal rays I,15; pectoral rays 18-19; body very elongate and compressed, tapering posteriorly; eye very small, interorbital space broad; mouth moderately large and oblique; *distinct from all other gobies treated here by its eel-like body and single dorsal fin*; pale tan with golden sheen on gill cover and side of body; upper back brown with ventrally directed, bar-like extensions reaching middle of side, these becoming obscure on posterior third of body. Honduras to Peru; inhabits brackish tidal rivers and fresh water, lives in muddy burrows. Attains 45 cm.

ESTUARY GOBY
Gobionellus microdon (Gilbert, 1891)

Dorsal rays VI-I,12; anal rays I,12-13; *midlateral scales 63-72*; scales ctenoid; head scaleless; pelvic disc present; *caudal fin lanceolate, much longer than head*; teeth conical; head pores extending to above posterior end of operculum; a vertical row of papillae posteriorly on operculum; light tan with silvery sheen on opercle and side of body; irregular, brown, saddle-like markings on back; first dorsal and caudal fin with small dark spots; second dorsal fin with fine, wavy, dark stripes. Gulf of California to Panama; inhabits estuaries and brackish mangrove creeks. To 13 cm. Two other *Gobionellus* are known from the region: *G. daguae* (Eigenmann) from Colombia, and *G. liolepis* (Meek & Hildebrand), ranging from Costa Rica to Ecuador.

SONORA GOBY
Gobiosoma chiquita (Jenkins & Evermann, 1889)

Dorsal rays VII+I,10 (rarely 9 or 11); anal rays I,9 (rarely 8 or 10); pectoral rays 18-21 (usually 19-20); pelvic disc present; *scales of middle of side extending forward from caudal fin base to pectoral fin axil, dorsally the scales extending forward to level between base of fourth and sixth dorsal spines; preopercular pores 3; mandibular papilla rows parallel*; first 2 dorsal spines of males typically prolonged; brown with small white spots on cheek, operculum, and body; a series of diffuse white bars (sometimes fragmented) on side of body; *diagnostic markings include 8-10 small, blackish spots along middle of side, and blackish patch just behind cheek*; dorsal fins and especially caudal fin with small white spots. Gulf of California; common on sand bottoms of tide pools and rocky reefs to 9-10 m. Grows to 6.5 cm.

Gobiosoma chiquita Male ▲ Female ▼

KNOBCHIN GOBY
Gobiosoma nudum (Meek & Hildebrand, 1928)

Dorsal rays VII+I,11-12; anal rays I,9; pectoral rays 18-20; pelvic disc present; *scales usually absent on head and body* (sometimes 1-4 scales in pectoral axil); tip of *snout with a small, median, fleshy protuberance; first dorsal spine of males not prolonged; chin with 2 small barbels*; side of head brown; body with alternating gray-brown and white bars; a midlateral row of small dark spots (in middle of each brown bar); first dorsal fin frequently with large black spot on basal half. Baja California to Panama; inhabits tide pools and shallow rocky reefs. Reaches 4.5 cm. Other species of *Gobiosoma* (not illustrated) from the eastern Pacific include *G. homochroma* (Ginsburg) (Panama; head distinctly depressed; scales of back reaching forward to just behind second dorsal fin origin); *G. etheostoma* (Jordan & Starks) (southern Mexico to Panama; scales of back reaching forward to first or second D spine; 4-5 longitudinal rows of black spots on side; anal rays I,10); *G. paradoxum* Günther (Sonora, Mexico to Ecuador; second dorsal fin rays usually I,11; usually an isolated patch of scales under pectoral fin); *G. hildebrandi* (Ginsburg) (Panama; body with well-defined, slightly oblique brown bars); and 3 undescribed species: "A" (Baja California and Gulf of California south to Chacala, Mexico; similar to *nudum*, but caudal peduncle usually with some scales, and no fleshy protuberance on tip of snout); "B" (southern Mexico to Panama; similar to *chiquita*, but scales of back extend only to origin of second dorsal fin, second dorsal rays I,9, anal rays I,8); "C" (Panama; scales of back reaching forward to near end of second dorsal fin; body with gray-brown bars).

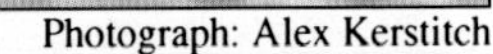
Photograph: Alex Kerstitch

SANDTOP GOBY
Gobulus hancocki Ginsburg, 1938

Dorsal rays VII+I,11; anal rays I,10; pectoral rays 16; *no pelvic frenum, pelvic fins fused to form flat, plate-like structure rather than typical disc*; scales entirely absent on head and body; *head noticeably flattened; head pores absent*; dark brown on lower half of head and body, mottled whitish above. Gulf of California to Panama; inhabits sand-rubble bottoms close to reefs in 1-20 m. Attains 2.5 cm. *G. crescentalis* (Gilbert) is similar in appearance, but has a greater proportion of white coloration on the head (usually the upper ½-⅔ of the head is white compared to upper ⅓ in *G. hancocki*) and grows to a larger size (6 cm). It occurs in the Gulf of California and outer coast of Baja California.

SPLITBANDED GOBY
Gymneleotris seminudus (Günther, 1864)

Dorsal rays VII+I,10 (rarely 9); anal rays I,9 (rarely 8); pectoral rays 18-19 (rarely 20); *pelvic fins separate, not forming disc*; head scaleless; *body with cycloid scales on posterior three-fourths, becoming reduced and embedded anteriorly; overall brown with striking pattern of oblique white bands on head, and white bars on body*; median fins with dark membranes and white rays. Baja California to Ecuador; inhabits algae-covered rocky or rubble reefs in 1-23 m. Maximum size 5 cm.

BLUEBANDED GOBY
Lythrypnus dalli (Gilbert, 1890)

Dorsal rays VI+I,15-19; anal rays I,11-16; pectoral rays 17-20; pelvic disc present; head and nape scaleless; body covered with small ctenoid scales; first 3-4 dorsal spines prolonged; *bright red, grading to orange posteriorly; a pair of narrow blue bars on head, and 3-5 similar bars on body.* Southern California to Ecuador; inhabits rocky slopes to at least 75 m, often seen with the Purple Sea Urchin (*Echinometra vanbrundti*). Attains 5.7 cm. According to Bussing (1990) *L. crinitus* Ginsburg (Galapagos) and *L. latifasciatus* Ginsburg (Southern California) are synonyms of *L. dalli*.

GILBERT'S GOBY
Lythrypnus gilberti (Heller & Snodgrass, 1903)

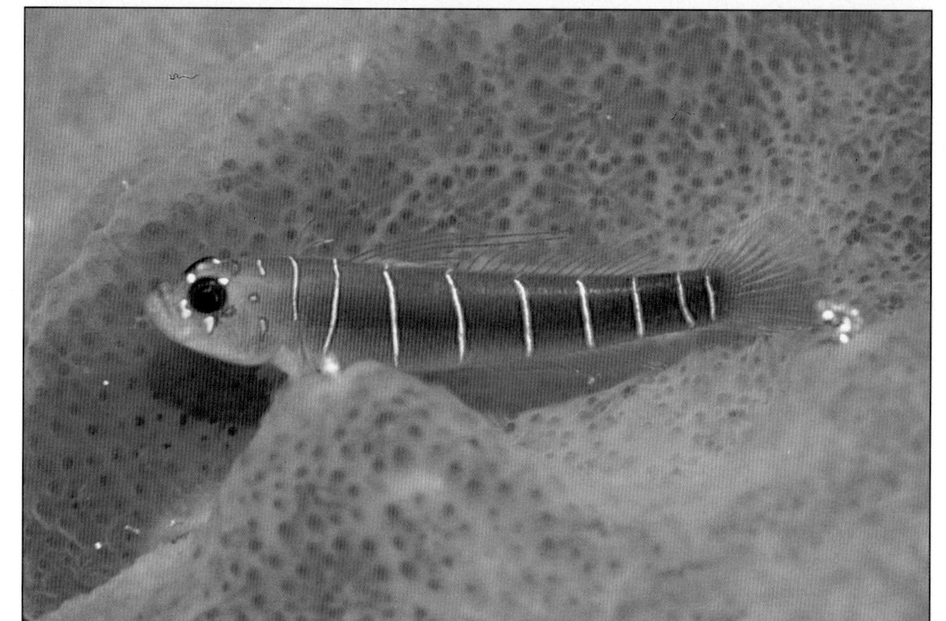

Dorsal rays VI+I,14-15; anal rays I,11; pectoral rays 18-19; pelvic disc present; head and nape scaleless; body covered with small ctenoid scales; first 3 dorsal spines prolonged; *head pinkish orange to red; body purplish brown with 8-9 narrow, dark-edged white bars.* Galapagos Islands; inhabits rocky reefs, often seen on sponges. To 4.5 cm.

ISLAND GOBY
Lythrypnus insularis Bussing, 1990

Dorsal rays VI+I,10-12 (usually 11); anal rays I,8-10 (usually 9); pectoral rays 17-20 (usually 18-19); pelvic disc present; head and nape scaleless; body covered with small ctenoid scales, about 29 between gill opening and base of caudal fin; first and second dorsal spines prolonged in males; upper half of head red, whitish below with red spots; 4 narrow blue bars dorsally on head; *body purplish brown with 13-15 narrow blue bars.* Revillagigedo Islands; inhabits rocky reefs in 3-18 m. To 2.5 cm. *L. solanensis* Acero (known from 1 specimen collected at Colombia in 57 m) is similar, but the body and head are red, and it has 21-22 pectoral rays.

GORGEOUS GOBY
Lythrypnus pulchellus Ginsburg, 1938

Dorsal rays VI+I,10-12 (usually 11); anal rays I,8-10 (usually 9); pectoral rays 17-21 (usually 17-19); pelvic disc present; head and nape scaleless; body covered with small ctenoid scales; first dorsal spine prolonged in males; *reddish orange head with 4 narrow whitish bars, and 11-14 (usually 13) similar bars on side of body; usually a thin, dark, vertical line in middle of each bar.* Outer coast of Baja California and Gulf of California; inhabits rocky reefs in 2-70 m. Reaches 4.5 cm. *L. lavenbergi* Bussing (known from 1 specimen collected in 137-146 m at Isla del Coco) is similar, but has 14 total rays in the second dorsal fin, and broader dark, vertical lines in the middle of each pale bar.

BANDED GOBY
Lythrypnus rhizophora (Heller & Snodgrass, 1903)

Dorsal rays VI+I,10-12 (usually 11); anal rays I,8-10 (usually 9); pectoral rays 17-20 (usually 17-18); pelvic disc present; head and nape scaleless; body covered with small ctenoid scales; first dorsal spine of males prolonged; *brownish orange; small brown spots on cheek and opercle; 3-4 bluish gray bars on head, and 9-14 similar narrow bars on body; caudal fin with vertical rows of small brown spots.* Isla del Coco and Galapagos; inhabits rocky reefs in 1-20 m. Grows to 3.4 cm. *L. alphigena* Bussing (known from 1 specimen collected at Isla del Coco in 91 m) is somewhat similar, but has white spots on the lower head, and the pattern of bars is reversed (narrow dark bars on head and body, rather than narrow pale bars).

Lythrypnus rhizophora ▲ *Microgobius cyclolepis* ▼

ROUNDSCALE GOBY
Microgobius cyclolepis Gilbert, 1891

Dorsal rays VII+I,14-16 (usually 15); anal rays I,15-16 (usually 15); pectoral rays 22-24; pelvic disc present; *midlateral scales 46-55; scales mainly cycloid, except patch of ctenoid scales beneath pectoral fin*; head, nape, breast, belly, and base of pectoral fin scaleless; *caudal fin 32-33 percent of standard length; dorsal spines 2-6 produced to form moderate filaments in both sexes; fleshy nuchal crest well developed in females, poorly developed in males*; generally pale tannish brown with cream-colored belly; an ovoid to crescent-shaped smudge or spot with bluish posterior border below origin of spinous dorsal fin. Baja California to Panama; inhabits flat sand bottoms in 10-35 m. Grows to 6.5 cm. *M. curtus* Ginsburg (Costa Rica to Peru) is similar, but has 62-78 midlateral scales, and yellowish spots on the body. Also, the dark blotch below the spinous dorsal is faint or absent.

EMBLEM GOBY

Microgobius emblematicus (Jordan & Gilbert, 1882)

Dorsal rays VII+I,15-17 (usually 16); anal rays I,15-17 (usually 16); pectoral rays 18-23 (usually 20-21); pelvic disc present; *midlateral scales 50-75; scales mainly cycloid, no ctenoid scales beneath pectoral fin*; head, nape, breast, belly, and pectoral fin base scaleless; *caudal fin 27-28 percent of standard length; dorsal spines 4-7 produced into elongate filaments in males; fleshy nuchal crest low and poorly developed in both sexes*; generally light brown with diffuse, dark, midlateral stripe; belly cream-colored; a dark smudge below origin of spinous dorsal fin; side of head with 2-3 orange stripes separated by blue stripes. Gulf of California to Panama; inhabits sand or mud broken-shell bottoms near shore. To 5 cm. *M. brevispinis* Ginsburg is similar, but usually has 17 segmented dorsal and anal rays, and males have a yellow-margined dark bar below the anterior dorsal spines (females with yellow bar above pectoral fins). Its range overlaps that of *M. emblematicus*.

ERECT GOBY

Microgobius erectus Ginsburg, 1938

Dorsal rays VII+I,13-14 (*usually 14*); anal rays I,13-14; pectoral rays 20-23 (usually 22-23); pelvic disc present; *midlateral scales about 35; scales mainly cycloid, except patch of ctenoid scales beneath pectoral fin*; head, nape, breast, belly, and pectoral fin base scaleless; *caudal fin lanceolate, about 45 percent of standard length; dorsal spines not produced into filaments in either sex; a low, fleshy nuchal crest in both sexes*; pale tannish brown with whitish belly, often a silvery sheen on opercle and side of body; sometimes a hint of 3-4 diffuse whitish bars on upper side. Gulf of California to Panama; common in the Gulf of Panama on mud and broken-shell bottoms in 15-30 m. Maximum size 6 cm.

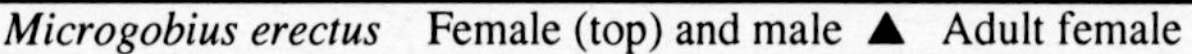

Microgobius erectus Female (top) and male ▲ Adult female ▼

MIRAFLORES GOBY
Microgobius miraflorensis Gilbert & Starks, 1904

Dorsal rays VII+I,15-17 (usually 16); anal rays I,15-17 (usually 16); pectoral rays 20-23 (usually 21-22); pelvic disc present; *midlateral scales 40-48; scales mainly ctenoid*; head, nape, breast, belly, and base of pectoral fin scaleless; *caudal fin lanceolate, 36-37 percent of standard length; dorsal spines 2-6 produced into filaments that may reach as far as caudal base in males, those of females only slightly produced; fleshy nuchal crest absent in both sexes*; light tan to nearly whitish; a narrow (wider in females) dark streak or bar below front of spinous dorsal fin, and several faint, narrow brown bars usually evident on side. Gulf of California to Colombia; inhabits sand, silt, or mud bottoms of bays, estuaries, and tidal creeks. Attains 5 cm. *M. tabogensis* Meek & Hildebrand (Baja California to Colombia) is similar, but both sexes have a diffuse spot (rather than narrow bar) below the spinous dorsal, and females have a well-developed nuchal crest. Also, the pelvic fins are dusky or darkly pigmented.

LUCRETIA'S GOBY
Parella lucretiae (Eigenmann & Eigenmann, 1888)

Dorsal rays VII-I,9; anal rays I,11; pectoral rays 17; pelvic disc present; scales relatively large and mainly ctenoid, midlateral scales 28; head scaleless; caudal fin lanceolate, much longer than head; head and back brown, whitish on sides; scales of body with narrow blackish margins; *5-6 diffuse blackish bars on side, and large diffuse blackish spot at base of caudal fin; a large, prominent black spot on upper half of pectoral fin base; lower margin of caudal and anal fins white.* Known only from Panama; inhabits shallow bays and estuaries. To at least 7 cm. *P. fusca* Ginsburg (Panama) is a poorly-known species that is similar in appearance, but is reported to have a more slender body (depth at anal origin 11-12 percent of standard length vs 16-17 percent for *P. lucretiae*).

DOUBLESTRIPE GOBY
Parella maxillaris Ginsburg, 1938

Dorsal rays VII+I,10-11; anal rays I,10-11; pectoral rays 19-20; pelvic disc present; scales relatively large and mainly ctenoid; midlateral scales 32; head scaleless; caudal fin lanceolate, longer than head; generally whitish, grading to brown on upper part of back and head; *a pair of narrow brown stripes originating on upper pectoral base and extending to middle of side below origin of second dorsal fin, the upper stripe continuing farther posteriorly; middle of side with 4-5 brown blotches and black, crescent-shaped spot at base of caudal fin*; upper margin of caudal fin white. Gulf of California to Panama; inhabits flat sand bottoms in 8-30 m. Grows to at least 5 cm.

SECRET GOBY
Pycnomma semisquamatum Rutter, 1904

Dorsal rays VII+I,10-11; anal rays I,9-10; *pelvic fins separate*; head and anterior part of body scaleless; *posterior part of body with larger ctenoid scales, anteriorly scales are smaller and have few or no ctenii*; caudal fin base with 4 enlarged scales; 3 *small pores present on top of each side of head*; generally mottled brownish; side of body with about 4-5 diffuse blackish, hourglass-shaped bars, and black bar at base of caudal fin; a pair of creamy yellow bars below spinous dorsal fin; head with white spots, and diagonal brown bars radiating from eye; *a small black spot at bottom of gill cover; median fins with white spots on rays, and white margin on both dorsal fins.* Gulf of California; inhabits crevices and ledges of weed-covered rocky reefs in 2-20 m. Reaches 6.3 cm.

Microgobius miraflorensis ▲

Parella lucretiae ▼

Parella maxillaris ▼

Pycnomma semisquamatum ▼

WORM GOBIES (GOBIOS GUSANOS)

FAMILY MICRODESMIDAE

Microdesmidae is divisible into two subfamilies, the Microdesminae (worm gobies) and the Ptereleotrinae (hover gobies). The latter group was formerly included in the goby family, but recent research indicates it is more closely allied to the worm gobies. Ultimately it may prove deserving of separate family status. These fishes have elongate bodies with small embedded cycloid scales (except ctenoid posteriorly on some hover gobies), no lateral line, and the lower jaw usually protrudes. The very slender wormfishes have a long-based dorsal fin that is continuous with X-XXVIII flexible spines and 28-66 soft rays, and an anal fin with 23-61 soft rays. Hover gobies usually have a divided dorsal fin, the first part containing IV-VI flexible spines. Microdesmids live over sand and mud bottoms, taking refuge in burrows when danger threatens. Most species are small, usually under 15 cm total length. They feed mainly on zooplankton and tiny benthic crustaceans. The subfamily Ptereleotrinae contains five genera and approximately 35 species. The eastern Pacific and western Atlantic fishes formerly classified as *Ioglossus* are now thought to belong to the mainly Indo-West Pacific *Ptereleotris*. The subfamily Microdesminae is composed of five genera with 19 species.

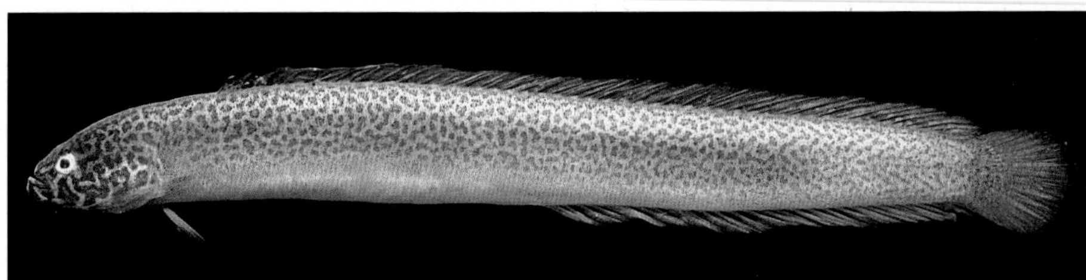

Cerdale ionthas ▲

Microdesmus dipus ▼

Ptereleotris sp. ▼

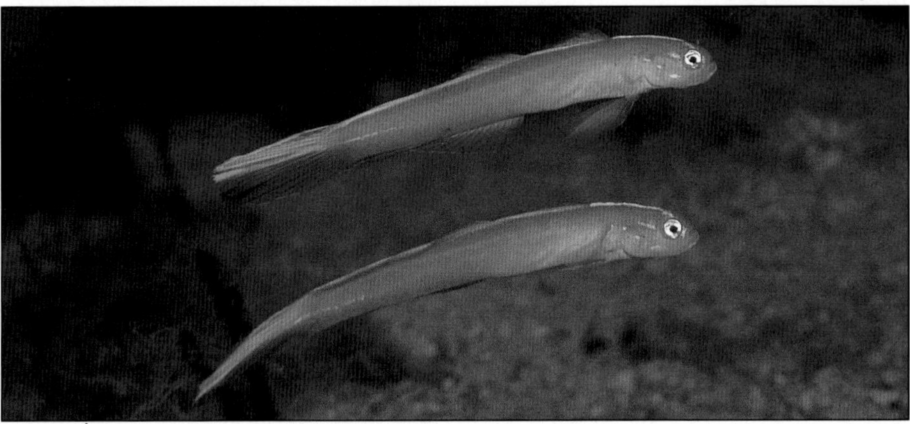

SPOTTED WORM GOBY

Cerdale ionthas Jordan & Gilbert, 1882

Dorsal rays XI-XIII,31-34 (usually XII,32-33), total dorsal elements 43-46 (usually 44-45); anal rays 27-31 (usually 29-31); pectoral rays 13-15 (usually 14); dorsal fin origin in front of level of pectoral fin tips; *body long and slender, somewhat eel-like; lower jaw protruding; overall whitish with heavy brown to blackish spotting and mottling.* Costa Rica to Colombia; occurs in very shallow water on rocky coasts, usually 1 m or less, including tide pools. Grows to 7.5 cm. Two additional species (not illustrated) occur in the region; both are plain brownish: *C. prolata* (Panama) has 51 and 37 dorsal and anal elements respectively, and *C. paludicola* (mangrove habitat; Costa Rica to Colombia) has the dorsal origin above or behind the level of the pectoral fin tips, and has a notch in the lower lip.

BANDED WORM GOBY

Microdesmus dipus Günther, 1864

Dorsal rays XVI-XVIII,36-41 (usually XVII,37-39), total dorsal elements 53-58 (usually 54-56); anal rays 32-37 (usually 33-36); pectoral rays 11-13 (usually 12); *body long and slender, somewhat eel-like; lower jaw protruding; overall whitish with brown, saddle-like markings on back, and numerous short brown bars on side.* Gulf of California to Panama; usually found on mud bottoms close to shore. To 8 cm. A similar species, *M. dorsipunctatus* (not illustrated) ranges from Baja California to Panama; it is mainly devoid of dark markings on the posterior half of the body, and usually has 16 dorsal spines, 36 anal rays, and 13 pectoral rays.

PACIFIC HOVER GOBY

Ptereleotris sp.

Dorsal rays VI+I,21-22; anal rays I,21-22; pectoral rays 21; pelvic rays I,14, *pelvic fins separate*; total gill rakers on first arch 26; head scaleless; *scales of body very small, appproximately 140-150 in midlateral row; scales ctenoid posteriorly on body, cycloid elsewhere*; generally pale gray or slightly bluish; scattered blue spots and/or lines on head; a silvery white to blue-white stripe from above eye to origin of dorsal fin; fins pale except lower part of caudal dusky reddish. Gulf of California to Panama; inhabits sand-rubble bottoms in depths between about 8-25 m. Reaches about 8 cm. This fish, which remains undescribed, is possibly divisible into 2 species, 1 that is endemic to the Gulf of California, and the other ranging outside the Gulf as far south as Panama.

SURGEONFISHES (CIRUJANOS, BARBEROS, SANGRADORES)

FAMILY ACANTHURIDAE

Surgeonfishes are one of the primary algal feeding groups on tropical reefs. Large grazing schools of *Prionurus* surgeons are frequently sighted in our area. Schooling behavior is an adaptation that serves to thwart the aggressive attacks of territorial bottom dwellers such as damselfishes. It is virtually impossible for the latter to repel the invading horde of surgeons. A few species of *Acanthurus* and most *Naso* (an Indo-West Pacific genus), that characteristically swim high above the bottom in areas of strong currents, feed on planktonic organisms. Their common name is derived from the sharp, scalpel-like spines located on the sides of the tail base. There is a single collapsible spine on each side in *Acanthurus* and *Ctenochaetus*. However, in *Prionurus* there are 3-6 fixed keel-like spines or knobs on each side. Other family characteristics include a deep, compressed body with the eye placed high on the head; a single unnotched dorsal fin with IV-IX spines and 19-33 rays; an anal fin with II or III spines (only *Naso* with II spines) and 18-28 rays; very small ctenoid scales; and a small mouth with close-set teeth which may be spatulate with denticulate edges, or numerous and comb-like with expanded incurved tips. Worldwide there are 77 species belonging to nine genera. The genera *Acanthurus* and *Ctenochaetus* were revised by Randall (1955 and 1956), and Smith (1966) presented a synopsis of eastern Pacific *Xesurus* (now *Prionurus*).

Chanco Surgeonfish (*Prionurus laticlavius*), Isla Pinzon, Galapagos

ACHILLES TANG
Acanthurus achilles Shaw, 1803

Dorsal rays IX,29-33; anal rays III,26-29; pectoral rays 16; body depth 1.7-1.9 in standard length; gill rakers on first arch 16-20; caudal fin emarginate; *overall bluish black with large, elliptical, orange patch posteriorly on body*; a white slash or short bar covering median portion of opercular membrane; caudal fin with broad orange bar and white posterior margin. Mainly islands of the tropical northern Pacific, including the Hawaiian Archipelago; in the eastern Pacific it has been seen near Cabo San Lucas (Baja California) and at offshore islands; usually inhabits shallow surge areas. To 22 cm.

Photograph: John Randall

WHITECHEEK SURGEONFISH
Acanthurus nigricans (Linnaeus, 1758)

Dorsal rays IX,28-31; anal rays III,26-29; pectoral rays 16; body depth 1.7-1.9 in standard length; gill rakers on first arch 17-19; caudal fin slightly emarginate; *overall blackish with large white blotch below eye*; narrow white band encircling mouth; a yellow band, becoming broader posteriorly, at base of dorsal and anal fins; caudal fin white with narrow yellow bar posteriorly. Mainly island areas of the tropical Pacific; in the eastern Pacific it is known from Baja California and offshore islands including the Revillagigedos, Isla del Coco, and the Galapagos; usually seen in shallow water of exposed coral reefs or rocky shores, but ranging to 45 m. Largest specimen 21.3 cm. *A. glaucoparieus* is a synonym.

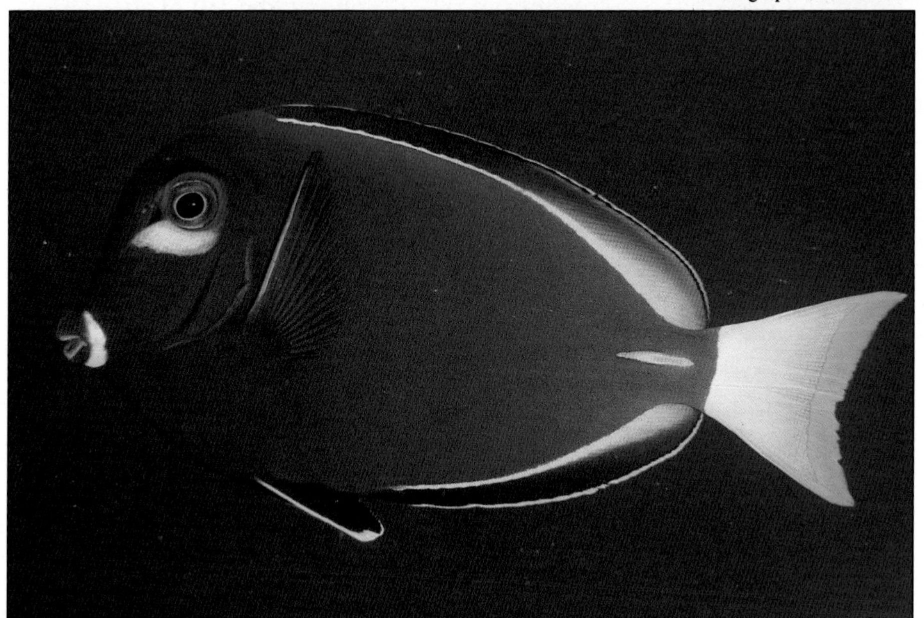

CONVICT SURGEONFISH
Acanthurus triostegus (Linnaeus, 1758)

Dorsal rays IX,22-26; anal rays III,19-22; pectoral rays 15-16; gill rakers on first arch 18-22; body depth 1.8-2.0 in standard length; caudal spine relatively small, about 10 in head; caudal fin slightly emarginate; *overall whitish to light greenish gray, with 6 narrow black bars on head and body.* Widespread in the tropical Indo-Pacific, from East Africa to the Americas; known in the eastern Pacific from the lower Gulf of California to Panama and the Galapagos; often seen in aggregations that graze on algae over rocky or coral reefs in 1-10 m; common around offshore islands; the young sometimes occur in tide pools. Maximum size 26.3 cm.

YELLOWFIN SURGEONFISH
Acanthurus xanthopterus Valenciennes, 1835

Dorsal rays IX,25-27; anal rays III,23-25; pectoral rays 16-17; gill rakers on first arch 16-24; body depth 1.9-2.3 in standard length; length of caudal spine 4.4-5.7 in head; caudal fin lunate; *purplish gray to brown, outer third of pectoral fin yellowish; can quickly assume a very pale gray coloration with numerous dark wavy lines on the side; yellow area around and in front of eye.* Widespread in the tropical Indo-Pacific, from East Africa to the Americas; in the eastern Pacific it ranges from the lower Gulf of California to Panama and the Galapagos; generally seen in sandy areas, often well above the bottom, usual depth range 12-20 m. Attains 56 cm; the largest species of *Acanthurus.*

Acanthurus xanthopterus Purple variety ▲ Dark brown variety ▼

BLUE-SPOTTED SURGEONFISH
Ctenochaetus marginatus (Valenciennes, 1835)

Dorsal rays VIII,27-28; anal rays III,25; pectoral rays 16-17; gill rakers on first arch 26-29; body depth 1.8-2.0 in standard length; teeth numerous, in a single row in jaws, movable, elongate, with tips expanded, incurved and denticulate on lateral margins; upper and lower teeth with 3-4 denticulations (including tip); adults with as many as 44 upper and 60 lower teeth; caudal fin lunate; *dark brown with numerous close-set, small blue spots covering head and body*; brightness of spots can be quickly intensified. Scattered localities in the tropical western and eastern Pacific; in the latter region known from Isla del Coco and Panama (Gulf of Chiriqui); inhabits rock and coral reefs in shallow wave-affected areas. To 22 cm. *C. cyanoguttatus* Randall is a synonym.

Ctenochaetus marginatus Freshly captured adult ▲ Live adult ▼

CHANCHO SURGEONFISH

Prionurus laticlavius (Valenciennes, 1846)

Dorsal rays VII-VIII,27-28; anal rays III-IV,23; this species and *P. punctatus* differ from other eastern Pacific surgeonfishes in lacking a single, scalpel-like spine on each side of the caudal peduncle, instead there are 3 bony knobs; *generally gray with bright yellow caudal fin*; a dark-edged, whitish bar on head behind eye; small juveniles are mainly yellow. Galapagos Islands and Panama; often occurs in large schools at depths between 3-15 m. Maximum size about 60 cm.

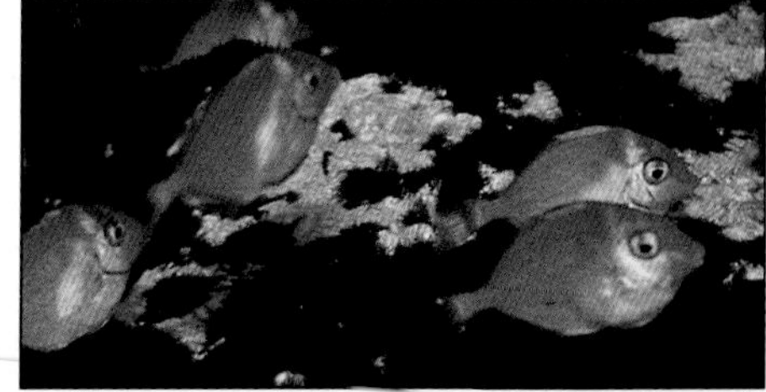

Prionurus laticlavius Adults ▼ Juvs. ▲

Prionurus punctatus Grazing school ▼

YELLOWTAIL SURGEONFISH

Prionurus punctatus Gill, 1862

Dorsal rays VII-VIII,25; anal rays III,23; caudal peduncle with 3 bony white knobs; *overall gray with numerous small black spots covering head and body; caudal fin bright yellow*; a dark-edged, whitish bar on head behind eye; juveniles with 2 color varieties: overall yellow, or similar to adults. Gulf of California to El Salvador; common along rocky shores in 6-12 m; usually seen in schools. Grows to 60 cm.

Prionurus punctatus Solitary adult ▼

Photograph: Mark Conlin

MOORISH IDOL (ÍDOLOS MOROS, MORICHAIRES)
FAMILY ZANCLIDAE

See discussion of the single species of the family given below.

Zanclus cornutus

MOORISH IDOL
Zanclus cornutus (Linnaeus, 1758)

Dorsal rays VII,40-43; anal rays III,33-35; pectoral rays 19; body very deep, 1.0-1.4 in standard length, and compressed; *third dorsal spine extremely long and filamentous, normally longer than standard length*; snout strongly pointed; mouth small, the teeth slender and slightly incurved; adults with a small bony projection in front of each eye (larger in males); white anteriorly, yellow posteriorly, with 2 broad black bars, 1 from nape to thorax and abdomen (enclosing eye), and the other curving across posterior body and entering dorsal and anal fins; a black-edged orange saddle on snout; caudal fin largely black. Indo-Pacific and tropical eastern Pacific; occurs over a large depth range from the shallows to at least 180 m; omnivorous, but feeds more on benthic animal life, such as sponges, than algae. *Z. canescens* (Linnaeus) is a synonym based on the large postlarval stage (to 8 cm).

CUTLASSFISHES (PECES CINTA, SABLES)
FAMILY TRICHIURIDAE

These fishes are relatives of the tunas and mackerels, but are easily recognized by their thin, ribbon-like body, and barracuda-like mouth that is equipped with large fangs. The dorsal and anal fins are extremely long-based and low; the anal fin is sometimes reduced to a series of projecting spinules. The pelvic fins are absent, or represented only by a scale-like spine. Scales are lacking. Trichiurids are voracious fish predators, occurring in all temperate and tropical seas. They generally inhabit deeper waters of the continental shelf and slope, but the single species treated below is frequently caught by trawlers in relatively shallow inshore waters. The family contains about 17 species in nine genera.

CUTLASSFISH
Trichiurus lepturus Linnaeus, 1758

Dorsal rays III,124-138; anal rays II,105-108; pectoral rays 10-12; *body very elongate and strongly compressed, ribbon-like, tapering to a fine point; mouth large with strong canine teeth, those at front of upper jaw fang-like; anal fin inconspicuous*, composed of 2 minute spines and numerous soft-rays reduced to minute spinules; overall silvery, bluish on upper back; pectoral fins dusky blackish. Circumglobal distribution in tropical and temperate seas; benthopelagic on continental shelf and slope to 350 m depth, but frequently enters shallow coastal waters. To 200 cm. *T. nitens* Garman is a synonym.

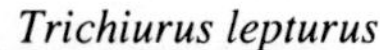

Trichiurus lepturus

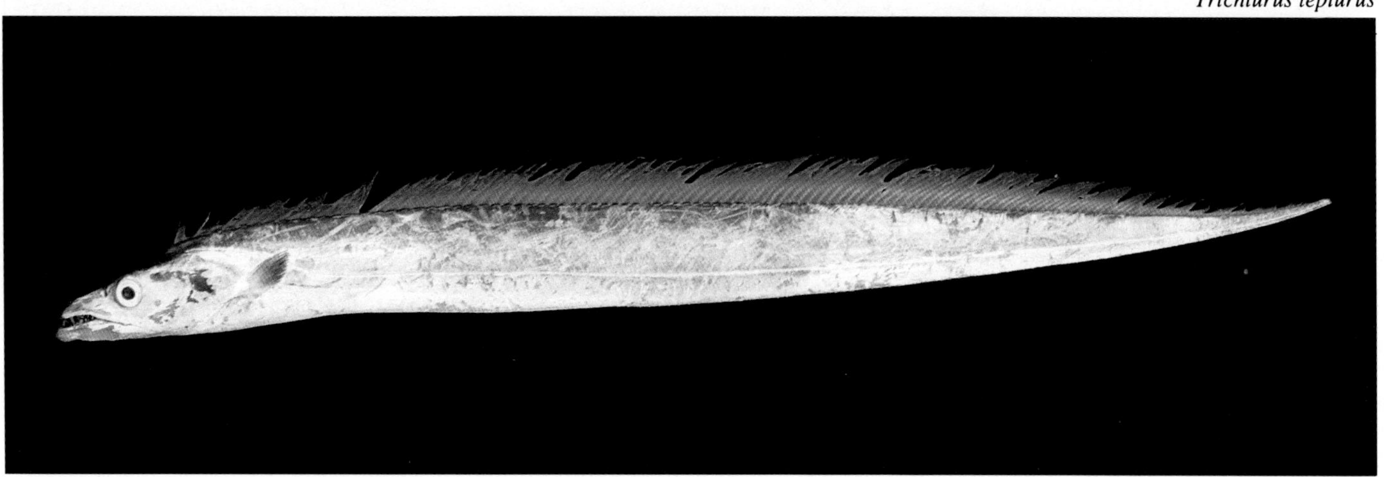

TUNAS AND MACKERELS (ATUNES, BONITOS, SIERRAS, CABALLAS, BARRILETES, MACARELAS)

FAMILY SCOMBRIDAE

The scombrids, or tunas, mackerels, and bonitos as they are commonly named, are well-known fishes that form the basis of valuable commercial fisheries in many regions. Worldwide the group contains 49 species. Over the past 10 years, world catches have generally fluctuated between about five and six million tons annually. These fishes are also highly prized by recreational anglers, and throughout much of the world they support important subsistence fisheries. All species are powerful swimmers, and some undertake extensive annual migrations. Scombrids are characterized by two dorsal fins that fold into grooves, and they have distinct finlets behind the second dorsal and anal fins; the pelvic fins have six rays and are situated below the pectoral fins; the scales are small and cycloid; the slender caudal peduncle has at least two small keels on each side, sometimes with a larger keel in between. Unlike typical "cold-blooded" fishes, some tunas have body temperatures several degrees warmer than the surrounding sea.

WAHOO

Acanthocybium solandri (Cuvier, 1831)
(Plate XV-1, p. 283)

Dorsal rays XXIII to XXVII-12 to 16 + 8 or 9 finlets; anal rays 12 to 14 + 9 finlets; body covered with small scales; no anterior corselet developed; iridescent bluish green on back; sides silvery with 24-30 cobalt-blue bars, some doubled or Y-shaped; a large, elongate tuna with a very long snout; teeth strong, triangular, compressed and finely serrate; *distinguished by its elongate shape, numerous spines in the first dorsal fin, and banded pattern.* All tropical seas; inhabits oceanic waters, generally well offshore. Reaches 210 cm; all-tackle world record 67.6 kg.

Sierra Mackerel (*Scomberomorus sierra*), Isla Bartolomé, Galapagos

BULLET MACKEREL
Auxis rochei Risso, 1810
(Plate XV-11, p. 283)

Two well-separated dorsal fins, the first with 10 or 11 tall spines; second dorsal fin followed by 8 finlets; pectoral fins short, not reaching a vertical line from anterior margin of scaleless area on back; anal fin followed by 7 finlets; body scaleless, except for the well-developed corselet; back dark bluish; *15 or more fairly broad, nearly vertical dark bars in scaleless area above lateral line*; belly white. Worldwide in tropical and subtropical seas; inhabits coastal and oceanic waters, forming large schools. Maximum size 50 cm.

FRIGATE MACKEREL
Auxis thazard (Lacepède, 1800)
(Plate XV-5, p. 283)

Two well-separated dorsal fins, the first with 10-12 spines; second dorsal fin followed by 8 finlets; pectoral fins short, not reaching a vertical line from anterior margin of scaleless area on back; anal fin followed by 7 finlets; body scaleless except for the well-developed corselet; back dark bluish; *15 or more narrow, oblique to nearly horizontal, dark, wavy lines in scaleless area above lateral line; belly white; similar in color pattern to* Euthynnus lineatus *(see below), but differs by having a much wider space between the dorsal fins, a lower spinous dorsal fin, and more slender shape.* Worldwide in tropical and subtropical seas; inhabits coastal and oceanic waters, sometimes forming large schools. To 58 cm and 4.5 kg.

BLACK SKIPJACK TUNA
Euthynnus lineatus Kishinouye, 1920
(Plate XV-8, p. 283)

Dorsal fins close together, the first with 10-15 spines, the anterior spines much higher than those posteriorly, giving the fin a strongly concave outline; second dorsal fin much lower than first, and followed by 8-10 finlets; pectoral fins short, reaching to about level of middle spines of first dorsal fin; dark blue on back, with *a patch of wavy black stripes on back, extending forward to below anterior part of first dorsal fin; lower sides and belly silvery white; several dark spots usually present on lower sides between pelvic and pectoral fins.* Eastern Pacific Ocean from California to Peru, including Galapagos Archipelago and other offshore islands; also 2 stray specimens captured in the Hawaiian Islands; forms schools in coastal waters and near offshore reefs. To 100 cm; all-tackle world record (Clarion Island) 9.12 kg.

SKIPJACK TUNA
Katsuwonus pelamis (Linnaeus, 1775)
(Plate XV-4, p. 283)

Dorsal fins close together, the first with 14-16 spines, the anterior ones relatively tall, giving fin a falcate outline; second dorsal fin falcate, shorter than first dorsal, followed by 7-9 finlets; pectoral rays 26-27; gill rakers numerous, 53-63 on first gill arch; body scaleless except for corselet and lateral line; dark purplish blue on back; *lower sides and belly silvery, with 4-6 conspicuous wavy stripes or discontinuous lines of dark blotches.* Circumglobal in seas warmer than about 15°C; makes up about 40 percent of world tuna catch. Attains 110 cm; all-tackle world record 18.93 kg.

ORIENTAL BONITO
Sarda orientalis (Temminck & Schlegel, 1844)
(Plate XV-7, p. 283)

Dorsal fins close together, the first without prolonged spines anteriorly, its outline sloping in a straight line posteriorly, with 17-19 spines; second dorsal fin lower than first with 15-16 rays followed by 7-8 finlets; anal fin with 14-16 rays followed by 5-6 finlets; pectoral rays 23-26; gill rakers on first arch 8-13; lateral line wavy; body entirely covered with minute scales except on the well-developed corselet; *caudal peduncle with a prominent lateral keel between 2 smaller keels on each side; back and upper sides steel-blue, with 5-11 dark, slightly oblique stripes*; lower sides and belly silvery; first dorsal fin entirely black. Indo-Pacific and eastern Pacific. Grows to 102 cm; all-tackle world record 10.65 kg.

CHUB MACKEREL
Scomber japonicus Houttuyn, 1782
(Plate XV-2, p. 283)

Body elongate and rounded; dorsal fins well-separated, the first triangular with 8-10 spines; second dorsal and anal fins with 12 rays, followed by 5 finlets; pectoral fins very short with 18-21 rays; gill rakers shorter than gill filaments, 25-35 on lower limb of first arch; *front and hind margins of eye covered by adipose eyelid; scales behind head and around pectoral fins larger and more conspicuous than those covering rest of body, but no well-developed corselet; 2 small keels on each side of caudal peduncle; generally silver, back with blackish, oblique wavy lines.* Circumglobal mainly in subtropical and temperate seas; in the eastern Pacific it ranges from Alaska to the Gulf of California, and from Panama and the Galapagos Archipelago to Chile; occurs in schools and undergoes extensive summer migrations into temperate regions. Feeds on plankton, fishes, crustaceans, and squids. Attains 50 cm.

SIERRA MACKEREL
Scomberomorus sierra Jordan & Starks, 1895
(Plate XV-10, p. 283)

Dorsal fins close together, the first relatively low with 15-18 spines; second dorsal higher than first, with 16-19 rays followed by 7-10 finlets; anal fin similar in shape and height to second dorsal, with 16-21 rays followed by 7-10 finlets; gill rakers on first arch 2-4+9-14; lateral line gradually curving down toward caudal peduncle; bronze-green to greenish blue on back, silvery white on lower sides and belly; *series of yellow-brown spots on sides; anterior half of first dorsal fin jet black*, posterior half white; pectoral fins dusky. Southern California to Peru, including Galapagos and other offshore islands; occurs in schools in coastal seas. Feeds on small fishes, especially anchovies and herrings. Attains 100 cm and at least 5.4 kg.

YELLOWFIN TUNA
Thunnus albacares (Bonnaterre, 1788)
(Plate XV-3, p. 283)

Dorsal fins close together, the first with prolonged spines anteriorly, giving a strongly concave outline; first dorsal with 13 or 14 spines; *second dorsal and anal fins becoming extremely tall in large specimens, well over 20 percent of fork length*; 7-10 finlets behind second dorsal and anal fins; *pectoral fins moderately long, usually reaching beyond second dorsal fin origin, but not beyond end of its base, usually 22-31 percent of fork length*; body with very small scales; corselet of larger scales present, but not very distinct; back metallic dark blue, becoming yellow to silver on sides and belly; belly may be crossed by about 20 broken, nearly vertical whitish lines; *dorsal and anal fins, including finlets, bright yellow, the finlets with a narrow black border.* Worldwide in tropical and subtropical seas; an important commercial species. Reaches 210 cm; all-tackle world record 176.4 kg

BIGEYE TUNA
Thunnus obesus (Lowe, 1839)
(Plate XV-9, p. 283)

Dorsal fins close together, the first with prolonged spines anteriorly, giving a strongly concave outline; first dorsal with 13-15 spines; second dorsal and anal fins falcate, each followed by 8-10 finlets; *pectoral fins moderately long, 22-31 percent of fork length*, extending to end of first dorsal fin or beyond; very small scales on body; corselet of larger scales well-developed, but not conspicuous; back metallic dark blue, lower sides and belly whitish; a lateral, iridescent blue band running along sides of live fish; *first dorsal fin yellow; second dorsal and anal fins pale yellow; finlets bright yellow, edged with black*. Circumglobal in tropical and temperate seas; an important commercial species. Maximum size, 240 cm; all-tackle world record 197.2 kg.

NORTHERN BLUEFIN TUNA
Thunnus thynnus (Linnaeus, 1758)
(Plate XV-6, p. 283)

Dorsal fins close together, the first with prolonged spines anteriorly; first dorsal with 11-14 spines; second dorsal and anal fins falcate, each followed by 7-10 finlets; *pectoral fins very short, 16-21 percent of fork length*; small scales on body; corselet of larger scales developed, but indistinct; back metallic dark blue; lower sides and belly silvery white; first dorsal fin yellowish or bluish, second dorsal reddish brown; *anal fin and finlets dusky yellow, edged with black*; Worldwide in tropical and temperate seas; the Pacific population is sometimes recognized as a separate subspecies, *T. thynnus orientalis*; an important commercial species. Attains 300 cm; common to 200 cm; all-tackle world record 679 kg (Nova Scotia).

SWORDFISHES (PEZ ESPADA)
FAMILY XIPHIIDAE

This family contains a single species, the swordfish, that is closely allied to the billfish family Istiophoridae. The main differences are its lack of jaw teeth, absence of pelvic fins, no scales, and only one keel on the side of the tail base. The swordfish is mainly a warm-water species that migrates to temperate and cold waters for feeding in summer, and back to warm waters in autumn for spawning and overwintering. Like its billfish cousins, the swordfish uses its beak to stun prey that consists of a variety of fishes, squids, and cuttlefishes. Major sportfishing areas for trolling are located off the eastern United States, western central America, and off eastern Australia and New Zealand.

SWORDFISH
Xiphias gladius Linnaeus, 1758
(Plate XIV-1, p. 281)

Bill extremely long, its cross-section flat; no teeth in jaws; adults with a single median keel on each side of caudal fin base; a very tall, short-based dorsal fin with pointed apex; first dorsal rays 34-49; second dorsal and second anal fins very small, second dorsal rays 4-6; first anal fin moderately elevated, falcate, with 13-14 rays; second anal rays 3-4; pelvic fins and pelvic girdle absent; pectoral fins situated low on the sides, with 16-18 rays; blackish brown on back, fading to light brown or silvery on sides; first dorsal fin dark blackish brown; other fins brown or blackish brown. Circumglobal in tropical, temperate, and adjacent cold seas. Maximum size 450 cm; all-tackle world record 100.24 kg.

BILLFISHES (MERLINES, PECES VELA)
FAMILY ISTIOPHORIDAE

This family is well known to gamefishermen, and contains the marlins, spearfishes, and sailfishes. These groups are typified by a combination of features including a long, bill-like snout composed of the elongated premaxilla and nasal bones; elongate, narrow pelvic fins with three or fewer rays; jaws with teeth; caudal peduncle of adults with two keels on each side; and dorsal fin, with a very long base, that folds into a groove. Billfishes are primarily oceanic, epipelagic animals that inhabit tropical and temperate seas, periodically entering colder waters. They are active, voracious predators that use the extended "bill" to attack and stun prey, generally fishes and cephalopods. Because of their large size and renowned fighting ability, billfishes are highly prized by anglers. However, the flesh is not particularly tasty. Charter-fishing boat operators are encouraged to promote tagging and release of captured fish rather than participating in their needless slaughter.

INDO-PACIFIC SAILFISH
Istiophorus platypterus (Shaw & Nodder, 1791)
(Plate XIV-2, p. 281)

A tall, long-based, sail-like dorsal fin with 42-49 rays, and small second dorsal fin with 6 or 7 rays; 2 anal fins, the first with 12-17 rays, and the second with 6-7 rays; pectoral rays 18-20; pelvic fins extremely long, almost reaching anus, depressible into a groove; dark blue dorsally, light blue blotched with brown laterally, and silvery white ventrally; *about 20 rows of vertical bars on sides, each composed of many light blue, round spots*; membrane of first dorsal fin dark blue or blackish blue, with scattered, small black spots; remaining fins blackish brown to dark blue. Circumglobal in tropical and temperate seas; undergoes extensive seasonal spawning migration. Grows to 360 cm; all-tackle world record 100.24 kg.

BLACK MARLIN
Makaira indica (Cuvier, 1832)
(Plate XIV-3, p. 281)

Anterior dorsal rays elevated into triangular peak, remainder of fin very low; height of anterior lobe of first dorsal fin smaller than greatest body depth; first dorsal fin rays 34-43; second dorsal fin rays 5-7; first anal fin rays 10-14; second anal fin rays 6-7; pectoral rays 12-20; *pectoral fins rigid and cannot be folded close to body*; pelvic fins filamentous, shorter than pectoral fins; dark blue dorsally and silvery white ventrally; usually no blotches or dark bars on body in adults; first dorsal fin blackish to dark blue; other fins dark brown tinged with blue. Tropical and temperate Indo-Pacific and eastern Atlantic. Feeds mainly on various tunas and other pelagic fishes. To 500 cm; all-tackle world record 707.61 kg.

INDO-PACIFIC BLUE MARLIN
Makaira mazara (Jordan & Snyder, 1901)
(Plate XIV-4, p. 281)

Anterior dorsal rays elevated into triangular peak, remainder of fin very low; height of anterior lobe of first dorsal fin less than greatest body depth; first dorsal fin rays 40-45; second dorsal fin rays 6-7; pectoral fin rays 20-23; *pectoral rays not rigid, and can be folded against side of body*; pelvic fins filamentous, shorter than pectoral fins; blue-black dorsally and silvery white ventrally with *about 15 cobalt-colored rows of round spots or narrow bars*; first dorsal fin blackish or dark blue; other fins blackish brown, sometimes tinged with dark blue. Circumglobal, in tropical and temperate seas; undergoes seasonal north-south migrations. Attains 500 cm; all-tackle world record 498 kg.

SHORTBILL SPEARFISH

Tetrapturus angustirostris Tanaka, 1915
(Plate XIV-5, p. 281)

Bill very short, usually less than 15 percent of total length; anterior dorsal rays elevated into triangular peak, most of remaining part of fin relatively high; height of anterior lobe of first dorsal fin exceeding greatest body depth; first dorsal fin rays 45-50; second dorsal fin rays 6-7; first anal fin rays 12-15; second anal fin rays 6-8; pectoral rays 17-19; pelvic fins slender, filamentous, about twice the length of the pectorals; dark blue dorsally, blue blotched with brown laterally, and silvery white ventrally, without dots or bars; first dorsal fin dark blue; remaining fins dark brown. Tropical and temperate Indo-Pacific, entering eastern Atlantic (via Cape of Good Hope), but not spawning there. To 200 cm.

STRIPED MARLIN

Tetrapturus audax (Philippi, 1887)
(Plate XIV-6, p. 281)

Anterior dorsal rays elevated into triangular peak, remainder of fin very low; *height of anterior lobe of first dorsal fin about equal to greatest body depth*; first dorsal fin rays 37-42; second dorsal fin rays 5-6; first anal fin rays 13-18; second anal fin rays 5-6; pectoral fin rays 18-22; pelvic fins long and slender, about equal to or slightly shorter than pectoral fins in large specimens, and slightly longer than pectorals in small individuals; blue-black dorsally and silvery white ventrally, with *about 15 cobalt-colored bars or vertical rows of spots*; first dorsal fin dark blue; other fins dark brown, sometimes tinged with blue. Tropical and temperate seas of the Indo-Pacific region. Reaches 420 cm; all-tackle world record 498.95 kg.

Indo-Pacific Sailfish (*Istiophorus platypterus*), Gulf of California, Mexico

Photograph: Mark Conlin

PLATE XIV

1 **SWORDFISH** (*Xiphias gladius*)

2 **INDO-PACIFIC SAILFISH** (*Istiophorus platypterus*)

3 **BLACK MARLIN** (*Makaira indica*)

4 **INDO-PACIFIC BLUE MARLIN** (*Makaira mazara*)

5 **SHORTBILL SPEARFISH** (*Tetrapterus angustirostris*)

6 **STRIPED MARLIN** (*Tetrapterus audax*)

1
2
3
4
5
6

PLATE XV

1 **WAHOO** (*Acanthocybium solandri*)

2 **CHUB MACKEREL** (*Scomber japonicus*)

3 **YELLOWFIN TUNA** (*Thunnus albacares*)

4 **SKIPJACK TUNA** (*Katsuwonus pelamis*)

5 **FRIGATE MACKEREL** (*Auxis thazard*)

6 **NORTHERN BLUEFIN TUNA** (*Thunnus thynnus*)

7 **ORIENTAL BONITO** (*Sarda orientalis*)

8 **BLACK SKIPJACK TUNA** (*Euthynnus lineatus*)

9 **BIGEYE TUNA** (*Thunnus obesus*)

10 **SIERRA MACKEREL** (*Scomberomorus sierra*)

11 **BULLET MACKEREL** (*Auxis rochei*)

1
2
3
4
5
6
7
8
9
10
11

DRIFTFISHES (PECES MEDUSA, PECES AZULES)
FAMILY NOMEIDAE

This family is generally not associated with reefs or inshore habitats; rather they are inhabitants of the high seas, or on the bottom in deep water on the outer continental shelf. Our coverage includes two examples because the juveniles are found swimming alongside or among the tentacles of pelagic coelenterates (jellyfish and siphonophores) near reef areas. Evidently they utilize the protection of their host (in much the same fashion as reef-dwelling anemonefishes of the Indo-West Pacific) until large enough to fend for themselves. General features include an ovate to elongate, laterally compressed body; a dorsal fin that is deeply notched or divided into separate portions, and has the longest spine at least equal in height to the longest soft dorsal ray; and the second dorsal and anal fin bases of equal length. Both the first dorsal and pelvic fins fold into grooves at their bases. Distribution is worldwide in tropical and subtropical seas. The family contains three genera and about 15 species. The group was reviewed by Haedrich (1972).

Nomeus gronovii [Photograph: Alex Kerstitch] ▲ *Psenes cyanophrys* ▼

MAN-OF-WAR FISH
Nomeus gronovii Gmelin, 1789

Dorsal rays IX-XII+ I,24-28; anal rays I or II,24-29; pectoral rays 19-24; pelvic fins of juvenile greatly enlarged; gill rakers on first arch 8+15-19; *juveniles silvery with black bars and spots, pelvic fins black with white blotches*; adults uniformly dark brown. Worldwide in warm temperate and tropical seas; inhabits surface layers of the high seas; young are associated with drifting Man-of-War siphonophores (*Physalia*). Reaches 39 cm.

FRECKLED DRIFTFISH
Psenes cyanophrys Valenciennes, 1833

Dorsal rays IX-XI,24-28; anal rays III,24-28; pectoral rays 17-20; pelvic fins present; gill rakers on first arch 8-9 + 20; snout blunt; second dorsal and anal fins higher than spinous dorsal; caudal fin forked with somewhat rounded lobes; scales very tiny; *young are translucent whitish with silvery head and abdomen*; adults yellowish with dark longitudinal lines on side. Worldwide in temperate and tropical seas; inhabits surface layers of the high seas; young are associated with drifting jellyfish and floating *Sargassum*. Grows to at least 15 cm. The identification of the illustrated species is problematical, as no specimens were collected. We provisionally identlfy it as *P. cyanophrys*. According to Haedrich (1967) *P. pacificus* Meek & Hildebrand, described from Panama Bay, is a synonym of *P. cyanophrys*. The illustrated fish were approximately 5 cm total length, and were photographed in association with an unidentified jellyfish at the Perlas Islands, Panama.

BUTTERFISHES (PALOMETAS, PAMPANITAS, PECES MANTECA)

FAMILY STROMATEIDAE

Stromateids are distinctive fishes with a relatively deep body, long-based dorsal and anal fins each usually with more than 30 rays, and no pelvic fins. They frequent continental shorelines of North and South America, West Africa, and southern Asia. The family contains about 13 species in three genera. These fishes are good eating, and are often seen in fish markets throughout the region. The family was reviewed by Haedrich (1967).

LONG-FINNED BUTTERFISH
Peprilus medius (Peters, 1869)

Dorsal rays III,42-48; anal rays II,38-46; pectoral rays 22-23; pelvic fins absent; caudal fin strongly forked; total gill rakers on first arch 23-27; *body depth 1.6-2.1 in standard length; dorsal and particularly anal fin long and falcate*; overall silvery white. Gulf of California to Peru; frequently captured by trawlers in the Gulf of Panama in depths of 10-40 m. Grows to at least 18 cm.

SHORT-FINNED BUTTERFISH
Peprilus snyderi Gilbert & Starks, 1904

Dorsal rays III,41-47; anal rays III,41-42; pectoral rays 22-23; pelvic fins absent; caudal fin strongly forked; total gill rakers on first arch 19; *body ovate, usually not as deep as in* P. medius, *the depth 1.9-2.7 in standard length; dorsal and anal fins moderately falcate and shorter than in* P. medius; silvery white to bluish on back. Baja California to Peru; inhabits soft bottoms of coastal areas, including river mouths. Attains 25 cm.

FLOUNDERS (LENGUADOS, LENGUADOS IZQUIERDOS)

FAMILY BOTHIDAE

Flounders are recognized by their flattened shape, with both eyes located on the left side of the body. The dorsal fin originates either above or ahead of the eye, and dorsal and anal fins both contain numerous segmented rays and are separated from the caudal fin. They typically inhabit soft-bottom areas, and are frequently a common component of trawler catches. Their coloration blends remarkably well with the bottom, and they possess the ability to rapidly change their coloration to match the substratum. The diet of flounders consists mainly of fishes and crustaceans. Worldwide the family contains about 215 species in 37 genera. Many of the eastern Pacific species occur in deep offshore waters. We include only a representative sample of the species that are commonly captured by trawlers, or that frequent inshore areas.

STARRY FLOUNDER
Bothus constellatus (Jordan, 1889)

Dorsal rays 85-95; anal rays about 65; pectoral rays 12; body strongly rounded; body depth about 1.5-1.8 in standard length; lateral-line scales about 75; *lateral line strongly arched above pectoral fin; eye diameter much less than interorbital width; snout profile with slight notch*; light to dark brown, with numerous stellate, pale (often blue) spots, some of these with darker centers or grouped together to form small rings. Baja California to Peru; inhabits flat sandy bottoms. Grows to at least 20 cm.

LEOPARD FLOUNDER
Bothus leopardinus (Günther, 1862)

Dorsal rays 86-92; anal rays 64-70; pectoral rays 10; body slightly less rounded than *B. constellatus*; body depth about 1.9 in standard length; lateral-line scales about 80; *lateral line strongly arched above pectoral fin; eye diameter about equal to or only slightly less than interorbital width; snout profile distinctly notched*; generally brown with pale blotches and numerous ocellated spots. Gulf of California to Panama; inhabits flat sandy bottoms. Reaches at least 15 cm.

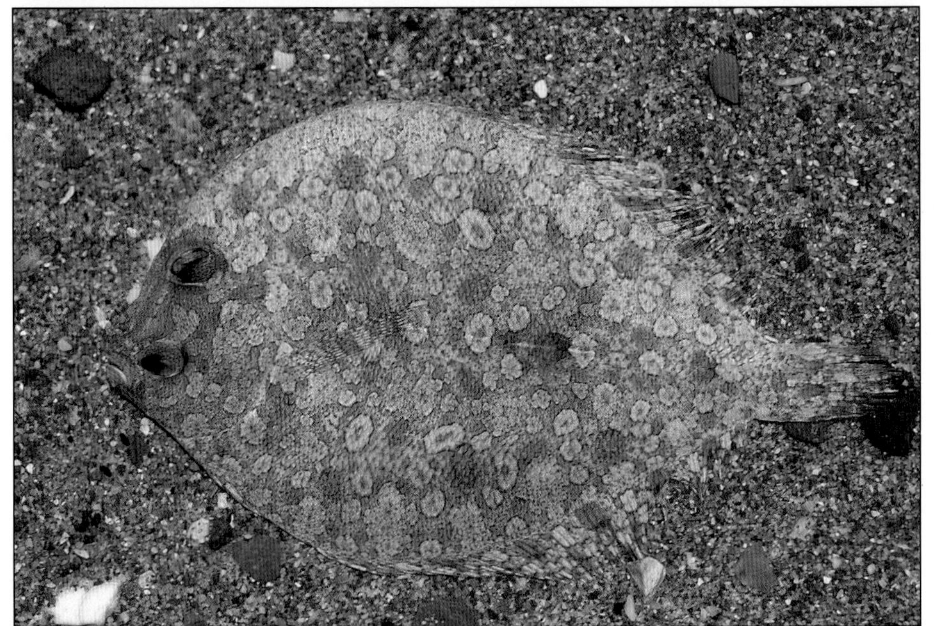

Photograph: Alex Kerstitch

FLOWERY FLOUNDER
Bothus mancus (Broussonet, 1782)

Dorsal rays 96-104; anal rays 74-81; pectoral rays 10-13; body ovate, its depth about 2.0 in standard length; lateral-line scales 85-90; *lateral line strongly arched above pectoral fin*; sexual differences develop at about 23-25 cm, *characterized by the male having very elongate pectoral fin on eyed side*, tentacles on eyes, wider interorbital space, and spines on snout and above eye; brown with numerous variable-sized white to pale blue spots, some forming small partial circles; also scattered dark brown spots present. Widely distributed in the tropical Indo-Pacific from East Africa to offshore islands of the eastern Pacific; known in the latter area only from the Galapagos. Reaches 42 cm.

GILBERT'S FOUNDER

Citharichthys gilberti Jenkins & Evermann, 1889

Dorsal rays 77-89; anal rays 58-68; lateral-line scales 40-43; body depth 2.0-2.3 in standard length; *lateral line straight, without arch above pectoral fin; teeth in upper jaw in a single row; scales thin, more or less deciduous, weakly ctenoid on both eyed and blind side*; overall brown with darker brown mottling, sometimes with scattered whitish flecks and blotches. Baja California to Peru; inhabits soft bottoms of trawling grounds, and also silt or mud bottoms of bays and estuaries; sometimes enters fresh water. Attains 26 cm.

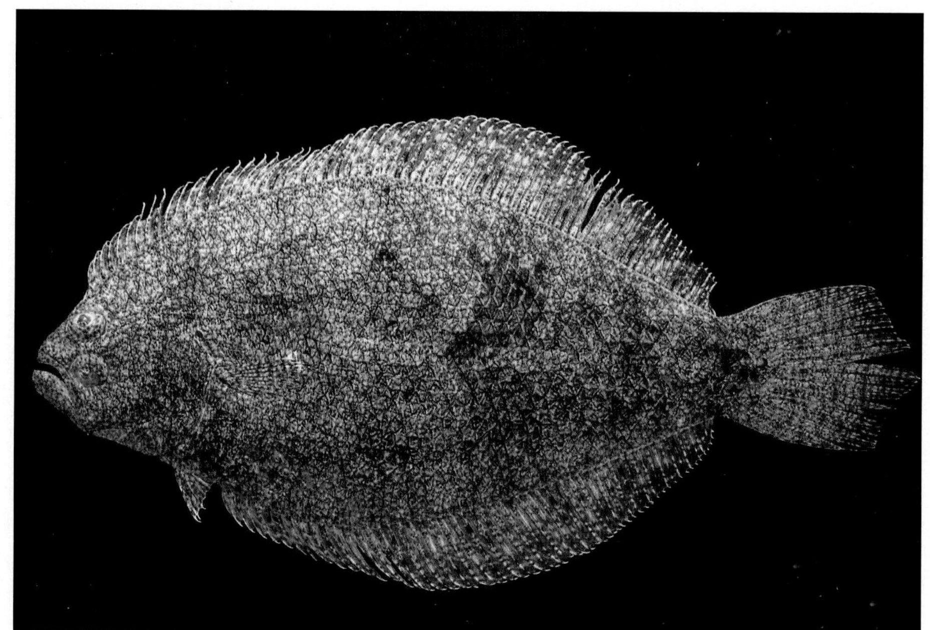

Citharichthys gilberti Mottled variety ▲ White-flecked variety ▼

PANAMANIAN FLOUNDER

Cyclopsetta panamensis (Steindachner, 1875)

Dorsal rays 90-99; anal rays 70-78; lateral-line scales about 75; body depth 2.2-2.3 in standard length; *lateral line straight, without arch above pectoral fin; teeth in upper jaw in a single row; gill rakers very short and thick (not longer than broad); scales ctenoid on eyed side, smooth on blind side*; brown with more or less ocellated pale spots on head and body; dorsal and anal fins with several large dark spots. Baja California to Panama; inhabits soft mud or sand bottoms. Grows to 25 cm.

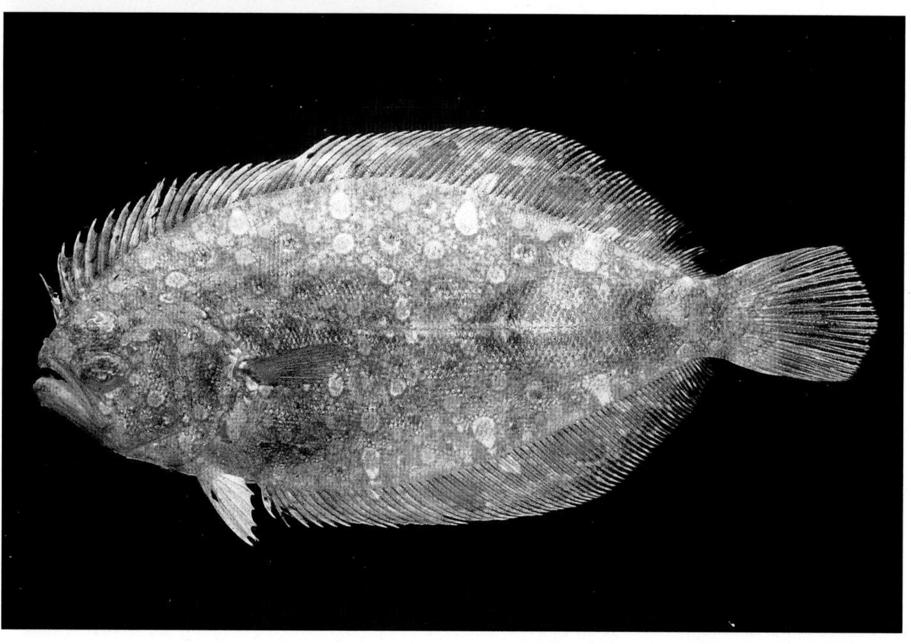

CHOCOLATE FLOUNDER
Cyclopsetta querna (Jordan & Bollman, 1890)

Dorsal rays 89-93; anal rays 70-74; lateral-line scales 92-98; body depth 2.2-2.3 in standard length; *lateral line straight, without arch above pectoral fin; teeth in upper jaw in a single row; gill rakers very short and thick (not longer than broad); scales cycloid on both eyed and blind side*; dark brown with faint indication of lighter crossbars; 2-3 large blackish spots on dorsal and anal fins. Gulf of California to Peru; often caught by trawlers on sand or mud bottoms. Reaches at least 33 cm.

SMALLMOUTH FLOUNDER
Etropus crossotus Jordan & Gilbert, 1881

Dorsal rays 75-87; anal rays 58-68; pectoral rays 10; body depth 1.7-2.0 in standard length; *lateral line straight, without arch above pectoral fin; mouth very small, length of maxillary less than one-third that of head; teeth small, in a single series, but absent on eyed side of upper jaw; interorbital space very narrow, the orbital ridge of each eye coalesced*; overall dark brown with narrow black scale margins; median fins dusky blackish. Baja California to Colombia; inhabits soft mud and sand bottoms. To 22 cm.

FOURSPOT FLOUNDER
Hippoglossina tetrophthalmus (Gilbert, 1891)

Dorsal rays 76-83; anal rays 58-62; pectoral rays 11; body depth 2.3-2.4 in standard length; lateral-line scales about 95-100; *lateral line strongly arched above pectoral fin; teeth in a single row in both jaws; gill rakers very short and thick; scales cycloid on both eyed and blind side*; generally brown with *2 pairs of ocellated black spots on rear half of body*. Gulf of California to Peru; inhabits sand-mud bottoms to at least 140 m. To at least 32 cm.

DAPPLED FLOUNDER
Paralichthys woolmani Jordan & Williams (*in* Gilbert, 1897)

Dorsal rays 70-80; anal rays 52-64; pectoral rays 11-12; body depth 1.9-2.3 in standard length; *lateral line strongly arched above pectoral fin; pelvic fins in lateral position (i.e. neither is situated on the median [abdominal] ridge); origin of dorsal fin in advance of upper eye; mouth large, jaws with a single row of teeth*; dark brown with blackish blotches and pale (blue or whitish) round spots; usually with 3 large blackish blotches on middle of body along lateral line; median fins blackish with scattered pale spots. Baja California to Peru; inhabits flat sand or mud bottoms. Reaches 35 cm.

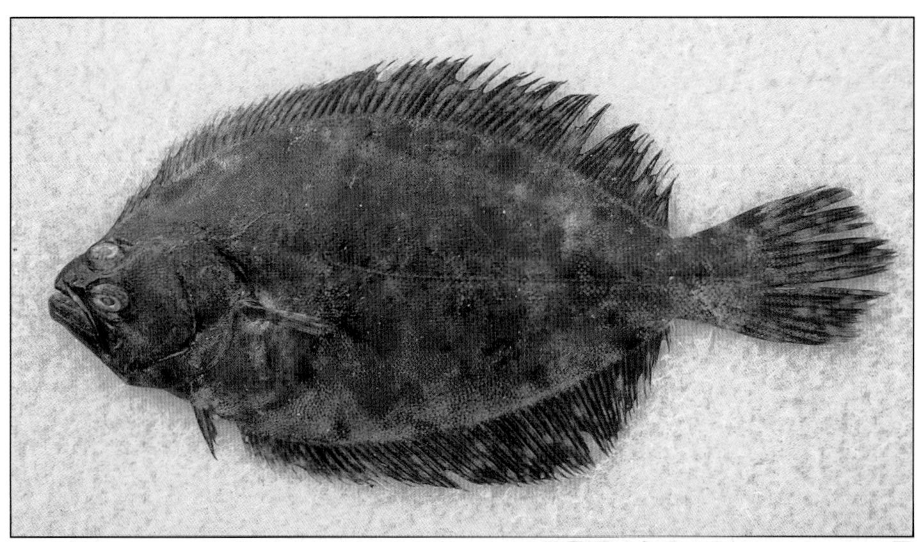

Syacium ovale Spotted variety ▼

OVAL FLOUNDER
Syacium ovale (Günther, 1864)

Dorsal rays 80-88; anal rays 63-68; pectoral rays 12, the 2 uppermost rays of males prolonged and filamentous; body depth 2.1-2.3 in standard length; *lateral line straight, without arch above pectoral fin; teeth in upper jaw biserial; gill rakers shorter than pupil width*; scales strongly ctenoid on eyed side, cycloid on blind side; color variable, from more or less uniform brown, to brown with numerous ocellated white spots which may extend onto median fins. Gulf of California to Colombia; inhabits flat sand bottoms. To about 20 cm.

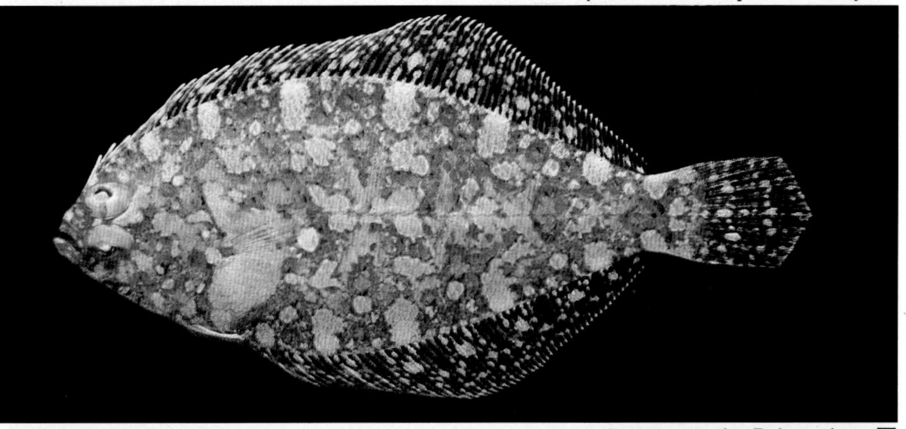

Syacium ovale Pale variety ▼

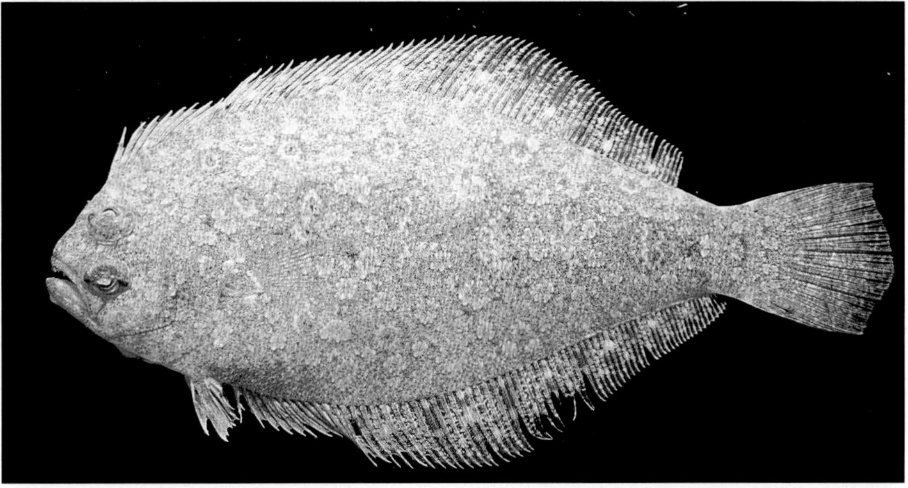

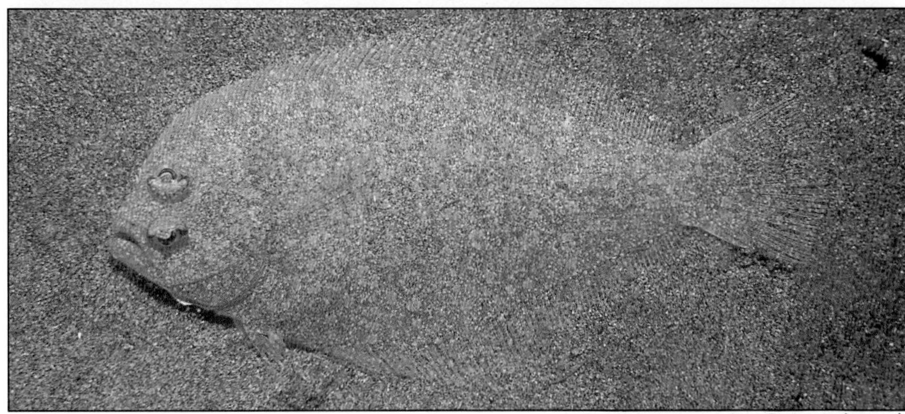

Syacium ovale Pale variety ▲

TONGUE SOLES (LENGÜETAS)
FAMILY CYNOGLOSSIDAE

Tongue soles are flattened, flounder-like fishes that are characterized by a somewhat elongate shape, with both eyes on the left side of the body. The many-rayed dorsal and anal fins are confluent with the pointed caudal fin; only the left pelvic fin is developed, and the pectoral fins are absent (except for a fine membrane in *Symphurus*). These fishes occur worldwide in tropical and subtropical seas (also a few species in fresh water). They are residents of flat, soft bottoms, often of sandy areas near reefs, or of more open stretches of mud, silt, or sand. The family contains three genera and approximately 110 species. We include coverage of a small representative sample of several of the more common shallow-water species in our area.

BAR-FINNED SOLE
Symphurus atramentatus Jordan & Bollman, 1889

Dorsal rays 88-98; anal rays 74-82; a single pelvic fin with 4 rays; pectoral fins absent; scales in longitudinal series 102-109; *tan with about 6 brown crossbars, the 2 halves of the first bar offset and sometimes separate; dorsal and anal fins with black bars, those at rear of fins particularly intense*; peritoneum and isthmus unpigmented. Gulf of California to Peru; inhabits flat sand bottoms to at least 20 m depth. Grows to at least 12 cm.

BLACK-TAILED SOLE
Symphurus fasciolaris Gilbert, 1891

Dorsal rays 85-98; anal rays 72-82; a single pelvic fin with 4 rays; pectoral fins absent; scales in longitudinal series 86-105; *generally tan with dark brown spotting on ocular side of head and body; about 6-8 brown crossbars; dorsal and anal fins with blackish rectilinear bars and narrow white outer margin; tip of tail black*; inner opercular linings of both sides of body darkly pigmented; peritoneum unpigmented. Gulf of California to Peru; inhabits flat sand bottoms to at least 20 m depth. Grows to at least 12 cm.

Photograph: Alex Kerstitch

BARRED SOLE
Symphurus chabanaudi Mahadeva and Munroe, 1990

Dorsal rays 98-109; anal rays 82-92; a single pelvic fin with 4 rays; pectoral fins absent; scales in longitudinal series 91-104; small ctenoid scales present on blind-side dorsal and anal fin rays; *light brown with 5-8 darker brown crossbars*; a blackish blotch on lower opercle; peritoneum unpigmented; isthmus pigmented. Gulf of California to Peru; inhabits soft sand or silt-mud bottoms in 2-59 m; usually in less than 28 m. Reaches about 24 cm.

LEE'S SOLE
Symphurus leei Jordan & Bollman, 1889

Dorsal rays 90-100; anal rays 70-79; a single pelvic fin with 3-4 rays; pectoral fins absent; scales in longitudinal series 80-90; *tan, sometimes with indications of grayish bars*; peritoneum heavily spotted to darkly pigmented; isthmus unpigmented; inner opercular lining of eyed side heavily pigmented. Costa Rica to Colombia; inhabits flat sandy bottoms to 90 m depth. Size to 13 cm.

DARKTAILED SOLE
Symphurus melanurus Clark (1936)

Dorsal rays 97-101; anal rays 79-85; a single pelvic fin with 4 rays; pectoral fins absent; scales in longitudinal series 88-91; *light brown with dense covering of small irregular, pale tan blotches*; dorsal and anal fins pale with blackish rectilinear blotches; inner linings of both opercles darkly pigmented; peritoneum unpigmented; isthmus darkly pigmented. Gulf of California to Peru; inhabits soft sand or silt-mud bottoms in 1-25 m. Grows to about 18 cm.

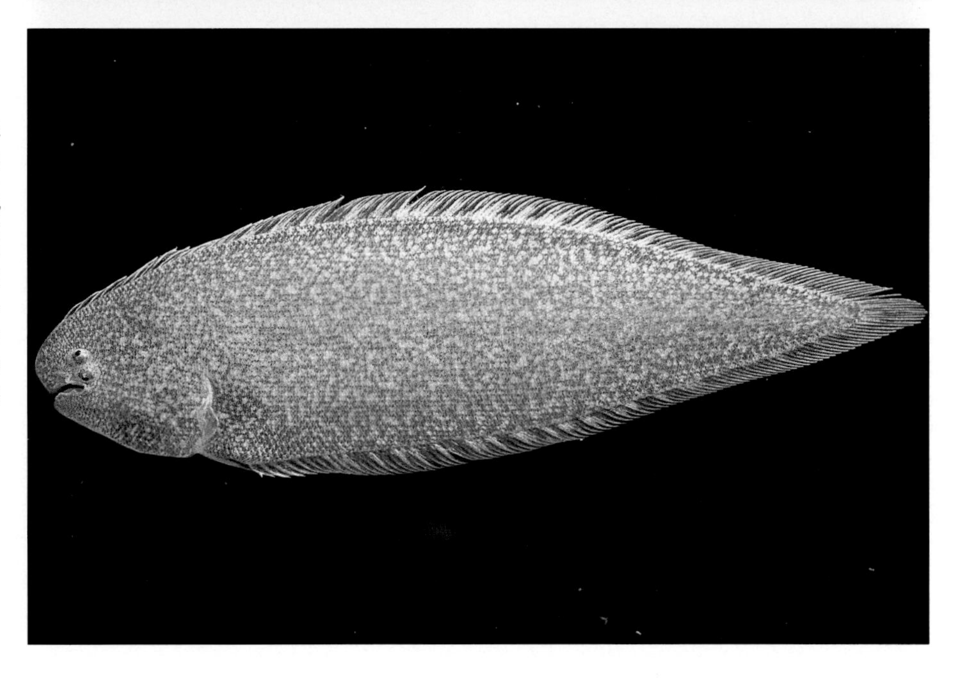

NEW WORLD SOLES (SUELAS, LENGUADOS REDONDOS, LENGUADOS DERECHOS)
FAMILY ACHIRIDAE

Soles differ from the bothids and cynoglossids (see above) in having the eyes on the right side of the body rather than the left. The family contains 9 genera with approximately 30 species. It is characterized by a superficial groove along the edge of the preoperculum, a caudal fin that is never connected to the dorsal and anal fin, and the right pelvic fin joined to the anal fin. By contrast, the closely related (and considered by many previous authors to be in the same family) Soleidae of the Old World (primarily Europe to Australia and Japan) has a concealed preopercular margin, a caudal fin that is either free or joined to the dorsal and anal fins, and free pelvic fins. Like other flatfishes, soles are bottom inhabitants of sandy or muddy areas.

Achirus klunzingeri ▲

TAN SOLE
Achirus klunzingeri (Steindachner, 1880)

Dorsal rays 59-63; anal rays 46-48; pectoral rays 2-5; anterior part of head on blind side with short fleshy tentacles; lower lip on eyed side fringed; *scales on eyed side with black, hair-like filaments; more or less uniform brown* with faint indication of dark spotting. Costa Rica to Peru; inhabits sand or mud bottoms. Grows to about 30 cm.

Achirus mazatlanus Pale variety ▲ Dark variety ▼

MAZATLAN SOLE
Achirus mazatlanus (Steindachner, 1869)

Dorsal rays 55-57; anal rays 41-44; pectoral rays 1-4; anterior part of head on blind side with very short fleshy tentacles; lower lip on eyed side with short fringes; *scales on eyed side with cluster of fine, black, hair-like filaments*; dark gray or brownish gray to light tan, with *about 8 vertical dark lines across eyed side, and scattered clumps of black, hair-like filaments.* Baja California to Peru; inhabits soft mud or sand bottoms of bays and estuaries. To 20 cm.

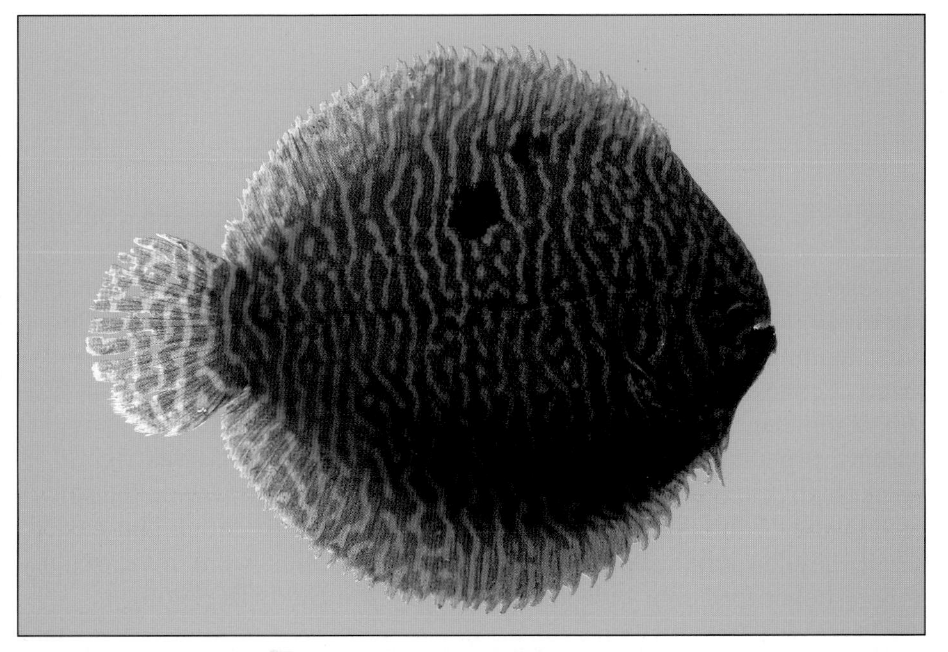

BANDED SOLE
Achirus scutum (Günther, 1862)

Dorsal rays 53-59; anal rays 42-44; pectoral rays 3-4; blind side of head with short fleshy tentacles; lower lip of eyed side fringed; no hair-like filaments on scales; *dark brown with tan or whitish bars, these generally wavy and becoming branched or joined with adjacent bars to form a maze-like pattern*; median fins with pale, wavy bands. Mexico to Peru; inhabits flat sand or mud bottoms of bays and estuaries. Reaches 28 cm.

WHITE-SPOTTED SOLE
Trinectes fimbriatus (Günther, 1862)

Dorsal rays 46-49; anal rays 33-35; pectoral rays absent; scales in longitudinal series 70-74; body ovate, nearly round, its depth 1.4-1.5 in standard length; *overall dark brown with distinct, pale yellow spots*. Guatemala to Peru; inhabits flat sand or mud bottoms. Reaches at least 8 cm. Reported to possess tiny white filaments on the scales, but the 2 individuals we have examined lack this feature.

FONSECA SOLE
Trinectes fonsecensis (Günther, 1862)

Dorsal rays 58-60; anal rays 43-46; pectoral rays 1-3 (usually a single small ray); blind side of head with fleshy long tentacles; lower lip of eyed side fringed; *no hair-like filaments on scales*; generally *tan with numerous dark brown bars (alternate bars usually interrupted)*; median fins with dark brown spots. Mexico to Peru; inhabits flat sand bottoms; frequently captured by trawlers. Reaches at least 25 cm.

TRIGGERFISHES (PECES PUERCO, PECES CHANCHO, CALAFATES)
FAMILY BALISTIDAE

Triggerfishes are characterized by a football shape, leathery skin, and small mouth with powerful jaws. They are closely related and similar in appearance to the filefishes (Monacanthidae), but generally are more robust and have three dorsal spines (filefishes have only two). Triggers are primarily inhabitants of the tropical Indo-Pacific region, but also occur in other warm seas. There are an estimated 40 species in 11 genera worldwide; six species are encountered in the eastern Pacific. Some of the larger triggers in the Indo-West Pacific region are a hazard to divers when tending nests, which are recognized as cone-shaped depressions in rubble-bottom areas. The female parent is especially vicious, and may aggressively charge other large fishes or humans. Although the mouth is small, they can deliver a painful bite. Triggerfishes feed on a wide variety of items. A few species consume mainly plankton, but most feed on a combination of such items as algae, live coral, urchins, crabs, shrimps, brittle stars, sponges, tunicates, polychaete worms, cephalopods, hydrozoans, and fishes. At night, triggerfishes wedge themselves tightly in crevices by locking the dorsal spines into an erect position. The eastern Pacific species were reviewed by Berry and Baldwin (1966).

FINESCALE TRIGGERFISH
Balistes polylepis Steindachner, 1876

Dorsal rays III + 26-28; anal rays 24-26; pectoral rays 13-15; *teeth in jaws deeply notched*; anterior rays of second dorsal and anal fins much longer than posterior rays; caudal fin lunate or double emarginate, with prolonged lobes; scales very small; *olive-brown to pale blue-gray; no distinguishing marks on body or fins*. Northern California to Chile; inhabits rocky reefs, often seen over adjacent sand or rubble areas; sometimes found in aggregations. Attains a length of 76 cm.

OCEANIC TRIGGERFISH
Canthidermis maculatus (Bloch, 1786)

Dorsal rays III + 23-27; anal rays 20-27; pectoral rays 14-15; *body of adults relatively elongate, depth of body about 2.3 in standard length*; dorsal and anal fins somewhat falcate, the anterior rays well elevated, much longer than posterior rays; longitudinal groove present in front of eye; no grooves on cheek; caudal fin double emarginate, rounded in juveniles; *blue-gray to nearly black, usually covered with small white spots*. Circumglobal in tropical seas; in our region it is usually seen at offshore islands, or far out to sea around floating debris, such as logs or *Sargassum*. Grows to 35 cm.

BLACK TRIGGERFISH
Melichthys niger (Bloch, 1786)

Dorsal rays III + 30-35; anal rays 28-31; pectoral rays 15-17; anterior rays of second dorsal and anal fins moderately elevated, much longer than posterior rays; no grooves on cheek; caudal fin deeply emarginate to lunate; *easily distinguished by the overall blackish coloration of the body and fins*; head may have slight yellowish hue; thin blue lines radiating dorsally and anteriorly from eye. Circumglobal in tropical seas; eastern Pacific records include the Revillagigedo Islands, Isla del Coco, Clipperton, Malpelo, and Panama (Gulf of Chiriqui); inhabits rocky reefs and coralline areas in 4-20 m. Reaches 29 cm.

BLUNTHEAD TRIGGERFISH
Pseudobalistes naufragium (Jordan & Starks, 1895)

Dorsal rays III + 27; anal rays 24; pectoral rays 15-16; a deep groove in front of eye; anterior rays of second dorsal and anal fins moderately elevated, longer than posterior rays; rear margin of caudal fin double emarginate or rounded, with slightly prolonged and pointed lobes; no scales around snout; *pale blue-gray to brownish gray; several alternating light and dark bars visible on side; gill opening white.* Baja California and Gulf of California to Ecuador; inhabits rocky reefs, sometimes seen in adjacent rubble areas; ranges to at least 30 m. The largest triggerfish in the region, reaching 100 cm.

ORANGESIDE TRIGGERFISH
Sufflamen verres (Gilbert & Starks, 1904)

Dorsal rays III + 30-33; anal rays 27-30; pectoral rays 14-15; anterior rays of second dorsal and anal fins only slightly longer than posterior rays, these fins more or less uniform in height; caudal fin truncate; *dark brown on upper half of body, light brown to yellowish orange on head and lower side of body; a thin, oblique, yellow stripe from rear corner of mouth across lower part of head; cheek with narrow black stripes*; juveniles with dark brown spots and broken lines on upper half of body. Baja California and Gulf of California to Ecuador; inhabits rocky reefs in 3-35 m. Grows to 38 cm.

Sufflamen verres Adult ▼ Juv. ▲

CROSSHATCH TRIGGERFISH
Xanthichthys mento (Jordan & Gilbert, 1882)

Dorsal rays III + 29-32; anal rays 26-29; pectoral rays 12-14; groove present in front of eye; 5 slightly diagonal, darkly pigmented grooves on cheek; anterior rays of second dorsal and anal fins elevated, forming triangular peak, much longer than posterior rays; lower jaw projecting slightly; caudal peduncle relatively narrow; *pale blue-gray to yellowish brown or tan; scale outlines narrowly blackish giving crosshatch appearance; caudal fin of males red, that of females yellow*; outer edge of second dorsal and anal fins of males broadly yellow. Scattered localities in the Pacific Ocean, including Ryukyu and Izu Islands, Marcus Island, Hawaiian Islands, Pitcairn Island, and Easter Island; in the eastern Pacific it is known from Southern California to the Galapagos, being more common at offshore localities such as the Revillagigedos. Reaches 28 cm. Also known as Redtail Triggerfish, an inappropriate name because of the female's bright yellow tail.

Xanthichthys mento ▼

FILEFISHES (LIJAS, PECES LIMA, PECES UNICORNIO)
FAMILY MONACANTHIDAE

Filefishes are close relatives of triggerfishes (Balistidae), but differ from them by having more compressed bodies, generally a more pointed snout, a longer first dorsal spine, a very small second spine (sometimes absent), and no third spine; the teeth are similar, but not as stout, and there are six instead of eight in the outer row, and four in the inner row; the scales, which are not conspicuous, have small setae that give the skin its coarse texture. Unlike the triggerfishes, most species are able to change color to match their surroundings, and some develop small skin flaps or tassels which further enhances their camouflage. They tend to be secretive, often hiding in seagrass, thick algal cover, gorgonians, or coral. Most exhibit omnivorous food habits, feeding on a great variety of benthic animal and plant life. There are approximately 85 species worldwide; the majority are confined to cooler waters of temperate and subtropical seas. In Japan and Australia they are important commercial fishes.

SCRAWLED FILEFISH

Aluterus scriptus (Osbeck, 1765)

Dorsal rays II+43-49; anal rays 46-52; pectoral rays 13-15; body elongate and very compressed; snout long, the dorsal and ventral profile concave; mouth small and upturned; *caudal fin long and rounded, the posterior edge often ragged; olive-brown to gray, with irregular blue spots and short lines and small black spots.* Circumtropical distribution, ranges from the Gulf of California to the Galapagos in the eastern Pacific; inhabits coral and rocky reefs; the young are sometimes seen around floating debris, at a considerable distance away from land; feeds on algae, hydrozoans, gorgonians, anemones, and tunicates. Grows to 75 cm.

Photograph: John Randall

YELLOWEYE FILEFISH

Cantherhinus dumerilii (Hollard, 1854)

Dorsal rays II-34-39; anal rays 28-35; pectoral rays 14-15 (usually 15); dorsal profile of snout nearly straight; *2 pairs of blade-like spines on caudal peduncle (males have longer, stouter spines)*; grayish brown, often with series of incomplete darker bars on posterior half of body; iris and caudal fin orange; males with caudal fin and peduncular spines more orange; juveniles and subadults with white spots scattered on body. East Africa to tropical eastern Pacific; inhabits coral and rocky reefs; feeds on tips of branching corals, and also on algae, sponges, sea urchins, and molluscs. Reaches 35 cm.

BOXFISHES OR TRUNKFISHES (PECES COFRE, PECES CAJA)
FAMILY OSTRACIIDAE

Boxfishes are easily recognized by their rigid external armor composed of sutured bony plates. Their fins are much reduced, and they are usually slow swimmers, but capable of short rapid bursts. When feeding, boxfishes sometimes squirt a jet of water into the sand to uncover small plants and invertebrates. The diet includes tunicates, sponges, soft corals, crustaceans, worms, and algae. Boxfishes have a harem social structure, with three to four females per male. Spawning occurs in pairs after dusk. Pelagic eggs are released and fertilized after the pair rises well off the bottom and hovers in place for several seconds. Boxfishes produce a toxic slime that can kill other fishes, or themselves, in a confined space such as an aquarium. About 30 boxfishes are known worldwide, but only a single, wide-ranging species is encountered on reefs in our area.

Ostracion meleagris Male ▲ Female ▼

SPOTTED BOXFISH
Ostracion meleagris Shaw, 1796

Dorsal rays 9; anal rays 9; pectoral rays 11; caudal fin rounded in females, truncate with rounded corners in males; juveniles and *females brown with small brown-white spots; males dark brown with small white spots dorsally on carapace, blue elsewhere, with dark-edged, orange-yellow spots on side*, these sometimes coalescing along upper lateral ridge to form an irregular band. Wide-ranging in the tropical Indo-Pacific, from East Africa to the Americas; inhabits coral and rocky reefs. Grows to 18 cm.

PUFFERS (TAMBORILES, BOTETES)
FAMILY TETRAODONTIDAE

The puffers or blowfishes and the related porcupinefishes (family Diodontidae) can inflate their body by swallowing water (or air if out of water). This adaptation no doubt is a deterrent to potential predators, although large sharks sometimes consume them. Puffers are further characterized by having tough, scaleless skin (often with small spinules); beak-like dental plates with a median suture; a slit-like gill opening in front of the pectoral fin base; no spines in the fins; a single, short-based dorsal fin; a similar anal fin below; no pelvic fins; and no ribs. They produce a powerful poison, tetraodontoxin, in their tissues, especially the liver and ovaries. Eating them can cause serious illness, and frequently death. The degree of toxicity varies greatly, depending on the species and also according to geographic area and season. However, some puffers are considered a great delicacy in Japan, where they are prepared by specially trained and licensed cooks. The diet of puffers consists of urchins, sponges, coral, starfishes, crabs, molluscs, worms, tunicates, and algae. The family is distributed in all tropical and subtropical seas (also some freshwater species); they are frequently associated with soft-bottom habitats. Worldwide there are 16 genera with about 120 species.

SPOTTED GREEN PUFFER
Arothron hispidus (Linnaeus, 1758)

Dorsal rays 10-11; anal rays 10-11; pectoral rays 17-19; small spinules on head and body except snout and posterior caudal peduncle; nostril consisting of 2 fleshy flaps from a common base; caudal fin rounded; grayish to greenish brown, with small white spots dorsally, shading to white below with curved dark stripes; pectoral base and gill opening alternately circled by narrow white and black bands. Widespread in the tropical Indo-Pacific; in the eastern Pacific it ranges from the Gulf of California to Panama; inhabits shalllow coral and rocky reefs. Reaches 48 cm.

Arothron hispidus Juv.

Arothron meleagris Spotted variety ▲ Yellow variety ▼

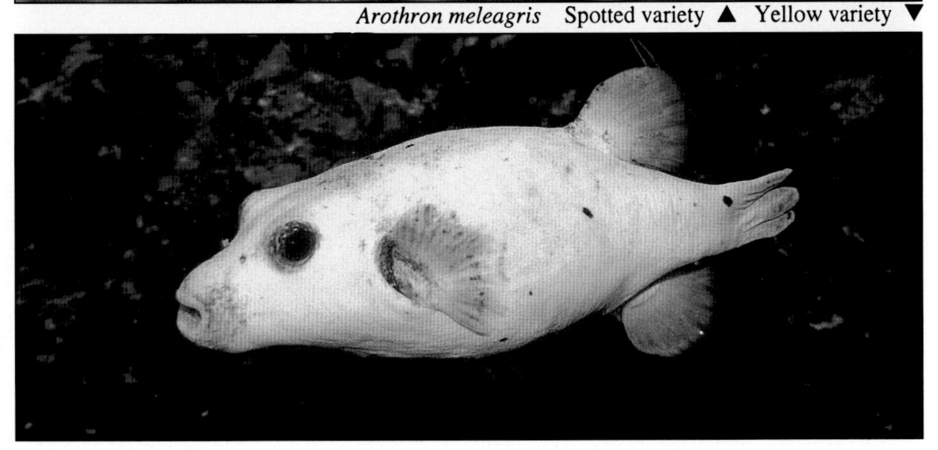

GUINEAFOWL PUFFER
Arothron meleagris (Bloch & Schneider, 1801)

Dorsal rays 11-12; anal rays 12; pectoral rays 17-19; body covered with prickles, more conspicuous when inflated; nostril with pair of fleshy flaps formed by bifurcation of a single base; 2 color phases are commonly encountered: *overall blackish with numerous small white spots, or bright yellow*. Widespread in the tropical Indo-Pacific; in the eastern Pacific it ranges from the Gulf of California to Ecuador; inhabits coral areas and rocky reefs in 2-30 m. Grows to 30 cm.

SPOTTED SHARPNOSE PUFFER
Canthigaster punctatissima (Günther, 1870)

Dorsal rays 8-10 (usually 9); anal rays 8-10 (usually 9); pectoral rays 15-18 (usually 17); a ridge of skin middorsally; *snout relatively long and pointed; overall dark brown; often yellowish brown on snout region; numerous small white spots covering head and body*; belly whitish. Gulf of California to Panama, also Galapagos and other offshore islands; common on rocky reefs in 2-15 m. Grows to 8.5 cm.

SKINFLAP PUFFER
Sphoeroides angusticeps (Jenyns, 1842)

Dorsal rays 7-8; anal rays 7; body relatively elongate, the depth about 3.5 in standard length; a pair of small black cirri straddling median line of back, just behind level of gill opening; *sides usually with scattered cirri; overall tan or light brown, with pale mottlings; white on belly; iris and fins yellowish, basal portion of dorsal fin usually brown.* Baja California to Panama and the Galapagos; usually encountered near the bottom in sandy areas. Attains 28 cm.

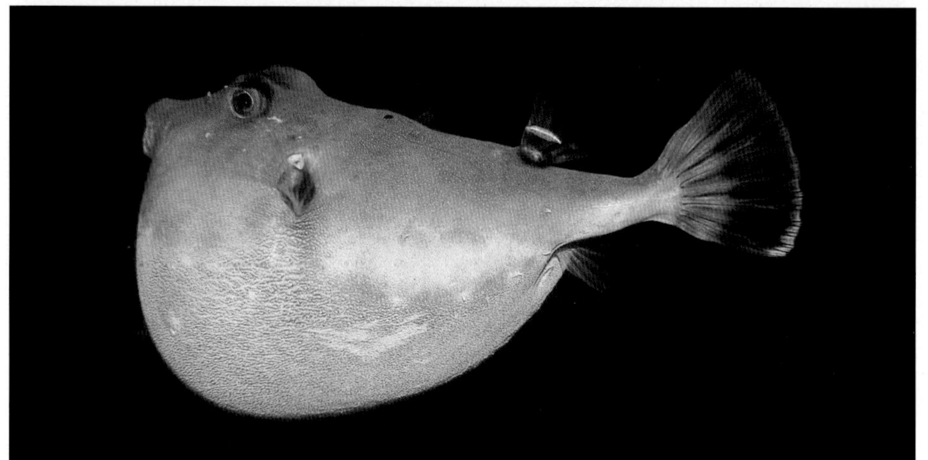

Sphoeroides angusticeps Inflated adult ▲ Deflated adult ▼

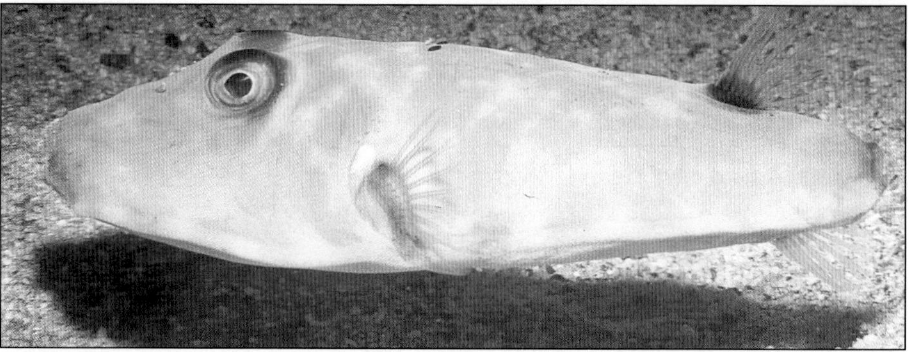

BULLSEYE PUFFER
Sphoeroides annulatus (Jenyns, 1843)

Dorsal rays 7-9; anal rays 6-9; body relatively elongate, the depth about 3.3-4.0 in standard length; small prickles on forehead, sides, and belly; blackish to olive-brown on upper half, white below; *"maze" pattern of narrow white to yellow concentric lines, bars, and oblique bands on back; head and sides with numerous small dark spots*; iris yellow. Southern California to Peru, including Galapagos; common on rocky reefs and adjacent sand patches; often seen in midwater high off the bottom, or at the surface; huge numbers may be attracted to anchored boats, feeding on garbage that is thrown overboard. To 38 cm.

Sphoeroides annulatus ▼

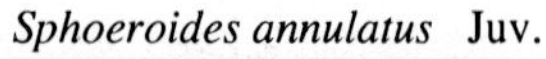

Sphoeroides annulatus Juv.

LOBESKIN PUFFER
Sphoeroides lobatus (Steindachner, 1870)

Dorsal rays 8; anal rays 6; body moderately elongate, the depth about 3.0-3.5 in standard length; small prickles covering most of body; *small triangular flaps of skin (cirri) scattered along side*; olive on back and upper sides, with brownish mottlings and numerous small white spots; ventral parts white; *a horizontal row of brown blotches or short bars between dark upper part of body and white lower part.* Gulf of California to Peru, including Galapagos; inhabits sandy or weedy areas in 1-20 m; has an effective camouflage coloration, and is therefore difficult to detect on weed or rubble bottoms. Reaches at least 25 cm.

PANAMANIAN PUFFER
Sphoeroides trichocephalus (Cope, 1870)

Dorsal rays 8; anal rays 6-7; body relatively robust, the depth about 2.5-3.0 in standard length; small prickles on forehead, breast, and belly, remainder of body smooth; caudal fin truncate or slightly emarginate; *olive or brownish, with small whitish spots and several large dark blotches or saddles on back; silvery white on lower sides.* Costa Rica to Ecuador; inhabits turbid waters of bays; usually found on soft sandy or silty bottoms. Grows to at least 12 cm. *S. furthii* (Steindachner) is a synonym. The type locality of *S. trichocephalus* was given as the Atlantic Ocean off Newport, Rhode Island. However, it was later discovered that the type was mislabeled and actually originated from Panama.

PORCUPINEFISHES (TAMBORILES DE ESPINAS, PECES ERIZO)
FAMILY DIODONTIDAE

Porcupinefishes are similar to puffers (Tetraodontidae) in form and habits, but have the added protection of formidable sharp spines on the head and body. These spines are either fixed and immovable, as in *Chilomycterus,* or erectile, as in *Diodon.* Diodontids differ further from puffers in having broader pectoral fins, lacking a median suture on their very stout dental plates, and in having larger eyes. They are mainly nocturnal, usually hiding in caves or beneath ledges during the day. Their strong tooth plates and powerful jaws are well adapted for crushing urchins, molluscs, and crabs. They can also cause tetraodontoxin poisoning, but less frequently than puffers. Care should be exercised when handling these fishes because of their sharp spines and painful bite. The family occurs worldwide in tropical seas and contains 15 species in two genera.

SPOTTED BURRFISH
Chilomycterus reticulatus (Linnaeus, 1758)

Dorsal rays 12-14; anal rays 11-14; pectoral rays 19-22; *head and body with short, immovable spines, about 8-10 in an approximate row from snout to dorsal fin*; gray on back, grading to whitish ventrally; scattered black spots on head, body, and median fins; also several diffuse, dusky brown bars on head and body. Circumtropical and temperate distribution, including California to Chile; inhabits coral and rocky reefs. Grows to about 55 cm; the young lead a pelagic existence until about 20 cm. *C. affinis* Günther is a synonym.

Photograph: John Randall

FRECKLED PORCUPINEFISH
Diodon holocanthus Linnaeus, 1758

Dorsal rays 13-15; anal rays 13-14; pectoral rays 22-25; long erectile spines on head and body; *12-16 spines in an approximate row from top of snout to dorsal fin; anterior middle spines on top of head longer than longest spines posterior to pectoral fins; no spines on caudal peduncle; a pair of small barbels on chin*; light olive to pale brown, shading to white ventrally, with small black spots on upper two-thirds of head and body; a brown bar from above to below eye; a broad brown bar across occipital region, and another across middle of back; a large, oval, brown blotch above each pectoral fin and another around dorsal fin base; fins plain. Circumtropical distribution, including southern California to Colombia; inhabits reefs and open sand-rubble bottoms. Attains 29 cm.

Diodon holocanthus Deflated adult ▲ Inflated adult ▼

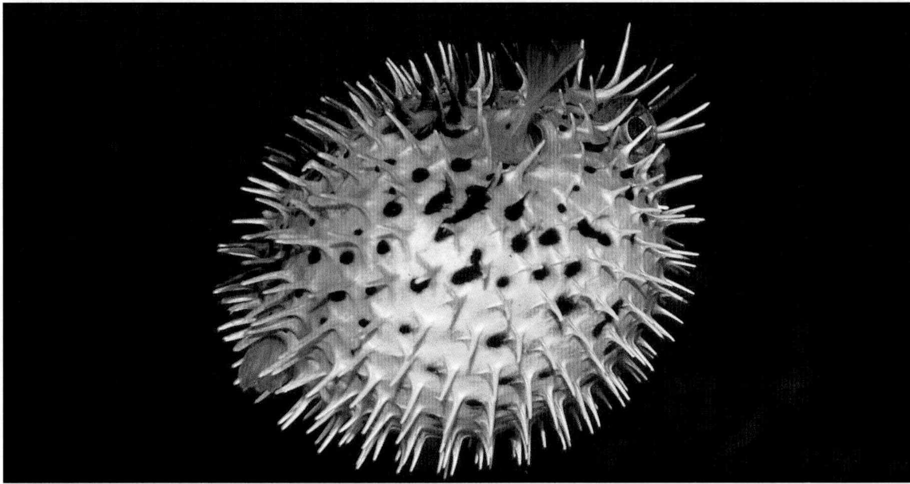

Diodon hystrix ▼

PORCUPINEFISH
Diodon hystrix Linnaeus, 1758

Dorsal rays 14-17; anal rays 14-16; pectoral rays 21-25; *16-20 erectile spines in an approximate row from top of snout to dorsal fin; spines anteriorly at front of head shorter than longest spines posterior to pectoral fins; 1 or more small spines dorsally on caudal peduncle; no barbels on chin*; olive to light gray-brown dorsally, with small black spots, shading to white ventrally; fins with small black spots. Circumtropical distribution, including southern California to Chile; inhabits coral or rocky reefs. Maximum length 71 cm.

APPENDIX

After this book was in press, during April 1994, the authors participated in a month-long expedition to Clipperton Island, including brief stops at several locations in the Revillagigedo Islands. We made a combined total of 82 dives at Clipperton, to depths of 55 m. Our objective was to thoroughly document the fish fauna by means of underwater observations and photography, and by collecting with rotenone and small spears.

The fish fauna of Clipperton Island is relatively small owing to its tiny size (approximately 4 km in length) and isolated position (approximately 1 000 km from the Mexican mainland). Although the results reported here are only preliminary, we recorded a total of 88 species. The faunal composition is a fascinating mix of elements from the eastern Pacific (39 species), Indo-West Pacific (37 species), and worldwide tropics (12 species). Of the eastern-Pacific species, there is a significant endemic element consisting of 7 species, only 3 of which have been described previously – *Myripristis gildi* (Holocentridae), *Holacanthus limbaughi* (Pomacanthidae), and *Stegastes baldwini* (Pomacentridae). Other elements we intend to describe in the near future include labrids of the genera *Thalassoma* and *Xyrichtys*, a pomacentrid (*Chromis*), and a muraenid (*Uropterygius*).

The most abundant reef fishes at Clipperton were as follows (family indicated in parentheses): *Gymnothorax dovii* (Muraenidae); *Myripristis berndti* (Holocentridae); *Paranthias colonus*, *Dermatolepis dermatolepis*, and *Epinephelus labriformis* (Serranidae); *Caranx melampygus* and *C. sexfasciatus* (Carangidae); *Lutjanus viridis* (Lutjanidae); *Mulloidichthys dentatus* (Mullidae); *Kyphosus elegans* (Kyphosidae); *Holacanthus limbaughi* (Pomacanthidae); *Stegastes baldwini* (Pomacentridae); *Thalassoma grammaticum* and *Thalassoma* sp. (Labridae); *Acanthurus nigricans* and *Ctenochaetus marginatus* (Acanthuridae); and *Melichthys niger* (Balistidae).

The Clipperton fauna is notable for several obvious omissions. It is the only tropical locality in which we have failed to record a damselfish in the genus *Abudufdef*. Gobiid fishes are frequently the most specious group on coral reefs, but except for a single intertidal species, they are absent from the island. In spite of a large, central lagoon, no fishes were present. The lagoon water is slightly brackish, and because of the lack of openings has poor circulation.

ADDENDUM

We also utilize this space to illustrate a common moray eel that was inadvertently omitted from coverage. The Tiger Reef Eel, *Uropterygius tigrinus* (Girard, 1859), ranges throughout our area of coverage. It grows to a length of about 140 cm.

Photograph: M. Kazmers/Sharksong

Tiger Reef Eel (*Uropterygius tigrinus*)

GLOSSARY

Adipose eyelid: an immovable, transparent outer covering or partial covering of the eye of some groups of bony fishes, such as mullets and jacks, which performs protective and streamlining functions.

Adipose fin: a small, fleshy fin without rays found on the back behind the dorsal fin of some primitive teleost fishes such as the lizardfishes.

Alcyonarian: an animal of the Subclass Anthozoa (corals, sea anemones) of Phylum Coelenterata; polyps with eight tentacles; includes soft corals and gorgonians.

Anal ring: body segment of syngnathids (pipefishes and seahorses), containing the anus.

Antrorse spine: a small, bony projection directed anteriorly (present on the lower preopercular margin of some serranids).

Anus: the posterior external opening of the digestive tract, from which wastes are voided; sometimes called the vent.

Axil: the acute angular region between a fin and the body; usually used in reference to the underside of the pectoral fin toward the base; equivalent to a human's armpit.

Band: an oblique or irregular marking (compare with 'bar', below).

Bar: an elongate color marking of vertical orientation, the sides of which are usually more or less straight (although they need not be parallel).

Barbel: a slender, tentacle-like protuberance of sensory function, which is often seen on the chin of some fishes such as goatfishes and some of the croakers.

Benthic: referring to the benthos, the fauna and flora of the sea bottom.

Bifid: having two tips; usually used in reference to the shape of teeth or cirri.

Bifurcate: divided or fork-shaped.

Biserial: arranged in two separate rows.

Branched tubules: refers to lateral-line scale tubules with one or more branches.

Branchial: referring to the gill region.

Branchiostegal: bony rays supporting the gill membranes behind the lower jaw.

Canine: a prominent, slender, sharp-pointed tooth.

Carapace: a rigid shield encasing the body.

Carnivore: a flesh-eating animal.

Caudal fin: the tail fin. The term 'tail' alone generally refers to that part of a fish posterior to the anus.

Caudal peduncle: the part of the body between the posterior basal parts of the dorsal and anal fins and the base of the caudal fin. The usual vertical measurement is the least depth; the length measurement herein is horizontal, generally from the rear base of the anal fin.

Cephalic flaps: the forward-directed projections on either side of the mouth of manta or devil rays (Mobulidae).

Ciguatera: a fish toxin, dangerous to humans, but only found in isolated tropical localities.

Ciliate: having a fringe of hair-like processes.

Circumnarial groove: a cleft or groove around the nostril openings, found in some sharks.

Circumpeduncular scales: the transverse series of scales that completely encircle the base of the tail.

Cirrus: a small, slender, flexible, fleshy protuberance; the plural is cirri.

Claspers: rod-like, grooved processes attached to the pelvic fins of male sharks, skates, rays, and guitarfishes. Used to transmit sperm during copulation.

Cloaca: cavity into which the intestinal, urinary, and reproductive canals open.

Coelenterate: an aquatic animal of the Phylum Coelenterata, which is characterized by a central mouth, usually surrounded by tentacles bearing stinging cells, and no anus; includes sea anemones, corals, and jellyfishes.

Compressed: laterally flattened; often used in reference to the shape of the body – in this case deeper than wide.

Cornea: transparent anterior part of the external coat of the eye.

Coronet: bony, crown-like structure on the top of the head of seahorses (Syngnathidae).

Corselet: area of large, thick scales covering the anterior part of the body in some scombrid fishes.

Crenate: having a scalloped or feebly incised margin.

Crustacean: an animal of the Class Crustacea of the Phylum Arthropoda; includes crabs, lobsters, shrimps, and copepods.

Ctenoid scales: scales of bony fishes, which have tiny, tooth-like projections along the posterior margin and part of the exposed portion. Collectively these little teeth (or ctenii) impart a rough texture to the surface of the scales.

Cuspidate: having lateral projections, as on the teeth of many sharks.

Cycloid scales: scales of bony fishes, the exposed surfaces and edges of which lack any small tooth-like projections; they are, therefore, smooth to the touch.

Deciduous: easily shed; refers to the scales of some fishes.

Demersal: living on the sea bottom.

Denticles: tooth-like projections, such as the scales which completely cover the bodies of cartilaginous fishes.

Depressed: dorsoventrally flattened. The opposite in body shape of compressed.

Depth: a vertical measurement of the body of a fish; most often employed for the maximum height of the body, excluding the fins.

Dorsal: toward the back, or upper part of the body; the opposite of ventral.

Dorsal fin: a median fin along the back, which is supported by rays. There may be two or more dorsal fins, in which case the most anterior one is designated the first.

Echinoderm: an aquatic marine animal of the Phylum Echinodermata; radially symmetrical, with a skeleton composed of calcareous plates (may be reduced to spicules); many move via their numerous tube feet; includes starfishes, brittle stars, sea urchins, and sea cucumbers.

Electrophoresis: a laboratory technique used for evaluating the protein similarities (or dissimilarities) of species.

Elongate: extended or drawn out.

Emarginate: concave; used to describe the posterior border of a caudal fin which is inwardly curved.

Epipelagic: pertaining to the surface layer of the open sea.

Esca: the bait or lure of lophiiform fishes (see illicium).

Falcate: sickle-shaped; e.g. some fins.

Family: a major entity in the classification of animals and plants, which consists of a group of related genera. Family words end in '-idae', an example being Gobiidae for the goby family; when used as an adjective, the '-ae' is dropped, hence gobiid fish.

Fimbriae: fine cirri (for example, on the lips of sand stargazers (Dactyloscopidae)).

Finlet: small individual fins in addition to the main fins (usually dorsal and anal) of some fishes (for example scombrids).

Fontanel: pertains to a space between the skull bones (the shape of the dorsomedian fontanel is diagnostic in certain ariid catfishes).

Forked: inwardly angular; used in describing the shape of a caudal fin which is divided into two equal lobes, the posterior border of each of which is relatively straight.

Frontal: bone on top of the skull, partially forming the anterior upper edge of the eye.

Gas bladder: a tough-walled, gas-filled sac lying in the upper part of the body cavity of many bony fishes, just beneath the vertebral column; the principal function is to offset the weight of the heavier tissues, particularly bone. The organ is also called the air bladder or the swim bladder.

Genital papilla: fleshy structure adjacent to the anus of some fishes, which contains the genital opening. Its shape may differ according to sex in some fishes, such as gobies (Gobiidae) and triplefins (Tripterygiidae).

Genus: a group of closely related species; the first part of the scientific name of an animal or plant. The plural is genera.

Gill arch: the bony support for the gill filaments and gill rakers. Normally there are four pairs of gill arches in bony fishes.

Gill opening: the opening posteriorly and often also ventrally on the head of fishes, where the water of respiration is expelled. Bony fishes have a single such opening on each side, whereas cartilaginous fishes (sharks and rays) have five to seven. The gill openings of sharks and rays are called gill slits.

Gill rakers: stout protuberances of the gill arch on the opposite side from the red gill filaments, which function in retaining food organisms. They vary greatly in number and length, and are important in the classification of fishes.

Gonads: reproductive organs.

Gonopodium: a specialized male reproductive organ that facilitates internal fertilization in some fishes. It is actually a modification of the anal fin.

Gorgonian: a sessile animal of the Subclass Alcyonaria, Class Anthozoa, Phylum Coelenterata; includes sea fans and sea whips.

Grammistin: toxic slime produced by soapfishes (Grammistidae).

Head length: the straight-line measurement of the head, taken from the front of the upper lip to the membranous posterior end of the operculum.

Herbivore: a plant-feeding animal.

Illicium: the 'fishing pole' and 'lure' of lophiiform (pediculate) fishes, which is used to attract prey close to the mouth of these fishes.

Imbricate: embedded or buried under the skin, referring to the scales of some fishes.

Incisiform: chisel-like; used to describe teeth which are flattened and truncate with sharp edges, like the front teeth of some mammals such as humans.

Inferior tail ridge: a bony ridge along the lower edge of the tail in pipefishes.

Inferior trunk ridge: a bony ridge along the lower edge of the body in pipefishes.

Infraorbital: below the eye, often referring to a series of small bones that form the lower edge of the eye.

Interdorsal ridge: a tough fold of skin that runs along the middle of the back between the dorsal fins of some sharks.

Internarial: space between the nostrils or nasal openings.

Interopercle: one of the bones comprising the operculum; bordered anterodorsally by the preopercle and posterodorsally by the opercle and subopercle.

Interorbital space: the region on the top of the head between the eyes; measurements may be taken of the least width, either fleshy (to the edges of the orbits) or bony (between the edges of the frontal bones which rim the orbits).

Inter-radial membrane: the soft membrane between fin rays.

Invertebrate: an animal lacking a vertebral column; includes the vast majority of animals on earth, such as the corals, the worms, and the insects.

Keel: a lateral strengthening ridge posteriorly on the caudal peduncle or base of the caudal fin; typically found on swift-swimming fishes with a narrow caudal peduncle and a broadly lunate caudal fin.

Labial furrows: grooves around the outer edges of the lips, which are prominent in some sharks and other fishes.

Lanceolate: lance-shaped, hence gradually tapering to a point; used to describe a caudal fin with very long middle rays. An unusual fin shape most often seen among the gobies.

Lateral: referring to the side, or directed toward the side; the opposite of medial.

Lateral line: a sensory organ of fishes, which consists of a canal running along the side of the body and communicating via pores through scales to the exterior; functions in perceiving low-frequency vibrations, hence provides a sense which might be termed 'touch at a distance'.

Lateral-line scales: the pored scales of the lateral line between the upper end of the gill opening and the base of the caudal fin. The count of this series of scales is of value in the description of fishes. Also of value at times is the number of scales above the lateral line (to the origin of the dorsal fin) and the number below the lateral line (to the origin of the anal fin).

Lateral trunk ridge: a bony ridge along the middle part of the anterior body of pipefishes.

Leptocephalus: the elongate, highly compressed, transparent larval stage of some primitive teleost fishes such as the tarpon, bonefish, and eels.

Lower limb: refers either to the horizontal margin of the preopercle, or to the number of gill rakers on the first gill arch below and including the one at the angle.

Luminous organ: light-emitting structure present on many deep-sea fishes and a few shallow-water species.

Lunate: sickle-shaped; used to describe a caudal fin which is deeply emarginate with narrow lobes.

Maxilla: a dermal bone of the upper jaw, which lies posterior to the premaxilla. On the higher fishes the maxilla is excluded from the gape, and the premaxilla bears the teeth.

Medial: toward the middle or median plane of the body; opposite of lateral.

Median fins: the fins in the median plane, hence the dorsal, anal, and caudal fins.

Midlateral scales: refers to the longitudinal series of scales from the upper edge of the operculum to the base of the caudal fin. Generally used for fishes without a lateral line.

Molariform: shaped like a molar, hence low, broad, and rounded.

Mollusc: an animal of the Phylum Mollusca; unsegmented, with a muscular foot and visceral mass; often protected by one or two shells; includes gastropods (snails and nudibranchs), pelecypods (bivalves such as clams and oysters), cephalopods (such as squids and octopuses), and amphineurans (chitons).

Multifid: having several tips or branches, usually in reference to cirri.

Nape: the dorsal region of the head, posterior to the occiput.

Nasal barbel: tentacle-like protuberance located close to the nasal opening.

Nasal fossa: cavity or pit containing the nasal opening.

Nasoral groove: a cleft or furrow between the nostril and mouth in some sharks.

Nictitating eyelid: a transparent fold of skin forming a protective inner eyelid.

Ocellus: an eye-like marking with a ring of one color surrounding a spot of another.

Occipital pit: depression on top of the head, just behind the eyes. Its presence or absence is diagnostic in some scorpionfishes.

Omnivore: an animal which feeds on both plant and animal material.

Opercle: the large bone which forms the upper posterior part of the operculum; often bears one to three backward-directed spines in the higher fishes.

Operculum: gill cover; comprises the following four bones – opercle, preopercle, interopercle, and subopercle.

Orbital: referring to the orbit or eye.

Order: a major unit in the classification of organisms; an assemblage of related families. The ordinal word ending in the animal kingdom is '-iformes'.

Origin: the beginning; often used for the anterior end of the dorsal or anal fin at the base. Also used in zoology to denote the more fixed attachment of a muscle.

Otoliths: 'ear bones'; small, calcareous structures found in the three main pockets of the inner ear. These often show growth rings, hence they are useful for determining age in fishes.

Oviparous: producing ova (eggs) that hatch after leaving the body of the mother; the mode of reproduction of the great majority of bony fishes.

Ovoviviparous: producing eggs which hatch within the body of the mother; the mode of reproduction of most sharks and rays.

Paired fins: collective term for the pectoral and pelvic fins.

Palatine: a paired lateral bone on the roof of the mouth, lying between the vomer and the upper jaw; the presence of absence of teeth on this bone is of significance in the classification of fishes.

Papilla: a small, fleshy protuberance.

Pectoral fin: the fin usually found on each side of the body behind the gill opening; in primitive fishes such as herrings, this pair of fins is lower on the body than in more advanced forms.

Pelagic: pertaining to the open sea (hence not living inshore or on the bottom); oceanic.

Pelvic disc: the disc-like structure that is composed of the fused pelvic fins of clingfishes (Gobiesocidae) and most gobies (Gobiidae).

Pelvic fin: one of a pair of juxtaposed fins ventrally on the body in front of the anus; varies from abdominal in position in primitive fishes, such as herrings, to the more anterior locations termed thoracic or jugular in advanced fishes. It is sometimes called the ventral fin.

Perinasal groove: cleft or furrow around the nasal opening in some sharks. Also called circumnarial groove.

Pharyngeal teeth: opposing patches of teeth which occur on the upper and lower elements of the gill arches. They vary from sharp and piercing to nodular or molariform; they may be modified into a grooved grinding apparatus (or pharyngeal mill), such as is seen in the parrotfishes.

Photophores: small light-producing organs that are scattered on the head and body of some (mostly deep-sea) fishes.

Pinnate: feather-like; usually used in reference to branching cirri of some fishes.

Plankton: a collective term for pelagic animals and plants that drift with ocean currents; though many are motile, they are too small, or swim too feebly or aimlessly, to resist the sweep of the current. By contrast, the animals of the nekton are independent of water movement.

Polychaete: an animal of Class Polychaeta of Phylum Annelida; a segmented worm with setae (bristles), which may move about freely or live permanently in a tube. Polychaete is from the Greek meaning many hairs or bristles.

Polyp: the sedentary form of coelenterate animals, consisting of a tubular body with one external opening (the mouth) rimmed with tentacles; may be one of a colony; the soft part of a living coral.

Precaudal pit: the dorsal depression or notch just in front of the caudal fin of sharks.

Predorsal scales: the series of scales along the mid-dorsal line between the snout and origin of the dorsal fin.

Predorsal plate: a bony exterior structure that is part of the head shield of ariid catfishes, and is located just anterior to the dorsal fin origin on the middle of the forehead.

Premaxilla: the more anterior bone forming the upper jaw. In the higher fishes it extends backward and bears all of the teeth of the jaw. It is this part of the upper jaw which can be protruded by many fishes.

Preopercle: a boomerang-shaped bone, the edges of which form the posterior and lower margins of the cheek region; it is the most anterior of the bones comprising the gill cover. The upper vertical margin is sometimes called the upper limb, and the lower horizontal edge the lower limb; the two limbs meet at the angle of the preopercle.

Preoral length: measurement used for sharks; taken between the snout tip and middle of the upper jaw.

Preorbital: the first and usually the largest of the suborbital bones; located along the ventroanterior rim of the eye. Sometimes called the lachrymal bone.

Principal caudal rays: the caudal rays which reach the posterior, terminal border of the fin; in those fishes with branched caudal rays, the count includes the branched rays plus the uppermost and lowermost rays which are unbranched.

Produced: drawn out to a point; lengthened.

Protrusible: capable of projection, as in some jaws.

Proximal: toward the center of the body; the opposite of distal.

Radii: small (often microscopic) grooves on the margin of scales.

Ray: the supporting bony elements of fins; includes spines and soft rays.

Rhomboid: wedge-shaped; refers to a caudal fin in which the middle rays are longest, and the upper and lower portions of the terminal border of the fin are more or less straight; essentially the opposite of forked. It is an uncommon fin shape.

Rostral plate: forward-projecting bony structure on the snout of triglid fishes.

Rounded: refers to a caudal fin in which the terminal border is smoothly convex.

Rudiment: a structure so deficient in size that it does not perform its normal function; often used in reference to the small nodular gill rakers at the ends of the gill arch.

Rugose: rough, usually referring to the bony outer surface of certain head bones.

Scute: an external bony plate or enlarged scale.

Segmented rays: the soft rays of the fins which bear cross striations, at least distally.

Serrate: notched along a free margin; like the edge of a saw.

Sexual dichromatism: a condition wherein the two sexes of the same species are of different color.

Simple: not branched.

Snout: the region of the head in front of the eye. Snout length is measured from the front of the upper lip to the anterior edge of the eye.

Soft ray: a segmented fin ray which is composed of two closely joined lateral elements. It is nearly always flexible and often branched.

Spatulate: flattened, sometimes used to describe tooth shape.

Species: the fundamental unit in the classification of animals and plants, consisting of a population of individuals which freely interbreed with one another. The word 'species' is both singular and plural.

Spine: an unsegmented bony process consisting of a single element which is usually rigid and sharply pointed. Those spines which support fins are never branched.

Spiracle: an opening between the eye and the first gill slit of sharks and rays which leads to the pharyngeal cavity.

Standard length: the length of a fish from the front of the upper lip to the posterior end of the vertebral column (the last element of which, the hypural plate, is somewhat broadened and forms the bony support for the caudal fin rays).

Striae: linear bony ridges, present on the head bones (often the cheek or gill cover) of some fishes.

Stripe: a horizontal, straight-sided color marking.

Subdermal spine: a sharp, bony protuberance covered by flesh and skin (present on the snout of some brotulas [Ophidiidae]).

Subopercle: an elongate, flat dermal bone which is one of the four comprising the operculum; lies below the opercle, and forms the ventroposterior margin of the operculum.

Suborbital depth: the distance from the lower edge of the eye to the nearest edge of the upper lip.

Suborbital stay: a bony ridge across the cheek, found in scorpaeniform fishes.

Subterminal: in reference to a mouth position that is behind and under the snout tip, in sharks for example.

Subterminal notch: indentation along the upper posterior edge of the upper caudal lobe of some sharks.

Supraorbital: the region bordering the upper edge of the eye.

Supraorbital ridge: a bony crest above the eye.

Supraorbital tentacle: a flap of skin situated above the eye.

Symbiosis: the living together in close association of two dissimilar organisms. This term includes commensalism, where one organism benefits from the association but the other does not (though it is not harmed); parasitism, where the association is disadvantageous or distractive to one of the organisms; and mutualism, where both organisms exist to mutual advantage.

Synonym: an invalid scientific name of an organism, proposed after the accepted name.

Tail ridge: bony dorsal and ventral edges on the body (posterior to the anus) of pipefishes (Syngnathidae).

Tholichthys larva: the pelagic stage of butterflyfishes (Chaetodontidae), characterized by external bony plates covering the head.

Thoracic: referring to the chest region.

Total length: the length of a fish from the front of whichever jaw is most anterior to the end of the longest caudal ray.

Transverse scales: series of scales in a vertical row, often counted between the dorsal and anal fin bases.

Trifid: having three tips, usually in reference to the shape of teeth or cirri.

Truncate: square ended; used to describe a caudal fin with a vertically straight terminal border and angular or slightly rounded corners.

Tubercles: small bony protuberances or knobs, sometimes present on the external surface of various skull bones.

Uniserial: arranged in a single row.

Ventral: toward the lower part of the body; the opposite of dorsal.

Vertical scale rows: see midlateral scales.

Villiform: like the villi of the intestine, hence with numerous small slender projections. Used to describe bands of small, close-set teeth, particularly if slender. If the teeth are short, they are often termed cardiform.

Viviparous: producing living young which develop from nourishment directly from the mother.

Vomer: a median unpaired bone toward the front of the roof of the mouth, the anterior end of which often bears teeth.

Zooplankton: the animals of the plankton.

BIBLIOGRAPHY

Acero, P.A. 1981. Two new species of gobiid fishes from the Colombian Pacific. *Japanese Journal of Ichthyology* 28(3):243-6.

Allen, G.R. 1980. A review of the damselfish genus *Stegastes* of the eastern Pacific with the description of a new species. *Rec. West. Aust. Mus.*, 8(2):171-98.

———. 1991. *Damselfishes of the World.* Mentor, Ohio: Aquarium Systems. 271 pp.

Allen, G.R. and D.R. Robertson. 1991. Quatre especes nouvelles d'Opistognathidae (Jawfishes) du Pacifique oriental tropical. *Revue fr. Aquariol.*, 18(2):47-52.

———. 1991. Descriptions of two new genera and four new species of triplefins (Pisces: Tripterygiidae) from the tropical Eastern Pacific. *Revue fr. Aquariol.*, 18(3):79-82

———. 1992. *Serranus socorroensis*, a new species of serranid fish from the tropical Eastern Pacific Ocean. *Revue fr. Aquariol*, 19(1-2):37-40

———. 1992. Deux nouvelles espèces de Girelles (Labridae: *Halichoeres*) du Pacifique oriental tropical. *Revue fr. Aquariol*, 19(1-2):47-52

———. 1992. Three new species of triplefins (Pisces: Tripterygiidae) from Malpelo and Socorro Islands, in the tropical eastern Pacific. *Revue fr. Aquariol*, 19(1-2):53-6

Alvarez-León, R. and O.B. Solano. 1983. Ictiofauna acompanante del camarón en aguas someras in el Pacifico colombiano. *Boletin Museo del Mar* 11.

Alvarez-Rubio, M., F. Amezcua-Linares and A. Yanez-Arancibia. 1986. Ecologia y estructura de las comunidades de pesces en el sistema lagunar Teacapán-Agua Brava, Nayarit, México. *Anales del Instituto de Ciencias del Mar y Limnologia de la Universidad Autónoma de México*, 13:185-242

Araya, H.A. 1984. Los sciaénidos (corvinas) del Golfo de Nicoya, Costa Rica. *Rev. Biol. Trop.*, 32(2):179-96.

Baldwin, W.J. 1963. A new chaetodont fish, *Holacanthus limbaughi*, from the eastern Pacific. *Contrib. Sci. Los Angeles Co. Mus.*, 74:8 pp.

Bath, H. 1977. Revision der Blenniini. *Senckenbergiana Biologica*, 57(4/6): 167-234.

———. 1987. *Hypsoblennius minutus* Meek & Hildebrand, 1928: ein Larvenstadium von *Parahypsos piersoni* (Gilbert & Starks, 1904) (Teleostei, Blenniidae). *Zool. Anz.*, 219(5-6):324-30.

Bean, T.H. 1895. Description of a new species of fish, *Bleekeria gilli. Proc. U.S. Nat. Mus.*, 17(1028):629-30.

Beebe, W. and J. Tee-Van. 1938. Eastern Pacific expeditions of the New York Zoological Society, XV. Seven new marine fishes from lower California. *Zoologica*, 23(3):299-312.

———. 1941. Eastern Pacific expeditions of the New York Zoological Society. XXVIII. Fishes from the tropical eastern Pacific. Pt. 3. Rays, mantas and chimaeras. *Zoologica* 26(3):245-80.

Berdegue, A.J. 1956. Peces de importancia comercial en la costa nor-occidental de México. Secretaria do Marina, México. 345 pp.

Berry, F.H. and W.J. Baldwin. 1966. Triggerfishes (Balistidae) of the eastern Pacific. *Proc. Calif. Acad. Sci.*, 34 (9):429-74.

Birdsong, R.S. 1981. A review of the gobiid fish genus *Microgobius* Poey. *Bull. Mar. Sci.*, 31(2):267-306.

Böhlke, J. 1953. A new stathmonotid blenny from the Pacific Coast of Mexico. *Zoologica*, 38(3):145-9.

Böhlke, J.E. and C.R. Robins. 1960. A revision of the gobioid fish genus *Coryphopterus. Proc. Acad. Nat. Sci. Philadel.*, 112(5):103-28.

———. 1968. Western Atlantic seven-spined gobies, with descriptions of ten new species and a new genus, and comments on Pacific relatives. *Proc. Acad. Nat. Sci. Philadel.*, 120(3):45-174.

Bortone, S.A. 1977. Revision of the sea basses of the genus *Diplectrum* (Pisces: Serranidae). NOAA Tech. Rept., National Marine Fisheries Service Circ. 404:1-49.

Breder, C.M., Jr. 1936. Heterosomata to Pediculati from Panama to Lower California. Scientific results of the second oceanographic expedition of the 'Pawnee' 1926. *Bull. Bingham Oceano. Coll.*, 2(3):1-56.

Briggs, J.C. 1955. A monograph of the clingfishes (Order Xenopterygii). *Stanf. Ichthyol. Bull.*, 6:1-224.

———. 1960. A new clingfish of the genus *Gobiesox* from the Tres Marias Islands. *Copeia*, 1960(3):215-17.

———. 1961. The East Pacific Barrier and the distribution of shore fishes. *Evolution*, 15(4):545-54.

———. 1964. Additional transpacific shore fishes. *Copeia*, (4):706-8.

———. 1969. The clingfishes (Gobiesocidae) of Panama. *Copeia*, 1969(4):774-8.

Briggs, J.C. and R.R. Miller. 1960. Two new freshwater clingfishes of the genus *Gobiesox* from southern Mexico. *Occ. Pap. Mus. Zool. Univ. Mich.*, 616:15 pp.

Brittan, M.R. 1966. A small collection of shore fishes from the west coast of Costa Rica. *Ichthyologica* 37(3):121-34.

Brock, V.E. 1943. Distributional notes on the fishes of Lower California and the West Coast of Mexico II. *Copeia*, (2): 130-1.

Bussing, W.A. 1972. *Halichoeres aestuaricola*, a replacement name for the tropical eastern Pacific labrid fish, *Iridio bimaculata* Wilson, with a redescription based on new material. *Brenesia (Nat. Mus. Nac. Costa Rica)*, 1:3-8.

———. 1981. *Elacatinas janssi*, a new gobiid fish from Costa Rica. *Revista de Biologia Tropical*, 29(2):251-6.

———. 1983. A new tropical eastern Pacific labrid fish, *Halichoeres discolor* endemic to Isla del Coco, Costa Rica. *Revista de Biologia Tropical*, 31(1):19-23.

———. 1983. *Evermannia erici*, a new burrowing gobiid fish from the Pacific coast of Costa Rica. *Revista de Biologia Tropical*, 31(1):125-32.

———. 1990. New species of gobiid fishes of the genus *Lythrypnus, Elacatinus* and *Chriolepis* from the eastern tropical Pacific. *Revista de Biologia Tropical*, 38(1):99-118.

———. 1991. A new genus and two new species of tripterygiid fishes from Costa Rica. *Rev. Biol. Trop.*, 39(1):77-85.

———. 1991. A new species of eastern Pacific moray eel (Pisces: Muraenidae). *Rev. Biol. Trop.*, 39(1):97-102.

Caruso, J.H. 1981. The systematics and distribution of the genus *Lophiodes* with the description of two new species. *Copeia*, 1981(3):522-49.

———. 1983. The systematics and distribution of the lophiid anglerfishes. II. Revisions of the genera *Lophiomus* and *Lophius. Copeia*, 1983(1):11-30.

Campos, J.A., B. Burgos and C. Gamboa. 1984. Effect of shrimp trawling on the commercial ichthyofauna of the Gulf of Nicoya, Costa Rica. *Rev. Biol. Trop.*, 32: 203-7.

Castro-Aguirre, J.L. 1978. Catálogo sistemático de los peces marinos que penetran a las aguas continentales de México con aspectos zoogeográficos y ecológicos. *Inst. Nal. de Pesca*. Serie Cient. No. 19 Mexico. 298 pp.

Castro-Aguirre, J.L., J. Arvizu-Martinez and Paez-Barrera. 1970. Contribucion al cococimiento de los peces del Golfo de

California. *Rev. Soc. Mex. Hist. Nat.*, 31: 107-81.
Castro-Aguirre, J.L. and F. de Lachica-Bonilla. 1973. Nuevos registros de peces marinos en la costa del Pacifico Mexicana. *Rev. Soc. Mex. Hist. Nat.*, 34:147-81.
Castro-Aguirre, J.L. and C. Villavicencio-Garayzar. 1989. Una neuva especie de Lonchopisthus (Pisces: Perciformes: Opistognathidae) del Golfo de California, Mexico. *Anales de la Escuela Nacional de Ciencias Biologicas*, 32(1-4):109-15.
Castillo-Campo, L.F. and E.A. Rubio-Rincón. 1987. Estudio de la ictiofauna de los esteros y partes bajas de los rios San Juan, Dagua y Calima, Departmento del Valle de Cauca. *Cespedesia* XIV-XV: 33-70.
Chavez, E.A. 1979. Análisis de la comunidad de una laguna costera en la costa suroccidental de México. *Anales del Centro de Ciencias del Mar y Limnologia de la Universidad Autónoma Nacional de México*, 6:15-44
Chirichigno, F.N. 1969. Lista sistemática de los peces marinos comunes para Ecuador-Perú-Chile. Conferencia sobre explotación y conservación de las riquezas maritimas del Pacifico Sur. Chile-Ecuador-Perú. Secretaria General, Lima. 108 pp.
———. 1974. Clave para identificar los peces marinos del Peru. *Inst. del Mar. (IMARPE)* Inf. 44. CALLAO (Peru) 387 pp.
Chirichigno, N., W. Fisher and C.E. Nauen. 1982. Catalogo de especies marinas de interes economico actual o potencial para America. *INFOPESCA – FAO*. SIC/82/2: 1-503.
Clark, H.W. 1936. The Templeton Crocker Expedition of the California Academy of Sciences. 1932. No. 29. New and noteworthy fishes. *Proc. Calif. Acad. Sci.*, 4:21(29):383-96.
Clemens, H.B. 1957. Fishes collected in the tropical eastern Pacific. 1954. *Calif. Fish & Game*, 43(4):299-307.
Collette, B.B. 1968. *Daector schmitti*, a new species of venomous toadfish from the Pacific coast of Central America. *Proc. Biol. Soc. Wash.*, 81:155-60.
Compagno, L.J.V. 1984. FAO Species Catalogue. Vol. 4. Sharks of the World. An annotated and illustrated catalogue of sharks species known to date. Part 1. Hexanchiformes to Lamniformes. *FAO Fish. Synop.* (125), Vol. 4, Pt. 1:249 pp.
———. 1984. FAO Species Catalogue. Vol. 4. Sharks of the World. An annotated and illustrated catalogue of shark species known to date. Part 2. Carcharhiniformes. *FAO Fish. Synop.*, (125), Vol. 4, Pt. 2: 251-655.
Cowan, G.I. and R.H. Rosenblatt. 1974. *Taenioconger canabus*, a new heterocongrin eel (Pisces: Congridae) from Baja California, with a comparison of a closely related species. *Copeia*, 1974(1):55-60.
Cuesta, T.C. 1932. Lista de los peces de las costas de la Baja California. *Inst. Biol. Mexico, Anal.*, 3(1):75-80.
Dawson, C.E. 1968. Eastern Pacific wormfishes, *Microdesmus dipus* Gunther and *Microdesmus dorsipunctatus* sp. nov. *Copeia*, 1968:512-31.
———. 1972. A new eastern Pacific wormfish, *Microdesmus knappi* (Pisces: Microdesmidae). *Proc. Biol. Soc. Wash.*, 85(15): 191-204.
———. 1974. Studies on eastern Pacific sand stargazers (Pisces: Dactyloscopidae). 1. *Platygillelus* new genus with descriptions of new species. *Copeia*, 1974(1):39-55.
———. 1974. A review of Microdesmidae (Pisces: Gobioidea). 1. *Cerdale* and *Clarkichthys* with descriptions of three new species. *Copeia*, 1974(2):409-48.
———. 1975. Studies on Eastern Pacific sand stargazers (Pisces: Dactyloscopidae). 2. Genus *Dactyloscopus*, with descriptions of new species and subspecies. *Nat. Hist. Mus. Los Angeles Co. Sci. Bull.*, 22:1-16.
———. 1976. Studies on Eastern Pacific sand stargazers. 3. *Dactylagnus* and *Myxodagnus* with description of a new species and subspecies. *Copeia*, 1976(1): 13-43.
———. 1977. Studies on eastern Pacific sand stargazers (Pisces: Dactyloscopidae). 4. *Gillellus, Sindoscopus* new genus, and *Heteristius* with description of new species. *Proc. Calif. Acad. Sci.*, 41(2): 125-60.
———. 1985. *Indo-Pacific Pipefishes*. Ocean Springs, Mississippi: Gulf Coast Research Laboratory. 230 pp.
D'Croz, L. and B. Kwiecinski. 1980. Contribución de los manglares a las pesquerías de la Bahía de Panama. *Rev. Biol. Trop.*, 28:13-29.
D'Croz, L., R. Rivera and E. Pineda. 1977. Observaciones sobre un arte de pesca fija en las costas de la Bahía de Panama. *Conciencia*, 14-17.
Dooley, J.K. 1978. Systematics and biology of the tilefishes (Perciformes: Branchiostegidae and Malacanthidae), with descriptions of two new species. *U.S. Nat. Ocean. Atmos. Admin. (NOAA) Tech. Rep.*, NMFS Circ. (411):1-78.
Ebeling, A.W. 1957. The dentition of eastern Pacific mullets with special reference to adaptation and taxonomy. *Copeia*, 1957(3):173-85.
———. 1961. *Mugil galapagensis*, a new mullet from the Galapagos Islands, with notes on related species and a key to the Mugilidae of the eastern Pacific. *Copeia*, 1961(3):295-305.
Eigenmann, C.H. 1917. Eighteen new species of fishes from northwestern South America. *Proc. Amer. Phil. Soc.*, 56:673-89.
Eigenmann, C.H. and R.S. Eigenmann. 1888. A list of American species of Gobiidae and Callionymidae, with notes on the specimens contained in the Museum of Comparative Zoology at Cambridge, Massachusetts. *Proc. Calif. Acad. Sci.*, (Series 2)1:51-78.
Evermann, B.W. 1898. Notes on fishes collected by E.W. Nelson on the Tres Marias islands and in Sinaloa and Jalisco, Mexico. *Proc. Biol. Soc. Wash.*, 12:1-3.
Evermann, B.W. and O.P. Jenkins. 1891. Report upon a collection of fishes made at Guaymas, Sonora, Mexico, with descriptions of new species. *Proc. U.S. Nat. Mus.*, 14:121-65.
Fowler, H.W. 1944. Results of the Fifth George Vanderbilt Expedition (1941) (Bahamas, Caribbean, Panama, Galapagos Archipelago and Mexican Pacific Islands). The Fishes. *Acad. Nat. Sci. Philadel., Monographs* 6:57-529.
Fritzsche, R.A. 1976. A review of the cornetfishes, genus *Fistularia* (Fistulariidae), with a discussion of the intrageneric relationships and zoogeography. *Bull. Mar. Sci.*, 26(2):196-204.
Garman, S. 1899. A species of goby from the shores of Clipperton Island. *New England Zool. Club*, 1:63-4.
———. 1899. Reports of an exploration off the west coasts of Mexico, Central and South America, and off the Galapagos Islands, in charge of Alexander Agassiz, by the U.S. Fish Commission Steamer "Albatross," during 1891. L.C.Z.L. Tannar, U.S.N. commanding XXVI. The Fishes. *Mem. Mus. Comp. Zool.*, 24: 432 pp.
Gilbert, C.H. 1890. XII – A preliminary report on the fishes collected by the steamer "Albatross" on the Pacific coast of North America during the year 1889, with descriptions of twelve new genera and ninety-two new species. *Proc. U.S. Nat. Mus.*, 13:49-126.
———. 1892. Scientific results of explorations by the U.S. Fish Commission steamer "Albatross." XXII. Descriptions of thirty-four new species of fishes collected in 1888 and 1889, principally among the Santa Barbara Islands and in the Gulf of California. *Proc. U.S. Nat. Mus.*, 14:539-66.
Gilbert, C.H. and E.C. Starks. 1904. The fishes of Panama Bay. *Mem. Calif. Acad. Sci.*, 4:1-304.
Gilbert, C.R. 1966. Two new worm fishes (family Microdesmidae) from Costa Rica.

Copeia, 1966:325-32.
Ginsburg, I. 1932. A revision of the genus *Gobionellus* (family Gobiidae). *Bull. Bingham Oceano. Collection*, 4:3-51.
———. 1933. Descriptions of new and imperfectly known species and genera of gobioid and pleuronectid fishes in the United States National Museum. *Proc. U.S. Nat. Mus.*, 82(20):1-23.
———. 1933. A revision of the genus *Gobiosoma* (family Gobiidae) with an account of the genus *Garmannia*. *Bull. Bingham Oceano. Collection*, 4(5):1-59.
———. 1938. Two new gobiid fishes of the genus *Gobiosoma* from Lower California. *Stanford Ichthyol. Bull.*, 1(2):57-9.
———. 1938. Eight new species of gobioid fishes from the American Pacific coast. *Allan Hancock Pacific Expeditions*, 2(7): 109-21.
———. 1939. Twenty-one new American gobies. *J. Wash. Acad. Sci.*, 29(2):51-63.
———. 1944. A description of a new gobiid fish from Venezuela, with notes on the genus *Garmannia*. *J. Wash. Acad. Sci.*, 34(11):375-80.
———. 1947. American species and subspecies of *Bathygobius*, with a demonstration of a suggested modified system of nomenclature. *J. Wash. Acad. Sci.*, 37: 275-84.
———. 1953. Ten new American gobioid fishes in the United States National Museum, including additions to a revision of *Gobionellus*. *J. Wash. Acad. Sci.*, 43: 18-26.
Gomon, M.F. 1974. A new eastern Pacific labrid (Pisces), *Decodon melasma*, a geminate species of the western Atlantic *D. puellaris*. *Proc. Biol. Soc. Wash.*, 87(19):205-16.
Gorman, G.C., Y. Kim and R. Rubinoff. 1976. Genetic relationships of three species of *Bathygobius* from the Atlantic and Pacific sides of Panama. *Copeia*, 1976(2):361-4.
Greenfield, D.W. 1974. A revision of the squirrelfish genus *Myripristis* Cuvier (Pisces: Holocentridae). *Nat. Hist. Mus. Los Angeles Co., Sci. Bull.*, 19:54 pp.
Grove, J.S., D. Gerzon, M.D. Saa and C. Strang. 1986. Distribucion y ecologia de la familia Pomacentridae (Pisces) en las Islas Galapagos. *Rev. Biol. Trop.*, 34(1):127-40.
Grove, J.S., S. Massay and S. Garcia. 1984. Peces de las Islas Galapagos, Ecuador. *Bol. Cient. Tech.*, 7(2):1-157.
Günther, A. 1862. On a collection of fishes sent by Capt. Dow from the Pacific coast of Central America. *Ann. Mag. Nat. Hist.*, (3)9(52):326-31.
———. 1862. On a collection of fishes sent by Capt. Dow from the Pacific coast of Central America. *Proc. Zool. Soc. London*, 1861(3):370-6.
———. 1864. On some new species of Central-American fishes. *Proc. Zool. Soc. London*, 1864:23-7.
———. 1864. Report on a collection of fishes made by Messrs. Dow, Godman and Salvin in Guatemala. Part first. *Proc. Zool. Soc. London*, 1864(1):144-54.
———. 1864. On some species of Central American fishes. *Ann. Mag. Nat. Hist.*, 1864:227-32.
Haedrich, R.L. 1967. The stromateoid fishes; systematics and a classification. *Bull. Mus. Comp. Zool., Harvard*. 135(2):31-139.
———. 1972. Fishes of the family Nomeidae (Perciformes: Stromateoidei). *Arch. Fisch Wiss.*, 23(2):73-88.
Hanna, G.D. 1926. Expedition to the Revillagigedo Islands, Mexico, in 1925, Pt. I, General Account. *Proc. Calif. Acad. Sci.*, (4)15(1):113.
Hastings, P.A. 1992. Phylogenetic relationships of *Tanyemblemaria alleni*, a new genus and species of Chaenopsid (Pisces: Blennioidei) from the Gulf of Panama. *Bull. Mar. Sci.*, 51(2):147-60
Heller, E. and R.E. Snodgrass. 1903. Papers from the Hopkins-Stanford Galapagos Expedition 1898-1899. XV. New fishes. *Proc. Wash. Acad. Sci.*, 5:189-229.
Herdson, D.M., W.T. Rodriquez and J. Martinez. 1985. Los recursos de peces demersales de la plataforma continental del Ecuador. Parte uno: Distribución, abundancia y variacones. Parte dos: Potencial y recomendaciones para la utilizacion del recurso de la pesca blanca en el Ecuador. *Boletin Cientifico y Técnico (Instituto Nacional de Pesca)*, 8(5), Guayaquil.
Herre, A.W. 1935. New fishes obtained by the Crane Pacific Expedition. *Field Mus. Nat. Hist., Zool. Ser. Publ. No. 335*, 18(12): 383-438.
———. 1936. Fishes of the Crane Pacific Expedition. *Fieldiana (Zool.), Field Mus. Nat. Hist. Chicago*, 21:1-472.
Hildebrand, S.F. 1946. A descriptive catalog of the shore fishes of Peru. *Bull. U.S. Nat. Mus.*, 189:1-530.
Hobson, E.S. 1965. Diurnal-nocturnal activity of some inshore fishes in the Gulf of California. *Copeia*, 1965(3):291-302.
———. 1968. Predatory behavior of some shore fishes in the Gulf of California. *Bur. Sports Fish. and Wildlife, Res.*, Rept. 73: 92 pp.
Hoese, D.F. 1973. *Gobius lucretiae* referred to the gobiid fish genus *Parrella*, with a review of the species. *Copeia*, 1973(4): 817-19.
———. 1976. Variation, synonymy and a redescription of the gobiid fish *Atuma histrio*, and a discussion of the related genus *Ophiogobius*. *Copeia*, 1976(2): 295-305.
Hoese, D.F. and H.K. Larson. 1985. Revision of the eastern Pacific species of the genus *Barbulifer* (Pisces: Gobiidae). *Copeia*, 1985(2):333-9.
Hong, S.L. 1977. Review of eastern Pacific *Haemulon* with notes on juvenile pigmentation. *Copeia*, 1977(3):493-501.
Hubbs, C.L. 1944. Species of the circumtropical fish genus *Brotula*. *Copeia*, 1944(3):162-78.
———. 1952. A contribution to the classification of the blennioid fishes of the family Clinidae, with a partial revision of the eastern Pacific forms. *Stanf. Ichthyol. Bull.*, 4(2):41-165.
———. 1953. Revision of the eastern Pacific fishes of the clinid genus *Labrisomus*. *Zoologica*, 38(3):113-36.
———. 1954. Additional records of clinid fishes, with the description of a new species of *Cryptotrema* from the Gulf of California. *Copeia*, 1954(1):17-19.
Hubbs, C.L. and A.B. Rechnitzer. 1958. A new fish, *Chaetodon falcifer*, from Guadalupe Island, Baja California, with notes on related species. *Proc. Calif. Acad. Sci.*, 29(8):273-313.
Hubbs, C.L. and R.H. Rosenblatt. 1961. Effects of the equatorial currents of the Pacific on the distribution of fishes and other marine animals. *Tenth Pacific Sci. Congress, Abstracts of Symposium Papers*, 340-1.
Hubbs, C.L. and L.P. Schultz. 1939. A revision of the toadfishes referred to *Porichthys* and related genera. *Proc. U.S. Nat. Mus.*, 86(3060):473-96.
Humann, P. 1993. *Reef fish identification. Galapagos*. Florida: New World Publishing.
Jenkins, O.P. and B.W. Evermann. 1889. Description of eighteen new species of fishes from the Gulf of California. *Proc. U.S. Nat. Mus.*, 11:37-158.
Jenyns, L. 1842. In C. Darwin, ed. *The zoology of the voyage of H.M.S. 'Beagle' under the command of Captain Fitzroy, R.N. during the years 1832 to 1836*. Part 4. 172 pp.
Jordan, D.S. 1884. Notes on fishes collected at Guaymas, Mexico, by Mr. H.F. Emeric, with a description of *Gobiosoma histrio*, a new species. *Proc. U.S. Nat. Mus.*, 7: 260-61.
———. 1886. A list of the fishes known from the Pacific coast of tropical America, from the tropic of Cancer to Panama. *Proc. U.S. Nat. Mus.*, 8(1885):361-94.
———. 1895. The fishes of Sinaloa. *Contrib. Biol. Hopkins Lab. Biol.*, 1: 377-514.
———. 1895. Description of *Evermannia*, a new genus of gobioid fishes. *Proc. Calif. Acad. Sci.*, (Series 2) 4:592.
Jordan, D.S. and C.H. Bollman. 1890.

Descriptions of new species of fishes collected at the Galapagos Islands and along the coast of the United States of Colombia, 1887-88, by the U.S. Fish Commission steamer 'Albatross'. *Proc. U.S. Nat. Mus.*, 12:149-83.

Jordan, D.S. and B.W. Evermann. 1896. The fishes of North and Middle America: a descriptive catalogue of the species of fish-like vertebrates found in the waters of North America, north of the Isthmus of Panama. Part I. *Bull. U.S. Nat. Mus.* No. 47:1-1240.

———. 1898. The fishes of North and Middle America: a descriptive catalogue of the species of fish-like vertebrates found in the waters of North America, north of the Isthmus of Panama. Part II. *Bull. U.S. Nat. Mus.* No. 47:1241-2183.

———. 1898. The fishes of North and Middle America: a descriptive catalogue of the species of fish-like vertebrates found in the waters of North America, north of the Isthmus of Panama. Part III. *Bull. U.S. Nat. Mus.* No. 47:2183-3136.

———. 1900. The fishes of North and Middle America: a descriptive catalogue of the species of fish-like vertebrates found in the waters of North America, north of the Isthmus of Panama. Part IV. *Bull. U.S. Nat. Mus.* No. 47:3137-313.

Jordan, D.S., B.W. Evermann and H.W. Clark. 1930. Check-list of the fishes and fish-like vertebrates of North and Middle America north of the northern boundary of Venezuela and Colombia. *Rept. U.S. Comm. Fisheries for 1928.* Part 2:1-670.

Jordan, D.S. and C.H. Gilbert. 1882. List of fishes collected by Lieut. Henry E. Nichols, U.S.N., in the Gulf of California, with descriptions of four new species. *Proc. U.S. Nat. Mus.*, 4(1881):273-9.

———. 1882. A synopsis of the fishes of North America. *Bull. U.S. Nat. Mus.*, 16: 1-1018.

———. 1882. Description of thirty-three new species of fishes from Mazatlan, Mexico. *Proc. U.S. Nat. Mus.*, 4(1881): 338-65.

———. 1882. Description of five new species of fishes from Mazatlan, Mexico. *Proc. U.S. Nat. Mus.*, 4:458-63.

———. 1882. Descriptions of nineteen new species of fishes from the Bay of Panama. *Bull. U.S. Fish Comm.*, 1:306-35.

———. 1883. List of fishes collected at Mazatlan, Mexico by Charles Henry Gilbert. *Bull. U.S. Fish Comm.*, 2(1882): 105-8.

———. 1883. Catalogue of the fishes collected by Mr. John Xantus at Cape San Lucas, which are now in the United States National Museum, with descriptions of eight new species. *Proc. U.S. Nat. Mus.*, 5(1882):353-71.

———. 1883. List of the fishes now in the museum of Yale College, collected by Prof. Frank H. Bradley at Panama, with descriptions of three new species. *Proc. U.S. Nat. Mus.*, 5(1882):620-32.

Kato, S., S. Springer and M.H. Wagner. 1967. Field Guide to Eastern Pacific and Hawaiian Sharks. *U.S. Fish. Wild. Ser. Circ. no. 271.*

Kendall, W.C. and L. Radcliffe. 1912. The shore fishes. Reports on the scientific results of the expedition to the eastern tropical Pacific, in charge of Alexander Agassiz, by the U.S. Fish Commission steamer ALBATROSS, from October, 1904, to March, 1905, Lieut. Commander L.M. Garret, U.S.N., Commanding. XXV. *Mem. Mus. Comp. Zool.*, 35(3): 77-171.

Krejsa, R.J. 1960. The eastern tropical Pacific fishes of the genus *Blenniolus*, including a new island endemic. *Copeia*, 1960(4):322-36.

Leis, J.M. 1984. Larval fish dispersal and the East Pacific Barrier. *Oceanogr. trop.*, 19(2):181-92.

Lopez, M.I. 1980. *Umbrina bussingi*, a new sciaenid fish from the tropical eastern Pacific. *Rev. Biol. Trop.* 28:203-8

———. 1981. Los "roncadores" del genero *Pomadasys (Haemulopsis)* [Pisces: Haemulidae] de la costa Pacifica de Centro América. *Rev. Biol. Trop.*, 29(1):83-94.

Lopez, M.I. and W.A. Bussing. 1982. Lista provisional de los peces marinos de la Costa Rica. *Rev. Biol. Trop.*, 30(1):5-26.

Madhu, M.N. and T.A. Munroe. 1990. Three new species of symphurine tonguefishes from tropical and warm temperate waters of the eastern Pacific (*Symphurus*: Cynoglossidae: Pleuronectiformes). *Proc. Biol. Soc. Wash.*, 103(4):931-54.

Massay, S. 1988. Revisión de la licta do los peces marinos del Ecuador. *Bol. Cient. Tech.*, 6(1):1-113.

McCarthy, L.V. 1979. Eastern Pacific *Rypticus* (Pisces: Grammistidae). *Copeia*, 1979(3):393-400.

McCosker, J.E. 1977. The osteology, classification, and relationships of the eel family Ophichthidae. *Proc. Calif. Acad. Sci.*, 41(1):1-123.

———. 1987. The fishes of the Galapagos Islands. *Oceanus*, 30(2):28-32.

McCosker, J.E. and R.H. Rosenblatt. 1975. The moray eels (Pisces: Muraenidae) of the Galapagos Islands, with new records and synonymies of extralimital species. *Proc. Cal. Acad. Sci.*, 40(13):417-27.

———. 1975. Fishes collected at Malpelo Island. In Graham, J.B. (ed.) The Biological Investigation of Malpelo Island, Colombia. *Smithsonian Contrib. Zool.*, 176:91-3.

———. 1987. Notes on the biology, taxonomy, and distribution of anomalopid fishes (Anomalopidae: Beryciformes). *Japan. J. Ichthyol.*, 34:157-64.

McPhail, J.D. 1958. Key to the croakers (Sciaenidae) of the eastern Pacific. *Inst. Fisheries, Univ. Brit. Colum. Mus. Contrib.*, 2:1-20.

———. 1961. A review of the tropical eastern Pacific species of *Pareques* (Sciaenidae). *Copeia*, 1961 (1):27-32.

Meek, S.E. and S.F. Hildebrand. 1912. Descriptions of new fishes from Panama. *Publ. Field Mus. Zool.*, 10:67-8.

———. 1916. The marine fishes of Panama. Part I. *Field Mus. Nat. Hist. Publ. Zool.*, Ser. v. 15:1-330.

———. 1925. The marine fishes of Panama. Part II. *Field Mus. Nat. Hist. Publ. Zool.*, Ser. v. 15:331-707.

———. 1928. The marine fishes of Panama. Part III. *Field Mus. Nat. Hist. Publ. Zool.*, Ser. v. 15:709-1045.

Meisler, M.R. and R.J. Lavenberg. 1995. A new species (Serranidae: *Serranus*) from Isla del Coco. *Pacific Science.*

Merlen, G.M. 1988. *A field guide to the fishes of Galapagos*. London: Wilmot Books.

Meyers, G.S. 1942. The fish fauna of the Pacific Ocean with especial reference to zoogeographical regions and distribution as they affect the international aspect of the fisheries. *Proc. Sixth Pac. Sci. Congr.*, 3(1939):201-10.

Miller, G.C. and W.J. Richards. 1991. Revision of the western Atlantic and eastern Pacific genus *Bellator* (Pisces: Triglidae). *Bull. Mar. Sci.* 48(3):635-56

Miller, G.C. and W.J. Richards. 1991. Nomenclatural changes in the genus *Prionotus* (Pisces: Triglidae). *Bull. Mar. Sci.* 48(3): 635-56

Miller, R.R. 1953. Second specimen of the eel *Gorgasia punctata*, an addition to the known fish fauna of Mexico. *Copeia*, 1953(4):236-7.

Miyake, T. and J.D. McEachran. 1988. Three new species of the genus *Urotyrgon* (Myliobatiformes: Urolophidae) from the eastern Pacific. *Bull. Mar. Sci.*, 42(3): 366-75.

Myers, G.S. and C.B. Wade. 1941. Four new genera and ten new species of eels from the Pacific coast of tropical America. *Allan Hancock Pacific Expeditions*, Vol. 9, No. 4:65- 111.

———. 1946. New fishes of the families Dactyloscopidae, Microdesmidae and Antennariidae from the west coast of Mexico and the Galapagos Islands with a brief account of the use of rotenone fish poisons in Ichthyological collecting. *Allan Hancock Pacific Expedition*, 9(6):151-79.

Nelson, J.S. 1984. *Fishes of the World.* 2nd ed. New York: J. Wiley and Sons. 523 pp.

Nichols, J.T. 1952. Four new gobies from the Eastern and Western Pacific. *Amer. Mus. Novit.*, No. 1594:1-5.

Pietsch, T.W. and D.B. Grobecker. 1987. *Frogfishes of the World.* California: Stanford University Press. 420 pp.

Poll, M. and N. Leleup. 1965. Un poisson aveugle noveau de la famille des Brotulidae provenant des iles Galapagos. *Bull. Cl. Sci. Acad. Roy. Belg.*, ser. 5, 51(4): 464-74.

Poll, M. and J.J. Van Mol. 1966. Au sujet d'une espece inconnue de Brotulidae littoral des iles Galapagos, apparentee a l'espece aveugle *Caecogilbia galapogensis* Poll et Leleup. *Bull. Cl. Sci. Acad. Roy. Belg.*, scr 5, vol. 52(11):1444-61.

Ramírez-Granados, R. and M.L. Sevilla. 1963. Lista preliminar de recursos pesqueros de México marinos y de agua dulce. *Nociones sobre Hidrobiologia applicada a la Pesca, Serie Trab. de Divul.* 5(42):325-61.

Ramírez-Hernandez, E. 1965. Estudios preliminares sobre los peces marinos de México. *Anales del Instituto Nacional de Investigaciones Biologico-Pesqueras*, 1: 257-92.

Ramírez-Hernandez, E. and J. Arvizu-Martinez. 1965. Investigaciones ictiologicas en las costas de Baja California. I. Lista de peces marinos de Baja California colectados en el periodo 1961-1965. *Anales del Instituto Nacional de Investigaciones Biologico-Pesqueras*, 1:293-324.

Randall, J.E. 1955. A revision of the surgeonfish genus *Ctenochaetus*, family Acanthuridae, with descriptions of five new species. *Zoologica*, 40(4):149-66.

———. 1956. A revision of the surgeon fish genus *Acanthurus*. *Pac. Sci.*, 10(2): 159-235.

———. 1963. Review of the hawkfishes (family Cirrhitidae). *Proc. U.S. Nat. Mus.*, 114:389-451.

———. 1964. A revision of the filefish genera *Amanses* and *Cantherhines*. *Copeia*, 1964(2):331-61.

———. 1987. Three nomenclatorial changes in Indo-Pacific surgeonfishes (Acanthuridae). *Pac. Sci.*, 41 (1-4):54-61.

Randall, J.E. and D.K. Caldwell. 1966. A review of the sparid fish genus *Calamus*, with descriptions of four new species. *Bull. Los Angeles Co. Mus. Nat. Hist.*, 2: 47 pp.

Randall, J.E. and J.E. McCosker. 1975. The eels of Easter Island with a description of a new moray. *Nat. Hist. Mus. Los Angeles Co. Contrib. Sci.* No. 264:32 pp.

Reid, E.D. 1936. Revision of the fishes of the family Microdesmidae, with descriptions of a new species. *Proc. U.S. Nat. Mus.*, 84(3002):55-72.

Ricker, K.E. 1959. Mexican shore and pelagic fishes collected from Acapulco to Cape San Lucas during the 1957 cruise of the "Marijean". *Univ. Brit. Columbia Inst. Fish., Mus. Contrib.*, 3:18 pp.

———. 1959. Fishes collected from the Revillagigedo Islands during the 1954-1958 cruises of the "Marijean." *Univ. Brit. Columbia Inst. Fish., Mus. Contrib.*, 4:10 pp.

Rivas, L.R. 1986. Systematic review of the perciform fishes of the genus *Centropomus*. *Copeia*, 1986(3):579-611.

Robertson, D.R. and G.R. Allen. In press. The Fishes of Clipperton Island, Eastern Pacific Ocean. *Smithsonian Contrib. Zool.*

Robins, C.R. and W.A. Starck II. 1961. Materials for a revision of *Serranus* and related fish genera. *Proc. Acad. Nat. Sci. Philadel.*, 113(11):259-314.

Rosenblatt, R.H. 1967. The zoogeographic relationships of the marine shore fishes of tropical America. *Stud. Tropical Oceanogr.*, 5:579-92.

Rosenblatt, R.H. and E.S. Hobson. 1969. Parrotfishes (Scaridae) of the eastern Pacific, with a generic rearrangement of the Scarinae. *Copeia*, 1969(3):434-53.

Rosenblatt, R.H. and G.D. Johnson. 1974. Two new species of sea basses of the genus *Diplectrum*, with a key to the Pacific species. *Calif. Fish & Game*, 60(4): 178-91.

Rosenblatt, R.H. and J.E. McCosker. 1970. A key to the genera of the ophichthid eels, with descriptions of two new genera and three new species from the eastern Pacific. *Pac. Sci.*, 24:494-505.

Rosenblatt, R.H., J.E. McCosker and I. Rubinoff. 1972. Indo-west Pacific fishes from the Gulf of Chiriqui, Panama. *Contrib. Sci. Nat. Hist. Mus. Los Angeles Co.*, 234:18 pp.

Rosenblatt, R.H. and L.W. Montgomery. 1976. *Kryptophaneron harveyi*, a new anomalopid fish from the eastern tropical Pacific, and the evolution of the Anomalopidae. *Copeia*, 1976(3):510-15.

Rosenblatt, R.H. and T.D. Parr. 1967. The identity of the blenny *Paraclinus altivelis* (Lockington) and the status of *P. sinus* Hubbs. *Copeia*, 1967(3):675-77.

———. 1969. The Pacific species of the clinid fish genus *Paraclinus*. *Copeia*, 1969(1):1-20.

Rosenblatt, R.H. and J.S. Stephen. 1978. *Mccoskerichthys sandae*, a new and unusual chaenopsid blenny from the Pacific coast of Panama and Costa Rica. *Contrib. Sci. Nat. Mus. Los Angeles Co.*, 293:1-22.

Rosenblatt, R.H. and L.R. Taylor. 1971. The Pacific species of the clinid fish tribe Starksiini. *Pac. Sci.*, 25:436-63.

———. 1972. A second record of *Starksia posthon* (Clinidae) with an expanded description. *Copeia*, 1972(3):599.

Rosenblatt, R.H. and B.W. Walker. 1963. The marine shore fishes of the Galapagos Islands. *Occ. Papers Calif. Acad. Sci.*, 44:97-106.

Rosenblatt, R.H. and B.J. Zahuranec. 1967. The eastern Pacific groupers of the genus *Mycteroperca*; including a new species. *Calif. Fish & Game*, 53(4):228-45.

Rubio, E.A. 1982. Estudio taxonomico de la ictiofauna asociada al ecosistema manglar-estero en la Bahia de Buenaventura. In: Cantera, J.R. (1982) Fauna asociada al ecosistema manglar estero en la Bahia de Buenaventura (Pac. Col.):76-122. *Cent. Publ. Div Cienc. Univ. del Valle. Cali.*

———. 1984. Estudios sobre la ictiofauna del Pacifico colombiano. I. Composición taxonomica de la ictiofauna asociada al ecosistema manglar-estuario en la Bahia de Buenventura. *Cespedesia*, Vol. XIII, No. 49-50:296-313.

———. 1984. Estudio taxonomico preliminar de la ictiofauna de la Bahia de Malaga. *Colombia Ann. Inst. Inv. Mar. de Punta de Betin.* Vol. 14.

———. 1986. Notas sobre la ictiofauna de la Isla de Gorgona, Colombia. *Boletin Ecotropica. Univ. Bog. Jorge Tadeo Lozano*, Vol. 13:86-112.

———. 1987. Composicion taxonomica de los peces del Golfo de Tortugas (Colombia). *Cespedesia*, 14-15:19-30.

———. 1987. Peces de importancia comercial para el Pacifico de Colombia. Contribucion Cientifica No. 1. Centro de Investigaciones Marinas y Estuarinas de la Universidad del Valle, Cali, Colombia.

Rubio, E.A., B. Gutierrez and R. Franke. 1987-88. *Pesces de la Isla Gorgona.* Centro de Publicaciones de la Faculdad de Ciencias de la Universidad del Valle, Cali, Colombia.

Rutter, C.M. 1904. Notes on fishes of the Gulf of California, with description of a new genus and species. *Proc. Calif. Acad. Sci. (Zoology)* (Series 3) 3:251-3.

Smith, C.L. 1971. A revision of the American Groupers: *Epinephelus* and allied genera. *Bull. Amer. Mus. nat. Hist.*, 146(2): 67-242.

Smith, J.L.B. 1966. Fishes of the sub-family Nasinae with a synopsis of the Prionurinae. *Rhodes Univ. Ichthyol. Bull.*, 32: 635-82.

Smith-Vaniz, W.F. 1976. The saber-toothed blennies, Tribe Nemophini (Pisces: Blenniidae). *Acad. Nat. Sci. Philadel.*, Monograph 19:1-196.

Smith-Vaniz, W.F. and F.J. Palacio. 1973. Atlantic fishes of the genus *Acanthemblemaria*, with description of three new species and comments on Pacific species (Clinidae: Chaenopsinae). *Proc. Acad.*

Nat. Sci. Philadel., 125:197-224.

Snodgrass, R.E. and E. Heller. 1905. Papers from the Hopkins-Stanford Galapagos Expedition, 1898-1899. XVII. Shore fishes of the Revillagigedo, Clipperton, Cocos and Galapagos Island. *Proc. Wash. Acad. Sci.*, 6:333-427.

Springer, V.G. 1958. Systematics and zoogeography of the clinid fishes of the subtribe Labrisomini Hubbs. *Pubs. Inst. Mar. Sci. Univ. Texas*, 5:417-92.

———. 1962. A review of the blenniid genus *Ophioblennius* Gill. *Copeia*, 1962(2): 426-433.

———. 1967. Revision of the circumtropical shorefish genus *Entomacrodus* (Blenniidae: Salariinae). *Proc. U.S. Nat. Mus.*, 122(3582):1-150.

Starnes, W.C. 1988. Revision, phylogeny and biogeographic comments on the circumtropical marine percoid fish family Priacanthidae. *Bull. Mar. Sci.*, 43(2): 117-203.

Stephens, J.S., Jr. 1963. A revised classification of the blennioid fishes of the American family Chaenopsidae. *Univ. Calif. Pubs. Zool.*, 68:1-133.

Stephens, J.S., Jr., E.S. Hobson and R.K. Johnson. 1966. Notes on distribution, behavior, and morphological variation in some chaenopsid fishes from the tropical eastern Pacific with descriptions of two new species, *Acanthemblemaria castroi* and *Coralliozetus springeri*. *Copeia*, 1966(3):424-38.

Sterling, J.E. 1977. Estudio taxonomico de peces marinos del Pacifico colombiano. *Mem. Sem. Pacif. Sud. Ame. Univ. del Valle. Div. de Ciencias*, 689-708.

Stiassny, M.L.I., J.E. McCosker and I. Rubinoff. 1972. Indo-west Pacific fishes from the Gulf of Chiriqui, Panama. *Contrib. Sci. (Los Angeles)*, 235:1-18.

Tarp, F.H. 1952. A revision of the family Embiotocidae (the surfperches). *Fish. Bull. Calif. Depart. Fish & Game*, 88: 1-99.

Thomson, D.A., L.T. Findley and A.N. Kerstitch. 1979. *Reef fishes of the Sea of Cortez*. New York: John Wiley and Sons.

Ulrey, A.B. 1929. A check-list of the fishes of Southern California and Lower California. *J. Pan-Pacific Res. Inst.*, 4(4):2-11.

Wade, C.B. 1946. Two new genera and five new species of apodal fishes from the eastern Pacific. *Allan Hancock Pacific Expeditions* Vol. 9, No. 7:181-213.

———. 1946. New fishes in the collections of the Allan Hancock Foundation. *Allan Hancock Pacific Expeditions*, Vol. 9, No. 8:215-35.

Wales, J.H. 1932. Report on two collections of Lower California marine fishes. *Copeia*, 1932(4):163-8.

Walford, L.A. 1937. *Marine game fishes of the Pacific coast from Alaska to the Equator*. Berkeley: Univ. Calif. Press. (T.F.H. – Smithsonian reprint, with emendations, 1974).

Walker, B.W. 1960. The distribution and affinities of the marine fish fauna of the Gulf of California. *Systematic Zool.*, 9(3-4):123-33.

Welander, A.D. and D.L. Alverson. 1954. New and little known fishes of the eastern Pacific. *Fish Res. Pap. Wash. Dept. Fish.*, 1(2):37-44.

Whitehead, P.J.P. 1985. FAO species catalogue. Vol. 7. Clupeoid fishes of the world. An annotated and illustrated catalogue of the herrings, sardines, pilchards, sprats, anchovies and wolf-herrings. Part 1. Chirocentridae, Clupeidae and Pristigasteridae. *FAO Fish. Synop.*, (125) Vol. 7, Pt. 1:303 pp.

Whitehead, P.J.P., G.J. Nelson and T. Wongratana. 1988. FAO species catalogue. Vol. 7. Clupeoid fishes of the world (Suborder Clupeoidei). An annotated and illustrated catalogue of the herrings, sardines, pilchards, sprats, anchovies and wolf-herrings. Part 2. Engraulididae. *FAO Fish. Synop.*, (125) Vol. 7, Pt. 2:305-579.

Williams, J.T. 1990. Phylogenetic relationships and revision of the blenniid fish genus *Scartichthys*. *Smithsonian Contrib. Zool.*, 492:1-30.

INDEX